I0066543

FRESHWATER FISH CULTURE

The Author

Dr. S.K. Sarkar obtained Bachelor and Master Degrees in Zoology from the Unviersity of Kalyani, West Bengal and did his Ph.D. Degree in Zoology in 1982 from the same university.

He joined as Technical Officer (Fisheries Science) in 1980 at the West Bengal Comprehensive Area Development Corporation and continued his service there upto 1985. Thereafter, he joined as Lecturer in Zoology at R.B.C. College, Naihati in 1985 and continued upto 1987. Dr. Sarkar is now working as Associate Professor in Zoology at Netaji Nagar Day College, Kolkata. His main research interests concern the biology of fishes, fish culture, limnology and toxicology. He has 23 years of teaching and research experience.

He has been invited to participate in many national and international conferences and symposia pertaining to the subject concerned. He has published six text books on life science at secondary level and 110 research papers in different Journals and edited books. He has also written several review articles pertaining to cultural and toxicological aspects. He has been Principal Investigator of the UGC Research Projects since 1997.

FRESHWATER FISH CULTURE

— Volume 3 —

S.K. Sarkar

Department of Zoology
Netaji Nagar College (Day), Kolkata

2014
Daya Publishing House®
A Division of
Astral International Pvt. Ltd.
New Delhi – 110 002

© 2014 AUTHOR

ISBN 9789351242925

Publisher's note:

Every possible effort has been made to ensure that the information contained in this book is accurate at the time of going to press, and the publisher and author cannot accept responsibility for any errors or omissions, however caused. No responsibility for loss or damage occasioned to any person acting, or refraining from action, as a result of the material in this publication can be accepted by the editor, the publisher or the author. The Publisher is not associated with any product or vendor mentioned in the book. The contents of this work are intended to further general scientific research, understanding and discussion only. Readers should consult with a specialist where appropriate.

Every effort has been made to trace the owners of copyright material used in this book, if any. The author and the publisher will be grateful for any omission brought to their notice for acknowledgement in the future editions of the book.

All Rights reserved under International Copyright Conventions. No part of this publication may be reproduced, stored in a retrieval system, or transmitted in any form or by any means, electronic, mechanical, photocopying, recording or otherwise without the prior written consent of the publisher and the copyright owner.

Published by   : **Daya Publishing House®**
                        A Division of
                        Astral International Pvt. Ltd.
                        – ISO 9001:2008 Certified Company –
                        4760-61/23, Ansari Road, Darya Ganj
                        New Delhi-110 002
                        Ph. 011-43549197, 23278134
                        E-mail: info@astralint.com
                        Website: www.astralint.com

Laser Typesetting   : **Classic Computer Services**, Delhi - 110 035

Printed at   : **Thomson Press India Limited**

PRINTED IN INDIA

Dedicated to
the Memory of My Mother
BINAPANI

to

Ratna (Sharmila), Kalyani, Swagata (Rimi)

UNIVERSITY OF CALCUTTA

Professor Samir Banerjee
Hiralal Chaudhuri Professor,

DEPARTMENT OF ZOOLOGY
AQUACULTURE RESEARCH UNIT
35, Ballygunge Circular Road
Kolkata-700 019 (India)
Phone: 24614981. (Ext. 306)
Fax : 91-033-2476-4419
E.mail- samirban@vsnl.net
20.11.06

FOREWORD

Dr. S.K. Sarkar has written in detail a book entitled "Freshwater Fish Culture" remarkably felicitous for fishery students. It symbolizes a great deal of effort and my consent goes to the author. Fish culture/aquaculture generally responds attentively to the conditions in the farmers' community. Fish culture provides not only low cost animal protein for the human society but helps foreign exchange earnings as well. Although the basic fundamental aspects offish culture are more or less similar, they also differ in many ways from country to country and even from region to region. Fish culture will differ in important respects depending on the development of farming strategy. Hence it is inevitable to have a freshwater fish culture book that responds commendably to the remarkable features of state-of-the-art culture system.

There is beyond doubt that this book will be more useful for students and farmers to learn fish culture and post-harvest lessons from a book that attempts to reflect the farming strategies for the nation in general and the farmers' community in particular.

This book uniquely affords an opportunity for the students, teachers and fish culturists to make a preference and to make it with the knowledge that the text concerned has the senility of region for culture that will definitely make it appropriate and useful to those who are actively engaged in the subject concerned.

20/11/06

(Prof. Samir Banerjee)

PREFACE

Although fish culture / aquaculture is a touch-and-go business, it is the life and soul of progressive fish culturists. Fish culture encompasses a very extensive water area. It accepts the study of both freshwater and marine fishery resources and their exploitation in full potential, the growth and yield of fish following the adoption of state-of-the-art technologies that have dealt with in different chapters of this book. It covers the relationship between fish cultivation and environmental factors and the manner in which farmers are actively involved in fish culture. It also embraces other kinds of economic activities and employment opportunities such as fishery management for sustained production, harvesting, processing and marketing as well as trade which have grown up as a result of technological development by exploiting aquatic resources. Fish culture activities no longer depend on the relationship between farmer and his aquatic ecosystem but must take into account of the economic interaction between different geographical regions. As a generalization, the success of aquaculture activities entirely depend, to a large extent, on the distribution of commercial fish species, availability of quality ecosystem, standard of living of the rural people, and the various kinds of aquaculture activities around this planet. At present, fish culture has, however, become so important that culture activities are undertaken by fish farmers and fishermen across the world.

In studying the various aquacultural activities for rural development, employment opportunities, and for human nutrition, several aspects nust be covered. First, the methods and means of production must be evaluated, the way in which fish are stocked and reared under any geographical locations such as from high altitudes to plains must be apprehended. Methods of production can transmute not only with the climate and soil but also according to the type of culture adopted, tecmological achievements, and the social characteristics of the farmer who accomplish them. Second, the allotment of aquacultural activities must be evaluated. There are several factors to be taken into attention in the location of aquacultural activities. Allotment may be affected by the techniques adopted which may alter with time as these techniques are gradually improved. Distribution of fish may depends largely on the availability of suitable culture ecosystem, but it may also hinge on comparative cost of production, availability of market for the fish, availability of transportation, economic and political conditions, market fluctuations and government economic policies.

Since population of aquatic species is so diverse and so unevenly distributed over the face of this planet that different regions have different aquacultural activities. Such regional differences lead to the trade between different regions and countries. Thus, fish marketing strategy and fish trade must be studied along with the culture-related activity.

Population explosion is one of the most important issues that must be kept in mind while fish culture and post-harvest technology are considered. Future projections made by the United States indicate that the world population will increase 41 per cent by the end of 2050 to 8.9 billion people. As a corollary, emphasis should be given to culture- and capture-related industries by adopting (i) genetic engineering, (ii) pollution control devices, and (iii) conservation strategies of fish germplasm and fish biodiversity to a large extent.

A text on the subject like Freshwater Fish Culture which is environmentally specific must be considered as great importance. The Second Volume is the continuation of the first one and is a comprehensive and full-fledged text on different major areas of the Freshwater Fish Culture that addresses specifically to the students of fisheries science. A student of fisheries science should have a proper idea of the several issues involved in fish culture.

Each chapter starts with an introductory panel, which summarizes the cotents. It is followed by flowing main text which discusses the main subject concerned. Supplementary informations have been given in panels and tables.

Illustrations play a key role in this book. Diagrams are a special feature of the book and have been authorized to further expand upon themes within the text. At the end of the chapters, some appendixes have been incorporated. Finally, the book ends with a glossary furnishing definitions and brief explanations of important technical terms. Through 26 Chapters (Vol. 2 & 3), however, the book comprehensively covers the aspect of Freshwater Fish Culture programs of the Graduate and Post-Graduate students of fisheries science.

Similar to Volume 1, this Volume has also been presented in a simple style. Clarity has been given top priority throughout. Utmost care has been taken to recite the most intricate ideas in a simple and easy-to-follow style.

Significant advances in fisheries science that are important in fish culture and fisheries continue to be emphasized. The book incorporated many informations motivated by the fact that students of fisheries science (and also fish culturists) desire for a systematic information and a shorter text. It is hoped that the book will be of yeoman's service to fishery students.

A large number of tables and panels have been provided to make the study easy and interesting. While panels have illustrated the various aspects noted in a chapter assembled from different sources such as books, magazines, and websites, tables and figures have been provided to further support the learning process.

This book would have not been possible without the co-operation extended to me time to time by Sharmila (Ratna) and Kalyani. Professor Samir Banerjee of the University of Calcutta has written the foreword of this Volume. In a few condensed lines he has highlighted the uniqueness of this Volume and commended it to teachers, students, progressive fish culturists, government and non-government organizations, planners, and entrepreneurs. The author is inmensely grateful to him. The Central Inland Capture Fisheries Research Institute (CICFRI), Barrackpur, West Bengal, has provided the library facility to the author in bringing out this book. The author is grateful to the CICFRI.

The author looks forward to receive more current informations from readers in the future.

Budge Budge (South 24 Parganas)   **S.K. Sarkar**
West Bengal, Kolkata

CONTENTS

13

Application of Recirculatory Systems in Fish Culture

Intensive, semi-intensive, and super-intensive fish culture systems are the three important strategies to be seriously considered for high production which involves stocking of fry/fingerlings at high rates, use of chemical fertilizers and organic manures, and supplementary feeds. The amounts used being substantial, an increase in the load of organic matter, nutrients, and waste materials which have an inimical effect on water quality variables. Reduction in the load of these materials from fish culture ponds/tanks through several treatments following recirculatory systems or by converting waste products into harmless compounds that are conducive to growth of farmed fish, is considered as a wise practice for reuse of water in fish culture.

13.1 General Considerations*

Ues of recirculatory systems has their irnportance in fish culture/aquaculture to a large extent (Figures 13.1.a & b). Decades of experience coupled with research on modren fish farming and on induced breeding have shown that while farming and breeding are relatively convenient, hatching and rearing of spawn/fry/fingerlings in commercial quantities face a lot of hassles by fish farmers. Also, there is a high mortality rate during the critical period of hatching.

Methods of fish farming and breeding in hatcheries differ considerably with the species under culture, the level of the technology, and the level of the mode of operation. At the same time, for better growth and reproduction, quality water must be used. Surface water from streams, rivers, lakes, and reservoirs may be less expensive to use. But in cases where there is insufficiency of quality water for use in fish culture, sophisticated filtration systems (See Chart 13.1) may be used to eliminate toxic substances from the pond water. As a generalization, sand or gravel filter with back-flushing will make the water suitable for use in aquafarming industries.

Recirculatory systems essentially help to recondition the pond water by decreasing or eliminating metabolic by-products and unconsumed food materials. Consequently, the water quality variables attain to an acceptable level before reuse in fish culture. Reconditioning facility involves (1) Biological filtration and/or solid filtration, (2) Biological filtration and/or ion-exchange filtration, (3) Water sterilization, and (4) Oxygenation by aerators.

---

* For further detail, see Staudenmann et. al. (1996).

CHART - 13.1
FILTRATION SYSTEM

Filtration of pond water is essential for intensive fish culture and hatchery units to rid the fish pond of toxic wastes, undesirable particles, and other dissolved chemicals. There are many different designs of filter and their mode of action may be biological, mechanical, or chemical, or a combination of these systems.

*   *Biological Filters*: Biological filters remove nitrite and ammonia from the water (See Chapter 17, volume 1 for a full description of the nitrogen cycle). They work by providing a home for bacteria that convert these waste substances into harmless compounds.

*   *Mechanical Filters*: Mechanical filters remove particles by forcing water through some kind of filter cartridges. These cartridges contain filter media that trap particles as small as 3 microns across and can be used periodically to scrub the water of algal blooms and bacteria.

*   *Chemical Filters*: Chemical filters remove dissolved substances from the water such as ozone, chlorine, heavy metals, and medications. These filters work by forcing the water through a filter medium of activated carbon (a manufactured form of carbon that is highly porous). It is a carbon that has been heated or treated to increase its adsorptive power.

*   *Undergravel Filter*: In many cases, undergravel filter which is a simple biological filter is used. Internal power filters also provide biological filtration, but can include additional media for purifying water by chemical and mechanical devices. In an undergravel filter system, a perforated corrugated or ridged plate is placed on the base of the tank and then covered by a substrate of gravel. The plate permits water to flow under the gravel, while the gravel particles (3 mm in diameter) help to ensure good water movement and establish the biological filter medium.

Biological filters such as underground and trickle designs are often teemed with external power filters which pump water through an external canister containing filter media such as filter wool or activated carbon. The filter water is then sprayed back into the pond through fine holes of a spray bar. It is a device that helps oxygenate the water.

*   *Ozonizers and Organic Waste Skimmers*: Ponds may well be equipped with ozonizers – units that generate bubbles of ozone gas to oxidize waste matter or organic waste skimmers. It is a more advanced method to remove potentially dangerous organic waste materials. Their use in hatcheries has, however, not been progressed very much beyond the experimental stage in some developed countries.

* *Water Sterilization*: Ultraviolet (UV) light is a powerful sterilizing agent, capable of destroying parasites, bacteria, and even tough algal spores. Progressive fish culturists use sterilizing apparatus that pass water from the filter over a UV lamp before returning it to the tank/pond. There are some evidences that use of these lamps reduces the risk of disease incidence considerably. It should be added as a note that the UV tube in the sterilizer unit should be replaced every six months, because output tails off quickly. Looking directly at this light source will damage the eye.

In different types of systems, there is the need for a regular supply of make-up water at the rate of at least 10 per cent of the total volume to compensate for the loss due to filter back-flushing and evaporation.

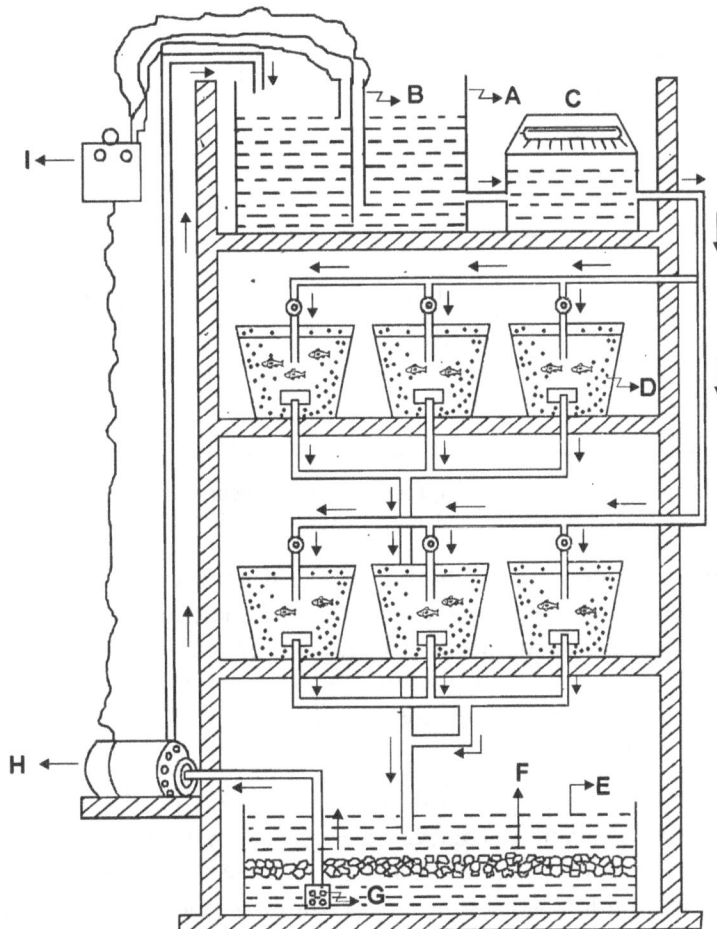

Fig. 13.1a : An indoor water recirculatory system : A. Upper reservoir; B. Probes; C. Ultra-Violet chamber; D. Experimental trough; E. Lower reservoir; F. Biological filter; G. Foot valve; H. Pump, I. Water level controller

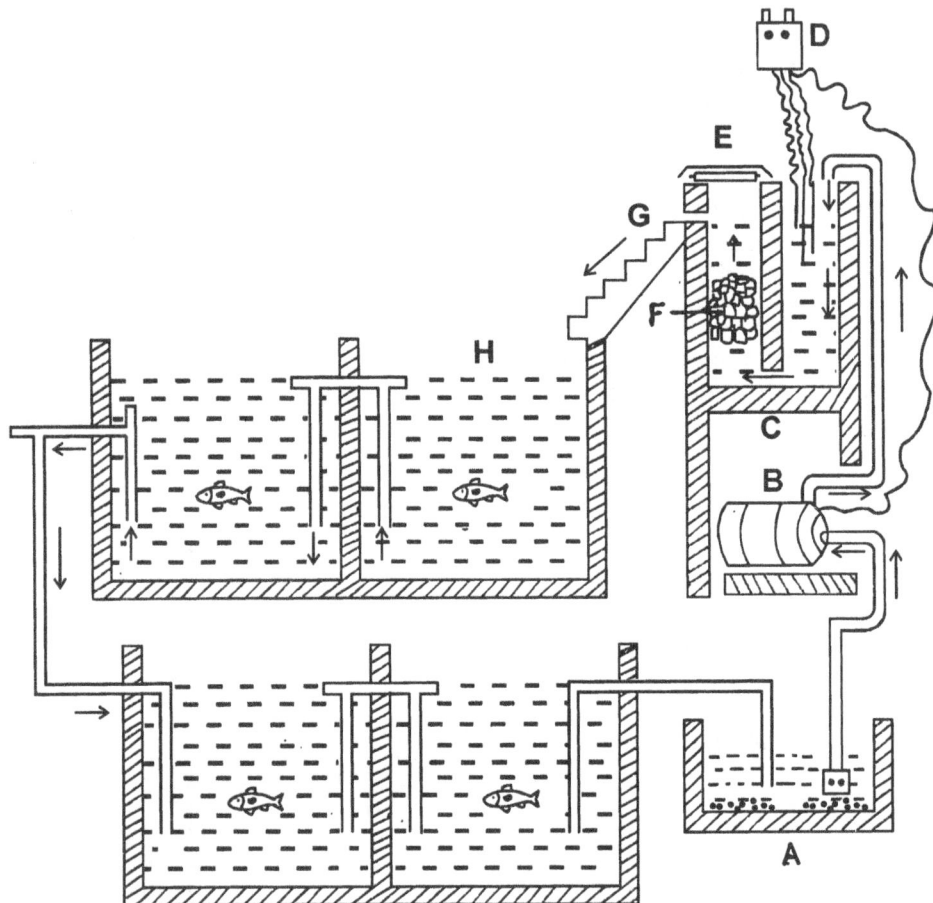

Fig. 13.1b : An outdoor water recirculation system : A. Lower reservoir; B. Pump; C. Upper reservoir; D. Water level controller, E. Ultra-violet light; F. Biological filter; G. Cascader; H. Fish tank

It is seen that static pond culture harbours a vast variety of disease-producing micro-organisms causing diseases of fish. Besides these, other toxic metabolites such as ammonia, nitrate, and nitrite are also generated in pond water. Gradual accumulation of these metabolites decreases the production potential of fish ponds. It is, therefore, inevitable to remove these unwanted materials from pond water to make the water body more suitable for further use in fish culture.

Effluent water of grow-out ponds that contains solid particles, is allowed to settle first to precipitate organic substances derived from faecal matters of fish and supplementary food. The sedimented effluent is then passed through ion exchange filter (Panel 13.1) or biofilter for reducing the level of anmonia in water. Although different types of filter materials are available, mostly a layer of sand and gravel is utilized to harbor heterogeneous microbial comnunity including bacteria, protozoa, fungi, and rotifers. The most frequent bacteria include *Zooglea ramigera, Beggiatoa alba, Sphaerotilus natans, Achromobactor* spp.,

PANEL - 13.1
ION EXCHANGE

Ion exchange is the reversible exchange of ions of the same charge between a solid and an aqueous solution in contact with it. The exchange of ions generally occurs without destroying crystal structure or disturbing electrical neutrality. The process widely occurs in nature particularly in the absorption and retention of water-soluble fertilizers by soil. If potassium salt, for example, is dissolved in water and applied to soil, potassium ions are absorbed by the soil and sodium or calcium ions are released from it. The soil, in this case, acts as an ion exchanger. Ion exchange is, however,accompained by diffusion and occurs most easily in crystals having one- or two-dimentional channel ways where ions are relatively weakly bonded.

Synthetic ion exchange resins consist of various co-polymers having a cross-linked three-dimentional hydrocarbon networks in which many ionizable groups are attached. An anionic resin has negative ions built into its structure and therefore exchanges positive ions. Conversely, the cationic resin has positive ions.

Ion exchange also occurs with inorganic polymers such as zeolites in which positive ions are held at sites in the silicate lattice. These are used for water-softening in which calcium ions in solution displace sodium ions in the zeolite. The zeolite can be regenerated with sodium chloride solution.

In most practical application, the co-polymer is used as an ion exchange resin. The entire polymer molecule, as its sodium salt, acts as a gigantic, insoluble anion which can exchange its cations with those in the solution that flows through it. Such resin soften water by exchanging sodium ions from the calcium and magnesium ions in hard water and are widely used in water treatment plants.

Ion exchange resins which are used in sediment effluents to remove hazardous substances, are synthetic organic polymers containing side groups such as $-NH_3^+$ to which anions 'X' are attached. The exchange reaction is one in which different anions in the solution displace the anion from the effluent substance. Similarly, cation exchange occurs with resins that have ionized acidic side groups such as $-COO^-$ or SO_2O^- with positive ions 'M' attached.

Flavobacterium spp., *Pseudomonas* etc. that plays a key role in the mineralization of dissolved organic matter in the effluent. Filters are of two types such as (1) slow sand filter and (2) rapid sand fiilter. While the slow sand filter contains several feet deep layer of fine sand particles, a rapid sand one contains coarse particles of gravel. Within the sand, however, there forms a layer of microorganisms which acts as an additional filter. This is termed as *dirty layer*. Most of the ammonia is removed from the effluent through nitrification within the filter. The filter provides a large surface area so that bacteria can be attached. Effluent may be passed through two or more filters or recirculated through the same filter with occasional changes of filter materials.

To remove the ammonia and carbon di-oxide from the water, a buffer material such as limestone is used as a part of the biofilter. After passing through the buffer medium, the water is then sterilized by using ozone or ultra-violet methods.

13.1 Sources of Waste Materials in Grow-out Ponds

The main source of pond water pollution includes food-derived waste materials, metabolic wastes, fertilizer-derived waste materials, and algal blooms. Pollution mostly stems from the combination of waste materials developed and feeds used in fish ponds.

Food-Derived Waste Materials

Organic matter and nutrients (particularly nitrogen and phosphorus) are developed in fish ponds during intensive and semi-intensive culture systems. More than 60 per cent of the total nutrient content in fish food are not utilized by fish. This excess amount of food contributes substantially to waste materials and becomes an integral part of the total waste load of the pond ecosystem. Accumulation of excess food materials makes the water quality unfit for fish culture.

Metabolic Waste Products

Faecal matter and excretory products of fishes significantly contribute to the production of toxic substances and their accumulation depends primarily on the stocking density of fish in ponds. Most of the nitrogen is excreted by fish in the form of ammonia. About 98 kg of N has been noted to excrete for every tonne of tilapia production. If the nitrogen content in faecal matter is assumed to be about 4 per cent, 18.4 and 54 kg of N are excreted respectively as faeces and urea as well as ammonia.

Fertilizer-Derived Waste Materials

Application of chemical fertilizers and organic manures is very important for intensifying fish production in grow-out ponds. But over-fertilization increases the levels of nitrogen and phosphorus and cause pollution primarily because they stimulate growth of water-loving plants and algae. Development of algae leads to plankton blooms. Blooms either generate obnoxious odors and tastes in waters or produce toxic metabolic products which can result in major fish kills. Development of algal blooms in pond water and their subsequent microbial degradation depletes the content of its oxygen.

13.3 Definition of Recirculation

The concept of recirculation of pond water in fish culture is unequivocally connected with the farming activities. Recirculation can be defined as waste water which has been influenced by farmers to such an extent that it has a higher concentration of dissolved or suspended constituents than the maximum permissible level recommended in national or international standards for safe use in fish culture.

13.4 Necessity of Recirculation

Waste-loaded pond waters require some degree of treatment which is then used in fish culture ponds. The considerations given below are intended to assist the entrepreneurs so as to enable to adopt precise measures to incorporate the system of recirculation to increase the rates of survival and growth of farmed fish.

1. The recirculatory system is necessary to reduce the risk of disease/parasite infections considerably.

2. Oxygen can be replineshed through aeration and most of the carbon di-oxide is dissipated, the removal of metabolic products especially ammonia involve more complex systems.

3. This system of farming highly improves survival and growth performance of fish due to high degree of control over the water quality.

4. This system eliminates water quality problems.

5. Recirculation of waste-loaded pond water reduces potential pollutants which assures the availability of quality water for fish farming where the source of fresh water is limited.

13.5 Advantage of Recirculatory System

Due to severe pollution of static intensive culture system, it is necessary to encourage progressive fish farmers by adopting closed recirculatory systems for fish production all the year round. The most important advantage for adopting such system is that relatively small areas are enough to yield commercially-viable fish biomass. Also, the requirement of water is very less which is very important particularly where there is scarcity of water. This system will definitely improve the quality of pond water.

13.6 Aeration Versus Pond Recirculatory System

Recirculatory system in fish culture is mostly confined to relatively smaller volume of water that ranges from a small aquarium (less than 200 L to 100 cu m) to an experimental fish culture pond. The conventional recirculatory system generally involves provisions for aeration, nitrification, water recirculation, solid waste removal, temperature control, denitrification, ozonation, and UV irradiation.

The aeration system differs markedly from the pond recirculatory system with respect to photosynthesis. The photosynthetic aspect in this system is very high which causes greater oxygenation and ammonia conversion in culture systems. At the same time, water replacement and energy use are essentially much less in pond recirculation. Table 13.1 shows the comparison between the two systems of recirculation.

Table 13.1 : Difference Between Pond Recirculatory System and Aeration Recirculatory System

Feature	Pond Recirculatory System	Aeration Recirculatory System
Size	Large pond (more than 0.5 hectare)	Small (Less than 200 litre capacity)
Culture system	Ponds	Cisterns
Aeration	Mechanical and photosyntheis	Mechanical
Denitrification	Pond bottom mud	Anaerobic component is necessary
Nitrtfication	Photosynthesis and pond surface	Inactive component
Exchange of water	0-2 per cent	0-10 per cent
Photosynthesis	Most important	Not or less important
Energy use	Less important	Most important

13.7 Types of Recirculatory System

Closed recirculatory system involves several methods by which water quality problems and the infection of fish with pathogens can be fully solved thereby improving growth performance and survival of fish. Although these systems are followed in finfish and shellfish farms all over the world, identical design is yet to be reported.

Zero-Water Exchange System

This system is characterized by recirculation, zero-water discharge, sludge removal, and fish culture. Each system includes the followings: three numbers of one hectare fish grow-out pond, one hectare water treatment pond for the culture of fish (such as carp and tilapia) and bivalves (*Lamellidens* spp), a cement cistern for sedimentation of phytoplankton, a sludge-settling pond, and a sludge-drying bed. Sludge is periodically removed from grow-out ponds through the settling pond. In this system, fish is stocked in each pond at the rate varying between 8,000 and 10,000 number/hectare. Since this system requires day and night attention, a software package may be used for water quality monitoring.

In-Pond Treatment System

In this system, a number of mussel rats with trays are suspended in the centre of the pond. Mussels help to remove excess algae and suspended solids. Four aerators are placed in such a way that the pond water is circulated and at the same time waste materials are concentrated at the centre of the culture pond.

Pond-in-Pond Recirculatory System

This system consists of two ponds: the first pond is utilized for intensive culture and the second one is for extensive farming. The intensive pond is somewhat deeper than the extensive one and is provided with paddle wheel aerators. Different species of fish such as carps, fresh water prawns, and bivalves are stocked in extensive ponds at lower densities. Effluents are discharged into the extensive pond from the intensive one where solid wastes are settled. Nutrients which are derived from the waste materials trigger to stimulate pond productivity that pushes the production capability of the extensive pond to a certain limit.

Drainage Canal System

In this system, a fish farm is circumscribed by a number of canals. These canals and farm ponds are filled up with water and the water is allowed to stand for three months. Water of the drainage canal is then circulated to the pond via the supply canal. The canals are provided with a number of aerators for aeration of the water. Canals are also used as sedimentation beds. Using this system, however, fish production to the tune of about 8 tonnes/ hectare/crop can be achieved.

Earthen Pond System

This system is often considered to be the most appropriate type of waste-water treatment for developing countries. The system consists of a series of ponds in which bacterial and algal growth can occur in a symbiotic manner. Bacteria utilize the organic matter from which they produce inorganic nutrients which are used by algae. Algae, in turn, produce oxygen through photosynthesis. This oxygen is consumed by bacteria. This co-existance of bacteria and algae for the benefit of each other is termed as *symbiosis*.

This system is characterized by the establishment of at least ten fish culture ponds each having 0.5 hectare area, four 0.5 hectare wastewater treatment ponds, two reservoir ponds, and four culture ponds. Water is stocked for several days (15-20 days at least) into the first reservoir pond for settlement of solids. This water is drained into the second reservoir pond for further settlement and then finally drained into the culture ponds. The drainage canal is stocked with about 200 stacks of mussel. The water is then entered step-by-step into a series of four treatment ponds. In the first pond, the water is aerated by four aerators. The next two are used as sedimentation ponds. After sedimentation, the water is allowed to enter into the fourth pond to enhance the level of nutrients and algal blooms. This pond is used to produce carp and tilapia. Development of algal blooms is very important since it absorbs carbon di-oxide and amnonia from the water. Under this culture, however, carp production to the tune of at least 8,000 kg/ha/crop in 4-6 months is possible in addition to the yield of tilapia at the rate of 5,500 kg/ ha/crop in 3 months culture.

Since earthen pond system has several advantages such as low technology and low operation cost, there are some disadvantages as the ponds have a long retention time (20 days) and consequently, require a large land area. This is a function of sunlight and temperature and hence these systems are most suited to lower latitudes. At the same time, effluent water in ponds generally contains different species of algae in large numbers which may affect the carbon di-oxide budget of receiving waters.

13.8 Filter System With Recirculation of Water for Fish Larval Rearing

Some developments in connection with the biological and mechanical treatments of water have been made for small-scale fish and prawn hatcheries to maximize the yield of fish hatchlings. For this purpose, two types of filters such as (1) rapid sand filters and (2) biological filters have been developed and the cost of these units is so cheap that rearing of fish hatchlings is specially suited for backyard and small-scale hatcheries.

Sand Filter

For filtration of contaminated freshwaters, a gravety filter is constructed on a cement platform with the following components: a one tonne capacity tank, one filter-in-filter device with coarse sand, one velong screen (20 mm mesh), coarse sand layer (1-5 mm in size), and gravel layer (5-15 mm in size) (Figure 13.2). At the bottom of the tank, a 10 inch deep sand layer is spread. Just above the sand layer, another deep coarse gravel layer of 10 inches in thickness is made which is followed by a velong net and this acts as separator. An outlet pipe with a tap-cock is fitted one feet above the bottom of the tank. A small basket filter with 20 nm mesh nylon cloth and coarse sand with a small perforated pipe is placed in the centre of the tank which acts as a filter-in-filter device. The coarse sand layer in the basket is periodically replaced.

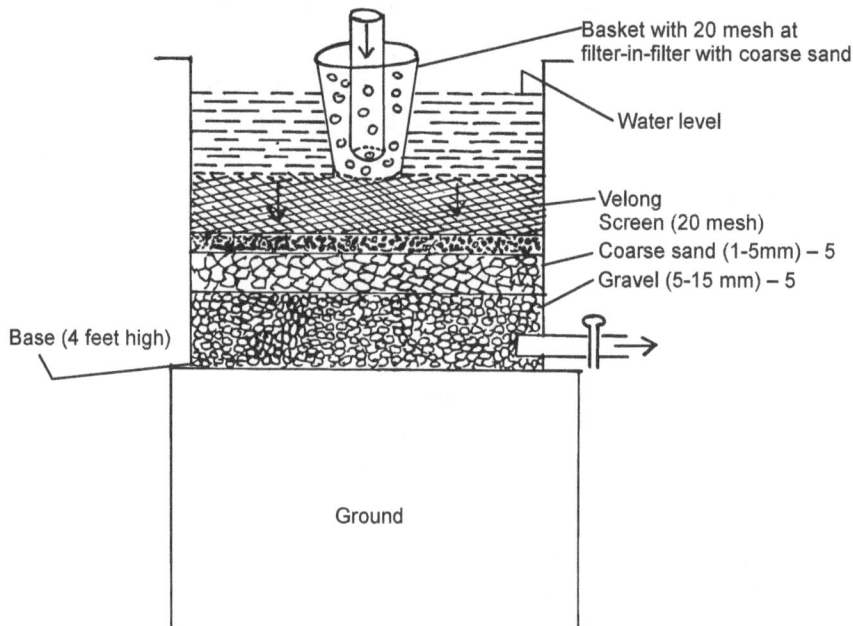

Fig. 13.2 : A Rapid sand-filter for filtration of contaminated freshwaters.

Biological Filter

Treatment of pond water with biological means involves the culture of bivalves and it is relatively a new application of ecological engineering (Figure 13.3). Bivalves can filter enornous quantities of water. Great productivity of filtration apparatus coupled with high efficiency of particulate matter retention determines such bivalves as natural powerful biofilters. Physiological features of common mussels such as *Mytilus edulis, Modiolus modiolus, Lameklidens* sp., *Dreissena* sp. etc., their world-wide distribution, and resistant to pollutants have attracted epoch-making attention to mussel culturists for solving the problems of fish culture in an ecologically sound manner. Choice of the biological species for ecological engineering is dependent on their physiological abilities that follows in pond water recycling.

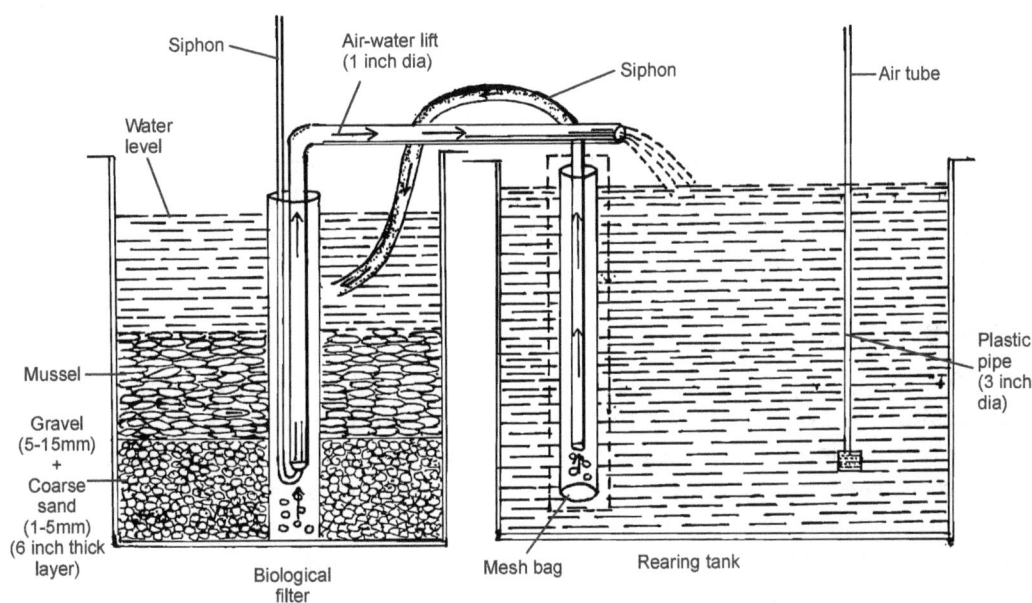

Fig. 13.3 : Biological filter with recirculation for larval rearing.

Filtration

The filtration rate generally reflects the level of feeding activity and is defined as the quantity of water cleared per unit time. Cirri of the gill act as filter organs that efficiently retain suspended particles in the water. The retention capacity in different species of bivalves varies from 85 to 3,900 ml/hour (Table 13.2).

Stimulation of Primary Production

Metabolites excreted by mussels such as dissolved organic matter possess certain biological activity in relation to producers. Metabolic products greatly influence the growth of phytoplankton and at the same time, control the development of bacterial populations. Thus mussels promote recycling some of their nutrients for the benefit of the pond ecosystem.

Table 13.2: Filtration Intensity of Some Bivalves per Gram of Dry Weight Tissue

Species	Filtration rate (ml/l)
Lamellidens sp.	85
Dreissena sp.	85
Mytilus edulis	386 - 3,900
Modiolus modiolus	240

Other Abilities

1. Mussels promote the circulation of water in ponds.

2. Continuous monitoring of water quality flowing in and out of the purification cycles is possible by bivalves due to their bioaccumulation features.

Though several major factors such as temperature, availability of food, suspended matter, and toxic substances limit the physiological activity and growth rate of mussels in different environments, such ecological engineering makes some contributions to sustainable development of fish culture ecosystems.

13.9 How Biological Filter is Established?

A 600 litre capacity of rectangular tank (8 X 4 X 4 feet) is filled with gravel (10 inches in thickness) at the bottom. Above this gravel layer, a four inch thick coarse (1-5 mm in size) sand bed is placed. This coarse sand layer is followed by a thick layer (10 inches thick) of fresh water bivalve. An air-water lift (3 inches diameter and one foot length) is fitted in the middle to withdraw the treated water into the rearing unit. The pipe is perforated in the region where it is embedded in the gravel and it is wrapped with a mesh bag. Water-lift plastic pipe having a diameter of 1.5 inch is inserted into the air-water pipe plastic siphon (1.5 inch diameter) is inserted into the rearing tank through which water is returned to the filter unit for purification. The siphon tube is inserted into the hard perforated plastic pipe which is covered by a nylon net so that hatchlings and their food organisms are not removed during the operation of the unit. Bivalve mussels completely remove toxic metabolites from rearing water developed during the metabolism of hatchlings. It should be added as a note that a single biological filter unit can able to purify the water of two rearing tanks at a time.

13.10 Ozone-Related Water Treatment

From the last 100 years or so, the use of chlorine in water and waste water-treatment has been in practice. To make the pond water suitable for fish culture and to treat the waste-waters generated from aqua hatcheries, aquafarms, and fish processing plants, chlorination is executed. But due to its inimical role in ozone layer in stratosphere, global warming, toxic and carcinogenic effects, the chlorine gas has recently been replaced by Ozone (Panel 13.2) which is cheaper, cleaner, safe, and eco-friendly.

PANEL - 13.2
OZONIZATION IN FISH CULTURE

What is Ozone? Ozone is a tri-atomic allotropic form of oxygen formed by recombination of three oxygen atoms in its molecule. It is an unstable, pungent, toxic form of oxygen and formed in electrical discharges or by ultra-violet light.

Characteristic Feature of Ozone :

(i)   It is a colorless gas with characteristic of pungent odor.

(ii) It is 30 times more soluble in water than oxygen.

(iii) It is a very strong oxidizing agent: 3,000 times faster disinfectant and 2 times stronger oxidizing agent compared to chlorine.

(iv) It is rapidly deodorized, oxidized, decolorized, and disinfected.

(v)   It is instantaneously destroyed bacteria.

How is Ozone Produced ?

(i)   *In Stratosphere* : Within the stratosphere, Ozone is produced from Oxygen by a photochemical reaction in which energy from the sun splits the Oxygen molecule to form atomic oxygen. In this process, short-wavelength Ultra-violet radiation (242 nano metre) is involved. Ozone formation in the stratosphere can be described by the following pair of reactions:

* In the first, atomix oxygen (O) is formed by photolytic decomposition of molecular oxygen (O_2)

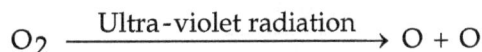

$$O_2 \xrightarrow{\text{Ultra-violet radiation}} O + O$$

* The atomic oxygen, in turn, reacts rapidly with molecular oxygen to form ozone:

$$O + O_2 + M \rightleftharpoons O_3 + M$$

Where M represents a third body (N_2 or O_2) necessary to carry away the energy released in the reaction.

Both these reactions are reversible, but the ozone content in the stratosphere is sufficiently constant. Ozone is being produced from oxygen as fast as it is broken down to molecular oxygen.

(ii) *In fish Culture*: In fish culture, ozone is produced through an ozonizer machine using a special device termed as 'Corona Discharging Technology'. The oxygen required to generate ozone is directly taken from ambient air. The atmospheric air is pre-filtered to eliminate excess humidity and dust particles. The filtered dry air is compressed and then passed through air-cooling systems. The system maintains the temperature of air. This dry-cool air is sucked by a thermetically-sealed ozone-generating cell charged with high voltage (5,000 - 20,000 volts) discharge electronic circuit. The charged ozone-generating cell excites the oxygen molecules of contained air and generates ozone molecules under the following chemical reaction:

$$3O_2 \longrightarrow 2O_3$$

Since normal air contains 19 per cent oxygen by volume, this entire amount of oxygen gets converted to about 3 per cent ozone by volume through ozonizer.

Features of an Ozonizer: Four characteristic features are noted in an ozonizer:

(i)   It is fully automatic and the maintenance cost is practically nil.

(ii) It is highly economical in use as ambient air is taken as feed gas.

(iii) It is easy and safe to operate.

(iv) It needs only electric current to operate the ozonizer and no other chemicals are required.

Benefits of Ozone Application

(i)   Ozone prevents the occurrence of water-borne diseases.

(ii) Ozone reduces biological oxygen demand, chemical oxygen demand, and toxicity load to a trace level.

(iii) It favors increased growth of plankton which, in turn, enhances the pond productivity.

(iv) It converts toxic nitrite into non-toxic nitrate through oxidation.

(v)   It reduces the turbidity of water and neutralizes obnoxious gases such as ammonia, carbon di-oxide, hydrogen sulfide, and methane.

Application of Ozone

The Ozonization not only effectively eliminates different organic compounds by neutralization and oxidation but completely destroys pathogenic micro-organisms as well. The recommended dose of ozone varies between 0.5 and 1.0 mg/l) and the contact time of 1-2 minutes is much effective that makes the water safe from contamination and pathogens. The safe limit of ozone exposure to fish culture ecosystem has been found to be 0.003 mg/l however.

13.11 Conclusion

Recirculatory system is indispensable for sustainable fish culture and as a principle that forces to develop ecological engineering design. This system creates greater efficiency and productivity of fish ponds. The application of ecological engineering principles to water pollution control in fish culture ecosystems can reduce treatment costs. Fish culture systems generate large volumes of nutrient-loaded water. Since nutrient mass loading is the critical factor contributing substantially to ecosystem degradation, treatment of pond water treatment is inevitable. Fish culture effluents are difficult to treat because they contain relatively dilute nutrients.

The use of some ecological engineering principles permits production of high-value fish crops while meeting stringent nutrient/toxicant discharge regulations. The cost of water treatment can be reduced through use of a less-expensive technology and this technology will definitely remove toxic metabolites to consistently less than toxic concentrations without drastic reduction in fish productivity or ecosystem quality.

Recirculatory system should be adopted in such a way that it would facilitate some of the important criteria such as (1) effective removal of dissolved organic matter, (2) cost-effective removal of suspended solids, (3) removal of ammonia, methane, hydrogen sulfide, and solid wastes, (4) high rate of water turn over, (5) adjustment of pH and feeding practices, and (6) management of the culture system to avoid the risk of fish kills in ponds. Adoption of recirculatory system will definitely transform the abandoned fish farm into highly productive one.

Reference

Staudenmann, J., A. Schonborn, and C. Etnier. 1996. Recycling the Resources. Transtec Publications, Switzerland.

Questions

1. What is recirculatory system? Why it is very important in fish culture industry ?

2. What are the source of waste materials in grow–out ponds? How waste materials are formed. What are the effects of waste materials on fish biomass?

3. State why the method of recirculation of pond water should be adopted in fish ponds? What are the advantages of such method?

4. Discuss different methods of recirculation generally adopted to treat effluent water for further use in fish culture. Which system is the most appropriate type of effluent treatment and why?

5. Write notes on the following: (a) zero-water exchange system, (b) sand filter, (c) biological, filter, (d) ion exchange, (e) algal blooms, (f) ozonization, (g) use of microbial community in recirculatory system.

14

Health Management in Fish Culture

Management of fish health is becoming an increasingly important strategy of sound sustainable fish culture. This involves the implementation of several management programs. Such programs not only increase the productive potential of ecosystems but simultaneously provides good health of farmed fish for sound economic returns.

Four major health management strategies to be primarily considered for sustained yield include (1) quarantine, (2) nutrition, (3) pollution, and (4) diseases. In the developing nations, significant progress has not yet been made towards these strategies to arrive at the desirable goals of fish health management whereas in developed countries, these strategies are most significant. This significant achievement is most notable in Japan, USA, and Canada. The lack of different planning protocols and reluctance to implement programs in most developing nations account for this constraint.

Disease outbreaks provide serious limitations in all fish producing countries of the world but are severe in most tropical and sub-tropical countries where the goal of improving health conditions of fish in particular and ecosystem in general has given very low priority. In addition to other fish culture activities, due attention must be given towards planning, programs, systems, and evaluation of health condition of fish culture ecosystems. With such challenge, fish health management may be referred to as *ecosystem health* . The problem of fish health management lies in the even more rapid rate at which farming activities are increasing. Health management programs are highly encouraging in several countries. In many other countries it is declining.

14.1 Rationale*

Most of the tropical countries produce finfish and shellfish from different aquatic resources through systems management. The maximum production depends upon a number of factors such as environmental conditions, availability of quality water, nutrition research information, state-of-the-art technology, and diseases. Incidence and degree of severity of aquatic pollution and disease outbreaks largely depend on the quality of environment and hence the most important step to be considered in fish culture health management is to afford the best quality ecosystem within the culture unit.

---

* For further detail, see Meyer et. al. (1983) and Langdon et. al. (1992).

The best quality aquatic ecosystem with respect to fish refers to a water area close to its natural ecosystan. This includes arrangement of good physical, chemical, and microbial quality of water, adequate space for swimming and feeding with a nutritionally-balanced diet. Though it is practically not possible to arrange a near-neutral ecosystem to a species under culture conditions, it is nonetheless obligatory that a quality ecosystem must be maintained as far as possible for sustained production.

In a culture system, several conditions involved in fish culture ecosystems include high stocking densities, incorrect feeding in both quality and quantity, inadequate quality of water and soil, disease outbreaks, and pollution which drastically affect the health status of farmed fish. The fish under these conditions are termed as *stress*. Once the fish are stressed, they are succumbed to primary and facultative pathogens to cause diseases. Thus, quality ecosystem is by far the most important constituent of an effective health management strategy in fish culture/aquaculture industry.

Effective health management programs involve (1) disease control, (2) systems management, (3) research and information, (4) training and extension, (5) control of pollution, (6) quarantine, (7) nutrition, and (8) management of pollutants. These programs should be strictly considered when present and future production potential is sustained. All these programs should be implemented and enforced after carefully considering the past activities, present status, future commands, and the institutional as well as manpower capabilities in the area concerned. In the following sections, we will discuss elaborately each of these effective health management programs one by one that should be adopted during fish culture operations.

14.2 Disease Control

There are a number of approach to disease control. These include chemotherapy, immunization, quarantine and environmental, nutritional and genetic manipulations. Prevention of fish diseases is far more economical and effective than other methods of control and must be included in health management programs. Fish diseases can, however, be controlled by maintaining correct stocking densities ans using quality water.

Chemotherapy, the use of chemicals and drugs for controlling fish diseases, is a widely practised management strategy; but ecosystem conditions must be evaluated before their extensive use is permitted. It should be noted that the hit-and-miss use of antibiotics, drugs, and chemicals without proper diagnosis has been the common incident among fish culturists. Though the short-term and immediate losses of fish stocks may be high and it takes time to conduct diagnostic tests and to ascertain adequate treatment procedure, irrational use of these ingredients is likely to cause more serious damage in the long run.

Prophylactic therapy, sanitation and maintenance of hygiene of culture systems are also important in fish health management. Prophylactic treatment before, during or after the management procedures such as stocking, grading, transport, and the like, can protect fish from subsequent disease outbreaks.

Vaccination

Vaccination, also termed as *immunization*, is another diseases-preventive measure. Although the knowledge on diseases and their etiological agents have significantly advanced, only a few effective commercial vaccines are available. Their use is mainly restricted to species in temperate regions though immunization is not the ultimate solution for prevention and control of fish diseases, due importance should be given to aquaculture as it could be a crucial part of an overall planned health management strategy. It is expected that more commercial vaccines would be available in the market in future to maintain the health status of fish in tropical and sub-tropical regions.

Immunostimulation

Non-specific defence mechanisms in many species of fish are more resistant to environmental stressors. Thus as an alternative to vaccines, immunostimulators offer potentials for prophylactic use in fish culture. A number of immunostimulators have been noted and are known to enhance the phagocytic activity of fish thus protecting them from certain pathogenic bacteria. Unless the precise role of the irnmunostimulators are known, prospects for the development of effective vaccines for commercial fish is extremely limited.

14.3 System Management

In farming systems research, the interactions between different types of components of the sub-systems must be clearly understood if the system as a whole is to be understood and managed. In the context of fish health, the systems management approach implies an understanding of the ways in which the different components of fish culture systems interact with one another and the management of these components which provides conditions optimal for the farmed fish and least favorable for the occurrence of diseases. The approach relies on both environmental and non-environmental factors that follows.

1. Water, soil, pond location etc.
2. Cultured species, stress levels, feeding practices, absence or presence of pathogens, occurrence of plankton blooms, and genetic factors.
3. Management experience of the farmer, access to capital and socio-cultural aspects.
4. Components which influence the distribution of resources and institutional support, quarantine and access to training as well as expert assistance.
5. Economic components which determine the supply and demand situation for aquaculture products and influence farm management decisions which have an impact on fish health.

Need for Systems Approach

The introduction of myriad factors which are responsible for fish health management strategy may appear to complicate the management situation to a large extent. However, the followings are the reasons which consider the systems management rather than individual components.

1. Since fish disease is the end reesult of a series of linked events, disease treatment goes beyond consideration of the pathogen. Extensive research, diagnosis and treatment protocols can lead to inappropriate cures which may temporarily eliminate the symptoms but not the causes.

2. Recent approaches have had limited success in the prevention of fish diseases. Difficulties exist in dealing with some existing problems.

3. Large-scale outbreaks of fish disease demonstrate the problems easily spread beyond the management of individual farms which need ecosystem management approaches to curb environmental degradation at the farm level and adequate steps should be taken to abandon the introduction of pathogens.

Fish Health, Stress and the Ecosystem

The ecosystem where fish are cultured, plays a critical role in the degree to which fishes are prone to pathogens and the occurrence of diseases. At the same time, an understanding of the relationship among host, pathogen and ecosystem is inevitable for understanding the cause and prevention of fish diseases.

The relationship between the occurrence of fish diseases and the ecosystem is less clearly understood particularly when the relation between disease occurrence and the stress is established. Fishes, for example, are stressed by poor ecosystem conditions, but the interaction between disease occurrence and stress in many fish species is not distinct. Enough is known about the general conditions under which healthy fish stocks exist and hence the unavailability of research data is not a major impediment to improve fish health management using the systems approach.For instance, water and soil qualities, pond bottom conditions, food and plankton bloom management are widely considered as critical for disease prevention in fish culture.

Ecosystem Impacts and Risks

Deterioration of ecosystem condition of several aquaculture systems is of serious concern not only to farmers but also to nation as a whole and place an additional stress on farmed fish. The environmental impacts affecting fish culture systems include:

1. Domestic, agricultural, and industrial pollution.

2. Discharge of pond effluents and habitat changes due to pond construction.

3. Self-pollution of water supplies through discharge of pond effluents.

Self-pollution caused by over-stocking of ponds with fish is thought to have contributed to serious disease outbreaks in several Asian countries such as China, India, and Thailand. In many cases, the impacts of fish culture effluent are difficult to seggregate from ecosystem deterioration caused by non-fish culture sources. These problems point to an obvious conclusion that fish culture farms can never be regarded in isolation from the surrounding environment. It should be kept in mind, however, that environmental problems and disease outbreaks permit certain conditions to be identified when fish farms have increased the risk of environmental problems and as a result fish stocks are succumbed to diseases. The following three criteria should be considered in assessing environmental risks.

Fish Culture System and Management

The type of culture system and its management is beyond doubt an important aspect, to be considered and some factors that follows affect the risk of culure systems.

1. Intensification of farming through increased stocking densities and feed inputs leading to (i) increased risk of water and soil quality deterioration and (ii) increased need for on-farm ecosystem management.

2. *Open versus closed systems*: While open systems such as cage farms and flow-through pond raceway systems are exposed to greater risk of external environmental impacts, closed systems offer greater opportunities for environmental control and management.

3. *Degree of reliance on external inputs* : Increased reliance on external inputs such as fish food and seed is related to intensification and also increased risk of imparting pathogens onto farms.

Water Quality Standards for Ecosystem Health

Measurement of water quality essentially involves a multi-disciplinary approach, wherein the required water quality is related to various categories of requirements such as agriculture, aquaculture, and fisheries. To adopt any management strategy stances, a cost-benefit ratio must be established between the benefit derived from certain water quality criteria and the cost of attaining the target. Such a program is invariably designed to establish the most economic long-term solution for achieving desired water quality for fish culture and other applications.

1. **Aspects of Quality Standards:** As a generalization, quality standards are based on three aspects such as (i) Chemical standards, (ii) Effluent standards, and (iii) Physical standards. Water Pollution Control Board of Federation has specified the criteria for more effective use of freshwater and accordingly, the freshwater bodies have been grouped into five categories: (1) Drinking water source with conventional treatment, (ii) Drinking water source without conventional treatment, (iii) Outdoor bathing, (iv) Water source for agriculture and industry, and (v) Developnent of aquaculture and fisheries.

(i) *Chemical Standards*: These standards are characterized by the presence of different chemical substances either at sub-lethal or at lethal levels in an aquatic ecosystem that cause severe damage to the ecosystem health and their hazardous effects on fish have become increasingly important around the world. Therefore, their presence in water must be ascertained and at the same time, the level of chemicals must be maintained following standard norms as prescribed by the competent authorities.These aspects have been described at length in Chapter 20 of this Volume.

(ii) *Effluent Standards*: Effluent standard is based on the maximum concentration of toxicants (always measured in terms of mg/l) or maximum pollution load (kg/day) discahrged into the receiving water. Since the legislation is a matter of regional concem, the Pollution Control Boards of each region have been empowered to promulgate their own standards but the optimum limits must be maintained below the standards prescribed by the Environmental Protection Agency, USA.

(iii) *Physical Standards*: Physical standards of water in fish culture ponds are characterized by the turbidity, odor, color, taste etc. Turbidity of water is mainly due to the presence of suspended matters such as clay, silt, finely-divided inorganic and organic matters, and plankton.

Odor is defined as preception of smell referring to the experience or one that is smelled referring to stimulus. Such perception is feasible through the receptors of olfactory epithelium. In almost all freshwater fish culture ecosystems, there do occur a number of biodegradable substances which are odorous and incessantly lose molecules to the environment. They are termed as *osmogenes* which generate odor or smell and the intensity of smell is expressed in absolute terms such as slight, moderate, strong, or extreme depending on the productivity of water bodies. Anaerobic decomposition of organic matter in the bottom mud mostly results in the production of various acids, alcohols, methane, hydrogen sulfide, ammonia, and a number of organic acids. The odor is, however, qualitatively described as medicinal (Phenols). fishy (due to blue-green algae), earthy (decaying matter) and chemical (chlorine, hydrogen sulfide, methane, and ammonia).

The color of water is due to the presence of humus materials in trace amounts developed from decaying vegetable matters. Some fish culture ponds contribute substantially to color of water. Water color has a marked influence on the survival and growth of farmed fish. The types of water color and their relationship with fish in polyculture systems are summarized below :

Type of color	Water quality	Fish growth	Fish mortality	Type of plankton	Instruction
Reddish brown	Excellent	Very good	Negligible	Diatom, Chaeto-ceros, Navicula	–
Light green	Good	Good	Low	Green algae and chlamydomonas	–
Dark green	Not detectable	Low	Low	Blue-green algae	Use fungicides under guidance; Check ferti-lization schedules, and use urea in proportion of phosphorus
Dark brown	Not good	Low	Moderate	Brown algae and Dinoflag-ellates	Exchange of water and check fertilization schedules
Yellowish	Bad	Very low	High	Crysophyta and green flagellate	Transform color through improved management
Clear water	Not good	Very low	–	Not found	Trace the reasons for lack of plankton growth

The color, turbidity, odor, and taste are extremely variable and depends very much upon the various factors. Temperature variations, micro-biological activities in bottom mud, management schedules, geographical conditions, and human interventions are apt factors. These parameters are based on approximation and may vary from region to region and even from pond to pond. It is, therefore, very difficult for any farmer to make a justifeable generalization on these aspects when fish culture operations are carried out.

14.4 System Approach

The emphasis of a system approach should be on the prevention rather than cure, which is likely to be the most cost-effective,involving both fish culturists and governments, and also involving on-farm as well as off-farm management. Whilst fish culturists are solely responsible for farm management , the government inputs are cardinal for rigorous control of the resource use, and for helping to provide legal arrangements which reduce resource use conflicts and ecosystem impacts.

On-Farm Ecosystem Management Options

These include site selection of farms, management of soil and water, quality and quantity of food, use of disease-free fish seed, fertilizers and manures.

Off-Farm Management

This essentially involves the management of ecosystems which support fish culture and consideration of environmental impacts such as efficient management, controlling risks associated with the introduction of exotic species and pathogens, pollution control strategies, and reducing user conflicts through equitable allocation of resources.

Advantages of System Approach for Fish Health Management

There are many advantages in promoting a system approach towards prevention and cure of fish diseases that follows.

1 A more effective solving of fish disease problems, since the system approach considers management of the factors most involved leading to disease outbreaks.

2. System approach can lead to increased self-reliance among farmers and less reliance among experts and extension workers. Given the lack of adequate manpower resources, an important consideration for fish health practitioners should be to reduce reliance of farmers on professional inputs for farm management and disease control.

3. System approaches develop solutions that are more likely to appropriate and contribute more to sustainable development.

4. System approaches can be more cost-effective as they do not depend on costly inputs.

14.5 Implication for Training and Extension

Since substantial amounts of aquaculture production (approximately 88 per cent of the world total) comes from Asia, the adoption of production strategy stances has raised some important issues for information on training and extension. The techniques developed for dealing with cultured species in one country can be applied to similar species for other countries. The most effective means of technology transfer is through training. The governments of various under-developed and developing countries realise the need to develop technology and to increase knowledge on fish health to achieve the goal of sustainable fish culture. Many institutes are now capable of conducting advanced research, and the number of experts in various fields of fish production is on the increase. Consequently, the present situation provides enormous possibilities for organizing various types of training within the region. This may lead to better cooperation among different countries.

In contrast to developed countries, developing ones require extensive training on fish health management. The number of trained staff is still limited in the region concerned and there is an urgent need to upgrade the existing staffs who are actively involved in fish health management programs. Since fish disease problems are similar in many countries, the knowledge acquired from training can easily be transferred from one country to another.

In many developing countries, training organized at an international level relies on donor funding. Without this support, the chances of setting up training programs are very slim. Therefore, donor agencies must be made to recognize the economic importance of fish health management and funding should be allocated for training in this field.

14.6 Research on Fish Health and Information*

Fish health research in many developing countries currently presents enormous challenges and opportunities. In many tropical and sub-tropical countries, research on fish disease and their correct diagnosis are not sufficient to meet the requirements of rapidly expanding aquaculture sectors. Relatively little is known about the infectious and non-infectious diseases affecting fish. It is essential that available research resources are focussed on important disease-related production constraints with the aim of affording cost-effective ways of reducing losses.

Since 1990, there have been three specific assessments of needs for fish health research such as (1) 1991 World Bank Study, (2) 1991 Asian Development Bank (ADB)/ Network of Aquaculture Centres in Asia-Pacific (NACA), and (3) 1993 Asian Fisheries Society (AFS). These assessments have meaningful aquacultural implications.

World Bank Study

In considering the potential contributions of science to fish culture/aquaculture, the world bank has recognized a major deficiency regarding knowledge on diseases of tropical

---

* For further detail, see Deomempo (1995).

and sub-tropical fish. The study has identified the following research needs and opportunities in the field of fish diseases.

1. Development of comprehensive monitoring schemes of fish health.

2. Development of disease-resistant strains of key species.

3. Development of specific pathogen-free cryopreserved gene pool of gametes.

4. Development of environment-friendly drugs.

5. Development of bio-engineered vaccines against parasites and viruses.

6. Description of the clinical pathology of fish species.

7. In-depth knowledge on the specific defence mechanisms of fish.

8. Development of nutritional pathology, particularly for fish meal substituted diets.

The study identified conditions necessary for conducting targeted research, assessed institutional requirements and suggested a framework for inter-regional cooperation in fish health research. Networks involving core teams of scientists, collaborative research programs and networks of research laboratories in developed and developing countries would definitely accelerate the development of aquaculture research.

ADB/NACA Study

Studies on the requirements for research concerned have acknowledged that the technical needs for research on fish health strategy within the region are diverse and that major distinctions exist between the requirements of technically-advanced countries and the requirements of countries with less-developed aquaculture production.

It is significant to note that adequate research personnel, better research and diagnostic facilities and information support are inevitable in countries with less-developed fish culture systems. Countries with more-developed fish culture systems, however, require more research on the (i) study of new diseases, (ii) virology, nutritional diseases and environmental factors in disease, and (iii) acceptable therapy and prevention methods.

AFS Study

Studies on the need for demanded research have recognized that a lot of factors significantly contribute to the optimization of freshwater fish production and that disease is one of the most serious constraints in this regard.

It seems quite plausible that the recommendations of the above-mentioned studies remain substantially valid and will continue to cater a basis for development of research capability in fish health. In general, this development is slowly proceeding and the starting points and rates of progress vary widely between countries. The following additional recommendations are, however, suggested as practical means to assist development of research capability in fish health.

1. Diagnostic and research efforts should be demanded and stressed on diseases causing major production constraints.

2. Information support to researchers is required on urgent basis and should include ready access to literature.

3. Expert suggestions on experimental design must be available to researchers so as to maximize the utility of their work.

4. An information package on all funding agencies which support work on fish disease diagnosis and research should be available to researchers.

14.7 Quarantine*

What is Quarantine?

Quarantine can be defined as the rearing or holding of fish under conditions which prevent their escape or the escape of organisms and potential disease agents infecting or associated with them into the natural environment. Quarantine involves the examination of fish for disease agents and certification, the issuing of a certificate indicating that a specific group of fish or a yield facility has been investigated and is devoid of any infection by a particular pathogen or pathogens.

Aim of Quarantine Program

Well-designed quarantine programs may be effective at several levels. Quarantine programs, whether nationally or internationally, are aimed at preventing the spread of exotic species or strains of parasites, bacteria and viruses into countries where they do not occur. While at the sub-national level, they assist to reduce within-country dispersion of pathogens whether they are native or introduced. At the regional level, quarantine of fingerlings, fry and broodstocks originating from an outside source helps protect fish farms from potentially devastating losses due to disease.

It should be kept in mind that quarantine programs form an integral part of much broader strategies aimed at protecting the aquatic ecosystem and native fauna from deleterious impacts of exotic species. They serve fish culture industries by protecting them from diseases which the farms themselves might introduce when fry or broodstocks are imported. Quarantine programs also protect human health by preventing the entry and possible establishment of exotic parasites infective to man which are transmitted by aquatic organisms.

International Codes of Practice for Quarantine Programs

Quarantine programs form an integral part of a first line of defence against possible adverse impacts resulting from the transboundary movement of fish and other aquatic

---

* For further detail, see Arthur (1987), Humphrey *et. al.* (2005).

species. Although codes of practice for the international movement of aquatic animals have been developed by various international organizations in many countries of the world, they have rarely been applied. Inspite of this, they cater a starting point for formulating regional and national fish health regulations as well as international agreements aimed at preventing the spread of disease-producing organisms among farmed fish. Several technical guidelines that follows have been developed on fish health certification and quarantine for the responsible transboundary movement of fish in particular in collaboration with the Network of Aquaculture Centres in the Asia-Pacific (NACA), the Aquatic Health Research Institute of the Thai (AAHRI), and the Australian Centre for International Agricultural Research (ACIAR). These organizations relate directly to the solution of fish quarantine problems. Their expert personnel act in concert to solve the problems and to make the program more effective.

Suitable measures taken by national governments to prevent the introduction of exotic fish diseases must be developed with the context of national and international policies dealing with the introduction and transfer of animals and plants. Several important organizations such as Office International Epizootics (OIE) and "European Inland Fisheries Advisory Commission of the FAO" (EIFAC) have developed revised recommendations and protocols for the prevention of the international spread of diseases of aquatic organisms. Several guidelines for the implementation of aquatic organisms including methods to reduce the possibility of disease transfer have also been developed by these organizations. The approach relating to disease control in case of exotic fish introduction as recommended by the EIFAC can be summarized as follows:

1. The possibility of introducing pathogenic organisms and parasites should be examined.

2. A brood stock should be established in the importing country by transfer of eggs to an approved quarantine facility for regular examination of pathogens.

3. If no pathogens become evident, first generation progeny should be transplanted to culture sites or the natural ecosystems.

4. Disease studies on these transplanted consignments should be continued.

5. Total prohibition on import can be justified where a total ban impose no great inconvenience and/or where the disease is potentially so serious that risks cannot be taken into consideration.

6. A certificate of health issued by appropriate authorities in the exporting country may be required before entry is permitted. Certificates must be mentioned specifically that the entire consignment is free from pathogenic organisms. It is unfortunate, however, that certificates from some countries are not dependable and therefore other measures need to be adopted.

7. Chemical treatment of imported fish consignments, either before or immediately after entry, must be undertaken as an added precaution.

8. Isolation and inspection after entry may be required so that imported fish can be grown initially in isolation and can also be inspected before distribution is permitted. This may cover a period of six months or more. If diseased fishes are found, they may have to be destroyed.

9. Different legislative measures are issued by governments of different countries, prohibiting or restricting the import of fish species from other countries into the receiving ones (foreign quarantine). Suitable prohibitions also exist from preventing the spread of fish diseases from one state to another within the country (domestic quarantine).

Such measures, though stringent, are justified to protect existing fish culture development and avoid harmful effects on indigenous species, both those important to recreational and commercial fisheries and those not currently exploited. The latter may be vulnerable to introduced pathogens as not having been previously exposed to infection. They often lack natural resistance.

The methods for fish quarantine recommended in the code were essentially formulated among developed nations, though have not been widely implemented. And a few developing nations have the necessary resources to adopt such stringent measures. Implementation of the code has sometimes been burdensome.

Problems in Implementation of Quarantine Programs

Quarantine program cannot be undertaken by individuals but must be the responsibility of governments to establish suitable legislation and necessary steps must be taken for its implementation.

A basic problem in the attempt to prevent the introduction of parasites and diseases into a country is that international trade and travel cannot be completely shut down. It is not even practical to totally prohibit the importation of fish. Therefore, any quarantine action must be, to some extent, a compromise with restrictions on trade and imports being commensurate with the risk of introduction of a new disease problem. Such measures might include total prohibition on import of fish stocks which could act as carriers.

The implementation of effective quarantine programs by developing countries is difficult due to financial supports and the need for efficient technical personnel. Some important requirements are noted below.

1. Effective legislation and adequate enforcement.

2. Adequate diagnostic and quarantine facilities.

3. Highly skilled personnel.

4. Detailed knowledge on potential pathogens.

5. Establishment of a communication network of experts.

Knowledge on fish diseases is highly restricted to a very few species, less than 5 per cent of the total number known to fisheries science. In many tropical and sub-tropical countries, little or no information exists for the vast majority of aquatic organisms. Only when basic work on the identity, biology and distribution of pathogens has been accomplished, reliable diagnostic techniques are developed and at the same time lists of certifiable pathogens are compiled.

Fish importers and fish culturists often regard quarantine and inspection programs as unnecessary corruptions imposed upon them by government bureaucracies to embarrass their constitutional business activities. Efforts need to be made to show these groups that these programs are *sine qua-non* that helps protect aquaculture industries and make the farmers more profitable by reducing losses due to disease outbreaks.

Progress Towards Quarantine in Some Tropical/Sub-tropical Regions

This sub-section briefly focusses on some regions regarding the status of quarantine and certification programs.

1. **Australia:** Australia is the only country which has a fully operational quarantine system for imported fishes. The system strictly prohibits the entry of live fish unless the species to be imported is listed by the Australian Nature Conservation Agency. The entire consignment may be freely imported or can be imported under special permit for scientific purposes. The fish stocks to be imported are held for a minimum of 14 days, during which time they are examined regularly for clinical symptoms of disease by technical personnel. The current Australian quarantine systems, however, provide a good practical measure of prevention against the introduction of diseases along with imported aquatic species. And this encouraging result should definitely provide a useful model for quarantine development in many developing countries of the world.

2. **Southeast Asia:** Among different Southeast Asian countries,only three countries such as Indonesia, Malaysia, and Thailand have recently made substantial advances and have established a quarantine and inspection services. Importation of fish stocks is done under permit and a health certificate issued by the country is required. Fish are supposed to be quarantined for a period of at least one month, inspected and if found infected with communicable diseases must be destroyed. However, due to lack of infrastructure, routine quarantine and inspection on imports have not yet been fully implemented in these countries and fish generally receive only visual inspection for signs of disease.

    Considerable health expertise also exists in Singapore and the Philippines. In these countries, sanitization and quarantine procedures are routinely practised in many farms but are not compulsory. Other Southeast Asian countries such as Vietnam, Myanmar, Laos, and Kampuchea have little expertise or infrastructure related to fish health and none of these countries is currently developing quarantine programs for aquatic animals.

3. **East Asia:** No countries in this region have implemented programs for the routine quarantine and inspection of imported fishes. Although some countries such as

China, Japan and Hong Kong have experience and expertise in fish health programs, there are no control on the introduction of aquatic animals for culture purposes and consequently can be freely imported.

Taiwan, Korea, and Hong Kong consider strict quarantine to be unworkable due to the nature of the local aquaculture industry, which depends heavily on imported fry. The governments desire to promote free trade and the perception that a quarantine system would not be put in force and not cost-effective.

4. **South Asia:** The situation in this region is more or less similar to that of East Asia. India, Sri Lanka, Nepal, Bangladesh, and Pakistan have no legislation on the importation of introduction of aquatic organisms.

To sum up, the effective control of the international spread of fish pathogens remains far away and difficult to achieve goal for much of the Asia-Pacific, and it is elshwhere in many developed and developing countries. Effective national quarantine programs, coupled with a reliable program of certification of exported aquatic animals at source, would definitely reduce the risk of disease introduction considerably. Given the high costs of disease control in fish culture, quarantine programs would undoubtedly result in a net profit and financial savings to the area concerned.

The improvement made by such countries as Australia, Indonesia, Malaysia, and Thailand to circumscribe the international spread of fish pathogens deserves praise and commands. Effective systems take considerable time and effort and protocols must be developed and modified on the basis of regional experience. New trends toward more effective disease control strategies include a more ecologically-based management approach.

For several countries, lack of political will at the national level seems to be the major problem that prevents an effective quarantine system to be established. Several other constraints such as lack of trained personnel,poor position in infrastructure, administrative apathy and the like will be more readily vanquished. In this regard, several international organizations such as OIE, NACA, FAO, and regional professional societies (such as Asian Fisheries society) should continue to play a key role, highlighting the hassles faced by policy makers and bringing groups of experts together who are capable of working in concert towards solutions.

14.8 Nutrition in Relation to Fish Health*

The term nutrition generally refers to the study of the qualitative requirements and amounts of food materials that promotes the growth, health, and reproduction. It has become increasingly apparent, however, that under several conditions of intensive fish culture, proper nutrition plays a critical role in maintaining normal growth and health of fish. All of the essential nutrients noted in the sub-sections that follows and also in Volume 1, should be provided in the diet in adequate quantities to sustain the health status of

---

* For further detail, see Hepher (1988), Halver and Hardy (1989).

fish. A number of nutrients and other dietary ingredients as well as various feeding practices, have been shown to influence the susceptability of fish to various non-infectious and infectious diseases. In addition to this, prepared diets also serve as a primary method of administering chemotherapeautics and immunostimulants to fish that are infected with certain pathogenic organisms.

As cultivation techniques for a given fish species are established, investigations on nutritional requirements and feeding to yield robust fish have become more comprehensive. Hair-splitting knowledge about nutritional requirements is particularly important for high-value species held in intensive culture because they are dependent upon formulated feeds for their primary source of nutrients. However, the formulation of diets for farmed fish has been stressed on account of their potentiality in fish production. Thee diets are now common for fish culturists and are referred to as larval diets, juvenile diets, food for on-growing, broodstock diets and other diets formulated for specific purposes such as medicated food, pigmented food, etc.This section has been structured to describe, in brief, general aspects of fish health and mention the various roles in sustaining the health status of farmed fish.

Issues on Fish Nutrition

A number of issues on fish nutrition, largely related to intensive culture practices, have been focused during the eighties and these can be briefly noted that follows.

1. Most of the formulated diets that contain 40-50 per cent protein fed to intensively-reared carnivorous fish, are supplied by incorporating large proportions of fish meals in the formulated feed. The fish meal production is not expected to increase in the future and the increase in fish meal prices is foreshadowed. It is unlikely therefore, that expansion of intensive fish farming will depend very much upon the use of fish meal-based diets. For this reason, alternative sources of protein for use in formulated feeds for intensively-farmed fish must be taken into consideration.

2. In any fish culture system, the food must be designed and absorbed prior to utilization by fish and nutrients that are not absorbed are voided as faeces. Faeces and soluble wastes together with wastes arising from unconsumed food, constitute the wastes associated with fish production strategies. As a consequence, to reduce fish farm wastes, attention has been focussed on diet formulations and feeding schedules to improve digestibility and food utilization and to develop feeding regimes which lead to reduction in losses from uneaten food.

3. Development of broodstock nutrition and nutrient-enriched food items are two prerequisites for successful fish culture.

4. The economic viability of intensive fish culture systems depends upon the ability of the farmer to deliver a product at a price that is acceptable to the consumer and it has been traditional to maximize yield at minimum costs.

Nutritional Requirements

The quantitative requirements of essential nutrients for fish are generally determined by feeding suitable diets containing graded level of nutrients. Feeding response to fish is then measured in terms of growth, whole-body accumulation of the nutrient or by using an enzyme activity. In general, however, growth responses are recorded and dose-response curves are constructed by plotting growth against the amount of nutrient added in the diet. A variety of regression techniques and the ANOVA are used to analyze the dose-response relationships.

The use of supplementary feeds in fish culture is an utopian view and is regarded as an essential practice throughout the world. In intensive or semi-intensive systems of fish culture where the availability of natural food is very much limited, feeding the fish stock with nutrient-dense balanced diets and the use of cost-effective diets with degree of regularity are inevitable. Such a diet is regarded as a complete when it contains balanced levels of all essential nutrients such as protein, carbohydrate, lipid, energy, minerals, and vitamins. These promote the biological and physiological processes influencing survival, growth, and prodcuction. Most practical diets contain all these ingredients. Vitamins and minerals are important to proper nutrition, but the requirements for many of them are less clearly known for many freshwater fish species. The same is true for fatty acids, amino acids, starches, and sugars. However a considerable information is known about the nutritional requirements of trout, salmon, and a few species of carps.

1. **Carbohydrates:** Carbohydrates are simple sugars (such as glucose, galactose, and fructose), compound sugars (such as sucrose and lactose), and complex sugars (such as starches and celluloses).However, although carbohydrates provide a significant amount of energy in mammals, they are less important to some species of aquatic animals including fish. This is due to the fact that the carbohydrates (particularly polysaccharides) present in many aqua feeds are poorly digested. For this reason no specific recommendations for carbohydrate levels have been put forth for most species of fish. The use of carbohydrates in excess than requirements can lead to toxic levels of liver glycogen in trout. Glycogen problems, however, do not arise at all in case of channel catfish.

   In contrast to the soluble and fibrous carbohydrates associated with common food ingredients included in fish diets, some structural carbohydrates associated with the cell-wall of yeast and fungi (such as beta-glucans and oligosaccharides) have been shown to enhance the immune response of several fish species. Their mode of action has, however, been considered in the section 9.21 of Volume 1.

2. **Proteins:** The minimum dietary requirement for protein or a balanced mixture of amino acids is of primary concern to fish culture because this requirement is inevitable to ensure adequate growth, energy, and health status of fish. But the use of these energy-generating nutrients in excessive levels is generally not economical, since protein is the most expensive dietary component used in fish

culture. Though studies have been concerned with determining minimum dietary protein requirements, very limited investigations have been noted concerning the manipulation of dietary protein and ten indispensable amino acids to maintain fish health.

Though the quantity of protein required for different species of fish is extremely variable, the most primary concern is to provide nutrition for sustaining fish health. However, the optimum protein level for a fish species alters due to age, condition, reproductive state, and fluctuation of environmental parameters. Young and rapidly-growing fishes, however, require higher protein levels. Tilapia and trout fry generally depend on diets containing protein that varies between 35 and 50 per cent whereas fingerlings may receive only 20-25 per cent protein.

3. **Lipids:** Lipids are very important components of fish diets because they provide a concentrated source of energy that is typically well-utilized. In addition to this, dietary lipids supply essential fatty acids which cannot be synthesized by the fish. Adequate quantities of these essential fatty acids must be supplied in the diet to sustain normal growth and health of fish, but excessive levels have been shown to have growth-suppressing effects. Many fish diets contain relatively high levels of poly-unsaturated fatty acids, which are susceptible to oxidation. Oxidation products of lipid may react with proteins, vitamins and other dietary components and limit their nutritional value. For example, vitamin E is susceptible to destruction by oxidized lipid due to its antioxidant properties. Thus, if fish are fed with oxidized lipids, it may results in signs of vitamin E deficiency such as membrane fragility, liver degeneration, and anemia.

   For the maintenance of fish health, the principal source of dietary lipids in prepared diets is triglycerides. Most of the commercial aquaculture diets, however, contain lipids that vary between 4 and 80 per cent, though it is hard to recommend the use of lipids at optimum level in fish diets. Excessive use of lipids may lead to nutritional diseases such as fatty liver. Excess lipid which is accumulated in viscera may be a function of dietary lipid levels. During processing, this fat is eliminated.

4. **Minerals:** Fish generally require more than 14 essential mineral elements and dietary requirements for several of these minerals have been quantified. Dissolved minerals in fish culture ecosystems may contribute substantially to satisfy the metabolic requirements of fish and interact with dietary requirements. Dietary supplementation of certain minerals (particularly calcium, phosphorus, iron, potassium, iodine and selenium) at concentrations above those required (as shown in Volume 1, Table 9.8) to support normal growth and health of fish, but below those causing deficiency signs (Table 14.1) have been noted for several species of fish especially trout, tilapia, salmon, and channel catfish.

   Most practical diets contain suitable levels of minerals for increased growth and mortality, good health, skeletal formation, and for various metabolic functions.

In general, freshwater fish diets are primarily supplemented with dicalcium phosphate, sodium chloride, and some other essential minerals. Significant amounts of these minerals are, however, available in some artificial feeds particularly fish meal.

Table 14.1 : Mineral Deficiency Signs in Fish

Mineral	Deficiency sign
Calcium	Reduced growth and poor feed conversion, anorexia, and reduced bone mineralization
Phosphorus	Reduced growth and poor feed conversion, reduced bone mineralization, skeletal deformity and increased visceral fat
Potassium	Anorexia, convulsion, mortality and titany
Magnesium	Anorexia, reduced growth and bone formation, sluggishness, cataract, degeneration of muscle fibres and gill filaments
Iron	Reduced growth and poor conversion efficiency, anemia
Zinc	Reduced growth, anorexia, skin and fin erosion
Manganese	Reduced growth, loss of equilibrium, cataract, mortality, poor hatching of fish eggs
Copper	Reduced growth and cataract
Iodine	Thyroid hyperplasia
Selenium	Reduced growth, anemia, cataract and muscular dystrophy

5. *Vitamins*: Similar to minerals and trace elements, vitamins are also important in fish diet. But unfortunately, except for trout, catfishes, and salmonoids, vitamin requirements have not been well established for many other species of fish. If natural food organisms are available to farm fishes, artificial food may require no vitamin supplementation, but in case of intensive culture when natural food organisms contribute insignificantly to their intake, a complete prepared diet fortified with vitamins is inevitable. Further informations on these nutrients used in formulating fish feeds with respect to their nutritional potential have been summarized in Volume 1 (See Chapter 9).

Of 15 vitamins and vitamin-like compounds established as being essential for fish, vitamins C and E are the most noteworthy in their ability to influence immunity and disease resistance to fish when these vitamins are fortified with diets. Though these vitamins have several distinct metabolic functions, they have antioxidant properties. However, the effective dose of these vitamins is extremely variable for different species of fish: it varies from 1, 000 to 4,000 and 600 to 1,200 mg/kg of food, respectively for vitamins C and E. Other vitamins, however, that have been noted with regard to their effects on immune functions and disease resistance include folic acid, pyridoxine, riboflavin, and pantothenic acid. These vitamins, to a greater or less degree, have fish health sustaining potential.

14.9 Use of Alternative Protein Sources for Fish Health Management*

Previous discussions remind us that proteins are essential components of the diet needed for survival and growth of fish. The quality of a protein depends very much upon its amino acid composition, the availability of amino acids to fish, and effective utilization of amino acids following digestion and absorption.

As a general rule-of-thumb, it can be said that the greater the proportion of the essential amino acids (EAAs) in a protein source the better is the quality. Although a wide range of ingredients have been used in the formulation of feeds for fish, several of the plant protein sources used may be deficient in one or more of EAAs. Soyabeans and oilseed meals are, for example, deficient in methionine and lysine, maize is deficient in tryptophan and lysine, and cereals are deficient in lysine. Therefore, fish feeds are usually formulated to include quite a large proportion of the protein from animal sources and, traditionally, fish meal has been the major source of animal protein in formulated feeds. Quality fish meals are produced by processing fresh whole fish under low temperature conditions (60-70°C). This is due to the fact that high processing temperatures (100-150°C) result in meals of lower digestibility. Reduced digestibility may, in turn, have several negative effects such as nutrient loading of pond water will be increased, the utilization of protein for growth will be reduced, and the overall reduction in growth rates of fish fed on diets containing fish meals produced under acrimonious processing conditions.

Differences in fish meal quality due to variability in the freshness of raw materials should be considered before the production of fish meal is permitted. If the processing of raw fish is delayed, there is a tendency to increase the level of biogenic amines and the loss of freshness will be more intense at high storage temperatures.

Source of Plant Protein

The most important protein sources of plant origin are the oilseed meals. They are the dried residues remaining after the oil has been extracted from cotton seed, soyabeans, rape-seed, peanuts, and sunflower seed. A certain proportion of cereal meals may be used in food formulations as they act as binders. In addition to this, single-cell proteins (See Section 7.13 of Volume 1) are sometimes used in fish food formulations. Some plant products are too expensive to form the main protein source in fish food, and serious concern has been expressed about the possibility of the presence of undesirable compounds such as heavy metals. Thus, although there is availability of plant protein sources, many of them may be discarded as major ingredients in fish feeds due to lack of general availability, expensive, or potential nutritional problems. It should be kept in mind that soyabean products are highly economic as ingredients in formulated feeds and the well-balanced amino acid profiles make the soyabean meal nutritious and considered as the best widely available plant protein source.

---

* For further detail, see Jobling (1994), Higgs et. al. (1995), Friedman (1996), Aksnes and Mundhein (1997), Hendricks and Bailey (1989), and Liener (1994).

Protein Concentrate

Partial destruction of anti-nutritional factors (to be discussed in section 14.10) under normal processing may reduce the potentiality of conventional oilseed meals in diets for several species of fish and plant protein concentrates (such as soyabean, potato, cereals, maize concentrates etc.) may be a better alternative. The use of protein concentrates surpasses many of the problems associated with the presence of anti-nutritional factors, and affords the fish with a source of high protein (60-65 per cent) having an essential amino acid profile. The protein concentrates resemble to those of fish meals and that would seem to fulfil the known requirements for maintaining the health status of fish. Obviously, it may be possible to replace much of the fish meal in formulated feeds by protein concentrates without revealing an embarrassing situation on the growth and health performance of farmed fish.

14.10 Feed Management for the Maintenance of Fish Health

Immediately after the production of feed, it is essential to store the product properly to retain the quality. Semi-moist and moist feeds and feed ingredients should be used immediately after preparation. However,the use of binders and proper storage of feeds is very important to prevent food items from (1) physical and chemical damages and (2) infestation by insects, rodents, and micro-organisms. Unless the quality changes of food stuffs are taken into consideration, economic losses will be more intense. Feed should be kept on wooden logs so that they do not touch the ground to prevent moisture absorption from the floor. Fungal and bacterial growth produces toxins and makes the feed ineffective. Their growth is very rapid when temperature exceeds from 24 to 28ºC and humidity remains above 80 per cent. Obviously, temperature control must be made in storage room before their extensive use in fish culture is permitted. At the same time, the use of cost-effective preservatives such as sodium and calcium propionates, potassium sulfate, propionic acid, calcium sulfate, and sodium benzoate are also recommended.

The use of screens in the warehouse ventilations prevent the entry of insects and rodents. Vitamins and vitamin premixes should be kept in cool and dry places as far as possible in air-tight containers and should be used within 4-6 months. Antioxidants such as Vitamin E, ascorbic acid, butylated hydroxyanisole etc. should be added to prevent unsaturated fatty acids from oxidation. If feeds are not properly processed nutrients can never be utilized in full potential.

Anti-nutritional Factors

Ingredients of different plant protein sources contain several anti-nutritional factors such as lectins, anti-vitamins, saponins, gitogens, protease inhibitors, gossypol (a phenolic pigment), phyto-oestrogens, trypsin and chimotrypsin inhibitors, malvalic acid, antigenic proteins (allergens), tannins, cyanogens, and various oligosaccharides. All these anti-nutritional factors can give rise to inflamation of the alimentary canal, thereby result in major disturbances in the digestive processes and consequent to fish health deterioration.

Although various plant protein sources are extremely valuable, their full nutritional potentiality may only be attained following processing techniques to destroy, remove, denature or inactivate the various anti-nutritional factors. Processing techniques for improving the nutritional value of plant ingredients involve a range of chemical and physical treatments. Processing techniques are, however, very complex and beyond the scope of the present discussion.

14.11 Health Management of Fish Larvae and Broodfish

Health management of fish larvae through use of high quality finished feeds continue to be one of the most critical aspects in the evaluation of candidate species for culture, and the ability to cater nutrient-rich food items in adequate amounts holds the key to successful culture of fish larvae. The culture of fish larvae depends primarily upon the provision of natural food organisms (particularly Artemia, Daphnids and Rotifers) that may be enriched with specific nutrients. Among different planktonic organisms, however, *Artemia nauplii* has been the most important live food used across the world. Constraints on the availability of Artemia and some other lucrative aquatic organisms coupled with their increasing demands and ascending prices cause serious operational problems to hatcheries.

Several problems which are encountered in the preparation of formulated microdiets for extensive use in the culture of fish larvae relate to the difficulties involved in making scientifically-formulated, nutritionally-stable, and homogeneous particles of suitable size. Poor acceptance and digestion of formulated feeds by the larvae may be considered as an additional problem.

It should not be over-estimated that production and viability of eggs largely depend on the health of broodfishes and the type of the formulated feed offered to them. Essential nutrient deficiencies in the diet of maturing fishes are associated with the reduced fecundity, egg size, and hatchability of many farmed fish. Since multiple breeding (especially carps) is widely practised in several tropical countries, it is essential for rematuration of broodfishes by adopting appropriate nutritional strategy. Feeds having adequate protein: energy ratio, carbohydrate : lipid ratio and feeding strategy have high health potential of broodfishes.

14.12 Use of Quality Feeds for Fish Health Management

Any product that contains crude protein, unsaturated fatty acids and a complete vitamin as well as mineral package to (1) enhance rapid growth of fish, (2) keep the fish in good health, and (3) help in disease resistance may be designated as a *high quality* feed. Some important internationally-accepted refined and powdered feeds include black algae powder, spirulina powder, tropical fish flakes/pellets, etc. These products are highly effective in producing robust fish from tropical and sub-tropical ponds and are most practical for high-value species and commercial hatcheries. These feeds are used in more than 35 countries around the world to supply fish nutrients. However ,there is a wide range of protein content of these feeds from 50 to 60 per cent. Fat content also varies between 10 and 20 per cent. Intensive fish farming definitely justifies the use of these nutrient-dense materials for sustainable health management.

The food quality attribute to farmed fish must be manipulated. Attributes that are extremely important in regulating the acceptance of fish products include flavor, general health, texture, nutritional value, and consumer preference. These attributes are, however, important in the marketing of fish.

It should be added as a note that high quality feeds are scientifically formulated, water-stable, and eco-friendly. It is geared to fight environmental stress conditions with the aid of minerals, vitamins, essential amino acids, phospholipids, carotenoids, glucans, and other immunostimulants. These feeds have definitely immense practical significance in relation to fish nutrition and health.

14.13 Guarantee of Formulated Feeds

In the preparation of feeds and for their use in fish culture, strict control measures should be adopted so that every ingredient has a guarantee so far their quality and quantity are taken into consideration. Quality control analysis of aquafeeds involves the regulation of (1) aflatoxin (the toxin produced by *Aspergillus flavus* — a mould commonly found in groundnut cake, (2) physical quality (pellet texture, shape, size, water-stable, etc.), (3) chemical quality (contents of nutrients, amino acids, fatty acids, calcium, phosphorus, vitamins, minerals, etc.), and (4) microbiological quality (mould count, total bacterial count, salmonella count, etc.). Therefore, in order to improve the health status of farmed fish, the quality control analysis must be followed. Since most of the global aquaculture production is still based on the farming activity using formulated diets, their guarantee must be ascertained.

14.14 Immunostimulation and Fish Health Management

Any compound which is used in feed to enhance the humoral and immune response at cellular levels in both specific (vaccination) and non-specific (upgradation of immunological capabilities at a time when fishes are exposed to pathogens) ways may be designated as an *immunostimulant*. A variety of heterogeneous biochemical substances including some nutrients have been established as immunostimulants. These compounds are widely used for impaired immune functions and to stabilize the improved immune status of fish. The most extensive means of producing healthy and disease-free fish from freshwater ecosystems is the use of immunostimulants. In general, however, these compounds essentially comprise a group of biological and synthetic compounds that enhance the non-specific and specific defence mechanisms in fish.

On the basis of the origin and mode of action, immunostimulants can be classified as (1) Bacteria and bacterial products, (2) complex carbohydrates, (3) Vaccines, (4) Immunity-enhancing drugs, (5) Nutritional factors, (6) Animal extracts, (7) Plant extracts, and (8) Cytokines.

Benefits of Immunostimulation

Immunostimulation of fish promotes many conditions which are favorable to fish health and provides greater stability of building the protective functions of fish. Unless the process of Immunostimulation is routinely followed in fish culture, health status of fish will seriously interfere by disrupting their immune systems.

The process of immunostimulation plays a critical role in maintaining the productivity of farmed ponds. The provision of this process in fish culture permits the prevention of diseases if they are already present in land-based aquaculture and in hatcheries and prevents their development in the ecosystem. Unfortunately, the essential need for adequate health status of fish is yet to be recognized in most fish producing nations. As a result, over time the outbreak of fish diseases has substantially increased to the extent that fish production has been seriously curtailed. Some tropical countries have been so affected.

Features of an Ideal Immunostimulant

An ideal Immunostimulant has many features. Its feature is determined primarily by its structure. The characteristic features of an immunostimulant may account for many functions so important in fish health management. Some important features of an Immunostimulant are noted below.

1. It should be non-toxic.
2. It should not be carcinogenic or having long-term side effects.
3. At therapeutic levels, it should have a short withdrawl period.
4. It should stimulate a wide range of non-specific immune responses against bacteria, viruses, helminths, protozoa, and fungi.
5. It should be orally active and should be stable after incorporating into the feed.
6. It should be compatible with a range of drugs such as antibiotics.
7. It should be capable of amplifying immune responses to infectious agents.
8. It shouid be inexpensive and either palatable or tasteless.

Some Common Immunostimulants*

Though many chemical substances are used as immunostimulants, the most important ones that follows are widely used in fish culture.

1. **Glucans:** Glucan products such as vetregard, vitastim, macrogard etc. ate the most popular immunostimulants used in fish culture. These products exhibit excellent immunostimulatory properties and operate well when injected or fed to fish. They are derived from yeast cells and from certain higher plants. Treatment of fish with these products has shown to increase the activity in non-specific defence mechanisms and in protection against pathogenic microorganisms.

---

* For further detail, see Muzzarelli (1977), Knorr and Klein (1986).

2. **Chitin and Chitosan:** Next to glucan, chitin and chitosan (See Panel 14.1) are the most extensively used Immunostimulants in many countries of the world. Chitosan is a deacetylation product of chitin. Being non-specific immunostimulants, they have great practical significance in relation to protection against bacterial diseases in fish and (ii) controlled release of vaccines. They are also used in the formulation of fish diets.

3. **Bacterial Products:** Bacteria as a group, almost without exception, participate vigorously in all the activities so vital if a productive fish culture ecosystem is set up. Certain bacteria such as *Bordetella pertuosis*, *Brucella abortus*, *Bacillus subtilis* and *Klebsiella Pneumoniae* have immunostimulating activities. These bacteria possess some components that can be readily taken up by fish. The activity of these components includes (i) enhancement of antibody production and (ii) activation of macrophages.

4. **Complex Carbohydrates:** In recent years, a growing body of research has heightened the interest in the capability of beta-glucans, polysaccharide derivatives from yeast and fungi to act as immunostimulants in fish. These compounds are

---

PANEL - 14.1
CHITIN AND CHITOSAN

Chitin is a non-toxic and biodegradable compound, with high molecular weight, and a naturally-occurring polymer of N-acetylglucosamine. Though chitin is widely distributed in nature, the most practical source of commercially preparing these polymers is shellfish processing waste materials. This chitin is firmly combined with proteins, calcium deposits, and pigments. Therefore, it is essential to remove these components before the process of separation of chitin from shellfish wastes is carried out. Removal of these components essentially involves demineralization (removal of calcium carbonate with dilute acid or chelating agents), deproteinization (removal of protein with dilute alkali), and decoloration (removal of pigments with acetone or hydrogen peroxide). After removal of these components from shellfish wastes, chitin is produced. Chitin is further processed into chitosan. This process involves deacetylation that proceeds to about 80 per cent within the first 60 minutes of treatment with sodium hydroxide solution at 100°C. Further deacetylation can rarely extend beyond 80 per cent unless the alkali fusion procedure is applied along with fractionation.

Chitosan may also be prepared from chitin using deacetylating enzymes which are derived from microbes such as *Aspergillus niger*, *Absidia butleri*, etc. The preparation may be accomplished by two ways. *First*, by fermentation whereby culture of chitin deacetylase-producing microorganisms are innoculated in a medium containing chitinous substrate and *second*, by a direct treatment of chitin polymers with extracts of microbial chitin deacetylase enzyme.

insoluble polysaccharides consisting of repeatable glucose units which can be joined through beta 1,3 and beta 1,6 linkages when derived from yeast and mycelial fungi or beta 1,3 and beta 1,4 linkages when derived from barley. The results with respect to an improvement in specific immune responses of fish are highly encouraging. When vaccinated rainbow trout, for example, was fed with glucan, antibody production significantly increased.

14.15 Stimulation of Non-Specific and Specific Immune Responses

Non-Specific Immune Response

Responses of macrophages such as phagocytosis, bacterial activity, and changes in the number of leucocytes, activation potential of cells on stimulation, etc. are generally used to assess non-specific immunity. Improvement of non-specific immune responses are, however, considered as the first line of defence against invading organisms. Immunostimulants which help to increase the non-specific immune responses include glucans and livamisole. These compounds have been shown to enhance non-specific immune responses of various species of fish by increasing their phagocytosis, macrophages and resistance to infection by pathogenic organisms.

Specific Immune Response

A specific immune response requires the fish to have prior exposure to an antigen. Vaccination is the best-known method of specific immunostimulation in humans including fish. The most common vaccines used in aquaculture are preparations consisting of live attenuated bacteria, killed cells, or cellular extracts which are administered to fish by injection or immersion. Intra-peritoneal injection of antigen is most effective in stimulating the immune response which is labor-intensive and may impose excessive stress on fish. Immersion vaccines consisting of either long-term exposure to dilute suspensions of vaccine or short-term exposure to a concentrated solution of antigen have been shown to be effective in immunizing fish. Use of vaccines to fish via the diet is also practised mainly to provide periodical booster doses of vaccine in conjunction with the initial vaccine exposure by immersion or injection. Since destruction of oral vaccine in the gastro-intestinal tract may influence the effectiveness of oral vaccination, various means of protecting the vaccine (such as antigen encapsulated in alginate) have been devised with encouraging results.

There is potential for incorporation of certain types of immunostimulants in diets especially beta glucans, livamisole, and chitosan in health management of fish through disease prevention and control regimes in particular. However, a number of factors that need to be well understood more elaborately before these substances can be routinely used in fish culture. Many of these factors such as the type of immunostimulant, dosage level, species of fish, and disease organism(s) contribute substantially to determine the success or failure of an immunostimulation strategy. Their proper dosage is extremely critical to assess because failure to use these immunostimulants in adequate amounts may not result in enhancement of response to fish, while their use in excess than requirements can lead to immuno-suppression. Considerable variations in the effectiveness of immunostimulants have been noted with different channels of administration, and with injection strategies producing best results. Oral application of immunostimulants through diets has been found to be the most convenient means of administration, however.

14.16 Feeding Schedules Affecting Fish Health

Certain deviations from feeding schedules on a daily basis may affect fish health and resistance to diseases. Most of the commercial fish such as channel catfish have shown to increase mortality and disease resistance following at least 10-weeks dispossession of food. Use of limited amount of food with or without nutrients for increasing yield efficiency concomitant with the improvement of health status of fish deserves additional consideration.

Manipulation of Nutritional Condition

Previous sections (sections 14.8 and 14.9) remind us that feeding practices definitely influence the nutritional status of fish which may, in turn, affect their disease resistance by reducing the growth of pathogenic organisms or by altering the immune function of fish. However, dietary composition of energy-generating nutrients largely influence the nutritional condition and health of fish. In general, the distribution and storage of protein, carbohydrate and fat vary considerably in fish body. Such variations in nutrient utilization may influence the health of some species of fish such as rainbow trout and channel catfish when artificial diets are added to culture systems. The proper balance of energy-generating nutrients in diet formulations is essential not only for sustaining the growth of fish but also for maintaining the body composition within limits to permit normal health.

Seasonal Feeding Regimes

Some seasonal factors particularly temperature, pH and dissolved oxygen concentration may influence feeding regimes in fish culture where environmental conditions fluctuate seasonally. It should be mentioned that the preponderance of certain infectious diseases varies with the season, because most of the pathogenic microorganisms have specific temperature ranges in which their virulence is greatest. Also, the immune system of fish is altered as a result of changes in seasonal temperatures. Therefore, modification of feeding schedules and diet compositions in relation to different seasons also deserve serious consideration.

Use of Medicated Feeds

Control of fish diseases is a primary concern to fish culture industry. A very limited number of government-approved chemotherapeutics are used in several developed countries to treat fish diseases. Most of the countries have less stringent rules concerning the use of medicated food items in fish culture industry. Use of antibiotics in the diet is highly effective in treating numerous fish in large volume of water. The most common and widely used antibiotic is Terramycin. Though it is a broad-spectrum antibiotic that is effective against a variety of bacteria, its use has been restricted to diets manufactured by compression pelleting since the antibiotic is heat-labile.

Antibiotics are also widely used as growth promoters for several species of fish. Increase in digestibility of some unsaturated fatty acids has been noted in fish (rainbow trout, for example) diets containing chloramphenicol, oxytetracycline, and oxolinic acid.

Due to the incompatible improvement in the performance of fish-food fortified with antibiotics, the practice of antibiotic treatment has not been widely followed in commercial fish culture. At the same time, the development of resistance by certain strains of bacteria and residues in fish flesh are the most potential limitations to antibiotic treatment in fish diets. Spinal deformities due to prolonged administration of antibiotics in fish are the additional complications. Therefore, caution must be taken before their extensive use in fish culture is recommended.

14.17 Management of Persistent Pollutants

As discussed in Chapter 20 of this Volume, a vast variety of pollutants exist that are resistant to most of the decaying and decomposing materials of the eocsystem. Most of the pollutants are non-biodegradable. Non-biodegradable pollutants generally persist in the ecosystem for several years and become harmful to fish and other aquatic species to a large extent. During this period of time, pollutants are taken up by the ecosystem and accumulated in the food chain. The accumulation of pollutants in the biota is termed as *bioaccumulation*. During the transfer of accumulated toxicants in the food chain by eating and being eaten processes, the concentration increases at each successive trophic level, thus concentrating highly to become potantially toxic to fish. The enhancement of toxicant concentration is termed as *biological magnification*.

Though designated as persistent, many of these chemicals are subject to decomposition by natural processes. However, such decomposition requires suitable conditions and well-integrated biological as well as non-biological processes. Such processes definitely require long enough period to degrade these pollutants at a rate much slower than the rate of their entry into the ecosystem.

Most of these pollutants are pervasive and fatal to ecosystem in general and fish in particular. In general, pollutants exert a notable influence on ecosystem health, causing serious disturbance in the health status of farmed fish. As discussed in Chapter 12 of Volume 1, the two terms such as bioaccumulation and biomagnification are the most fatal effects of persistent toxic materials.

It should be kept in mind that prevention is the best way to control any problem, especially those related to health management in fish farms. In this sense, prevention of biomagnification of persistent organic pollutants with respect to pesticides, perhaps, constitute the key step that follows to arrive at the goal.

Detoxification of Pollutants

Detoxification of pollutants involves biotic or abiotic transformation of persistent pollutants into relatively harmless substances. Since the toxicity of a compound depends on its chemical structure and its ability to alter its properties radically, detoxification on the other hand, involves the detachment of its active group and alter in its structure. Compounds having highly-branched structure and increased substitution are difficult to degrade and to detoxify. Hence, suitable devices for their degradation have to be worked out.

14.18 Control of Pollution

Though biodegradable pollutants alone are not responsible for water pollution, they indicate the level of pollution. Besides these, a substantial pollution load is contributed by slow degradable or non-degradable pollutants such a heavy metals, biocides, plastic materials and the like. To keep the fish stocks in healthy condition throughout the culture period, pollution must be controlled at source by treatment. The non-degradable substances can be removed from water by suitable devices.

In addition to these methods, some standards, conditions and requirements have to be legally enforced by the government through legislations. The various methods suggested for control of water pollution are summarized below.

Stabilization of the Ecosystem

This is the most scientific way to control water pollution. The main principles involve the reduction in waste input, harvesting and removal of biomass, trapping of nutrients, ecosystem management and aeration. Various methods may be used (physical and biological) to restore fish health and ecological balance in the water body to prevent pollution.

Reutilization and Recycling of Waste Water

Various kinds of waste water may be judiciouly recycled for further use in fish culture. For this purpose, suitable technology must be followed which will maintain the health status of fish.

Use of Zeolite

One of the valuable properties of zeolite is to exchange its sodium for cations such as Ca^{+2}, Mg^{+2}, Cu^{+2}, Cd^{+2}, Zn^{+2}, Fe^{+2}, etc. in water. Zeolite can bind with metal ions forming ion-exchanged zeolite complex which is deposited in the sediment. The ion-exchanged zeolite complex can easily enter into the fish body than raw metal ions. The complex does not accumulate in the body but remove through faeces thus reduces the metal burden in fish body which in tum enhance both biochemical and growth parameters of fish. This phenomenon is more pronounced in fish ponds having low water hardness that varies between 120 and 200 mg $CaCO_3$ per litre.

1. **Mechanism of Action of Zeolite on Toxicants :** The explanation regarding effects of zeolite on metal toxicity, for example, generally involves (i) the interaction of metal and zeolite and (ii) the influence of water hardness on zeolite. Zeolite has extra framework ion (Na^+) and ions (Al^{3+} and Si^{4+}) which are exchangeable and non-exchangeable, respectively. The ionic radii of Na^+, AL^{3+}, and Si^{4+} are 0.95, 0.50, and 0.54 A° respectively; and that of the metal Cd^{2+}, for example, is 0.97 A° which corresponds to the ionic radius of Na^+ (0.95 A°) in zeolite and as a corallary, both ions can be easily exchanged with each other than that of Al^{3+} and Si^{4+} ions. Cadmium ions can easily bind with exchangeable extra framework Na^+ ion in zeolite.

Hardness of the pond water generally results from cations dissolved in water. Along with increase in the hardness of water, there are more cations competing with metals for a fixed number of exchange sites on the zeolite. Consequently, less the water hardness, more metal ions could be removed from the pond water. The metal-removing potential of zeolite is highly affected by water hardness, the maximum potential being reported as the least water hardness (120 mg $CaCO_3$ per litre). The application of low-cost ion-exchange agent zeolite and also synthetic zeolite in fish culture results in the immobilization of toxic intermediates of nitrogen-containing fertilizers (such as ammonia and nitrite) and heavy metals and hence recommended for their use over extensive water areas before stocking them with fish or during the pond preparation.

Removal of Pollutants

Various pollutants present in different water bodies can be effectively removed by several methods such as absorption, ion-exchange, reverseosrosis, and solar power. Mercury and ammonia are, for instance, removed by ion-exchange technique. Phenolic compounds and sodium salts are, on the other hand, eliminated respectively by use of polymeric absorbents and reverse osmosis. Decoloration of water is carried out by electrolytic decomposition technique by electrolytic decomposition technique.

The use of solar power for cleaning up polluted waters to maintain the quality of water is very cheap. A combination of sunlight and a catalyst such as titanium dioxide can break down chemical toxicants of water. Such photo-catalytic reactions can destroy pesticides, polychlorinated biphenyls, cyanides, and dioxines.

14.9 Conclusion

Effective health management in fish culture plays a major role in increasing the yield of robust and diesease-free fish. Several methods are followed by many developed countries to maintain the health status of fish and to satisfy the demand of farmers as well as consumers.

Proper health management strategies essentially involve dietary immunostimulation, use of quality food, vaccination, precise determination of the effects of dietary modulation on immune response and disease resistance, diet composition, feeding schedules, quarantine programs and other management strategies described in this chapter. However, before implementing the culture programs of high-value species, these factors would have to be kept in mind. At the same time, to meet the consumer preference and the marketing of quality fish products, health management programs must be seriously considered.

Different research institutions and government organizations form an administrative system of management that are responsible for planning fishery development and the scale of culture, circulating informations on epidemic diseases in various regions, organizing experts to formulate a policy for quarantine and nutrition in particular, and to tackle key problems pertaining to fish health management.

Most of the farmers do not understand the relationship between the culture ecosystem and disease outbreaks nutritional requirements for fish health management. Therefore, it is most important to increase their level of knowledge by offering intensive training courses on fish health management in general.

References

Aksnes, A. and H. Mundhein. 1997. The impact of raw material freshness and processing temperature for fish meal on growth, feed efficiency and chemical composition of Atlantic halibut (*Hippoglossus hippoglossus*) *Aquaculture* 149 : 87-106.

Arthur, J. R. 1987. Fish quarantine and fish diseases in South and Southeast Asia. *Asian Fish. Soc Spec.* Publ No 1 86 P.

Deomempo, N. R. 1995. Farming systems, marketing, and trade for sustainable aquaculture. 1n Report of the ADB/NACA Regional Study and Workshop on Aquaculture Sustainability and Environment. Network of Aquaculture Centres in Asia-Pacific (NACA), Bangkok, Thailand.

Friedman, M. 1996. Nutritional value of protein from different food sources. A review.*Journal of Agricultural and Food Chemistry.* 44, 6-29.

Halver, J. E. and R. W. Hardy. 1989. Fish Nutrition. Academic Press California USA.

Hepher, B. 1988. Nutrition of Pond Fishes. Cambridge University Press, Cambridge.

Hendricks,J. D. and G. S. Bailey. 1989. Advantitious toxins in fish nutrition (*Ed* J. E. Halver). Academic Press, London, pp. 605-651.

Higgs, D. A., J. S. Macdonald, C. D. Levings and P. S. Dosanjh. 1995. Nutrition and feeding habits in relation to life history stage.*In* Physiological Ecology of Pacific salmon (*Ed C. Groot L.* Margolis and W. C. Clarke), UBC Press, Vancouver, pp. 160-315.

Humphrey, J. and *et al* 2005. Aquatic Animal Quarantine and Health Certification in Asia. FAO, Rome, Italy. 153 p.

Jobling, M. 1994. Fish Bioenergetics. Chapman and Hall, London.

Langdon, J. S., G. L. Enriquez and P. Sukimin (Eds.). 1992. Proceedings of the Symposium on Tropical Fish Health Management in Aquaculture. Biotrop. Spec. Publ. No. 48, 172 PP.

Liener, I. E. 1994. Implications of antinutritional components in soyabean foods. Critical Review in Food Science and Nutrition. 34, 31-67.

Knorr, D. and Klein. 1986. Production and conversion of chitosan with culture of *Mucorrouxii* sp., *Phycomyces blakesleeanus. Biochem.* 8, 691-695.

Meyer, F. P., J. W. Waren and T. G. Carey. 1983. A guide to integrated fish health management in the Grate Lakes Basin. *Grate Lakes Fish. Comm. Spec. Publ.* No. 82, 262 PP.

Muzzarelli, R. A. A. 1977. Chitin. Pargamon Press, Oxford, Haworth Press, London.

Webster, C. D. 2002. Aquaculture.

Questions

1. What do you mean by ecosystem health? How it is related to fish culture strategy?

2. Discuss how health status of farmed fish is affected.

3. Discuss different aspects of fish health management programs that are considered as essential criteria during fish culture.

4. Discuss how system management is related to fish production.

5. What do you mean by system approach? How it is related to fish culture?

6. What are the needs that should be considered for fish health research?

7. What is quarantine? What are the codes and practice that should be considered for quarantine programs to protect aquatic life?

8. What are the problems which are related to quarantine programs?

9. Discuss the status of quarantine and certification programs taken in some tropical/sub-tropical regions.

10. What is nutrition? Discuss how nutrition is largely related to the maintenance of fish health.

11. State how health management of young fish and broodfish are maintained.

12. What is immunostimulation ? Discuss how it is related to fish health. What are the chemical substances that are used as immunostimulant ?

13. Discuss how feeding strategies affect fish health.

14. Highlight the importance of persistent pollutants that are pervasive and fatal to fish and aquatic ecosystems. How pollution is controlled in the context of fish health management?

15. Discuss the following terms:

    (a) Ecosystem health, (b) System management, (c) Disease control for fish health, (d) Quarantine, (e) Vaccination, (f) Immunostimulation, (g) System approach, (h) Training and extension programs for fish health management, (i) Use of alternative protein for fish health, (j) Protein concentrate, (k) Antinutritional factors, (l) Features of immunostimulants, (m) Chitin and chitosan, (n) Glucans, (0) Specific- and non-specific immune responses.

15

Genetics in Fish Culture

Genetics in fish culture is no doubt an applied subject. The growing interest on the subject has been due to two compelling reasons. *First*, to document and catalogue the genetic diversity in natural fish stocks for conservation and *second*, to increase fish production by using genetic technology. The genetic identity of commercially-important species of fish is of paramount importance in farming industries around the world. Sophisticated techniques have been developed in which biological and molecular alternatives to breeding by selection have come into view. The application of these technologies to fish has been rapid because eggs and embryos developed outside the body of the female and also perhaps because the ethical reservations are less with regard to fish than with other vertebrate groups. Here we shall discuss the issues on genetics that are specially related to fish culture and fish culturists in their endeavour at production level, and the possible strategies that fish culturists can employ for significant production. The discussion on genetics in fish culture will be briefly carried out under several major sections that follows.

* Qualitative traits
* Quantitative traits
* Selection, inbreeding, and cross-breeding
* Hybridization
* Gynogenesis and androgenesis
* Polyploidy
* Sex detemination and control
* Transgenic fish production
* Genetic threats to wild fish
* Future needs in genetic research

We shall take up these topics one by one.

15.1 Rationale*

The term genetics was first coined by W. Bateson in 1905, although beginning of the science of genetics was made in 1900. Genetics is the study of how the information needed

---

* For further detail, see Webster (2001), Reddy (2005).

to make an organism is stored and transmitted. It is often described as a subject which deals with two important terms such as *heredity* and *variation*. While the former includes those characters which are transmitted from generation to generation and hence fixed for a particular individual. The variation, on the other hand, describes the difference in characteristics shown by organisms belonging to the same natural population or species. Whilst Charls Darwin (1809-1882) recognized that particular characteristic features could be developed by selective breeding (See Section 1.8), it was Gregor Johann Mendel (1822-1884) who explained the mechanism by which selected characteristic features was passed from generation to generation. Variations are, however, two types such as (1) Hereditary Variations and (2) Environmental Variations. Hereditary variations generally refer to differences in inherited characters. Such variations which result due to sexual reproduction are found both in progenies of different parents and among progeny from the same parents. Characters which are present in parents can reassort and give rise to different combinations which are liable for hereditary variations within the same offspring. Environmental variations are those which are due principally to the environment. Hereditary variations are transmitted from generation to generation but environmental variations are transient.

Although freshwater fish farming in ponds, tanks, lakes, pens, cages, and in shallow areas of reservoirs is widely practised around the world, a wide range of technology levels can be observed across species, regions, countries, and farming operations. Developed countries are bringing an action against the potential gains of molecular genetics and other state-of-the-art technological approaches to the improvement of fish species, most of the farmers across the world have little interest in or access to these approaches.

Many cultured fish species represent unselected, semi-natural stocks and/or isolated populations suffering from inbreeding, indirect selection, and general inferiority to potentially available alternatives. However, extensive research has shown that practical application of traditional animal and plant breeding methodologies can produce substantial gains in productivity of many aquatic species.

Alarming gains have been achieved in poultry and dairy cattle production over the past century. In contrast, fish culture industry in every region of the world can generate substantial gains in productivity through application of simple principles of selection and breeding.

The high fecundity of fish admits for greater selection pressure than can be exercised in most forms of fowl or livestock. Production efficiency, profitability, and product attributes are all areas that can be definitely impacted through simple and practical genetic applications. More than 60 per cent of the salmon and trout produced in Norway have been found to be genetically-improved species resulting from intensive breeding programs.

In fish culture and production systems, growth is one of the most important production traits which is closely correlated with the profitability. Although growth is the principal trait for most of the fish species when genetic resource improvement (see Panel 15.1) is taken into consideration. Disease resistance, food conversion efficiency, resistance to environmental extremes of heat, cold, salinity, poor oxygenation, high demand for fish

PANEL - 15.1
IMPROVEMENT OF GENETIC RESOURCES

The application of genetic principles to aquaculture is still far behind the agriculture or animal husbandry sectors. Some of the reasons for this condition are due to only very recent intensification of aquaculture practices and the low level of domestication of fish.

While significant research on selection, genetic management and population genetics has been conducted for several species of fish such as carp, trout, and mahseer, the results have not yet been widely applied in breeding programs to benefit fish culturists in rural areas particularly in several fish producing countries of the world. The aquaculturists have encouraged by recent research on cytogenetical studies on several species of warm wate fish but it should be kept in mind that this research has not yet reached at a stage where it has had an influence on both cold- and warm water fish culture.

Determination of the conservation value of varieties and strains in freshwater aquaculture requires both quantitative trait and evaluation of population genetics. Both the private and the public sectors should contribute data to evaluate conservation value. Private sector attempts to meditate economic concern and will be directed towards producing and conserving breeds for particular purposes, whereas public sector endaevours may consolidate ecological concern and will be commanded towards conservation and re-establishment of endangered species.

It should be borne in mind that selective breeding, hybridization, and genetic management are the main strategies for genetic improvement. These strategies should, however, be evaluated during breeding programs for sustained aquaculture development in farming regions. The choice of strategy for fish genetic improvement can have effect on the structure of fish hatchery operations and farming systems. A system excited to yield F_1 hybrids, for example, will have diverse structure and management requirements than those needed for propagation of fish breeds developed by selective breeding. All those concerned with fish breeding programs should prosecute strategies to preserve biodiversity of fish. And the impact of such strategies should be monitored both as to their impact on socio-economic structures of the local community and on biodiversity.

The potential to maintain genetic diversity in the context of the control of transfer of fish species by localized seed production should be evaluated. Breeding programs, taking due attention in respect of ecological conditions and its implication for conservation of biodiversity should be considered. It should be recommended, therefore, that the planning and operation of breeding programs must consider the likely expansion of the construction of fish seed supply sectors and must endorse strategies for genetic improvement that will not only conserve genetic and biological diversities but will also bring genetic earn to benefit farmers.

Since the gene pool has been extensively infected because of inexorable and unplanned hybridization among the Indian major carps, a scrutiny should be encouraged to recognize pure stocks of economically-important coldwater species such as Tor, Schizothorax, Salmon, etc. to conserve their genome in the original form and that recognition of the importance of this sector can facilitate the conservation of aquatic genetic resources. Gene conserving efforts must be superior to fish production. The lack of adequate data on such activities promotes an attitude that are not valuable. Fishery scientists are claimed as one's own to devise systems that will alter this situation.

of a single sex, and several qualitative traits are some of the essential criteria which must be improved through practical approaches. It has been suggested that drastic genetic improvement can likely be brought to bear to reduce the impact of fish diseases in future. Genetic improvement definitely has, at the same time, a cardinal role to play in efforts to reduce the potential environmental impacts on fish production.

The term genotype and phenotype in genetic study is highly important. The particular alleles at specified loci present in fish is termed as *Genotype*. The physical characteristics of fish as determined by the interaction of its genotype are, on the other hand, referred to as fish *Phenotypes*. The phenotype in terms of appearance or performance, essentially reflects to both the environment the fish has been exposed to and the genetic composition of the organism itself. In fish culture, the study of qualitative phenotypes, quantitative traits, and selection are the three major goals in fish breeding as will become apparent in sections 1 through 3.

Genetic improvement of aquaculture stocks is expressed in the same quality with the higher yields or increased profitability. Genetic changes in many fish species can result in increased yields and profitability and have been discussed in sections 4 through 6.

Control of sex is inevitable for monosex stock production and various techniques which address this need are presented in section 7. The extension of these concepts to the development and production of transgenic fish are summarized in section 8. Genetic threats to wild fish and needs in genetic research have been briefly noted in sections 12 and 13.

Scope and Significance of Genetics

Genetics may be alleged as one of the most important subjects of biology. For thousands of years, human societies have adopted the techniques of genetics in the betterment of agricultural crops, vegetables, and domestic animals without any knowledge of the mechanisms which exist beneath the superficial aspect of these practices. Fragmentary reports on these aspects dating back to 7,000 years have come into person's mind that human society understood that some physical factors could be transmitted from one generation to another. Through selection of particular organisms from the wild stocks and allowing them to inbreed, humans have been able to generate high-yielding varieties of animals and plants to suit their needs.

In the beginning of the 20th century, scientists have begun to esteem highly the principles and mechanisms of heredity. Development of microscope has illuminated the fact that the ova and the sperm are definitely responsible for transmission of the hereditary characters from one generation to another. But the hassle, nonetheless, remained of how very small particles of biological matter could sustain the vast number of features that constitute an individual organism.

The scientific progress on the study of heredity based on the recurrence of characteristics transmistted as genes, as formulated by the 19th century Moravian Botanist Gregor Johann Mendel and his work has laid to foundations for the present-day science of genetics. He reported that characteristic features do not mix but transmit from parents to offspring as separate units. These units disclose in the offspring in pairs, remain separate and are transmitted to subsequent generations by the male and female gametes. The Danish Botanist Johannsen called these units as *genes* in 1909 and the American geneticist T. H. Morgan in 1912 reported that genes were possibly arranged in a linear fashion on chromosomes. Since the beginning of the 20th century, the study of genetics has made substantial contributions to enumerate the nature of inheritance at both the level of the organism and at the level of the gene. Therefore, it is indisputable that the science of genetics has a tremendous impact in applied areas such as aquaculture and fishery, forestry, ecology, agriculture, and biotechnology and has become a subject of discussion on food production through genetic manipulations. From aquaculture point of view, however, an understanding of genetics is cardinal for producing robust and disease-resistant high-yielding fish through selective breeding for mass consumption and marketing. Different aspects of genetics in fish culture will receive an exhaustive treatment in different sections of this chapter.

15.2 Why the Study of Fish Genetics is Essential?

In general, fish need conducive environment to reproduce and require minimum effective population size to maintain intra-populational genetic variability. The gene pools of a species bear all the gene properties. Human activities (such as over-exploitation, pollution, etc.) drastically affect the gene pools and genetic diversity of fishes at large. Also, artificial breeding is performed without concern towards the genetic consequences. Mixed spawning of Indian major carps in hatchery, for example, leads to hybridization between the genus. These inter-genetic hybrids are fertile and find opportunity to back-cross with the parental species in natural waters. This causes a threat to the genetic integrity of these species by genetic introgression. Also, the use of spawners in low numbers at the time of artificial breeding lead to inbreeding resulting in intra-populational variability. Therefore, the study of fish genetics is inevitable that will address these issues and it is also concerned with the approaches by which genetic potential of culture stocks can be improved.

15.3 Qualitative Phenotypes*

Qualitative phenotypes are characters which can be defined as *discontinuous categories*. Distinct color variants, variations in fin size or shape, characteristic scale patterns or distinct flesh coloration of fish particularly common carp, guppy, channel catfish, rainbow trout, gold fish, etc. are apt examples. Since these phenotypic characters are controlled by only one or two loci, they tend to themselves conveniently as illustrations of the basic principles of inheritance as well as the relationship between genotype and phenotype.

Ascertaining Qualitative Inheritance

When an attractive color variant is found in a pond, it is necessary to isolate the fish and ultimately allow it to spawn with one or more normal-colored, or wild-type individuals (usually unrelated to the color variant of each other). The potential benefit of spawning color variant to a number of individuals is, however, to generate offspring in large numbers that sustain the allele(s) responsible for the phenotype (if the variation is not the result of some environmental factors. Though the first generation (F_1) of offspring from these matings may provide some insight into the genetic control of the phenotype, it is inevitable to allow these F_1 offspring to cross among themselves. This effort reduces the essential inbreeding that arises when trying to increase numbers of a rare phenotype.

The F_1 offspring produced by crossing a normal individual with one or more normal individuals present normal phenotype. The F_2 generation subsequently produced by crossing F_1 individuals will generally provide easily-characterized phenotypic categories in proportions that can afford insight into the alleles generating the specific phenotypes generally observed within both offspring and parents.

Relating Phenotype to Genotype

When the genetic control of a particular novel trait is known, the breeding outcomes can be easily predicted. The following examples will illustrate this sub-section.

1. **First Example:** Suppose a minnow culturist has identified a fish in his stock that is golden in color, as opposed to the normal gray-green phenotype of the species he produces. Appropriately, the fish happens to be a male. The farmer places the male with a number of females and collects the eggs. The F_1 offspring produced are all gray-green in color. Subsequently, the farmer produces the F_1 fish and allows them to spawn among themselves. As the F_2 generation grows, it becomes conspicuous that some golden individuals are present. Sampling reveals that there are soughly 3 gray-green individuals for every golden one. These results would suggest that the golden phenotype results from a single locus being homozygous for a recessive allele (Fig. 15.1). The golden allele can be regarded as *recessive* because those individuals that carry both golden and normal (gray-green) colors are phenotypically normal. The color phenotypes of gray-green and golden would occur in a 1 : 2 : 1 ratio.

---

* For further detail, see Tave *et. al.* (1989), Thorgaard *et. al.* (1995).

Gg male   X   Gg female

Sperms / Eggs	G	g
G	GG	Gg
g	Gg	gg

GG = Gray-green (1/4)

Gg = Gray-green (2/4)

gg = Golden (1/4)

Fig. 15.1 : Progeny classes from mating two Gg heterozygotes

2. **Second Example:** Suppose fish culturists identified a minnow that is pink and has elongated fins. If this fish spawns, the F_1 offspring produced are all gray-green in color with normal fin. In the F_2 generation, roughly 3 gray-green normal-finned fish for every one will be produced. Therefore, it is suggested that the pink coloration and long fin phenotypes have resulted from a single controlling locus being homozygous for a recessive allele.

Though the offspring produced in the F_1 generation would all be gray-green in color with normal finnage, the subsequent generation would reveal a number of phenotypes such as gray-green with normal fins, gray-green with long fins, pink with normal fin, and a few pink individuals with long fins. These phenotypes would, however, occur in a ratio of roughly 9 : 3 : 3 : 1 (Fig. 15.2).

Pp Ff female x Pp Ff male

Sperms / Eggs	PF	Pf	pF	pf
PF	PP FF	PP Ff	Pp FF	Pp Ff
Pf	PP Ff	PP ff	Pp Ff	Pp ff
PF	Pp FF	Pp Ff	pp FF	pp Ff
pf	Pp Ff	Pp ff	pp Ff	pp ff

PPFF, PpFF, PPFf and PpFf = Gray, normal fin (9/16)

Ppff and Ppff = Gray, long fin (3/16)

ppFF and ppFf = Pink, normal fin (3/16)

ppff = long fin (1/16)

Figure 15.2 Progeny classes from mating between two double heterozygotes, Pp Ff.

Color Inheritance in Tilapia

Though the color inheritance and hybrids of tilapia have been given serious considerations over several decades, some breeding works on tilapia represent complex examples of qualitative inheritance in a food fish species. Some unusual white Nile tilapia (*Oreochromis niloticus*) fingerlings have been reported. Out of about 15,000 fingerlings, between 20 and 30 individuals exhibited white coloration.

In some developed countries, markets for live tilapia place extra value on silvery or light-colored fish and consequently, the market potential for these white Nile tilapias among culturists was discernible. Such light-colored tilapias are derived from crosses of certain red hybrids. However, light-colored Nile tilapia fingerlings were set aside for grow-out on site. As fingerlings grew, they developed accessory coloration. (Fig. 15.3). Additionally, their coloration became more intense and their opalescence, the phenotype was termed as *pearl*.

Fig. 15.3 : The pearl strain of *Oreochromis niloticus* (Nile tilapia) generally exhibits the accessory pigments associated with mood and behaviour but most of the skin pigments are absent normally found in this species.

When a number of the pearl-colored fish are crossed with each other, the offspring displayed a 50 per cent white: 50 per cent normal color ratio. But Chi-square tests of a number of broods from these fishes indicated the ratio was statistically close to 9 : 7 (white to normal). This indicates that two loci (designated as A and B) are involved in the control of this trait. Each locus has two alleles (a and b). The a and b alleles in combination, and only in combination, would result in pearl coloration. The pearls colored individuals produced from different genotype parents are shown in the following checker boards. (Figures 15.4 through 15.7).

AABB female × aabb male

Sperms \ Eggs	AB
ab	AaBb

AaBb = Pearl

Fig. 15.4 : Expected progeny when crossing a double-homozygous pearl Nile tilapia

AaBB female × AABb male

Eggs \ Sperms	AB	aB
AB	AABB	AaBB
Aa	<u>AABb</u>	<u>AaBb</u>

AABB, AaBB, AABb = Normal
AaBb = Pearl (1/4)

Fig. 15.5 : Outcome of crossing normal-colored AaBB and AABb Nile tilapia

aaBb female × aaBb male

Eggs \ Sperms	aB	ab
aB	aaBB	aaBb
ab	<u>aaBb</u>	<u>aabb</u>

aaBb, aabb, Pearl (3/4)
aaBB = Normal (1/4)

Fig. 15.6 : Outcome of crossing partially heterozygous pearl-colored Nile tilapia

AaBb × AaBb

Eggs \ Sperms	AB	Ab	aB	ab
AB	AABB	AABb	AaBB	<u>AaBb</u>
Ab	AABb	<u>AAbb</u>	<u>AaBb</u>	<u>Aabb</u>
aB	AaBB	<u>AaBb</u>	aaBB	<u>aaBb</u>
ab	<u>AaBb</u>	<u>Aabb</u>	<u>aaBb</u>	<u>pabb</u>

AABB = Noral (1/16)
AABb = Normal (2/16)
AAbb = Normal (1/16)
AaBB = Normal (2/16)
AaBb = Pearl (4/16)
Aabb = Pearl (2/16)
aaBB = Normal (1/16)
aaBb = Pearl (2/16)
aabb = Pearl (1/16)

Fig. 15.7 : Progeny classes produced from crossing two first generation pearl Nile tilapia

As a generalization, for production of fry in large numbers through breeding commercially, congregation of large numbers of broodfish (generally varies between 15 and 50) in a breeding pond/tank is necessary. Consequently, it is very arduous to isolate pearl-colored brood stock from fry production unless a few percentage of normal-colored fish are produced. The presence of multiple genotypes exhibited as similar pearl phenotypes, the testing of progeny is difficult other than those individuals (aabb) harmoniously generate all-pearl offspring. Production of pearl heterozygotes on commercial basis require breeding operations which is similar to those of mirror/leather/scale carp (Figure 15.7a)

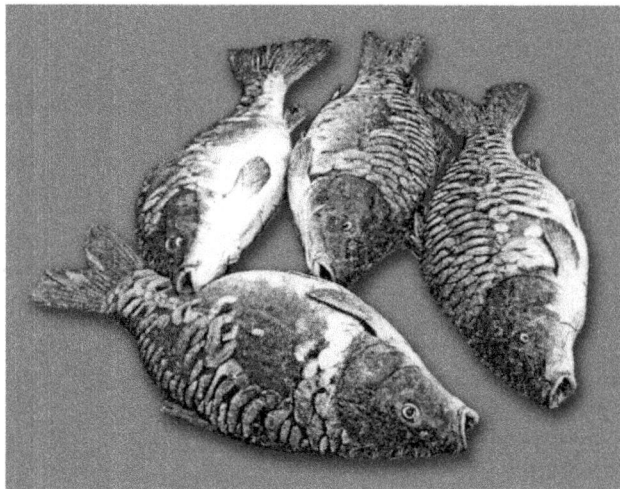

Fig. 15.7a : The fish *Cyprinus carpio* (mirror form) sometimes emerges in normally-scaled courses and it can be obtained by mating female and male leather carp. The phenotype of mirror form results from a homozygous recessive condition and breeds genuine.

Many distinct color variants such as blond (a faded Khaki-like appearance), syrup (orange-yellow), red and gold have been noted in different species of tilapia populations. Gold individuals in *Oreochromis mossambicus* are homozygous for the recessive allele (g) while wild-type (normal-colored) individuals are homozygous for the normal (G) allele. Interestingly, heterozygotes (Gg) individuals displayed a bronze color.

The syrup and blond colors were determined to result from recessive alleles at single autosomal loci, each with complete dominance for the wild type (normal) coloration. A third colour variant in this stock (red color) also resulted from a single autosomal gene, but the mutant allele (red color) was dominant to the wild type. Selection against any carrier of such a mutant allele would serve to eliminate it as quickly as it arose in a normal population, since both heterozygotes and homozygotes exhibit the red phenotype. It should be added as a note that these color variants result from a single autosomal gene.

When light-colored *O. niloticus* are crossed at random, the first generation progeny produced normal (wild-type) coloration. Inbreeding between these light-colored individuals, produced light-pink individuals, with a smaller numbers of bright-red, orange, and a few albino offspring.

1   **Significance of color Variants in Tilapia:** The importance of the emergence of many colorful phenotypes is extremely significant to the global tilapia industry. The development of stable variety of red tilapia (*O. mossambicus*) and production of hybrids with *O.niloticus* is a boon to fish farmers because it has great commercial significance. The following examples will illustrte how different types of colors in tilapia can be produced in aquaculture farms.

* Offspring derived from *O.niloticus* by *O.mossambicus* are normally white or pink which are homozygous and red individuals are heterozygous. When black-flecked red-gold hybrids are, however, allowed to mate randomly,F_2 generations produced phenotypes with different color variants such as wild type, reddish-brown with darkly pigmented dorsal area, reddish-gold with black patches, pink and reddish-orange mottled with white, continuous pink with no black markings, and continuous off-white. Also, it is interesting to note that some phenotypes were predominantly male (reddish-gold with patches and continuous pink) while others were predominantly female (reddish-brown, wild type, pink-orange with white mottling).

* A number of color variants have been produced from random breeding of Philippine red tilapia. This stock has resulted from crossing of *O. niloticus* with red tilapia. When 'red-orange gold with black spots' founder stocks were allowed to spawn, a number of phenotypes have been noted in the offspring such as red gold-orange without spots, red gold-orange with spots, uniform pink without spots, uniform pink with spots, albino with black eyes and uniform gray. Selective breeding within red gold-orange without spots generated uniform pink without spots, uniform pink with spots, albino with black eyes, red gold-orange without spots, and with no normal-colored fry. But breeding within uniform pink without spots produced the gold-orange without spots, uniform pink without spots, albino with black eyes phenotypes, also without the spotted pink and with no normal-colored progeny.

* A cold-tolerant strain of red tilapia has been developed through hybridization and recurrent back crossing with a cold-tolerant line of *O.aureus*. In this case, F_1 hybrids and back progeny ratio has been to be 1 : 1 (50 per cent red and 50 per cent normal color offspring). Crosses between red- and normal-colored hybrids of *O aureus* and between similar *O. niloticus* hybrids also produced red- and normal-colored offspring in a 1 : 1 ratio. But when male white tilapia is mated with female *O.aureus*, or with female *O.niloticus* or when male dominant red phenotypes of *O niloticus* is crossed with normal *O. niloticus*, consistently produced all-red progeny.

2. **Albino Strains of Fish:** In general, the term albino refers to an animal having a congenital absence of pigment in the skin and hair (which are white) and the eyes

(which are usually pink). Albinism is the most familiar type of recessive trait and many albino ornamental fish can be obtained across the world. However, albinos occur in a variety of cultured fish including salmonoids, cichlids, and cyprinids. They are often put to use in studies of gynogenesis and androgenesis (See section 15.7).

Albino refers to a genetic condition that does not allow the production of melanin, but many other white- or light- colored genetically-based variants can be found in a variet of fish and these are sometimes referred to as *albinos*. Through induced spawning, a stock of albino *Catla catla* has been developed from a few initial fingerlings. Similarly, when albino channel catfish (*Ictalurus punctatus*) were crossed with each other, all-albino offspring were produced. All-albino progeny of grass carp (*Ctenopharyngodon idella*) were produced when they are crossed with each other.

It should be added as a note that albinism in commercial fish can result in more marketable products, but if predation occurs in culture ponds or tanks, production potential significantly reduced. Survival is lower for albinos and this was the result of increased vulnerability to predation due to the high visibility of these fish at the time of feeding.

To sum up, in contrast to red tilapia where the red phenotype exhibited higher mortality than their normal-colored counterparts (37 per cent versus 15 per cent), the pearl tilapia exhibited high survival potential and at the same time, the white phenotype can apparently play an important role in the mass production of red tilapia in some populations. At present, a number of other red tilapia varieties have arisen or have been developed, some of which display much better survival. Two independant genes, the Mm locus determining melanism pigmentation and the Rr locus determining red coloration might have resulted in the production of pink, normal, red, or white individuals. However, the production of red tilapia and albino fish in grow-out ponds or tanks have good demand in the global market.

15.4 Quantitative Characters

Quantitative characters can include traits measured in the form of dimensions or counts, such as fecundity, dress-out precentage, metabolic activity, length, weight, behavior, or other various morphometric dimensions. However, several to many genes (perhaps ten or more) as well as environmental variations (See Panel 15.2) are generally involved in the expression of quantitative traits and direct patterns of inheritance are dissembled. Each gene contributing such a small amount of phenotype that their individual effect cannot be detected by Mendelian methods but by only statistical methods. Such genes which are non-allelic and affect the phenotype of a single quantitative trait, are termed as *polygenes* or *cumulative genes*. The inheritance of polygenes is called *quantitative* or *multiple factor inheritance*. The quantitative traits are also termed as *metric traits* (characters). Metric traits exhibit clearcut differences between individuals and forms a spectrum of phenotypes which

---

PANEL - 15.2

INFLUENCE OF ENVIRONMENTAL VARIATIONS OVER GENE ACTION

A large number of gene loci are unnecessary for continuous variations because all biological characters are influenced to some degree by the environment and genotype. The environment will definitely interact with the genotype to produce phenotype. Thus if a single genotype is exposed to a range of environments, a range of phenotypes are observed which is described as the *norm of reaction*. This illustrates much of the phenotypic variance in the population where each individual has the same genotype termed as *isogenic population* as all individuals are not exposed to identical environments. The degree to which a phenotype can be shaped by the emvironment is termed as its *phenotypic plasticity*.

---

blend imperceptibly from one type to another to cause continuous variations. It is practically very difficult to determine exactly how many genes (loci) are actively involved in the phenotypic expression of such traits. Rather, it is possible to estimate the types of genetic effects that exercise the most control over a trait and formulation of improvement programs to make the best use of these sources of variation. Ascertaining the genetic basis of observed variation and determining methods for improving a trait require precise examination of variations among various types of relatives and with the aid of statistical methods.

Genetic Effects and Phenotypic Variation

Genetic control of metric trait discloses two phenomena which are involved in animal. and plant-breeding strategies such as: (1) heterosis often referred to as *hybrid vigor* or the alternative, termed as inbreeding *inbreeding depression* and (2) resemblance between relatives. These phenomena are based on dominant genetic effects and additive genetic effect, respectively. For genetic improvement of any trait, the influence of dominance and additive effects will resolve the importance of selection, cross-breeding, or a combination of these practices in a breeding program.

1. **Additive Effects:** Additive effects are related to the heritability of various characters and consequently are the basis for selection. These effects are the basis of an individuals breeding value which can be rendered as the expected performance of its offspring when it mated at random within a population. The observed variation of these breeding values is termed as *additive genetic variance*.

2. **Dominance Effects:** These effects are the result of specific combinations of alleles and accordingly, combinations of individuals, or species that contribute substantially to those alleles. This influence of parental combinations relates to the manifestation of dominance effects through inbreeding depression and or hybrid vigor.

3. **Epistasis :** The term epistasis describes a situation in which expression of one gene obscures the phenotypic effects of another gene. Epistatic genes are sometimes called *inhibiting genes* because of their effect on the other genes which are described as *hypostatic*. Simply defined, epistasis refers to variation resulting from the interreactons of alleles occupying at different loci. Variation, on the other hand, describes the difference in characteristics shown by organisms belnging to the same natural population or species. In fact, most of the breeding facilities available to fish culturists are not adequate to access epistasis with exactness. It is, however, fortunate that it rarely influences phenotypic variation.

4. **Maternal Effects:** These effects frequently contribute substantially to the phenotypic variation found in fish species. These influences have a genetic basis, and may reflect characteristic features such as fertility, egg size or quality, survival and care of hatchlings/fry, and synchrony of ovulation as well as spawning. Maternal effects are also observed in the application of gynogenesis.

Attributing Observed Variation to Genetic Effects

To access individual genetic potential in most fish species, easily-calculated values are taken into consideration. Definitely, the phenotype is the most measureable physical expression of most continuous features such as growth rate, food conversion efficiency, and tolerance to extremes of temperature or dissolved oxygen. In most statistical analyses of variation within cultivated populations, environmental influences are supposed to have a mean of zero, at least within a given generation. The result of this assumption is the phenotypic mean for a character should approach the genotypic mean if the sample is large.

To evaluate the relative importance of different sources of genetic variation for a given trait, it is inevitable to separate random environmental variation from genetic variation and at the same time must ascribe genetic variation to additive, dominance, or maternal effects. Separation of the genetic portion of the observed variance among various types of relatives such as half siblings (progeny with only a common mother or father), full siblings (progeny with both parents in common), or unrelated individuals, estimates of the portions of phenotypic variance accounted for by dominance, additive and other genetic effects can be measured.

Since the genetic manifestation of similarity between relatives is associated with additive genetic effects, the greater the additive portion of genetic variance within a population for a particular character the greater the expected degree of similarity or covariance for that character between related individuals. In related individuals, alleles with additive effects will generate the some phenotypic influence. Similarly, covariances resulting from dominance effects are based on the combination of alleles producing a dominance effects at a particular locus. This covariance occurs only when individuals are related to each other. It is, however, unfortunate that since fish are raised under similar environmental conditions, it is obvious that similar individuals will be produced, and hair-splitting analyses must be designed to take this propensity into account.

Average Effects and Dominance Deviations

When a single-locus trait is considered, it is not impossible to determine a fish's genotype by the following observation

Species: *Oreochromis mossambicus*

Character: gold/bronze/normal (gray-green) body color

Phenotype:	Gold	Bronze	Normal
Genotype:	– a	o,d	+ a
	gg	Gg	GG

This condition is termes as *incomplete dominance* since heterozygote (Gg) individuals can be distinguished phenotypically from their normal (GG) counterparts. On examination of the additive attributes of alleles, the value of -a and +a are assigned to the two homozygous genotypes (GG, gg, for examples). Also, a value of '0' is measured at an intermediate point and some relative value of 'd' is assigned to the heterozygotes.

In some other cases, a heterozygote may be phenotypically indistinguishable from a homozygote for whichever of the alleles is dominant. This type of interaction is termed as *complete dominance*.

Species: *O. niloticus*

Character: blond/normal (gray) body color

Phenotype:	Blond	Normal	
Genotype:	– a	0 d	+ a
	blbl	BlBl	BlBl

The average deviation from the population mean of offspring which obtained a specific allele from one parent combined at randon with corresponding alleles at prevailing frequencies within the fish stock is termed as the *average effect* of that allele. As shown in the case of incomplete dominance, the g and G alleles will accordantly result respectively in light- and dark-colored offspring. These results represent additive genetic effects.

It should be kept in mind that allele frequencies can have some interesting effects within a fish stock. Let us consider the genotypes at the Nile tilapia locus (BlBl, Blbl, blbl) where the dominance is complete when a recessive allele (bl) is substituted at random. If the bl occurs at a low frequency, most of the individuals in the stock will be BlBl, and a substitution to produce Blbl will exhibit little or no change in phenotypic value. When the bl allele is very common, many of these substitutions will convert Blbl to blbl, generating a significant change in the population mean and a larger average effect. However, the average effect of an allele on the trait is not only additive but also persists from generation to generation. This is the key to a number of concepts developed in fish culture genetics.

Utility of Estimates of Genetic Variation

1. **Heritability :** The observed variation within a fish stock can be easily applied for genetic improvement. The term heritability (or h^2) is generally used to illustrate the contribution of additive variance to total observed variance. Simply defined, the term h^2 is that proportion of phenotypic variation in the expression of characteristics which is due entirely to genetic components. Fish breeders often prefer to estimate h^2 which is based on the knowledge of gene action. The details of gene action are beyond the scope of this book and the readers are advised to consult any text book on genetics. For this purpose, however, genetic variance is partitioned into additive genetic variance and dominance genetic variance. If A and a are two alleles at a locus controlling a quantitative trait, the additive genetic variance is associated with the average effect of substitution of A for a in a homozygote (aa → AA) and dominant genetic variance is associated with partial dominance of A over a in heterozygotes. The h^2 is used as a measure of relative proportion of additive genetic variance as follows:

$$h^2 = \frac{\text{additive genetic variance}}{\text{phenotypic variance}}$$

The estimation of $.h^2$ is the major goal of efforts to partition phenotypic variance. If 30 per cent of the observed variation in a trait (Vp) within a given fish population is attributed to additive variance (Va), the h^2 estimate of that trait is expressed as 0.30 (Va/Vp).

It should be remembered that h^2 is an integral concept in genetic improvement through selection. A large-sized fish may be large due to any number of causes. Some causes may have influence on its phenotype than the allelic average effects in its genotype. Relatively little additive genetic variance in relation to the total phenotypic variance indicates that little genetic improvement can be made from one generation to the next through selection since many other factors are responsible for phenotypic variation. Understanding the h^2 of a specific trait permits for the formulation of efficient breeding programs to improve performance in fish production stocks.

2. **Application of Heritability Estimates:** The study of partition phenotypic variation associated with a number of traits especially disease resistance, growth, survival, dress-out has great practical significance in relation to high fish production. These are, however, the most widespread areas of interest in fish culture genetics. If disease resistance were moderately heritable, the survivors of any given outbreak would pass along with superior resistance to their offspring, and the disease would cease to be a consideration. Genetic modification of the balance of power between the immune system of fish and a particular pathogen may or may not be feasible over several generations.

The study of the h^2 of resistance to infectious hematopoietic necrosis virus (IHNV) in rainbow trout (*Oncorhynchus mykiss*) and in Atlantic salmon (*Salmo salar*) for susceptibility to furunculosis disease has given encouraging results. When juveniles of rainbow trout are exposed to high or low dose solutions of IHNV, the h^2 of resistance of 0.50 and 0.05, respectively, has been noted. This indicates that under low dose conditions, several factors other than additive variance influenced individual susceptibility. At the high dosage level, these factors played a less important role compared to additive genetic attributes.

Heritability estimates of infected 5,000 Atlantic salmon with *Aeranonas salmonicida* (fish from 50 full-sib families within 25 parental half-sib groups) were 0.48 ± 0.17 and 0.32 ± 0.10, respectively. Though overall death rate was about 70 per cent, the family groups exhibited large variations in survival. Therefore, selection plays a very important role in improving resistance to furunculosis disease in fish.

3. **Growth, Survival and Dress-out:** Growth rate is one of the most important production traits in fish culture. If fish culture systems are successfully adopted, daily maintenance of body weight of fish is inevitable. Over time, this can be a tremendous cummulative feed use, complicating factors arise that have to be considered when heritability of growth rate is assessed. If excessive amount of food is given to any culture system and rapid growth requires more food consumption, the rate of stocking of fast-growing fish must be reduced while turnover within the system will increase.

Survival of fish in different months of age exhibited a heritability close to 0.0, but heritabilities for weight and length at 12 and 18 months of age for Arctic charr (*Salvelinus alpinus*), for example, ranged between 0.13 and 0.17. But when the relationship between weight and length was presented as condition factor, heritabilities for this trait were 0.54 and 0.32 respectively at 12 and 18 months. Other important production traits such as flesh coloration have become more important in selection programs. However, substantial variations in growth, survival, and flesh coloration have been noted in several commercial fish species and hence emphasizes the importance of utilizing superior populations in domestication and selection programs for farmed fish.

Genetic influence on growth trait is most notable in early life history. As time passes and fish grow and mature, its genetic merit may tend to come to forefront that permits better estimation of breeding values. It is, however, apparent that the type of aquatic environment results in increased environmental variations in the phenotype and subsequently reduces the partitions of additive genetic variance.

Selection for market-related traits is forecasted on procuring acceptable yield performance using available state-of-the-art technology. A continuous production process is necessary that enhances the end product to ascertain the consumer demand. When the fish coho salmon, for example, were fed a diet supplemented with canthaxanthin, the estimated heritability for carotenoid level in the flesh has been reported as 0.50; of course, the values of estimated heritability are highly variable and depends very much upon the age of brood fish.

15.5 Selection*

Selection is a process by which some individuals come to contribute more offspring than others to produce the next generation in natural selection through intrinsic differences in survival and fertility or in artificial selection through the choice of parents by the breeder. However, all of the progress made in genetic improvement of livestock and agricultural crops can be attributed to the application of selection pressure. The practice of selection principally relies on utilizing the most desirable individuals within a population to produce subsequent generations or on the omission of unwanted individuals from the breeding population.

When additive effects play an important role in production traits, selection becomes a practical means of genetic improvement in fish culture operation. The aim of a farmer is to obtain maximum profit feasible from the farmed species. This can be achieved by genetic improvement of fish stocks by selection. To implement the program on fish breeding and genetic improvement of stocks, maintenance of adequate broodfish is inevitable. Some important things for fish culturists may, however, wish to improve through selection include disease resistance, reproductive performance, market-related traits, and growth. Of course, setting up an experiment to estimate selection response in such traits are very arduous. When individuals are retained for breeding programs on the basis of their individual performance, the practice is referred to as *individual mass selection*. Other types of selection such as family selection, genotypic selection, progeny selection, etc. are often practised as well.

Selection relies on comparing different strains and utilizing the best one(s) for culture purposes. Selection can be recognized as the process of altering, generation after generation, the genetic make up of a population in order to produce more desirable phenotypes. In this sense, an inferior strain that possesses significant additive variance may be a better starting point for selection and domestication than a moderately superior population with little additive variance.

Pre-Requisite for Selection

The following pre-requisites should be kept in mind before a genetic selection program is undertaken.

1. The minimum number of characters (traits) that determine the value of a fish species should be specifically defined.

2. The complete life cycle of the fish should be under control that permits to identify the parents and progeny. Unless it is properly identified, it would lead to full-sib or half-sib mating resulting in inbreeding depression.

3. The relative economic value of the selected characters should be established.

---

* For further study, see Marian and Krasznai (1987).

4. The genotypic and phenotypic correlations of the selected characters should be known. Since fish have very high reproductive rates and hence phenotypic variations are large, large-sized brooder selection does not guarantee genetic improvements in offspring. And the economic and genetic values of a large and aged fish significantly reduced due to its greater maintenance cost and reduced number of eggs.

5. The environmental variation influences phenotypic expression of fish. Therefore, such variations should be well understood.

Type of Selection

1. **Individual selection:** This selection is made on the basis of phenotypic characteristic performance of the individual. If selection is known based only on one character (such as total weight of fish at one year of age), the individual can be selected. In this case, the performance of parents or offspring is neglected. Historically, this selection is the principal method followed by fish breeders but, in general, success has been limited for yield characteristics due to the following problems:

(i) Correlation between age and growth rate is very poor. Selection of large-sized fingerlings for stocking in ponds may not improve the overall growth rate for the table or selection of table-sized fish may not improve the growth rate at fingerling stages. This is due to the fact that fish growth at different ages is accelerated by different factors and the heritability.

(ii) The selection of aggressive individuals may results in insignificance of commercially-important fish. Hence, if a culture stock is dominated by the aggressive individuals, its overall performance will diminish.

(iii) In the absence of family selection (see latter), the full- and half-sib mating will result in inbreeding.

It should be pointed out that individual selection can be more effective if unrelated populations are considered for breeding so as to generate heterogeneous gene pool from which further selection can be done.

2. **Mass Selection:** This selection is principally based on traits of the fish under selection as hindered to selection based on the performance of their relatives. It is the most common method generally used in breeding programs where improved characteristic features can easily be measured. Mass selection can be used in selection for both growth rate and for age at maturity. However, the response to mass selection (R) is determined by the following equation.

$R = Sh^2$ Where S = selective differential (the difference in a certain character between the individuals selected and the population as a whole)

h^2 = Heritability of the differences

At present, mass selection is considered as a means of improving production stocks of many farmed fish especially tilapia. The genetic improvement of farmed tilapia involves the development of a synthetic base population including 8 strains of Nile tilapia. Combined family selection for growth was applied across five generations. The average response to selection per generation was 13 per cent and the accumulated response has resulted in 85 per cent increase in growth.

3. **Genotypic selection:** Two types of genotypic selection has great practical significance in fish culture strategy: (i) family selection and (ii) selection by progeny testing.

(i) *Family Selection*: Selection can be based solely on family averages. In this case, all (or a random sample of) members of of those families with the highest average are retained for breeding programs. For characters with low heritabilities, this type of selection is usually more efficient than mass selection. When families include siblings in high numbers, the efficiency of mass selection and family selection becomes roughly equal at heritabilities of 0.50 for full-sibs and 0.25 for half-sibs. An incapability to identify fish to particular families in many commercial settings often require growers to rely on mass selection even for characters with low heritabilities.

(ii) *Selection by Progeny Testing*: In this case, artificial fertilization of eggs is the main criterion for selection by progeny testing. The production of progeny should be repeated for several spawnings. In selection, the following three criteria should be taken into consideration.

* Correlated changes in other traits

* Genotypic influence with the environment

* Interaction between genetic variance (Gv) and environmental variance (Ev)

Genetic variance is the result of cummulated addition of variations caused by additive gene (Va), dominant gene (Vd), and by epistatic effect (Vi). Therefore, Gv = Va + Vd + Vi. Also, since Gv produces effect on phenotypic variance (Pv), such a Pv is not only due to Gv and Ev but also due to interaction that exists between the genetic and environmental variance (Vg - e). Hence, Pv = Gv + Ev + Vg - e.

Progeny testing is very useful when selection is made for traits with low heritability, traits that are expressed in only one sex, or traits that illustrate slaughter-related characters such as dress-out percentage. Progeny testing is highly significant to progressive fish-culturists since the required number of progeny can be produced within a short period.

4. **Pedigree Selection:** Simply defined, the term pedigree refers to as the record of descent of an animal, showing it to be pure-bred. Family selection is also termed as pedigree selection in which, on the basis of performance of the parents and grand parents, the progeny are selected to be considered as parents for the next generation. This has wide applications in farm animals where pedigree history of each individual is maintained. But in fish culture, records of pedigree of any

farmed fish are not available. However, on the basis of the performance of half-
or full-sibs of the individuals, selection can be made for breeding programs.

5. **Domestication Selection*:** The term domestication refers to a tame (an animal)
    which is kept as a pet or used for farm produce. The process of domestication
    and selection may occur simultaneously. Domestication can be interpreted as a form
    of selection in which only those wild individuals capable of surviving the stress
    and artificial conditions of a culture environment contribute to lines of captive-
    bred offspring. However, fish with shorter periods of domestication generally
    exhibits greater responses to selection.

6. **Tandem Selection:** The method of tandem selection is very simple for selecting
    one particular trait at a time for several generations. When the desired level of
    that trait is achieved in the population, selection for second trait is followed for
    further several generations. Once that desired level of trait is achieved, the others
    are followed. This method is rarely used since it takes very long time and the
    characters are greatly influenced by environmental factors.

To sum up, the broodstock management practices (such as choice of founder stock,
the number of breeders maintained, the type of mating system followed, stocking density,
etc.) exert a strong negative selection on the hatchery broodstocks. Some fishes such as
Indian major carps have an undesirable tendency to breed very slowly and exhibit maturity
towards end of season since fish breeders always use older broodfish at the time of
breeding season. Hence a careful and proper selection strategy for healthy fish is the need
of the hour which will definitely ensure the production of quality fish seed possessing all
the desired performance traits. Fish should be raised under hatchery conditions as far as
feasible that are more representative of natural selection forces. The hatchery-produced
fish seed can withstand the rigorous and often irregular conditions of fish farming in grow-
out regions.

Genetic selection might improve the level of production or slightly modify the quality
of fish stocks, but production of totally disease-free and robust fish is out of the question.
By genetic manipulation, it has paved the way for introducing any gene or set of genes
into the fish body and thereby causing it to produce immediate gene products.

The initial emphasis of those involved in the development of these capabilities has
been to produce fish that are useful to consumers. Already several species of fish that carry
foreign genes are commercially available (see section 15.10). It is hoped that other fish of
commercial importance will almost certainly beccome available in future. The capabilit of
this technology is, however, quite great.

15.6 Inbreeding, Cross-Breeding, and Hybridization**

Simply defined, the term inbreeding refers to the mating together of individuals that
are closely related by descent. In animal populations, mobility is restricted. As a result,

---

* For further study, see Dunham and Smitherman (1983).
** For further detail, see Argue and Dunham (1999), Falconer (1989), Kirpichnikov (1987),
    Makeyeve and Verigin (1992), Marengoni *et. al.* (1998).

mating between individuals of such populations is far from random, this being affected among individuals occurring in proximity with each other. Many small sub-gene pools are thus developed within a gene pool. These sub-gene pools disappear from overall characteristics of the entire pool. One of the consequences of restriction in population size is mating between relatives which is also termed as *inbreeding*. Inbreeding, though has very little influence on change in overall gene frequencies, have major effects on frequency of homozytes which is increased. However, inbreeding results in uniform sequence of genetic materials in the chromosome causing homozygosity thus reducing the genetic variance. The loss of genetic variance results in certain phenotypic losses of the individual. The degree of loss of the characters depends on the intensity of breeding. In general, many characters are lost due to intense breeding than mild inbreeding. Inbreeding has been associated with unfavorable biological effects, and the practice of cross-breeding (mating comparatively unrelated lines within a species) with an unrelated stock results in increased vigor. The consequences of inbreeding often, therefore, leads to loss of vigor and to other indications of deterioration and have wide implications in both natural and artificial selections.

Apart from the benefits of hybrid vigor, a cross-related species also termed as *hybridization* is sometimes undertaken simply to combine specific desirable traits of related species. Another important strategy in fish culture genetics is the production of sterile stocks not only for commercial culture but also for direct introduction into natural environments.

Inbreeding Depression

It is the gradual reduction of desirable characters such as growth rate, yield, fertility, etc. as a result of homozygosity produced by inbreeding. The loss or decrease in the phenotypic mean value of quantitative traits due to inbreeding is termed as *inbreeding depression* (See Panel 15.3).

Inbreeding depression, which is frequently associated with a general decline in fitness, can often be reduced from a practical standpoint by making special efforts to prevent closely-related individuals from breeding with each other. It can be offset by introducing new and unrelated individuals into a breeding program if care is taken to obtain high-performance stocks with adequate quarantine protocols.

Heterosis

The term heterosis was coined by Shull (1914) and in ordinary usage heterosis and hybrid vigor are virtually synonymous. Sometimes a distinction is made between hybrid vigor and heterosis. For example, according to Whaley (1944), 'hybrid vigor denotes to manifest effects of heterosis'. The practice of crossing is also ,observed in dominance effects and their manifestation through heterosis. However, hybrid superiority may results from a more efficient utilization of nutrient, increased rate of cell division, greater ability to synthesize growth substances, and others as yet unrecognized causes.

Heterosis is opposite to the phenomenon of inbreeding. When different inbred lines are allowed (or forced) to cross at random, the offspring exhibit an increase in mean

PANEL - 15.3
INBREEDING DEPRESSION

In animal populations, the mobility is highly restricted. As a result, mating between individuals of these populations is far from random, this being affected among individuals which occurs in proximity with each other in space. Thus, many small sub-gene pools are formed within a gene pool. These sub-gene pools are disappeared from overall characteristics of the entire pool. Crossing between relatives is one of the results of limitation in population size and it is called as *inbreeding*. Inbreeding, though, has little effect on change in overall gene frequencies, has major effects on frequency of homozygotes, which is increased. Inbreeding has been associated with unfavorable biological effects, and crossing with an unrelated stock results in increased vigour. The consequences of inbreeding may, however, have wide implications in both artificial and natural selections.

Three generalizations can be made with respect to inbreeding and hybrid vigour: (1) inbreeding often leads to loss of vigour and to other indications of deterioration, (2) hybridization between unrelated forms is often accompanied with great vigour and (3) cross-breeding must be biologically important.

Evidences have shown that the efficiency of selection is closely related to breeding system (outbreeder or inbreeder). For establishing superior homozygous genotypes, selection in a breeding system will be more effective which promotes homozygosity. At the same time, cross-breeding also maintains heterozygosity. When nornal outbreeding patterns are blocked to enforce inbreeding, the resultant effect is termed as *inbreeding depression*. If the frequencies of recessive homozygotes in inbred and random outbred populations are compared, a vivid picture will be observed. In a fish population at equilibrium, inbreeding alters the frequencies of genotypes (AA = P^2, Aa = $2pq$ and $aa = q^2$) to:

$$AA = p^2 + pqF$$

$$Aa = 2pq - 2\,pqF$$

$$\text{and} \quad aa = q^2 + pqF$$

Where F is the inbreeding coefficient. Therefore, it is evident that if a recessive disease with genotype *aa* occurs with frequencies q^2 in a random outbred population, its frequencies will be increased by pqF in an inbred population. The ratio of frequencies of recessive homozygotes in inbred and outbred populations will, therefore, be :

$$\frac{q^2 + pqF}{q2} = \frac{q(q + pF)}{q^2} = \frac{q + pF}{q}$$

It is clear from the above ratio that if q is large and p is small, the inbreeding increment PF will be small and the increase in frequency of recessive homozygotes will hardly be observed. But if q is small and is large, pF provides an increase in recessiveness even when F is small. Because different inbred families often produce homozygotes for various genes, paradoxically, one of the results of inbreeding will increase in variability between inbred families. The increase in homozygosity is escorted by a change in the means of quantitative traits towards that of the homozygous recessive. In case of incomplete domonance, the change in the value of heterozygotes will be zero but if the dominance is absent, the average quantitative value of heterozygote will be reduced to zero.

Features of Inbreeding: There are four principal features of inbreeding such as (1) increase in frequency of homozygotes, (2) increase in variability between different inbred families, (3) decrease in value of quantitative traits in the direction of recessive values, and (4) decrease in value depends upon dominance.

Effects of Inbreeding: The effects of inbreeding for several generations are as follows:

   (i)   Several lethal and sub-lethal forms appear in early generations.

  (ii)   The material rapidly separates into distinct lines which become uniform for differences in different morphological and functional features.

 (iii)   Several lines decline in vigour and fecundity until they cannot be maintained even under the most favorable cultural conditions.

 (iv)   The lines that survive exhibit a general decline in size, weight, and vigour of fish.

phenotypic values and can come to close to the foundation population. Therefore, decline in fitness as a result of inbreeding is revived by heterosis.

Heterosis has been exploited at commercial scale in fish. Intermediate phenotypes are preferred and in such cases, heterosis involves mating of parents having opposite phenotypes. The offspring generally produce an intermediate yield of fish biomass and dress-out percentage. Hybrids produced through heterosis may acquire disease resistance and successive rounds of selection combined with back-crossing (when F_1 individuals are crossed with one of the two parents from which they were derived) to the fish variety can eventually fix the gene or genes for disease resistance.

Dominance Effects

Dominance effects present interactions between pairs of alleles at the same locus. Since these interactions depend on the specific alleles involved at that particular locus, they terminate to exist during gamete formation through meiosis and are established anew in each generation.

In quantitatively measured traits, influence of dominance deviaions on all involved loci are reflected in dominance genetic variance.

These deviations can also influence the overall means of different traits which are manifested both through inbreeding depression and heterosis. The expression of heterosis is the foundation stone for hybridization and cross-breeding strategies. Since inbreeding depression and heterosis are the manifestation of dominance genetic effects, any discussion of one aspect generally refers to the other as well.

Population Genetics and Dominance Effects

Inbreeding and heterosis are the two important events that have definite impact on traits. These events are closely are associated with fitness for reasons that are related to molecular and population genetics. However, inbreeding is a great concern over small populations where random sampling in the formation of genotypes from an available pool of gametes can results in the eventual random exclusion of certain alleles and the fixation of others. Smaller populations have the tendency to inbred which is exacerbated when unaqual numbers of female and male broodstock are utilized for the purpose. The concept of an effective population number (Ne) takes gender ratios into account and ponders the relative contributions of maternal and paternal alleles to the subsequent generation (Table 15.1). The Ne for any number of broodstock is calculated by the following formula:

$$Ne = (4Nm\ Nf)/(Nm + Nf)$$

Where Nm = actual number of breeding males and Nf = actual number of breeding females

Table 15.1 : Relationship of Actual Numbers of Male and Female Broodstocks and Effective Population Size

Number of Broodstock			Effective Population
Male	Female	Total	Size
10	10	20	20.0
10	20	30	26.66
10	50	60	27.34
10	100	110	36.36
20	40	60	53.34
20	80	100	64.00
100	200	300	266.66
134	134	268	268.00

If the Ne for a commercial hatchery was originally 70 per cent heterozygotic and the inbreeding coefficient (F) of 0.3 has accumulated, then 0.3 X 70 per cent or 21 per cent of the original heterozygotic loci can be assumed to have become homozygous due to inbreeding. The resultant heterozygosity level is now 70 per cent- 21 per cent or 49 per cent. The increase in the F over one generation of random mating within a population size Ne can be expressed as :

$$F = \frac{1}{2\,Ne}$$

It is apparent that the smaller the Ne, the greater is the incremental increase in inbreeding per generation. Once inbreeding levels have accumulated within a population, they cannot be reduced.

In fitness-related traits, inbreeding depression results from directional dominance which helps improve the value. Breeding between more closely-related individuals tends to increase homozygosity of all alleles when their gametes unite to produce offspring. Therefore, as small populations reproduce over several generations, all alleles tend to become fixed. Consequently, inbreeding results in a reduction of the mean values of traits associated with fitness.

When inbreeding and dominance effects are discussed, hybrid vigor must be simultaneously taken into consideration. This phenomenon discloses itself when unrelated individuals produce offspring together. However, in case of directional dominance, fitness-related traits are superior in these offspring since over time certain alleles have already become fixed within parental lines or species.

Utilization of Dominance Effects for Genetic Improvement

Cross-breeding within species can be utilized as a practical alternative to selection to improve traits with low heritability. Nonetheless, since dominance effects are the basis for improvement of traits, progeny must be assessed on a cross-by-cross basis to determine how the alleles involved will interact to impact yield performance for the traits in question.

It is often possible to produce viable offspring by crossing closely-related species. This practice is referred to as *Hybridization* (see section 1.7 and Panel 1.4). The term hybridization is more appropriately used for crosses between species. It is unfortunate that the distinction between strains, races, and species is often obscured to a point where hybridization or cross-breeding are essentially not distinguishable.

1. **Alternative Goals in Hybridization Trails:** As noted earlier, the expression of heterosis is one of the consequences when species are crossed. Fish breeders are always try to set up a cross that will definitely produce sterile offspring to avoid unwanted reproduction in culture systems. In some cases, monosex population of some species of fish such as tilapia, some North American sunfish (*Lepomis* sp.) can be produced by crossing closely-related species that exhibits different sex-determining systems (For further discussion, see Section 15.9).

   Monosex culture is one of the methods in controlling wild spawning of some species of fish especially tilapias, and since male tilapia grows rapidly and attains a larger size, attention has been focussed on all-male stock (or all-female stocks as in the case where female stripped bass are crossed with male and yellow bass).

   The main approach to the production of monosex stocks is principally based on the use of steroid hormones. However, expectations wih desire using inter-specific hybrids is fish culture were provoked by the generation of all- male progeny by crossing *Tilapia mossambica* female with *T.hornorum* males. Also, monosex production strategy helps improve growth rate, body coloration, and temperature tolerance. These could enhance the value of tilapia for fish culture on commercial basis. At present, a number of all-male or only a slight preponderance of males has been successfully produced by crossing different species of tilapia that follows.

T. nilotica X T. hornorum (all male)

T. nilotica X T. aurea (all male)

T. nilotica X T. variabilis (predominantly male)

T. spilurus niger X T. hornorum (predominantly male)

T. vulcani X T. hornorum (all male)

T. vulcani X T. aurea (all male)

T. nilotica X T. mossambica (predominantly male)

The examples of inbreeding, heterosis, and hybridization in fish culture are predominated by studies on salmonoids, carps, tilapias, catfishes and others and some of the important achievements in this regard are noted below.

1. **Tilapia :** Improved performance in hybrid or crossbred offspring has been noted (Figure 15.8). Hybrid broodstocks were produced from a cross between female red tilapia and male Nile tilapia. Comparison of growth rates of first generation hybrids (F), second generation hybrids (F_1 X F_1:= F_2), and third generation hybrids (F2 X F2 = F_3) have resulted in improved spawning rates. The gains in fry production obtained by spawning hybrid parents were not substantially offset by any reduction in the performance of the F_2 progeny. This dictates the possibility of commercial application in fish culture industries around the world.

Fig. 15.8 : All strains of red tilapias are virtually synthetic lines derived from hybrids involving *Oreochromis mossambicus*.

2. **Catfish:** When *Clarias batrachus* was crossed with male *Heteropneustes fossilis* , offspring were viable through a 8-month grow-out period, but the reciprocal cross exhibited severe mortality at the time of hatching. In an example of heterosis,

however, it has been noted that African catfish hybrids resulting from a cross between *C.anquillaris* female and *C. gariepinus* male performed better than either parental species in growth during the first 5-weeks post-hatching. Different hybridization studies involving several species of freshwater fish (see Chapter 1) have been carried out. When hatching rate, growth and survival are statistically evaluated, offspring significantly performed better than either parental species.

Impacts of Inbreeding

When the differences in the relative proportion of gametes contributed by individual broodstock from generation to generation significantly increased, the potentiality for inbreeding is also increased dramatically. This is the concept of excessive reproductive contributions by dominant female and male broodstocks at the expense of genetic variation within a breeding population.

Common inbreeding impacts include gross deformities, reduced fecundity, survival, and disease resistance. A single generation of inbreeding in channel catfish (*Ictalurus punctatus*), for example, has been noted to slow egg development, decrease in hatching of eggs by over 15 percent, reduce booy weight and survival by over 15 and 5 per cent, respectively. Controlled inbreeding in conjunction with rigorous selection for fitness-related traits can be an effective approach to genetic improvement.

In general, a breeding stock can be divided into two or more sub-populations each under selection for fitness- and yield-related attributes. These lines should be crossed to generate fingerlings for culture purposes thus reducing the overall inbreeding load in the commercial stock.

Monosex and Sterile Hybrids

Hybridization can incidentally produce several benefits when particular stocks are segregated to produce all-female or all-male offspring. This is due to the result of closely-related species having different systems of sex determination. A number of sterile hybrid fish can be either produced artificially or occasionally found in nature. In sterile hybrids, male- and female-like gonads are partially developed. Though sterile, a dressout precentage is highly significant nonetheless.

Combination of Suitable Broodstock and Gametes

When cross-breeding or hybridization is carried out repetitively, the need to maintain populations of distinct species or lines within species is not avoidable. To avoid severe inbreeding, adequate facilities to house sufficient numbers of individuals are definitely needed. Also, care must be taken to avoid inadvertent mixing of distinct lines within species.

At times, handling stress, weather patterns, etc. can influence the spawning success of broodstocks and this condition permits obstruction of gametic effects. Therefore, repetition of treatments over several spawning seasons may be regarded as the result of hybridization.

The viability of hybrid offspring largely depends on the compatibility of the chromosome complements (also termed as *Karyotype*) of the parental lines. On examination of the physical features of chromosomes of grass carp, silver carp, and their hybrids, the diploid chromosome number has been reported as 48 for each parental species and the hybrid, but relative proportions of different types of chromosomes (such as metacentric, sub-metacentric, and sub-telocentric) in the hybrid differ from either species.

To sum up, a number of studies evaluating impacts of inbreeding, heterosis, and hybridization in several species of fish such as rainbow trout, brown trout, sunfish, tilapia, common carp, etc. have been carried out and their results are highly encouraging. The ultimate goal of all these efforts is the improvement of performance in commercial culture. Growth-related trait is one of the most important traits across all species and hybrids provide an added insight into the subject that has great practical significance. Other traits such as cold tolerance, disease resistance, and body confirmation are also equally important.

15.7 Androgenesis and Gynogenesis*

Selective breeding and hybridization are the classical methods of upgrdaing fish stocks but are long-term programs. Consquetly, the manipulation of entire sets or complements of chromosomes rather than individual strands of DNA is the key concept in this section, although the main title of this section refers to as chromosomal manipulation. Chromosomal manipulation generally includes androgenesis, gynogenesis, and sex reversal would definitely reduce the period for stock improvement to less than five years.

Viable offspring have been produced through androgenesis and gynogenesis in many farmed fish. Although these genetic manipulation techniques have little direct application in the production of commercial stocks they can be utilized to ascertain sex-determination mechanism as well as to produce (1) monosex populations, (2) genetically identical clones in large numbers, and (3) novel populations with far greater levels of homozygosity. Therefore, the contribution of androgenesis and gynogenesis to fish culture has been substantial.

Androgenesis

Simply defined, the term androgenesis referes to the development of an egg after entry of male germ cell (sperm) without the participation of the egg nucleus. In other words, it is the reciprocal of gynogenesis and involves the elimination of egg-derived DNA prior to fertilization. The phenomenon of androgenesis is extensively studied in tilapia, common carp, grass carp, trout and salmon.

Sometimes, it is possible to use irradiated eggs from one species as hosts for the production of androgens from a distinctly different species. Methods which are used for the production of androgenetic tilapia by the use of Nile tilapia (*Oreochromis niloticus*) eggs in combination with blue tilapia (*O.aurea*) sperm for successful production of androgenetic

---

* For further detail, see Kavumpurath and Pandian (1994), Nagoya *et. al.* (1996), Suwa and Suzuki (1994), Ihssen *et. al.* (1990).

tilapia, have proved quite successful. Similarly, irradiated common carp eggs and subsequently fertilized them with milt obtained from 3 varieties of goldfish (*Carassius auratus*) produced androgenetic common carp. Androgenetic diploid zygote is, however, developed through (1) dispermy, (2) fusion of second polar body and sperm, and (3) through suppression of first mitotic division.

1. **Production of Androgenetic Diploid Zygote:** Androgenetic diploid zygote production is generally carried out by thermal or hydrostatic pressure shocks. The induced androgenetic offspring are produced when gamma-irradiated (3.6 X 10^4 R) eggs are activated by normal sperm of desired species and diploidy is reinstated through application of shocks. Homozygosity level in androgenetic offspring is similar to that of homozygosity gynogenes. However, androgenetic progeny are all-males with certain parental phenotypes as they inherit most parental traits. High rate of mortality has been noted during embryonic and larval stages in androgenetic zygote when subject to thermal shocks, but hydrostatic pressure shocks exhibited about 50 per cent hatching rate. This suggest that the latter method is better than the former one. The amount and time of hydrostatic pressure shocks required to produce androgene (or gynogene) depends on the type of species to be considered (such as for tilapia: 7000 psi for 6 minutes, *Onchorynchus masou* 650 kg/cm^2 for 6 minutes to interrupt mitosis, 450 minutes after fertilization of egg, *Misgurnus angillicudatus* (loach) 800 kg/cm^2 for 1 minutes to interrupt mitosis and 35 minutes after fertilization). Similar to gynogenic individuals, the growth of androgenic ones is slow due to inbreeding effect. In cases where fish exhibit a system of sex determination, androgenesis can be used to produce novel YY male (to be discussed in section 15.9). Generally, sperm carries either an X or Y factor. Viable androgenes produced from Y-fertilized eggs will sustain two Y copies. when viable YY androgene in fish (common carp, for example) are crossed with a number of females from different genetic backgrounds, this fish will produce all-male progeny.

Gynogenesis

This genetic condition is induced in fish by three ways: (1) chemical treatment (such as colchicine, dimethyl sulfate, ethylene urea, etc.) (2) irradiation of sperm using Ultra-Violet rays, gamna rays etc. and (3) use of heat or pressure shock. In this case, unfertilised eggs are first activated with heat and pressure shocks to check the first cleavage division (mitotic gynogenesis) or to prevent the second polar body (meiotic gynogenesis) from elimination. Activated eggs are then allowed to fertilize with sperm. Consequently, the haploid (n) embryo that contains only maternal chromosomes is produced. During chemical treatment or irradiation, the chromosomes are broken into fragments and some fragments can persist in gynogenetic eggs and hence contribute to gene expression in the embryos. Since gynogenetic offspring inherit all maternal traits, it is the best way of producing all-female populations. Induced gynogenesis of many farmed fish such as tilapia, Common carp, grass carp, silver carp, Indian major carps, trout, salmon, etc. have been produced.

It should be pointed out that gynogenetic individuals are either haploid or restored to diploid condition by pressure shock at the expected time of formation of second polar body or prior to first mitotic division of the haploid zygote.

15.8 Polyploidy*

When an organism has more than twice the normal haploid number (IN) of chromosomes, such as 3N, 4N, 6N, etc., the condition is known as *polyploid*. They are also termed as *euploids*. Artificial polyploidy, which can be induced by the use of heat shock, pressure shock, or chemicals (notably colchicine),is of economic importance in producing hybrids with desired characteristics. In polyploidy, however, the genetic manipulation leads to the addition of chromosomes to the normal diploid complement (2N) resulting in an increase in the gynogene size. The increase in heterozygosity of the resultant individual is one of the best merits of polyploidy.

Sexual reproduction is the rule for virtually all species of fish and at the same time has a number of advantages, most notably is the formation of new genotypes from generation to generation through meiosis. Most of the organisms possess two sets of chromosomes in each cell which is designated as *diploid condition* (2N). The diploid cell generates haploid gametes that contain only 1N. Accordingly, as gametes unite to form new individuals, the resultant organisms possess the same number and type of chromosomes as their parents.

It should be rembered that this system of dividing and recombining pairs of chromosomes could potentially be disrupted if an animal possessed an odd number of chromosome set (say 3N) that could not be equally halved. Also, if organisms could be produced with multiples of their normal 2N karyotype (say 4N, for example), their gametes might be of particular interest in the production of novel offspring. The production of fish with greater than the normal number of chromosomal complements is generally termed as *induction polyploidy*.

Polyploidy has been reported to occur spontaneously in nature during hybridization between two distinctly-related species termed as *Natural polyploidy* and is criminated in the evolution of many groups of aquatic species; but for the purpose of fish culture, serious attention has been focussed on the development of techniques to produce triploid (3N) production stocks. The ultimate goal of developing 3N stocks includes the yield of fish that functions biologically in all respects other than gamete formation. In commercially-important species of fish, such sterile 3N stocks have great potential benefits that includes better growth rates and feed conversion efficiencies, higher survival, higher turnover within production systems, or reduction in potential ecological risks where introduced species, hybrids or transgenic stocks (see Section 15.10) are being cultured. Many species of fish gametes can be manipulated to produce large-scale viable triploid offspring.

---

* For further detail on the subject, see Maclean (1993), Maclean and Rahman (1994), Branick and Puckhaber (1995), Recoubrastky (1992), Rottmann *et. al.* (1991), Cherfas *et. al.* (1994).

Method of Ploidy Induction

Since polyploids are normally larger and more vigorous, their role in the improvement of fish stocks for high production has been realized and techniques have already been developed for artificial induction of polyploidy. However, induction of meiotic triploidy involves the use of thermal, pressure or chemical shocks to newly-fertilized eggs as meiotic gynogenesis, with the resultant disintegration of the mechanisms that would otherwise force the second polar body to retain within the egg (Figure 15.9). In contrast to gynogenesis where sperm cannot fertilize eggs due to irradiation, a motile sperm (1N) contributes to the meiotic triploid zygote, as do both the second polar body and the egg pronucleus. In this way, three sets of chromosomes (one of parental origin and two of maternal origin) combine within the triploid nucleus of the fertilized egg and all three sets replicate with each successive cell division as the zygote starts its development.

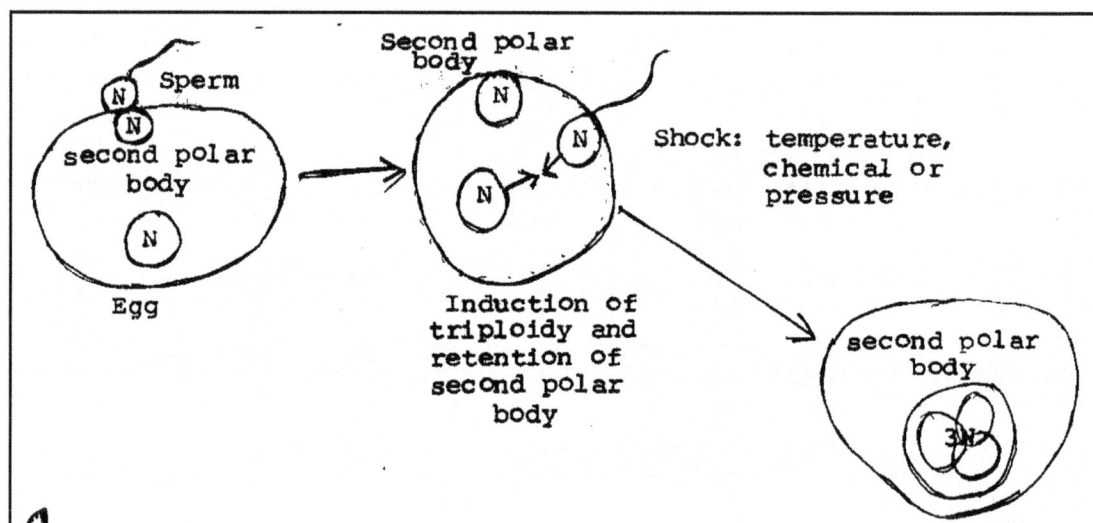

Fig. 15.9 : The induction of triploidy through retention of the second polar body

Tetraploids have been produced in a variety of fish such as *Tinca tinca, Perca flavescens, Hypophthalmichthys mobilis*, etc. In the production of tetraploids, however, normal (1N) eggs and sperm combine to form a diploid (2N) zygote. The 2N chromosomes then replicate in preparation for the first cleavage or first cell division (Fig. 15.10), but shocks are employed at the precise moment to prevent the first cleavage, leaving a 4N chromosomal complement in a single cell (2N from maternal origin and 2N from paternal origin, Figure 15.11). From this point on, cell division and chromosomal replication proceeds normally, but each cell contains a 4N complement of chromosome.

Viability, Growth, and Fertility of Ploidy Individuals

The viability of triploid individuals are highly variable in different species of fish. Triploid stickleback *Gasterosteus* sp. and blue tilapia exhibit less viable than diploids, but triploid common carp X grass carp and bighead carp X grass carp have better viability than diploids. Similarly, the survival of triploid heat shock-treated conmon carp and trout progeny has been noted to be as 70 and 95 per cent than that of diploid counterparts, respectively.

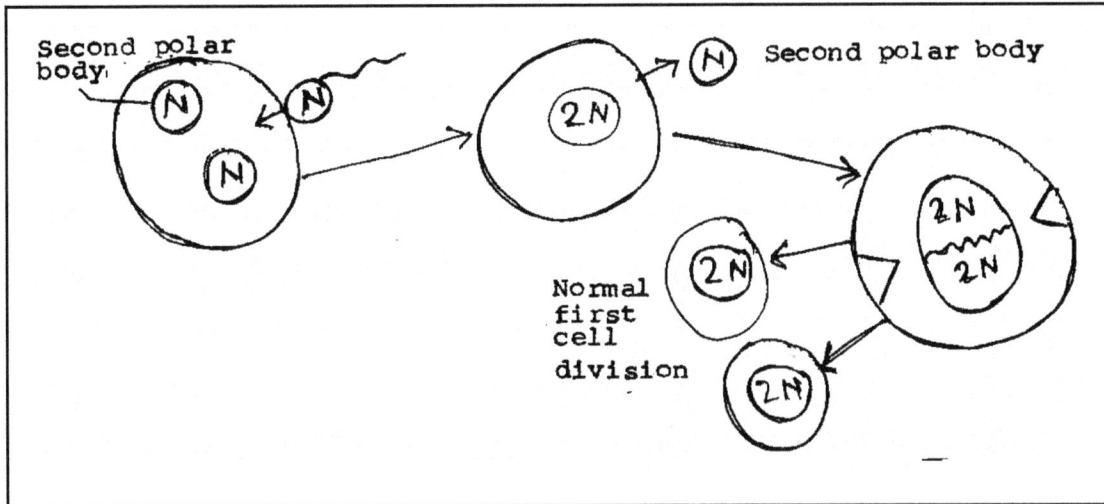

Fig. 15.10 : The normal course of events at the time of first cleavage

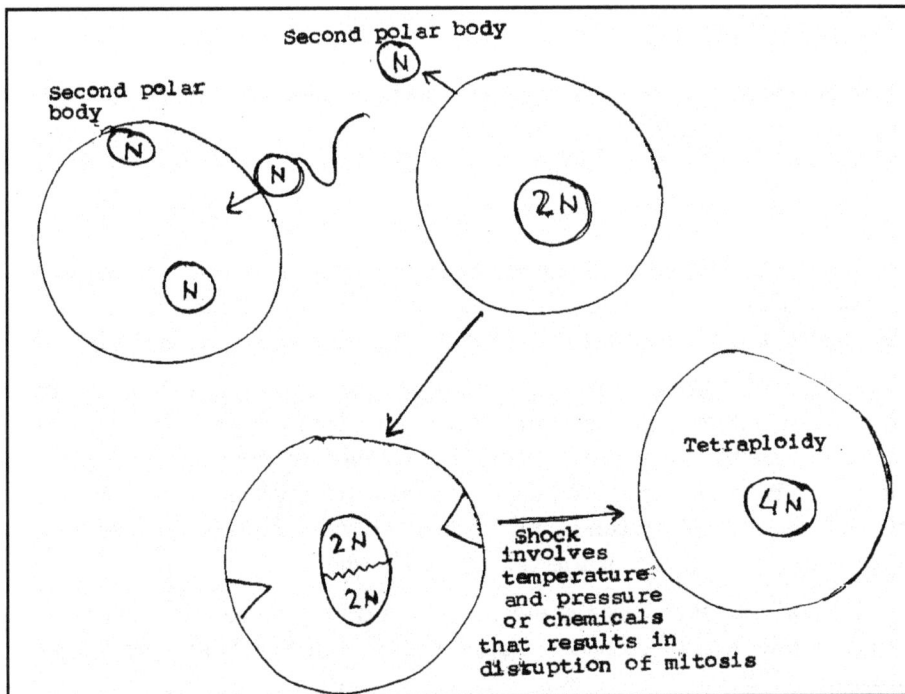

Fig. 15.11 : The interruption of first cleavage resulting in induction of tetraploidy

The increase in somatic growth in triploid tilapia has been found to vary between 20 and 30 per cent. Triploid male *O. niloticus* exceeded the body weight of diploid males by about 66 ± 15 per cent on an average. By weight, triploid females exceeded their diploid counterparts by about 95 ± 23 per cent. Similarly, mean body weight of triploid common carp has been noted to be 85 per cent to that of diploids. Generally, growth of triploids and diploids are more or less similar to the first several weeks or months, after which triploid fish exhibit superior growth. Although triploid fish stocks exhibit better survival and growth in culture conditions, they are functionally sterile.

The increasing interest in culture and demand of carps and catfishes throughout the Asian countries serve to emphasize the importance of developing production stocks through ploidy induction more efficiently. This interest and demand has been encouraged or forced to investigate polyploidy in potential fish stocks. It is hoped that such efforts could dramatically alter the freshwater fish culture scenario in the area concerned.

15.9 Sex Determination and Control*

In many farmed fish, production traits such as growth rate, age at maturity,and dressout percentage differ considerably between sexes. Consequently, it is more profitable to culture and market the more productive sex. It is also desirable to culture monosex stocks so that controlled or unwanted reproduction can be avoided. Several methods of monosex culture are widely practised. A growing practice involves phenotypic sex reversal through use of hormones and subsequent mating of phenotypically-reversed and normal broodstocks to generate either monosex populations or individuals which are themselves to produce monosex offspring.

The term sex refers to the sum total of the characteristics, structural and functional which distinguish male and female organisms especially with regard to the part played in reproduction. However, sex determination in fishes is concerned with the study of factors which are inevitable for making an individual male, female, or a hermaphrodite. Similar to other vertebrates, sex in fish is determined by the presence of a particular combination of chromosomes. In several cases, presence or absence of one special chromosome, known as *X-chromosome* which is considered as the determining factor. In the fish *Sternoptyx diaphana*, for example, males and females have 35 and 36 chromosomes, respectively.

In most common systems of sex determination, sex in fish is primarily controlled by the homozygous or heterozygous condition at one or two key loci. Inspite of the genetic basis for gender, several species of fish can be directed to develop into one phenotypic sex or another during their early life stages through the application of sex steroids. Thus, a fish that is genetically one sex can be made to function as the other. These methods result in reversal of some individuals of the sex being changed.

It is to be mentioned here that the mechanism of hormonal sex reversal are not justified for commercial production of monosex stocks and consumers in developed countries may

---

* For further detail, see Nair *et. al.* (1991), Melard (1995), Muller and Horstgen (1995), Shelton (1986).

be slanted to avoid steroid-treated products. Such manipulations are potentially of great benefit to a small farm. The motivation for such strategies is based on the heterogametic or homogametic nature of gender-determining in many species of fish. And the ultimate goal involves mass production of monosex stocks from novel and artificially-produced broodstocks.

Hybridization can have benefits when specific stocks can be isolated to produce all-female or all-male offspring and this is the result of closely-related species, sub-species or populations within species having evolved different systems of sex-determination. The commercial exploitation of such crosses could be profitable that results in monosex stocks and heterosis. .

Sexuality in Fishes

Similar to other animals, sexual dimorphism is also found in fishes. The expression of sexuality in fishes is diversified and this includes (i) various types of sex chromosomal mechanisms,(ii) gynogenetic reproduction,(iii) undifferentiated and differentiated types of gonochorism (See Panel 15.4), (iv) synchronous,protandrous and protogynous hermaphroditism (See Panel 15.5) (v) species-specific external and diversified ethological sexes, and (vi) environmental conditions. These diversifications are due to dual nature of the control mechanism, physiological and gonadal sexes as well as genetic sexes. In general, however, several differential characters in both sexes such as morphological, anatomical, growth rate, behavioral pattern, body color, and time of maturation provide an insight into the framework to understand the enlighted discussion pertaining to sex control and manipulation in fish.

1. *Physiological Sex and Manipulation*: Physiological sex is developed through biochemical processes under the control of genetic sex. The classification of physiological sex in fishes is shown in a flow-sheet diagram that follows.

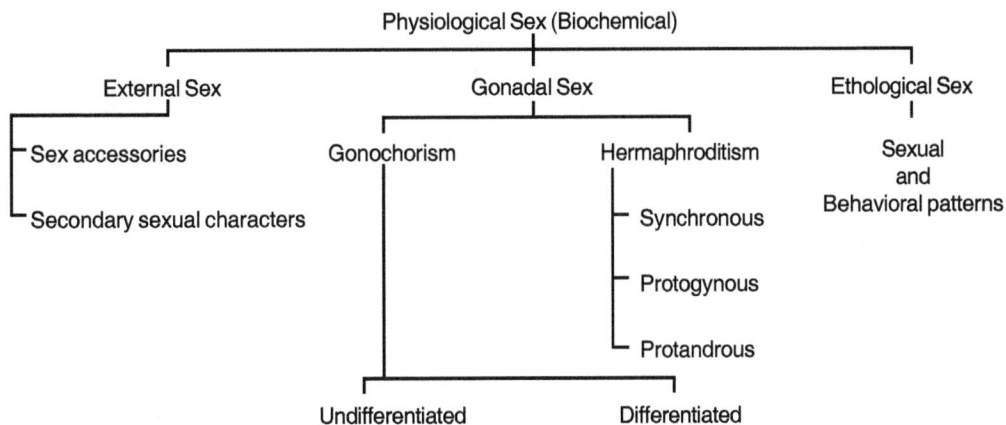

PANEL - 15.4
GONOCHORISM

It is the phenomenon in which either ovaries or testes are present in one individual fish species. Most of the farmed fish have this type of sexuality. As a generalization, two types of gonochorism exist in fishes. Such as (1) undifferentiated gonochoristic fish and (2) differentiated gonochoristic fish. In the former case, the undifferentiated gonad first develops into an ovary-like gonad and then about 50 per cent of the individuals are transformed into females and the other 50 per cent become male. In case of differentiated fish, the undifferentiated gonad differentiates into either an ovary or a testis. It should be pointed out that most of the species are regarded as an undifferentiated form. The artificial control of gonochoristic sex of differentiated fish is, however, successful only when sex steroid is administered during the specific stage of gonadal development.

PANEL - 15.5
HERMAPHRODITISM

Hermaphroditism refers to the ownership of both ovarian and testicular tissues in the same individual fish which functions both as male and female during its life history. More than 125 cases of hermaphroditism have been noted in different species of fish. Three types of hermaphroditism have been recognized such as synchronous, protogynous, and protandrous. The former hermaphrodites are those in which both spermatozoa and eggs mature at the same time. Protogynous hermaphrodite fish function first as females then as males by sex reversal at a certain stage of their growth. In case of protandous hermaphrodite fish, ovaries substitute the testes by natural sex reversal through a transitory inter-sexual stage. Reports on these phenomena have been published but the mechanism of natural sex reversal have not been explained. Spontaneous synchronous hermaphroditism is occasionally found in some differentiated gonochoristic fish such as *Oncorynchus keta*. Experimentally-induced synchronous hermaphroditism has been noted in rainbow trout by administration of estrone to fry at a dosage of 10-30 mg/kg diet following self fertilization. This suggests that the sex steroids are actively involved in naturally-occurring hermaphrodites.

The physiological sex can be manipulated by using sex steroids. Sex steroids are regarded as inducers of different reproductive phenomena such as gametogenesis, ovulation, spermiation, spawning or courtship behavior (the courting behavior of male fish), differentiation of gonads (make or become different in the process of growth or development of testes or ovaries), production of sex pheromone, and secondary sexual characters. Of these phenamena, gonadal sex differentiation is definitely significant and hence it is first included which is followed by the development of external and ethological sexes through the action of gonads.

2. **Gonadal Sex and Manipulation:** Gonad is a mass of tissue arising from the primordial germ cells (In the early embryo, cells which will later give rise to the gonads) and within which the ova or spermatozoa are developed. However, primary sex organs (testis or ovary) are the basis of the physiological sex and this principal sex is termed as *gonadal sex* which is further divided into two groups such as gonochorism and hermaphroditism.

In many cultured species, gonadal sex manipulation strategy involves artificial introduction of permanent sex reversal or the change of sexual phenotypes without change of genetic sex. The sex reversal through artificial means has been carried out in several species of fish such as *Carassius auratus* (gold fish); *Oreochromis mossambicus, O. aurea, O. nilotica, Salmo gairdneri, S. salar Poecilia reticulata* (Guppy), etc. by using androgens and oestrogens (acting as male and female inducers, respectively (Table 15.2). In differentiated fish, the artificially-reversed sex is possibly permanent since the action of sex genes is restricted to a short period during the early development of gonads and after the onset of gonadal sex differentiation, the sex genes become inactive. In undifferentiated species, on the other hand, sex reversal is very complicated since the sex genes can be active until later stages of the gonadal development.

Table 15.2 : Sex Inducers Used in Sex Reversal Strategy From Genotypic Female to Phenotypic Male or From Genotypic Male to Phenotypic Female

Male Sex Inducer	Female Sex Inducer
17-alphmethyl testosterone*	17-beta estradiol*
11-Ketotestosterone	Estrone
17-alpha-ethyltestosterone	Estradiol
Testosterone propionate	Diethylstilbestrol
Androstendione	Estradiolbutyryl acetate
Methyl androstendione	Ethynylestradiol

* The effective sex steroids and their use in more than 25 gonochoristic fish species is highly impressive. The results are highly encouraging.

(i) *Effective Dosage of Sex Inducer* : The treatment of fish to follow the strategy of sex reversal (Fig. 15.12) with sex steroid(s) is performed by feeding. Administration of male or female sex inducers to produce phenotypic feminization or masculinization have been successful by oral treatment in almost all farmed fish. (Table 15.3). This strategy of sex reversal is, however, an important step in the production of monosex fish populations. The production of monosex population of farmed fish would definitely eliminate unwanted rearing or reproduction. This will permit farmers to select fish for culture having growth-related traits and allow farmers to sale fish in live condition.

3. **Genetic Sex:** The genetic sex is generally evolved through the evolutionary process of each species of fish. This sex is determined at fertilization through the union

Table 15.3 : Effective Concentration and Treatment of 17-beta estradiol and 17-alpha methyl testosterone to induce sex reversal in Fish

Steroid / Species	Effective Concentration (mg/kg diet)	Percent Sex Reversal
17-alpha methyl testosterone		
Oreochromis mosambicus	30	100
O. niloticus	30	100
Tilapia aurea	30 - 60	100
Carassius auratus	25	100
Ictalurus punctatus	15 - 30	100
17-beta estradiol		
O. niloticus	100 - 200	more than 80 per cent
Salmo qairdneri	100	more than 90 per cent
Carassius auratus	100	100
Tilapia aurea	100 - 200	98

of chromosomes derived half from the sperm and half from the ovum. Therefore, the genetic sex is also termed as *chromosomal sex*. Each species has its own characteristic chromosomal sex.

(1) *Manipulation of Chromosomal Sex :* In many species there are generally two sex chromosomes, and the sex of fish depends on whether it has two identical chromosomes termed as *homogametic sex* (sex producing only one type of gamete such as X-producing eggs) or one of each of the two types-termed as *heterogametic sex* (sex producing two types of gametes such as X or Y-bearing sperm). Where the female fish is homogametic, the two chromosomes are designated as XX and the male chromosomes are referred to as XY. Where the male fish is homogametic, the male chromosomes are called ZZ and that of female chromosomes are known as ZW.

The manipulation of sex chromosomes has great possibilities for sex control in fish. Artificial gynogenesis in fish having female homogamety will produce all-female fry and the species having female heterogametic (ZW) will produce 1:1 of ZZ male and ZW female. Also, this super female will generate all-female offspring (ZW) when crossed with normal males (ZZ). This can be noted by using fish such as *Angiulla japonica* and *Gambusia affinis* having ZW types of sex chromosome. Similarly, artificial androgenesis in XY types of fish will also generate 1:1 of XX females and XY males. Hence, the males obtained by androgenesis in the XY type are expected to be XY homogametic males which generate an all-male population when mated with normal XX females. As noted earlier, the chromosomal manipulation for triploid induction by dissolution of spindle of the second meiotic division. This results in the inhibition of the second polar body formation. Different stock procedures are very common and hence frequently used for the purpose.

Mixed Population — Treated with Steroid

XY   X   XX

↓

Normal Life Cycle

Normal Male (XY)	Functional Male (XX)	Functional Male (Precocious) (XX)
↓	↓	↓
Reject or use for normal fertilization	Production of milt (female sperm only)	Reject or use if adult functional males are not available

↓

Functional male (XX)   ×   Normal female (XX)

↓   ↓

Milt   Eggs

↓

All-female Population (XX)

Treatment of fry with steroid	Culture in growout ponds for the table

↓

All functional
male population (XX)

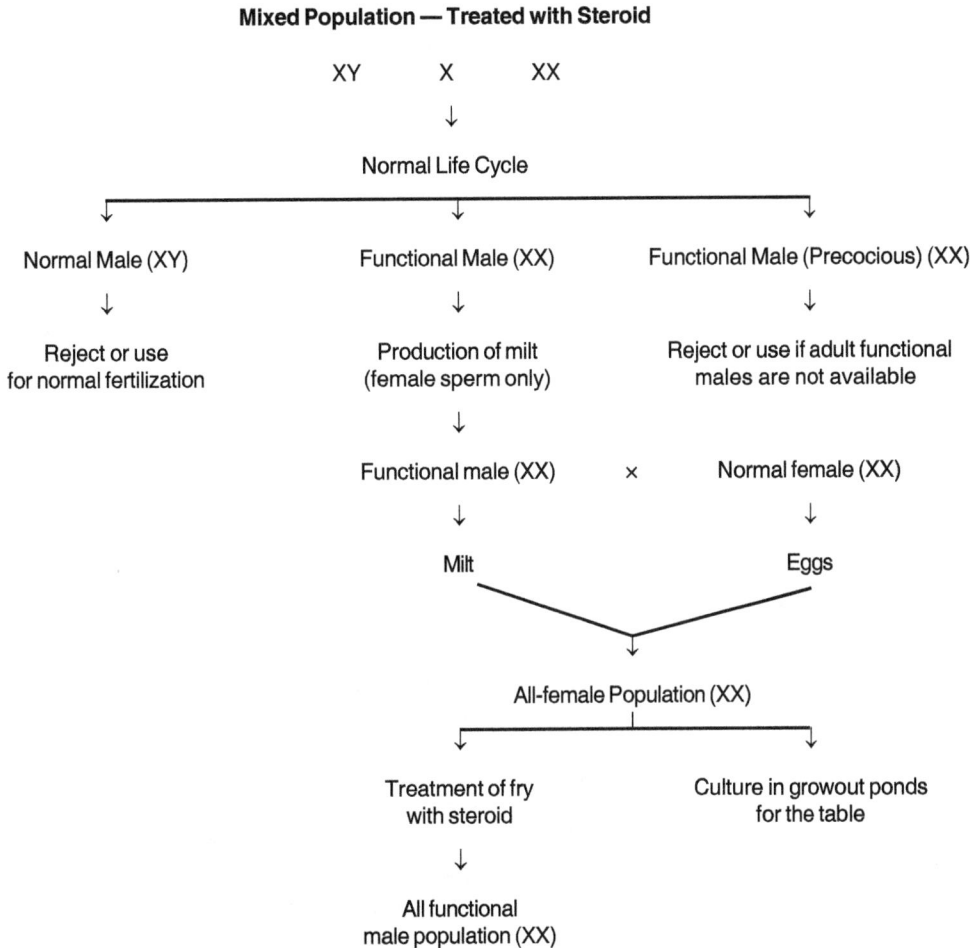

Fig. 15.12 : The Strategy of Sex Reversal in Fish

Generally, sex chromosomes could be morphologically identified in fish. The determination of sex by chromosomes has been noted in many species of fish which are listed in Table 15.4.

Homogametic and Heterogametic Monosex Stock Production

In many several salmonoid species, females are regarded as superior in production traits. Since female sex in these species is based on XX type of sex chromosomes, it is feasible to produce all-female stocks by crossing normal females with XX 'males'. These females with XX males are produced through steriodal masculinization of XX fry. Similar approach can be followed in the blue tilapia, *Oreochromis aureus*, with intial steriodal feminization rather than masculinization.

Table 15.4 : Determination of Sex by Chromosomes in Some Species of Fish

Species	Diploid (2n)	Sex Chromosome	
		Female	Male
Salmo gairdneri	58	XX	XY
Oncorynchus nerka	58 (57)	XXXX	CCY
Tor putitora	100	XX	XY
T. tor	100	XX	XY
Cyprinus carpio	100	XX	XY
Salmo trutta	80	XX	XY
Carassius carassius	50	XX	XY
Tinca tinca	48	XX	XY
Labeo rohita	50	XX	XY
Gambusia affinis	48	ZW	ZZ
Trichogaster fasciatus	48	ZW	ZZ
Anguilla japonics	38	ZW	ZZ
Sternoptyx diaphana	36 (35)	XX	XO
Lampanyctus ritteri	48 (47)	XX	XO

In some situations, use of gynogenesis to produce all-female salmonoid broods followed by sex reversal of these gynogenetic fry through application of male sex inducers has been quite successful in many salmon-producing farms.

In contrast to homogametic monosex stock production, mass production of heterogametic monosex ones requires more precise and complex manipulations. In cases where males are heterogametic (XY) such as *O. nilotica* and *Ictalurus punctatus*, it is not impossible to produce novel homozygous individuals. In an XX-XY system of sex determination, these novel individuals would be designated as YY. If these fish are mated with normal (homogametic) broodstocks, all-heterogametic offspring would be produced. This approach, however, involves sex reversal of hetergametic offspring and their crossing with normal heterogametic counterparts.

Production of Monosex Within Species

Large-scale production of all-male Nile tilapia generally involves sex reversal of normal male (XY) to functional female through admistration of female steroids. If these XY 'females' are crossed with normal XY males, some progeny will be YY males. When progeny of YY males are crossed with normal females (XX), about 96 per cent of male offspring will be produced. Ultimately, broods containing YY individuals were sex reversal after hatching to produce phenotypically female YY individuals. This allows mass production of YY male fish and permit them to cross with normal XX females.

On the basis of crossing between masculinized females and normal females, mass production of all-female grass carp has been possible. Masculinized individuals produced all-female offspring which dictates female homogameity.

Many species of fish are characterized by female homogamiety and male heterogameity, so that feminization of YY males might be used to produce YY offspring but the benefit derived from this effort to produce YY broodstock is yet to be realized by commercial fish producers.

Production of all-male or sexually undifferentiated stocks of channel catfish through use of male sex inducer (such as trembalone acetate) to undifferentiated fry has been found to be highly significant. But unfortunately, gonads of this species when exposed to this inducer for prolonged periods may fail to develop into either functional ovaries or testes. Therefore, it is necessary to find out the efficiency of other male sex inducers that would cause no harm and the process of gonadal maturation may proceed in an unhampered manner.

Possible Threats to Sex Reversal Phenomena

Sex reversal phenomena though significant in mass production of monosex stocks, have been noted to cause severe threats to fish culture industries around the world. However, possible threats include (1) ecological imbalance, (2) lack of natural competition, and (3) the possiblity of extinction of either male or female populations and failure of gonads to develop into either functional ovaries or testes. Moreover, use of steroids in sex reversal techniques for prolonged periods may deteriorate the human health condition. Therefore, all these threats to farmed fish must be carefully taken into consideration before the production of monosex fish stocks is permitted.

15.10 Transgenic Fish Production*

In 1990s, the use of novel methods for gene tranafer over a wide range of organisms, leading to the production of transgenic animals and plants, has been recognized as one of the thrust areas of biotechnology. The modern techniques of gene transfer using recombinant DNA technology (See Panel 15.6), allow transfer of genes from even bacteria to higher plants and animals, thus increasing manifold the possibility of improving the quality of animals and plants of commercial importance. The gene transfer methods, their use in production of a variety of desirable fish, significance of transgenesis for fish culture, future possibilities, contaminant and environmental impact and constraints to adoption of transgenic fish stocks in fish culture have been briefly noted below.

What is Transgenic Fish?

Transgenic fish can be defined as possessing within their chromosomal DNA., either directly or through inheritance, genetic constructs which have artificial organisms. Genetic constructs should be incorporated in such a way as to be expressed and passed along with subsequent generations.

---

* For further study, see Maclean (1993), Maclean and Rahman (1994), Purdan (1993), Rahman and Maclean (1992), Gupta (1996), Dubey (2002).

PANEL - 15.6
RECOMBINANT DNA TECHNOLOGY

Success in genetic engineering has been possible due to development of gene cloning techniques. It is essentially the insersion of a specific fragment of foreign-DNA into a cell, through a suitable vector, in such a way that the introduced DNA replicates independently and transferred to progenies as a result of cell division. The transformed cells containing DNA after their characterization and confirmation can be used commercially for various purposes.

1. **Plasmids and the Restriction Enzymes :** Two significant discoveries have contributed greatly to the development of recombinant DNA. These include (i) discovery of plasmids and (ii) restriction enzymes. In addition to the main chromosome (that contains double-stranded DNA), one or more pieces of circular DNA is floated inside the bacterial cell. These can replicate in the same way as the DNA from the chromosome and are termed as *plasmids*.

    Certain bacteria also contain what are known as *restriction enzymes*. These act as a fine chemical scalpel to break the long chain of DNA molecule into smaller units. Cells can employ them as defence against foreign DNA. There are, however, two types of restriction enzymes: type-I and type-II. While the former types break the DNA molecules at random site, the latter ones attack the specific sites only. One of the type II enzyme is the restriction endonuclease (EcoR-I) derived from the bacteria *Escherichia coil*. This enzyme recognizes the base sequence G↓AATTC on the DNA strands and cleaves at the site indicated by the arrow. The cleavage fragments carry short and single-stranded extensions at each end. These ends are sticky and can easily join with other fragments carrying complementary ends (Figure 15.13). Thus any two Eco R-I produced DNA fragments can be rejoined. This has made possible following the adoption of powerful technique of what is known as *recombinant DNA technology*.

2. **Steps Involved In Recombinant DNA :** Recombinant DNA involves breakage of a DNA molecule at desired places to isolate a specific DNA fragment and then insert it in another DNA molecule at a desired position. The product thus obtained is termed as *recombinant DNA* and the technique is often called *genetic engineering* or *recombinant DNA technology*. With this technique, genes can be separated, cloned and characterized so that the technique has led to significant progress in many areas of molecular biology. This technology helps isolate and clone single copy of a gene or a DNA segment into an indefinite number of identical copies. It has been possible because bacteria, phages, and plasmids reproduce in their usual style, even after insertion of foreign DNA so that the inserted DNA replicates with the parent DNA. This technique is termed as *gene cloning*.

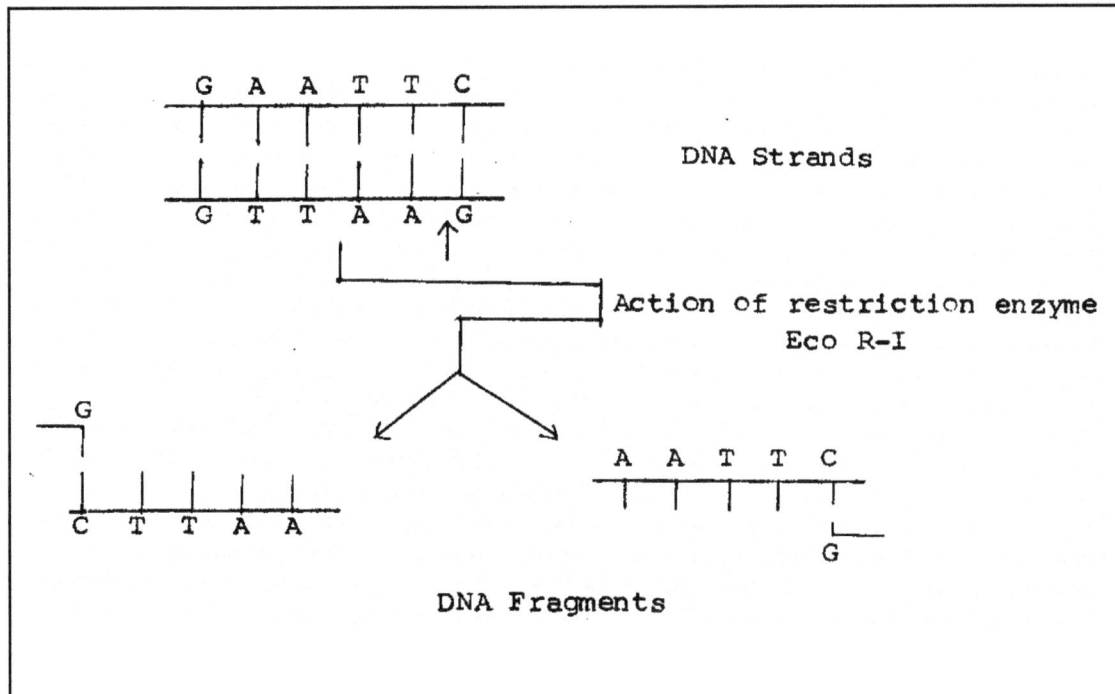

```
        G   A   A   T   T   C
        |   |   |   |   |   |                          DNA Strands
        |   |   |   |   |   |
        G   T   T   A   A   G
                                        Action of restriction enzyme
                                                  Eco R-I
    G
    |   |   |   |   |
    |   |   |   |   |                      A   A   T   T   C
    C   T   T   A   A                      |   |   |   |   |
                                                           G
```

Fig. 15.13: Restriction enzymes can cut the long chain of DNA molecules into several small pieces. There are several site specific DNA restriction enzymes which can splice the DNA molecules only at some specific places.

Genomic DNA is first extracted from any tissue of animal or plant species. It is then digested with a restriction enzyme, ECo R-I, also termed as *endonuclease*. This enzyme cleaves DNA molecules into fragments of different sizes which are then separated through gel electrophoresis. Electrophoresis is an analytical device generally used for separation and purification of DNA fragments. A gel is used in electrophoresis which is either agarose or polyacrylamide. The former is preferred for larger DNA fragments and the latter for smaller ones. Agarose is a purified powder isolated from agar which is a gelatinous material of sea weeds. This powder is dissolved in water and boiled that results into gel form. The gel prepared in a mixture of salt and water becomes a good conductor of electricity. The gel forms small pores that depends on its amount in a given water. These pores act as a molecular sieve. The gel is cylindrical or a slab, about 10 cm long and 0.5 cm thick. The gel is held in a electrophoresis chamber that consists of a negative and a positive electrode.

DNA fragments, after digesting with restriction enzymes, are mixed with sucrose and a dye (methylene blue) which altogether is known as *loading dye*. While dye increases the visibility of the preparation, sucrose increases the density of DNA preparation.

The preparation is loaded into wells at one end of the gel. One well is filled with DNA fragments of known length for comparison with those of unknown length. Electric current is applied at opposite ends of the electrophoresis chamber. A current is produced between the positive electrode at the bottom of the gel and the negative electrode at the top of loading end of the gel. This causes movement of DNA fragments through pores of the gel. Since the rate of movement of fragments is inversely correlated with the size of fragments, fragments which are heavy will retain closer to the size of loading and the lighter ones will move away. All DNA molecules of a given length move nearly the same distance into the gel and form bands. Each band represents many copies of DNA fragments having about the same length. The gel is removed from the chamber and then isolated by a technique called *Southern blotting technique** for furthel study.

The restriction enzyme cuts open the circular plasmid into a linear molecule. The DNA from a donar is similarly treated to produce fragments. The plasmid is now used as vector to carry a.piece of the donor DNA into the bacterial cell. When both are mixed under suitable nutrient conditions, a fragment of the donor DNA may join with the sticky ends of the linear molecule of the plasmid DNA. Another enzyme, *called ligase*, may be used to seal the two open ends of the plasmid DNA to make it again a circular molecule. Thus a piece of mammalian DNA gets incorporated into a plasmid to produce a hybrid plasmid.

In the next step, this hybrid plasmid is introduced into a bacterial cell where it can replicate. This process, called *transformation*, is achieved by mixing under suitable nutrition . In addition to plasmids, phagemid (phage + plasmid), virus DNA,and bacteriophage can also be used as vectors. The role of the donor and host cell can be interchanged thus creating hybrid plant or animal cells. The hybrid cells so produced are new organisms, their properties differ significantly from those of either the donor or the host cells. Another bacterial strain *Bacillus subtilis* as well as yeast, a higher organism, have been successfully employed as host.

An introduced gene construct includes a structural gene, the encoded DNA instructions for the production of a specific protein or similar gene product. Such DNA serves as a template from which proteins are assembled. However, the introduced constructs contain genetic sequences that serve as promoters and terminators for their transcription.

Fish Species and Genes Involved in Transgenesis

The main species of fish which have been subjected to transgenesis include *and* recommended for culture, *Oncorhynchus mykiss* (rainbow trout), *Salmo salar* (Atlantic salmon), *O.keta* (chum salmon), *O.kisutch* (coho salmon), *Esox lueius* (Northern pike), *Cyprinus carpio* (common carp), *Labeo rohita* (Rohu), *Ictalurus punctatus* (channel catfish), *Clarias batrachus* (Asian catfish), *C. gariepinus* (African catfish), *Heteropneustes fossilis* (Asian catfish), different species of tilapia and others. At the same time, a number of protein-coding genes used in transgenic fish production such as prolactin, growth hormones, antifreeze protein (AFF),

---

* For further detail, consult any Text Book of Biotechnology.

lysozyme, gonadotropin releasing hormone (GRH), bacterial/bovine beta-galactosidase, and the like have been widely used. These genes, called *reporter genes,* produce specific products and can be observed or detected through various assays to indicate whether genetic constructs have been incorporated, integrated and expressed in the target organisms. Different types of transgenes are used in different fish species. For examples, beta-galactosidase, chinook salmon GH plus a regulatory sequence from ocean pout, and beta-lactin proximal promoter are used in Asian catfish, tilapia, and common carp, respectively. These transgenes are inherited in a simple Mandelian fashion, indicating stable, permanent transformation of the germ-line.

Gene Transfer Methods

Gene transfer methods using recombinant DNA technology involve the injection of DNA into either the pronuclei of embryonic cells (or fertilized eggs) or the embryo stem cells. In animals, however, the uptake of genes by cells is termed as *transfection.* Transfection can be carried out at a cellular level and the transfected cells may be used for a variety of purposes such as (1) studies on the structure and function of genes, (2) production of transgenic animals for increased growth rate, and (3) resistance to certain diseases.

A number of variables play a critical role in the production of transgenic fish such as the number of construct copies incorporated on which chromosome incorporation takes place, epistatic interactions* with alleles at other loci, etc. As a result of incorporation, some sites of the chromosomal DNA strands appear to be much more conducive to gene expression than others. If the introduced construct can be successfully passed through subsequent generations, transgenic fish must be raised to maturity and permitted to reproduce.

1. **Microinjection :** It is the most commonly used method for the production of transgenic fish. This process typically involves the use of a powerful dissecting microscope with good lighting conditions and subsequent incorporation of multiple copies of genetic constructs through an extremely fine glass tube, called *micropipette,* into newly-fertilized eggs. Eggs/cells to be micro-injected are placed in a container. A holding pipette is kept on the field of view of the microscope. The holding pipette holds a target egg, cell, or embryo at the tip when gently sucked. The top of the micropipette is injected through cell membrane of the contents of the needle are delivered into the cytoplasm. At least one or a few introduced constructs would find their way onto the chromosome and at the same time the construct(s) would be replicated with each cell division and all the cells in the developing embryos would contain copies.

   In cases where chorionic eggs are very soft (such as carps), transgenic materials can be penetrated without any difficulty by fine needles. Salmon eggs are, on the

---

* Epistatic interaction means that one gene masks or changes the effect of another gene. This is a counterpart of dominance.

other hand, covered with tough chorions and injection of these eggs requires considerable skill and determination. A foot-pedal operated electronic pressure pump is used to force the required volume of transgenic suspension into the egg. With difficult eggs, the needle may have to be mounted in a micro-manipulator, but with easy eggs, the needle is held in hand. With tilapia eggs, injection needle can be held in readily passed into the egg via the chorionic micropyle. In others such as those of salmon, a method of boring a hole in the chorion is used and then passing the needle into the egg through this hole.

2. **Biolistics:** Biolistic involves the firing of microscopic particles that are coated with DNA construct into living cells through special devices termed as *gene gun*. These particles are made of tungsten or gold with different sizes, shapes, and chemical compositions that can be incorporated into the host genomes.

   As the term gene gun, it shoots foreign DNA into fish eggs, embryos or tissues at a very high speed. This technique is also termed as *particle bombardment, particle acceleration* or *micro-injectile bombardment*. This technique is most suitable for those eggs which do not exhibit adequate response to gene transfer.

   The equipment typically consists of a chamber fitted to an outlet to create vacuum (Fig. 15.14). At the top, a cylinder is sealed off from the rest of the chamber with a plastic rupture disk. A plastic micro-carrier is kept in close to rupture disk. The disk contains tungsten particles coated with DNA, called as *coated micro-projectiles*. Before working, the target cells/tissues are kept in the equipment. A screen is kept between the target cells and micro-carrier assembly. Helium gas is passed through the cylinder at high velocity. At this moment, plastic disk gets lacerated. Helium shock waves push forward the micro-carrier containing DNA-coated micro-pellets and the stopping screen permits them to travel through and deliver DNA into the target cells. These cells are then produced anew onto the nutrient medium. The embryos produced are then analyzed for expression of foreign DNA.

3. **Electroporation:** Electroporation is widely used to transfer the foreign DNA into the eggs. This approach, however, typically involves keeping the fertilized ova in a solution containing millions of copies of the gene construct of interest. Brief pulses of high voltage (about 350 V) of very short duration are then used to generate high-permeability of the egg membrane which permits for constructs to cross into the egg. The voltage can facilitate transport of the construct molecules through the egg membrane. By using this method, genes of choice have been successfully transformed in several species of carps, salmon, and trout. Also, the frequency of transformation can be improved by using linear DNA instead of circular ones and using 1.25 KV/cm voltage.

4. **Lipofection:** This approach is widely used to produce transgenic fish. Lipofection involves the encapsulation of DNA constructs in lipid vesicles and subsequently allowing the vesicles to come in contact with the plasma membrane of a target cell. This contact helps uptake the vesicles through fusion with the cell membrane

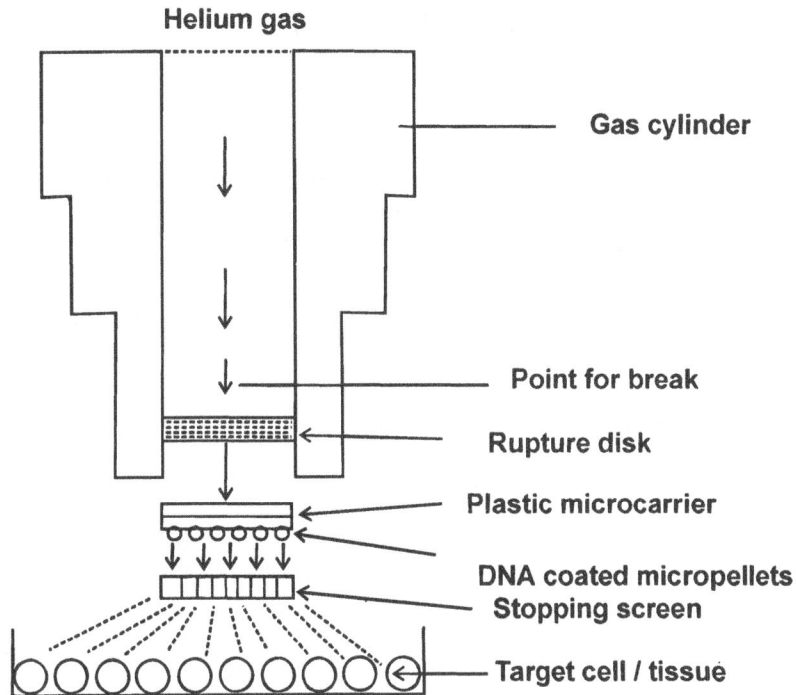

Fig. 15.14 : Working principle of particle bombardment gun. Using this technique, it is possible to succesfully transfer foreign DNA into the target tissue/cell/embryo

and is termed as *endocytosis*. And this approach can have disastrous consequences to target cells due to toxic action of lipofectin reagents which may alter cellular functions.

To sum up, the production of high-value fish and fish products through genetic selection provides an instructive lesson in the economics of industrial fish culture. Several such fish and their products are produced commercially by adopting gene manipulation technologies. Transgenic trout, catfish, carp and tilapia are being produced by several fish producing countries of the world. Though production of trout and catfish is low or moderately high compared to that of either carp or tilapia, the farming strategy remains competitive because of the very high price of the product. The production complexity of trout and other delicate species along with the precise control over the entire breeding and spawning periods, for example, virtually precludes the possibility of commercially-feasible production in many trout-producing regions of the world. But it is still trying to produce trout, catfish and their products for domestic consumption and export as a result of adoption of refined techniques. By further genetic selection and improvement of culture methods, strains have been developed that produce so much fish to meet the demand of food for ever-increasing populations.

Significance of Transgenesis in Fish Culture

The success of work with transgepic fish in terms of potential benefit to fish farming industry is highly encouraging. Expansion of transgenes may well be expected to produce four major phenotypic effects such as growth enhancement, cold tolerance, disease resistance, and induced sterility.

1. **Growth Enhancement:** It is clear that growth enhancement in fish has turned out to be much more successful than with mammals, though some species seem to be resistant to such improvement. The work definitely suggests that tilapia, channel catfish and common carp species respond well to transgenic technology and the growth enhancement of transgenic fish is quite impressive. However, responses of fish to growth enhancement seem to be variable with species and in some species the gains are modest. Also, the response to this technology is highly variable between different transgenic lines of the same fish species and the same construct. Therefore, it is essential to produce many lines and to select the optimal performers.

   A genteically-improved fish such as 'Jayanthi Rohu' (JR) has evolved in India. At present, this fish is being recommended for culture in ponds. This fish is a boon to farmers as the culture period could be brought down by 1 or 2 months without affecting its production potential and with a better output-input cost ratio. This excellent breed was developed for selective breeding of Rohu which over a period of about 10 years.

2. **Cold Tolerance:** In some species of fish such as carp and gold fish, cold tolerance could be enhanced by endowing them with extra copies of (delta-desaturase) gene. The protein prepared by this gene is cold-inducible. Also, the level of antifreeze protein production by use of anti-freeze gene in combination with its own promoter makes the Atlantic salmon effectively resistant to freezing.

3. **Disease Resistance:** Disease resistance in fish can be produced by using DNA vaccines. Vaccinated fish are resistant to a range of gram-negative bacteria such as *Aeromonas*, *Yersinia*, and *Vibrio* and infectious haematopoietic necrosis virus (IHNV). When the DNA fragment is injected into the nucleus of adult fish, it will lead to long-term expression of the relevant protein.

4. **Induced Sterility:** As noted earlier, sterile fish are most frequently produced by induced triploidy via the use of heat shock to the fertilized eggs. Hence it is feasible to produce sterile fish by exploiting transgenesis. Also, since fertile transgenic fish are prone to environmental hazards, there is much interest in producing double transgenics, namely fish which are transgenic for a growth enhancing gene but also transgenically sterile. This method may be compared with an advice like killing the goose that lays the golden eggs but there is a way at least theoretically.

   The proposed technique involves suppression of gonadotropin realising hormone (GnRH) by making transgenic fish for anti-sense against GnRH messenger RNA.

Such fish could, as desired for breeding, be rendered fertile by injection of GnRH. It should be kept in mind that apart from any additional transgenically- endowed trait, effective sterile strains of farmed fish would definitely be a desirable commodity to fish culture industry around the world.

Future Prospects

Field trials of transgenic growth-enhanced fish are underway in several fish-producing countries and that growth enhancement in commercial fish species such as tilapia, common carp, and salmonoids may well become one of the most definite benefits of the application of the transgenic technology. If growth enhancement can be combined with sterility, the advantages of fish farming could definitely be of considerable importance.

Another prospect of fish culture strategy involves the production of fish which are cold-tolerant or freeze-resistant or resistane to a wide range of bacterial or viral diseases. At the same time, DNA vaccination of young fish may offer improvements over conventional protein-based immunization. There is also great opportunity to use transgenic fish as biofactories for the production of valuable and desirable products.

The advent of transgenic technology opened completely new possibilities for the production of high-value products. Proper selection of techniques might improve the level of production to a considerable extent.

Environmental Impact

Transgenic animals in general, and fish in particular, are generally styled as *Genetically Modified Fish* (GMF). The problems faced by transgenic fish include (1) whether they pose an environmental hazards as a consequence of competition with existing species in natural ecological situations and (2) whether they interbreed with fish of the same or related species and so change the genetic structure of existing stocks.

It should be noted that many countries such as United States and United Kingdom have existing legislations to cope with the problems associated with contaminants and deliberate release of transgenic fish. Though the degree of control and the strictness of regulations are extremely variable from country to country, regulations require precise consideration for release of GMF, and divination of the effects of their proposed release in aquatic ecosystems could be substantial. However, production of transgenic sterile fish stocks is more safety. The likely environmental impact of the release or escape of sterile transgenic fish would possibly be inconsiderable. On the contrary, release of fertile transgenic fish would definitely lead to a reduction in overall fitness. In this regard, the observed hazards of release or escape of GMF have been greatly overestimated.

The question may arises whether eating transgenic fish offers any additional hazard to humans. This concern is highly significant. The potential benefits are so substantial that many farmers are interested to culture GMF. Any serious progress on the disease resistant front would be appreciated, in spite of the involvement of transgenic technology. The consumption of transgenic fish presents absolutely no hazard since their growth hormone, DNA, and protein contents are easily digested.

15.11 Genetic Threats to Wild Fish

Genetic risks are generally faced by wild fish populations only when they are escape from fish culture ecosystems. The most commonly cited threat is the loss of population identity. A number of direct effects on fish culture stocks can cause changes in the genetic structure of a wild population. The most dramatic situation is sometimes referred to as *swamping* when isolated populations are faced with overwhelming number of introduced individuals. There is a tendency to look at the genetic make-up of populations in terms of gene frequencies, that makes good sense within a population of constant size. But raw numbers come into play when gene frequencies for a mixed population of native, and introduced fish are examined.

A more real threat to swamping is the continuous introduction of small numbers of farmed breeding stocks into wild populations. And as a corollary, the transfer of genetic information generally occurs from one species to another through hibridization and repeated back-crossing. This is generally termed as *introgression*. *Introgression* from continuous introduction of non-native strains could eliminate unique genotypes within the species. Therefore, the potential for escaped production stocks to impose negative genetic and ecological impacts on wild populations should be preserved in memory.

Concern over genetic conservation of wild fish populations in terms of genetic variation both among populations of any particular species and within isolated populations is on the increase. The former issue is more concerned with future potential to maintain species survival. It should be added as a note that emphasis has already been given in many developed countries in shaping conservation policy for many farmed fish.

15.12 Future Needs in Genetic Research

Applied genetic research partaining to fish culture is lagging far behind than that of farm animals and crop plants. In view of the present importance to fish culture industry across the world and its future potential, this situation can never be tolerated. Some of the important high-priority problems associated with genetic research as applied to fish culture are briefly noted below.

1. Exploration and testing of wild genetic stocks to substitute culture strains that have been chosen from the wild.

2. Initiation of genetic studies of production-related traits in farmed fish are necessary for designing breeding programs.

3. Establishment of genetically-sound and controlled breeding projects for major fish species.

4. Transfer of fish from traditional, low-input fish culture to high-input intensive culture would be easier if parallel genetic changes are accompanied with the production strategy.

5. Since high stocking densities increase the severity of disease and parasite infections, breeding of resistant strains will definitely solve these problems.

6. Inter-specific hybridization programs should be expanded to yield new breeds that are well-adapted to new culture systems and for the production of sterile and monosex hybrids.

7. Cytogenetic work on chromosomal manipulations should be carried out.

8. Endangered genetic resources should be conserved at any cost.

15.13 Conclusion

The efforts of those involved in the developnent of genetic techniques around the world has been to produce robust and disease-resistant, and sterile fish that are highly significant to fish culture industry. Already genetically-improved variety of some farmed fish is now cultured commercially in developed countries. It is, however, hoped that in future, other commercially-important and genetically-engineered fish species will certainly become available to progressive fish culturists. The capability of these new state-of-the-art technologies is quite great.

Discussions have been made only to note that fish culturists will show greater awareness of the benefit of research-based studies on genetics. Several fish producing countries of Asia, Africa, and Latin America are applying genetic manipulation techniques for the growth of aquaculture sector. Expert personnel have taken the key responsibility to upgrade production stocks genetically. Genetic upgradation techniques have already acquired by developed countries.

References

Argue, B. J. and R. A. Dunham. 1999. Hybrid fertility, introgression, and back-crossing in fish. Review in Fisheries Science, 7 (Suppl. 3-4), 137-195.

Bramick, U., B. Puckhaber et al. 1995. Testing of triploid tilapia (*Oreochromis niloticus*) under tropical pond conditions. *Aquaculture* 137 (Suppl. 1-4), 343-353.

Cherfas, N. B., B. Gonelskyet al. 1994. Assessment of triploid common carp (*Cyprinus carpio*) for culture. *Aquaculture*, 127 (Suppl. 1), 11-18.

Dunham, R. A. and R. O. Smitherman. 1983. Response to selection and realized heritability of body weight in three strains of channel catfish, *Ictalurus punctatus*, grown in ponds. *Aquaculture*, 33, 89-96.

Falconer, D. S. 1989. Introduction to Quantitative Genetics. Longman, Esses.

Fast, A. W. 1998. Triploid Chinese Catfish. CTSA Publication No. 134. Centre for Tropical and Sub-Tropical Aquaculture, Makapu Point, Hawaii.

Ihseen, P. E., L. R. McKay et al. 1990. Ploidy manipulation and gynegenesis in fishes: Cytogenetic and Fisheries Applications. *Trans. Amer. Fish Soc.* 119, 698-717.

Kavumpurath, S. and T. J. Pandian. 1994. Induction of heterozygous and homozygous diploid gynogenesis in *Betta splendens* (Regan) using hydrostatic pressure. Aquaculture and Fisheries Management, 25 (Suppl. 2), 133-142.

Kirpichnikov, V.S. 1987. Fish culture and breeds of pond fish in the USSR J.*Ichthyology* (Suppl. 3), 88-99.

MaClean, N. and M. A. Rahman. 1993. Transgenic induction in salmonoid and tilapia fish *In* : Transgenesis Techniques (Eds : D. Murthy and D. A. Carter), Humana Press, New Jersey, PP. 95-110.

MaClean, N. and M. A. Rahman. 1994. Transgenic fish.*In*: Animals with Novel Genes (Ed: N. Ma Clean), Cambridge University Press, pp. 66-105.

Makeyeva, V. S. :and . V. Verigin. 1987. Morphological characters of age o+ reciprocal hybrids of grass carp, *Ctenopharynogodon idella* and black carp,*Mylopharyngodon piceus*. *J. Ichthyology*, 33 (Suppl. 3), 66-75; Voprosy Ikhtiologii, 32 (Suppl. 6), 41-49.

Marengoni, N. G., Y. Onoue, and T.Oyama. 1998. Offspring growth in a diallel cross-breeding with three strains of Nile tilapia, *Oreochrormis niloticus*. *J. World Aquaculture Society*, 29 (Suppl. 1),

Mair, G. C., A. Scottet *et. al.* 1991. Sex determination in *Oreochromis*. 1. Sex-reversal, gynogenesis and triploidy in *O. niloticus* and 2. sex reversal, hybridization, gynogenesis and triploidy in *O. aureus*. *Theoretical and Applied Genetics*. 82 (Suppl. 2), 144-160.

Marian, T. and T. Krasznai. 1987. Selection, hybridization and genetic engineering in aquaculture. FAO, EIEAC, Hamburg, PP. 253.

Melard, C. 1995. Production of high percentage of male offspring with 17 alpha-ethynylestradiol sex-reversed *O.aureas*.1. Estrogen sex-reversal and production of F2 pseudo-females. *Aquaculture*, 130 (Suppl. 1), 25-34.

Muller, B. A. and S. G. Horstgen. 1995. Sex determination in *O. niloticus*, sex-ratios in homozygous gynogenetic progeny and their offspring. *Aquaculture*, 137 (Suppl. 1-4), 56-65.

Magoya, H., H. Okamoto *et. al.* 1996. Production of androgenic diploids in salmon, *Oncorhynchus masou*. *Fisheries Science*, Tokyo, 62 (Suppl. 3), 380-383).

purdon, C. E. 1993. Genetics and Fish Breeding. Chapman and Hall, London.

Rahman, M. A. and N. MaClean. 1992. Production of transgenic tilapia (*Oreochrormis niloticus*) by one-cell stage micro-injection, *Aquaculture*, 105, 219-232.

Recoubratsky, A. V., B. I. Gomelsky et al. 1992. Triploid common carp produced by heat-shock with industrial fish farm technology.*Aquaculture*, 108 (Suppl. 1- 2), 13- 20.

Reddy, P.V.G.K. 2005. Genetic Resources of Indian Major Carps, FAO, Rome, Italy. 76p.

Rottmann, R. V., J. V. Shireman and F. A. Chapman. 1991. Induction and verification of triploidy in fish. SARC Publication No. 427. Southern Regional Aquaculture Centre, Stoneville, Mississippi.

Shelton, W. L. 1986. Broodstock development for monosex production of grass carp. *Aquaculture*, 57, 310-320.

Tave, D., M. Rezk and R. O. Snitherman. 1989. Genetics of body color in tilapia, .*Oreochromis mossambius*. *J. World Aquaculture Society*, (suppl. 4), 214-222.

Thorgaard, G- H., P. spruell *et. al.* 1995. Incidence of albinos as a monitor for induced triploidy in rainbow trout. *Aquaculture*. 137 (Suppl. 1-4), 112-130.

Webster, C. D. 2001. Aquaculture. Haworth Press, London.

Questions

1. Discuss the phenomenon of sex reversal in fish. How did the availability of XX males and XY females help in understanding and locating the genes responsible for sex determination?

2. Discuss different gene transfer methods in fishes.

3. State why the study of fish genetics is essential for fish production.

4. What is heterosis and what are the major manifestations? How has this phenomenon been utilized for fish breeding?

5. What do you understand from the term inbreeding depression? Discuss the consequences of inbreeding both in an inbred and a random outbred populations.

6. State why selection is essential in fish breeding programs? What are the pre-requisites for selection?

7. Discuss different types of selection generally adopted in fish culture strategies.

8. What is hybridization? Why it is so essential for improvement of fish stocks?

9. Distinguish between androgenesis and gynogenesis.

10. What is polyploidy? State the method of ploidy induction.

11. What are the advantages of ploidy induction?

12. How sex is determined in fish?

13. How chromosomal sex can be manipulated ?

14. How homogametic and heterogametic monosex stocks are produced?

15. Describe briefly the achievements already made in the   production of transgenic fish. Discuss their utility in fish culture.

16. State the importance of transgenesis in fish culture.

17. What is transgenic fish? Mention some important genes involved in transgenic fish production.

18. Write short notes on :

(i) Transfection, (ii) microinjection, (iii) promoter, (iv) genotype and phenotype, (v) epistatis, (vi) alleles, (vii) heritability, (viii) mass selection, (ix) sex inducers, (x) selective breeding, (xi) heterosis, (xii) additive and dominance effects, (xiii) cross-breeding, (xiv) inbreeding, (xv) inbreeding depression, (xvi) gynogenesis and androgenesis, (xvii) transgenesis, (xviii) introgression.

16

Regulation and Induction of Maturation and Spawning of Fish

Previous chapter dictates us that several approaches that are partinent to fish production require precise control over the entire spawning period from establishing specific combinations of brood-stock to correct timing of fertilization. Ascendancy over the timing of reproduction is inevitable for successful genetic improvement in fish species and various techniques which address this need are presented in a nutshell.

16.1 Necessity of Induction and Maturation of Spawning of Fish*

It seems quite plausible that on the basis of selection and propagation of production stocks the genetic improvement requires little more attention than other strategies such as good nutrition, suitable water quality and comfortable habitats for brood fishes. Also, it is virtually beyond possible under hatchery conditions to replicate the circumstances that cause to happen some species to release eggs. Unreliable spawning in captivity is a major stumbling block in the development of successful commercial production systems for many species of fish.

Artificial manipulation and regulation of both internal physiological processes (through hormonal intervention) and external stimuli is necessary for success of genetic manipulations and propagation of captive broodstocks. The degree of these techniques depends very much on the biology and life history of fish involved and the availability of expert personnel. Holding and conditioning of robust brood fish are the two important aspects to be reckoned with for restocking programs. The photothermal conditioning and hormonal intervention for obtaining quality eggs or fry in significant numbers from many farmed fish are, at present, highly encouraging.

16.2 Importance of External Stimuli

The sensory and reproductive physiology are adjusted to the frequency of the required stimuli which becomes essential for the action of maturing and volitional spawning. The spawning process involves the ability to become aware of the environmental conditions through senses that ultimately trigger a complex pathway of physiological developments.

---

* For further detail, see Barry *et. al.* (1995), Bates and Tiersch (1998), Chamberlain and Lawrence (1981), Faghenro *et. al.* (1992), Kobayashi and Stacey (1993), Roderick *et. al.* (1996), Tan-Fermin *et. al.* (1997).

Unfortunately, unsuitable stinuli or physiological stressors definitely rupture or halt this chain of events.

Some species of fish are able to produce seeds in captivity but others are virtually incapable to reproduce unless they are forced to do so. These behaviors reflect the range of environmental and seasonal conditions under which fish could normally spawn. External conditions can reflect seasonal patterns, whether events and other environmental factors that influence reproductive success.

Several factors such as photoperiod and lunar cycles, ambient temperature, water current and depth, pressure and behaviors of fish, presence of suitable spawning substrates and various changes in salinity, hardness, pH, and dissolved oxygen help to incite the reproductive process. Tremendous diversity of life history strategies among fish species generally reflects the chances for reproductive success. If farmers think seriously about the spawning-related acute environmental signals, particularly those with highly synchronous release of gametes, it becomes much easier to appreciate those species that spawn under ideal culture conditions.

16.3 Internal Processes

It should be kept in mind that the external stimuli are enough to trigger a series of physiological events resulting in maturation of gametes followed by ovulation and spawning. This series of events are intimately associated with various organs and sex steroids. Since external stimuli are interrupted at the time of spawning, hypothalamus of the brain responds well to generate substances such as gonadotropin releasing hormone (GnRH) and/or dopamine that inhibits gonadotropin secretion. The puitary gland which is situated below the brain just posterior to the optic nerve crossing or left behind on the floor of the cranium, is stimulated by GnRH and ultimately secretes gonadotrophic hormones (GtHs). The presence of these compounds in the blood stream activates the gonads to generate (i) steroids that, in turn, trigger final egg maturation or spermiation and (ii) prostaglandins that are involved in ovulation. If any link in the chain is broken as a consequence of injury, stress or other adverse effects, reproduction will not occur.

16.4 Artificial Induction

Under artificial situations, conditions may force fish seed producers to avoid external stimuli and internal mechanisms if spawning occurs successfully. Over the last several decades, studies have shown that it is feasible to strengthen links in the reproductive stimulus pathway through photothermal conditioning and hormonal intervention. Various methods have, however, already been developed through decades of scientific research for induced breeding that can be applied to the origins of the internal reproductive pathway.

16.5 Photochemical Conditioning and Maturation

Methods for photochemical manipulation to activate annual seasonal cycles have been well documented in a number of fish. Although not all the stimuli are necessary for full maturation and spawning, these manipulations are enough to stimulate the development of gametes and this refers to as *maturation*. Fish considered for seed production are kept in ponds and subject to controlled photoperiod and temperature to imitate annual cycles associated with maturation and spawning. These systems require recirculating and biological as well as mechanical filtrations. This permits continuous re-use of temperature-adjusted water. Depending on the temperature range encountered with the species normal life history, instrument such as chillers or heat-pumps is required to manipulate water temperatures. Although both temperature and photoperiod regimes may induce and maturation in some species, in others anyone regime is necessary.

16.6 Hormone-Induced Spawning

In stagnent waters where several species of broodfish (such as carps) do not spawn and hormonal intervention is generally undertaken where holding or conditioning facilities for broodstock are not available, wild broodfish are collected from natural spawning grounds. Hormonal intervention is necessary to prevent the rupture or complete shut-down of maturation processes as a response to transport and capture-induced stress.

When holding mature fish in ponds under ambient conditions or in tanks under controlled temperature and photoperiod conditions to induce spawning are not sufficient, hormones must be administered to reinforce or create internal links in the maturation chain. A single or double injection of hormone could induce ovulation and spawning. However, Hormonal intervention is very easy and most successful when it is used at later stages and conventional wisdom suggests that use of hormones should be considered when fish are on the verge of maturation.

Gonadotropins

The most common and widely-practised intervention in fish maturation involves administration of gonadotropic hormones (GTHs) generally produced by the pituitary gland. In addition to fish GTHs, purified gonadotropins such as human chorionic gonadotropin (HCG) are widely used. Both HCG and pituitary extracts are commercially available and the latter can be collected from almost any variety of fish with little effort. The collection of pituitary extracts involves sacrificing mature fish, removing the pituitary glands and extraction of GTHs. The pituitary tissue can be ground into a slurry and diluted with distilled water to produce a supernatant for injection. The solution can be used immediately or frozen for latter use. Whole pituitary glands can also be frozen or kept in acetone or absolute alcohol for later thawing, grinding and injection. However, pituitary extracts are highly variable in their performance depending on collection, storage methods and maturity of donar fish.

Steroids and Prostaglandins

In some situations, use of hormones other than gonadotropins can foster the reproductive process up to an advance stage, but the last links in the maturation chain must be reinforced to induce ovulation and spawning. Steroids and prostaglandins (See chart 16.1) have been criminated in the expression of spawning-related behaviors which serve as a sort of feedback to cater an external stimulus to other broodfish and reinforce the possibility of successful spawning.

CHART - 16.1
STEROIDS AND PROSTAGLANDINS

1. **Steroids :** The term steroid is applied to any of a class of natural or synthethic organic compounds with a molecular core, or nucleus, of 17 carbon atoms in a characteristic three-dimensional arrangement of four rings. The configuration of the nucleus, the nature of the groups attached to it, and their positions distinguish different steroids. The structures of the steroids are primarily based on the 1,2-cyclopentenophenanthrene skeleton. Hundreds have been found in plants and animals and thousands more synthesized or made by modifying natural steroids. Steroids are important in biology, medicine, and chemistry. Examples include many hormones (including sex hormones), bile acids, sterols, vitamin B, etc.

   The sex hormones belong to the steroid class of compounds, and are produced in the gonads (testes in the male and ovaries in the female). Their activity appears to be controlled by the hormones are sometimes called as *secondary sex hormones*, and the hormones of the anterior lobe of the pituitary (which are protein in nature) are called as *primary sex hormones*.

   The sex hormones are of three types such as (1) Androgens (male hormones such as Androsterone, Testosterone, etc.), (2) Oestrogens (female or follicular hormones such as Oestrone, Oestradiol, Oestriol, etc.), and (3) Gastrogenes (the corpus luteum hormone such as progesterone). They are either synthesized or to extract them from the urine of pregnant or post-menopausal women. But such methods are extremely time-consuming and expensive. The most cumbersome part of chemical synthesis of these hormones is the steroid backbone. As a corollary, if a steroid skeleton can be obtained from a relatively inexpensive and accessible source, the final synthesis of the hormone by addition of pitutiary components to the molecule can be relatively easily accomplished.

2. **Prostaglandins :** Prostaglandins (PGs) are locally known as *eicosanoids*. They are biologically active, carbon-20, unstaurated fatty acids that are produced by the metabolism of arachidonic acid through the cyclooxygenase pathway. They are extremely potent compounds that elicit a wide range of response, both physiologic and pathologic. They are local short-range hormones that are formed rapidly, act in the immediate area, and then decay or destroyed by enzymes. Prostaglandins are named as PG plus a third letter (for example, A, D, E, F), which designates the type and arrangement of functional groups in the molecule.

PGI$_2$ is known as prostacyclin. However, although PGs have many biological actions and clinical applications, effects on ovary and uterus are significant nonetheless. PGs are involved in Leutinizing Hormone-induced ovulation. Better fertilization rate is achieved with timely application of PGs.

Both steroids and prostaglandins whether they are natural or synthetic, are used in aquaculture production. But their extensive use has been limited due to their unavailability in the market and the cots of production may negate the benefit derived.

In addition to these, sufficient research data on this aspect are not available to farmers. As a corollary, this situation has been forced to use pituitary extracts in breeding programmes. Application of these hormones to induce breeding programmes should be recommended only after performing and effective risk assessment studies.

16.7 Some Important Synthetic Hormones Used in Spawning of Fish

Due to unavailability of natural hormones and some, other related problems, ,scientists have been compelled to develop certain chemical substances that would definitely help to induce spawning of fish within very short time interval. Decades of scientific research coupled with experiments on induced breeding of fish have helped to explore a large variety of synthetic hormones, mammalian hormones, and some other hormone-related substances which are solely responsible to ascertain induced breeding strategies of farmed fish.

Human Chorionic Gonadotropin (HCG)

This high molecular weight (30,000) glycoprotein hormone is secreted from the placenta of pregnant women and is excreted through urine. This hormone is an analogue to leutenizing hormone. Different commercial enterprises filter the HCG-containing urine and sale to the market in different trade names. The commercial name of HCG, for example, manufactured by Infar India Limited, Kolkata, is termed as' *Sumach*'.

Though this hormone is highly effective in case of exotic carps, continuous use of HCG in native carps is not recommended because this hormone may develop immunity within the body of fish resulting in depression of fertilization. For this reason, HCG is generally used after administration of pituitary extract as a primary dose.

Pregnant Mare Serum Gonadotropin (PMSG)

It is a serum derived from the blood of pregnant mares. The serum is rich in hormones especially prostaglandins. The hormone secreted by the placenta of pregnant female horse. It acts as a Follicle stimulating Hormone but due to the absence of characteristic features of Leutinizing Hormone, its wide use in induced spawning has been limited.

Leutinizing Hormone-Releasing Hormone (LHRH) and its Analogues (LHRHa) and Gonadotropin Releasing-Hormones (GnRH) and its Analogues (GnRHa)

1. The use of LHRH in combination with pituitary extract has been very successful in case of introduced carps.

2. The activity of gonadotropin releasing inhibitory factors can be inhibited by the use of LHRHa in combination with domperidon and pimozide at the rates of 5 and 10 μg respectively. This mixture can be conveniently used for successful sprawning of different types of carps.

Salmon/Murrel Gonadotropin Releasing Hormones (SgnRH/MGnRH)

These two hormones are more effective than other synthetic hormones. They are dissolved in propylene glycol and then mixed with domperidone. This mixture has almost 100 per cent efficient in the field of induced spawning programs which is available in the market as *Ovaprim*. This substance is highly effective than LHRHa. Ovaprim has, however, several advantages that follows.

1. It is available ready-to-use form.

2. Dependable results are obtained.

3. It can be stored for prolonged period of time.

4. Both male and female fish are injected simultaneously.

5. A single dose for female fish is enough.

6. Highly effective at any fish farming environment.

7. Can be successfully used even at the end of breeding season.

8. It is a germless substance.

9. The rate of egg production, their fertilization, and hatching is very high (the rate varies between 95 and 100 per cent).

Steroidal Hormones

Among different steroids, desoxycorticosterone acetate (DOCA), 11-desoxycorticosterol, 11-desoxycorticosterone, 21-desoxycortisol, cortisone, and hydrocortisone are very effective in case of catfish breeding but not carps. Of course, their extensive use must be restricted because these substances can easily enter into the human body through food fish and may cause hazardous effects.

Merits of Induced Spawning of Fish: Induced spawning of fish has several merits that follows.

(i) Pure seeds of desired species are obtained.

(ii) Fast-growing, tasty and disease-resistant high-yielding robust fish can be produced through hybridization.

(iii) Uncertainty of egg collection from rivers, streams, and the transport cost can be prevented.

(iv) Huge quantities of eggs can be obtained from every female.

(v) Superior quality of fry can be obtained from high-yielding varieties of fish that has tremendous market potential.

(vi) Preparation of ampules containing homoplastic extract* is a bold step towards induced breeding. As a result,'pituitary bank' has been set up with a view to produce quality fish seed from different hatchery systems.

16.8 Hormone Preparation and Injection Practices

As a generalization, the preparation, storing and use of hormones involve sanitation and sterilization procedures. Before they can be used, all materials used for mixing, storing and injection should be boiled for at least 15 minutes. This is due to the fact that bacterial contamination of hormone solution can results in partial or complete loss of potency. If hormone solutions are necessary to store for any period of time under frozen or refrigeration, the addition of anti-bacterial compounds may be warranted. Dosage calculation can at times be arduous since information received so far includes recommendation on the basis of weight, volume and international units (IUs). Recommendations may be highly variable within a single species. Consquently, it is necessary to perform range-finding trials on a small number of fish before deciding a dosage approximate for the particular fish and stocks encountered.

It should be kept in mind that all recommended dosages are measured in terms of key body weight of the fish to be injected. In general, it may be practical to obtain a correct weight for smaller fish (such as tilapia, catfish, etc) but when dealing with large broodfish, it is preferable to develop a weight estimate rather than exposing fish to the potential damage on handling stress involved in direct weighing. For this purpose, a rough length-weight relationship based primarily on dead fish can be used to provide a weight estimate.

Next to dosage calculation for a specific fish, the volume of the injection becomes an important consideration. Volumes should not exceed 1 CC injection. For smaller fish, however, the volume of the injection should be somewhere in the range of 0.1 to 1.0 CC. In general, the dose per kilogram body weight is multiplied by the body weight and the total dose is then divided by the volume of solution desired.

Pituitary Extracts

Once the pituitary extracts are collected, it must be chilled to prevent the breakdown of the hormonal components. If pituitary extracts is to be stored, bacteriostatic physiological solution may be used. Recommended dosages generally range from 2 to 8 mg/kg of body weight. If an injection dose of 5 mg/kg body weight of fish is prepared for a group of 3

---

* Pituitary extract collected from the same species of fish.

kg broodfish and a desired injection volume of 1 CC is determined, and once a known weight of pituitary material is crushed to small particles, distilled water must be added until a concentration of 15 mg/1 is attained.

HCG Solutions

The hormone HCG is generally available in quantities of 5,000 or 10,000 IUs. Before use, the freeze-dried powder is mixed with bacteriostatic water. Intramuscular administration is also the normal injection method for this hormone.

Other Injection

Ovaprim (See 11.28 of Volume 1) and dopamine blockers are generally injected intramuscularly in conjunction with an initial injection of releasing hormone. Procedures for handling and preparation of various other hormones, steroids, dopamine blockers, prostaglandins and the like should be determined from authorized suppliers.

16 9 Injection Protocols

Although intra-peritoneal injections have been recommended for certain species of fish, intramuscular injections are generally preferred for several reasons. An intramuscular injection when applied to the back of the fish, generally results in less chance of damaging the body and consequently gradual and sustained uptake of the injected hormone takes place. When an intra-peritoneal injection, on the other hand, is used at the base of the pelvic fins, injected materials are transferred into the body cavity, but risk for internal damages may increase.

In general, a series of two or more injections will definitely generate more consistent and reliable results for females. Males are generally injectad only once, and at the same time the last dose is administered to the females. Recommended intervals between injections will vary with species and material, but they fall in the range of 12-24 hours for cool- and warm-water species, 6-12 hours for tropical fish, and 48-72 hours for cold-water species. Some examples of fish, dose of injection, and combination of materials reported from different experimental studies are given in Table 16.1.

16.10 Determination of Maturational Status

A visual examination of sexual characters is perhaps the simplest method of determining the status of maturation in many species of fish. Sexual characters involve coloration, conformation, and behavior. Reproductive maturity is often easily distinguished in male fish with a superficial external examination. Some examples of reproductive maturity include the formation of pearl organs of the opercula and pectrol fins of certain cyprinids. Besides these features, the average diameter of eggs may be most practical characteristic and the relative positions of their nuceli can serve as indicators of development. Nuclei of mature eggs migrate to the end of the egg where the sperm will penetrate at the time of fertilization. The appreance of these characteristics can be considered as mature and the fish is considered as a good candidate for hormone injection.

Table 16.1 : Fish/Compound/Dose Combinations Reported to be Effective
in Injection-Induced Maturation and Spawning

Fish	Compound	Dose (kg body weight for each injection)			Number of injections	Injection interval (h)
		Males	Both sexes	Females		
Aristichthys	HCG and	-		100 & 850 IU	2	12
nobilis	Carp pituitary	0.1 mg	-	1.0 mg	1	After 24
Catla	Carp pituitary	-	1.3. & 2.7 mg	-	2	6
catla	and HCG	-	167 & 333 IU	-	2	6
Cirrhinus	Carp pituitary	-	0.7 & 1.3 mg	-	2	6
mrigala	and HCG	-	67 & 133 IU	-	2	6
Clarias spp.	-	-	4,000 IU	-	1	-
Ctenopharyn-	HCG and	-	-	100 & 850 IU	2	12
godon idella	Carp pituitary	0.1 mg	-	1.0 mg	1	After 24
Heteropneustes fossilis	sGnRHa	-	25g	-	1	-
Hypophthalmi-	Carp pituitary	0.1 mg	-	1.0 mg	1	After 24
chthys nobilis	and HCG	-	100 & 850 IU	-	2	12
Labeo rohita	Carp pituitary	-	1.3 & 2.7 mg	-	2	6
	and HCG	-	167 & 333 IU	-	2	6
Perca flavescens	HCG	-	-	150-660 IU	1	-

16.11 Holding and Handling of Broodstock

The most important drawback associated with the induction of spawning through hormone injections is the requisite confinement, handling and associated physiological stress faced by broodstock. Successful induction of spawning requires careful handling, good nutrition, and stress minimization.

The release of stress-meditated compounds such as cortisol into the blood stream during captive and handling has been associated with the disruption of the internal pathways that lead to final maturation, ovulation and spawning. When species such as cichlids, catfish, anabantids, etc. exhibit spawning-related hostile attitude are restricted in a very small area without opportunities for females to retreat or hide, reproduction may be reduced or prevented altogether.

Handling stress under deficiencies in water quality variables such as extremes pH, nitrites and ammonia may result in mass mortality of some delicate species. Brood fish may be injured during frequent capture associated with the captive spawning. Active, robust and large fish may be injured during collection from spawning grounds or holding tanks or ponds.

16.12 Collection of Gametes

Volitional spawning in holding tanks or ponds is the most practical approach to obtain offspring. When exact control is required, eggs and sperm must be taken from broodstocks.

While certain conditions may require surgical elimination of eggs (such as captive paddlefish and sturgeon) or milt (as in the case of certain catfish), the most common approach to collect eggs and milt involves normal stripping of broodfish.

Although some fish such as Nile tilapia ovulate in tanks even in the absence of males, hand-stripping of eggs and milt is associated with hormonal induction of spawning. Immediately after collecting broodfish from spawning grounds, they are segregated sex-wise and held separately in tanks for monitoring and subsequent stripping. This is very important for the timing of ovulation. The time of fertilization of eggs varies considerably from 10-20 minutes (as in the case of warm-water species) to several days (as in salmonoids).

It should be kept in mind that for successful ovulation and spawning, female broodstocks must be checked with utmost care several hours before the first ovulation would be expected. The adequate interval between examinations generally depends on the species, environmental temperature, and dosages of injection. For examples, for tropical, warm-water, temperate, and cold-water/cool-water species, the interval between examinations by applying thumb pressure is every 45 minutes, every hour to hour, a half, and every several hours, respectively.

The most common method of detecting ovulation involves physically preventing the fish from an inverted position and applying gentle pressure to the abdomen behind the pectoral fins. The ejection of a few eggs through the vent indicates that ovulation has begun. If the eggs flow freely from the vent, ovulation is at or near completion and the eggs must be collected immediately and are allowed to fertilize with sperms.

Eggs are collected without being contaminated by water or slime from the surface of the donar fish. Water and / or slime can be removed by the use of a damp towel and then the fish can be held for stripping, with the head slightly elevated and the vent down and slightly to one side or they may be placed on the table on a damp towel. When the fish is positioned, the abdomen is pressed gently to release eggs from the vent into a dry pan.

Milt can be collected by stripping males in much the same way as eggs are collected from females. Enough milt should be added to cover the eggs. The two contents (eggs and milt) are thoroughly mixed with a clear feather or spatula. Once the contents have been mixed, adequate water is then added not only to activate the sperm but also to ensure the eggs are completely covered.

Milt Storing

Milt can be collected for storage and later use. If milt is to be stored, contamination from water, urine, blood, slime or other sources must be avoided, and milt from individual fish should be stored separately to avoid the possiblity of cross-contanination. Milt can be stored under refrigeration in sterile plastic bags, preferably with oxygen and antibiotics.

For proper storage, the use of extenders is highly encouraging. The use of extender for tilapia milt, for example, consisting of 6.5 g/l NaCl, 3.0 g/l KC1, 0.30 g/l NaHCO$_3$,

0.30 g/l CaCl$_2$.6H$_2$O has given better results. Such storage permits high levels of fertility with more than 90 per cent survival.

Adhesive Eggs

Elimination of adhesiveness in fish eggs is the second important criteria for successful artificial fertilization. Silt-clay, bentonite, mixture of urea and common salt, dilute tannic acid solution, sodium sulfate, and Fuller's earth have all been used for the purpose. In case of carp eggs (*Cyprinus carpio* for example), the addition of urea-common salt solution (15 g urea + 20 g NaCl + 5l of water) is highly encouraging. In catfish that produces gelatinous mass of eggs, sodium sulfate solution (15 g/l) has been widely used to separate the eggs for incubation. However, saturated suspensions of these materials should be used during the activation and water hardening processes following mixing of milt and eggs.

16.13 Conclusion

Through use of suitable hormone extracts, successful ovulation and spawning are sufficient to permit the production of robust fry in large numbers. At the time of final oocyte maturation, ovulation, and spawning, careful observation (under conditions of high visibility) and frequent monitoring (with capture and handling) are inevitable to detect ovulation in time to avoid over-ripening of eggs.

In the practice of spawning induction, limited number of broodstocks can generate phenomenal number of eggs and fry. In India, however, induced spawning of the Indian major carps and catfish has been practised successfully for roughly 30 years with most hatcheries. Though techniques are constantly being refined, the ground-work for modern approaches to conditional and induced spawning of captive fish was being laid several decades ago and current investigations on the area involved have provided new insight into the operation of carp, catfish and tilapia spawning. This facilities to maximize the efficiency of seed production.

References

Barry, T. P. , J. A. Malison et al. 1995. Effects of selected hormones and male cohorts on final oocyte maturation, ovulation, and steroid profiles in walleye. *Aquaculture*, 138 (Suppl. 1-4), 331-347.

Bates, M. C. and T. R. Triersch. 1998. Preliminary studies of artificial spawning of channel catfish as male-female pairs or all-female groups in recirculatory systems. :*J* World *Aquaculture Soc* 29 (Suppl. 3), 325-334.

Chamberlain, C. W. and A. L. Lawrence. 1981. Maturation, reproduction, and growth of *Penaeus vannanei* and *P. stylirostris* fed with natural diets. *J.World Maricul. Soc* (Suppl. 1), 209-224.

Faghenro, A.O., A. A. Salami, and D. H. J. Sydenham 1992. Induced ovulation and spawning in the catfish *Clarias isheriensis*, using pituitary extracts from non-piscine sources.*J Appl. Aquaculture* (Suppl. 4),15-20.

Kobayashi, M. and N. Stacey. 1993. Prostaglandin-induced female spawning behavior in goldfish *Carassius auratus* appears independent of ovarian influence. *Hormones and Behavior, 27,* 38-55.

Roderick, E. E., L. P. Santiago et. al. 1996. Induced spawning in *Oreochromis niloticus. In* The Third International Symposium on Tilapia in Aquaculture. (Eds. R. S. V. Pullin, J. Lazard et. al.). p. 548, ICLARM, Makati City, Philippines.

Tan-Fermin, J. D., R. R. Pagader et al. 1997. LHRHa and pimozide-induced spawning of Asian catfish *Clarias macrocephalus* at different times during an annual reproductive cycle. *Aquaculture,* 148 (Suppl. 4), 323-331.

Questions

1. Discuss how maturation of gametes, ovulation and spawning of farmed fish takes place through external and internal stimuli.

2. Explain various techniques that are essential to control the timing of reproduction for successful genetic improvement of farmed fish.

3. Discuss the methods how gametes are obtained from farmed fish to produce phenomenal number of eggs and fry.

4. Write short notes on the following:

   (a) Prostaglandin, (b) Gonadotropins, (c) Volitional spawning.

17

Diseases and Parasites of Freshwater Fish

The term disease is used to incorporate any deviation from normal structure or function of the fish encompassing those states that result from activities of infectious agents, invesion of parasites, and environmentally- or genetically-induced abnormalities.

Fish diseases and their control are the most integral part of fish culture management strategy. These are the tools for ensuring that the programs for disease control and other management protocols of the farm always get directed towards the production activities of the farm. The study of fish disease and their control provides the means of the potentiality of fish producing regions whether the desired results are being achieved or not. And knowledge of the desired results involves adequate management of fish culture ecosystems. Symptoms and control of fish diseases are closely inter-related. In fact, disease identification, symptoms and control measures are the three inter-related phases of fish culture.

Fish diseases may be due to parasitic or non-parasitic causes. The former being the most numerous and from both pathological and economic newpoints, are the most important. In this chapter, we shall discuss about various aspects of fish diseases such as (1) the reason for occurrence of fish diseases, (2) environment and fish diseases, (3) techniques of investigation on fish diseases, (4) diagnosis of fish diseases, and (5) symptoms and control of diseases for effective performance of the hygienic conditions of ecosystems.

17.1 General Considerations

The study of freshwater fish diseases is until recently not scattered and adequate. Although knowledge of freshwater fish diseases has proliferated, particularly in pond culture systems, little or no incentive has existed for comparable investment in understanding the ills of many inland species. In contrast to ocean ecosystems, the smaller dimensions of the inland water areas, the simplicity of natural factors controlling the size of fish populations, and the availability of methods to control such factors, have tended to encourage extensive and systematic studies on such specific aspects of the freshwater environment as disease. Growing awareness that the human population is increasing enormously, that substantial quantities of freshwaters are used for agriculture, aquaculture and domestic purposes, and that robust as well as disease-free fish production is inevitable for domestic consumption and export has led to increasing scrutiny of those variables that might become set aside in their influence on population size. Disease is considered as an important variable.

At present, diseases of freshwater fish have logically received greatest attention. Diseases and parasites in addition to destroying the host, can materially reduce the value of fish as food for humans. This consideration serves as a further incentive to examine diseases of commercial species. Non-commercial species may also be considered due to academic interest; of course, more importance should be given concerning to food fishes.

17.2 Reasons for Occurrence of Fish Disease

For high productton of fish in freshwater impoundments, excessive feeding and over-crowiding is followed by farmers without any scientific management options. Consequently, accumulation of toxic metabolites such as ammonia, nitrite, etc. occurs beyond the threshold limits. These substances stinulate to develop blue-green algae in water which causes production of off-flavor in fish culture systems. Development of algal blooms and macrophytes in fish culture ecosystems deplete the content of its oxygen. Moreover, stressful aquatic environment, pollution of fish culture ecosystems by pesticides and heavy metals, drainage of untreated wastewater in fish culture ecosystems and accumulation of pathogenic micro-organisms and helminth eggs are responsible for outbreak of diseases in fish stocks.

In general, fish diseases occur through interactions of the pathogens, the fish and the environmental conditions. The environmental conditions include poor water quality, overstocking, etc. The disease-producing agents may be non-pathogenic and pathogenic. The former is characterized by extreme temperature, chemical poisoning, and acidic conditions of water and soil which cause non-infectious diseases that results in sudden mass mortality. The pathogenic organisms include bacteria, protozoa, helminths, and micro-crustaceans. These organisms cause infectious diseases and results in fish mortalities. Micro-organisms are always present in fish culture ecosystems and some of them cause disease when immune system in fishes is lost or the fish has been weakened to a significant extent.

17.3 Environment and Fish Disease

The impression of high production potential through state-of-the-art technologies established firmly on fish biomass of exchanging inputs is chemical and biological in nature. These components produce waste substances that produce regressive impact on the ecosystem under different farming strategies. These factors trigger some other factors that influence parasites to infect fish stocks. A number of fish diseases recognized until now include ulcerative dropsy, columnaris disease, trichodiniasis, white gill spot, white scale spot, black spot, lernaeosis, dactylogyrosis, gyrodactylosis, lingulosis, ergasilosis, argulosis, branchiomycosis, saprolegniasis, and viral, bacterial as well as fungal diseases. Environmental diseases generally encountered are gas bubble disease, unionized ammonia stress, and algal toxicosis. The most dangerous freshwater fish disease in Asian countries has been epizootic ulcerative syndrome (EUS) affecting wild and cultured fishes.

Water and Soil Quality, Versus Fish Disease

Good quality of aquatic environment is very important for fish health. Any deterioration in the water and soil qualities create stress to fish and favor multiplication of pathogens.

In fish culture, the dictum 'prevention is better than cure' should be kept in mind. However, water quality management practices essentially involve aeration and water exchange that helps prevent fish stocks from disease outbreaks. In cases where such practices are not feasible, application of some chemicals is urgently needed. The most commonly used and recommended chemicals include formalin (@15-25 mg/l), malachite green (@ 0.3-10 mg/l), potassium permanganate (@2-4 mg/l), copper sulfate (@ 0.3-10, mg/l), chelated copper (@ 0.2 mg/l), aquazine (@ 0.25 mg/l), filter alum (@10-20 mg/l), gypsum (50-1, 000 kg/ha) and hydrated or slaked lime.

Good water chemistry is the guiding principle of fish culture and hence better fish production. Water is considered as a binder in which inorganic and organic substances as well as dissolved gases are present. The presence of micro-organisms stimulate chemical exchange between micro-organisms and in some situations, may have adverse effects on fish.

Five parameters such as temperature, hardness, dissolved oxygen, pH, and ammonia are closely inter-related. So long as these parameters exist in balanced conditions, a desirable state of equilibrium is achieved. This means that fish will remain healthy and disease-free in culture systems and grow well.

Pond Bottom

The condition of pond bottom is very important so far the outbreak of diseases is considered. To make the pond bottom healthy and clean, it should either be completely dried to get rid the pond of toxic gases or be applied waste digesters. The waste digesters are harmless micro-organisms and enzymes that consume organic matter from the pond bottom. Following the application of digesters, disinfectants such as organic iodine and organic silver are used at the rate of 18 litres/ha when the depth of water becomes 0.5 m. Seven to 10 days after the application, the organic silver disintegrates and therefore flushing the pond is not necessary. At the same time, this chemical also prevents the development of algal blooms.

Dissolved Organics

Dissolved organic matter is the main stress factor in culture pond water. Increase in dissolved organics results in increase the level of ammonia and micro-organisms. To prevent the accumulation of dissolved organics, frequent total or partial water exchange is necessary or the pollution can be reduced by eliminating the organics by adsorption using activated carbon. Dissolved organics are developed through various ways such as by decaying excess feed, dead larvae and algae. Stress includes discontinuous feeding of fish, weakness and susceptible to diseases.

17.4 Techniques of Investigation of Fish Disease

Techniques of investigation of fish diseases involve proper diagnosis of the infected fish. Diagnosis requires a method of investigation which in the great majority of cases by making a clinical history of the disease and following external exploration. Many progressive fish culturists may have an idea of the condition of fish health by applying simple techniques for investigation and by the use of an ordinary student's microscope to facilitate minute examination. To serve the purpose, however, the following instruments are necessary: (1) One or two pairs of stainless steel scissors, (2) one pair of forceps, (3) two curved forceps, (4) pipettes (5) one spatula, (6) two or more petridishes, (7) slides and coverslips, (8) a wooden fish dissecting board, (9) a ordinary student's microscope, (10) a simple microscope, and (11) a high power hand lens (Fig. 17.1).

Methods of Examination of a Diseased Fish

When any fish is suspected to be infected with disease, it should be carefully examined with external observation. Since monthly netting is a routine work in composite fish culture programs, there is every scope to examine almost all the fishes whether exhibiting any abnormal behavior and for this purpose special attention must be given to the following points:

1. **General Symptoms:** The general symptoms of a diseased fish are characterized by the abnormal swimming, swimming around in circle, frequent surfacing behavior, scrapping against the bottom of the pond when fishes swim, rapid and frequent respiratory movements, and frequent jumping out of the water. Moreover, the opercular movement of a diseased fish is not frequent and the fish consumes less quantity of food than the healthy one and in acute cases, the feeding rate drastically reduced. Healthy fish always remain in the bottom of water. As soon as these symptoms are noted, proper care must be taken so that the entire fish stock is kept in good condition.

2. **External Examination of an Infected Fish:** If fish exhibits any abnormal behavior in culture ponds, special attention should be given to the following points:

   (i) Whether the color of the body of fish has changed (such as darkening, blackening, decoloration, etc.).

   (ii) Cloudiness of the skin.

   (iii) Elevated skin and damage of fins.

   (iv) White spot or parasites visible to the naked eye.

   (v) Reddening.

   (vi) Magnifying glass should be used to examine the external parasites.

   (vii) The texture of the muscle should be determined whether there is any ulcer or inflamation over the skin.

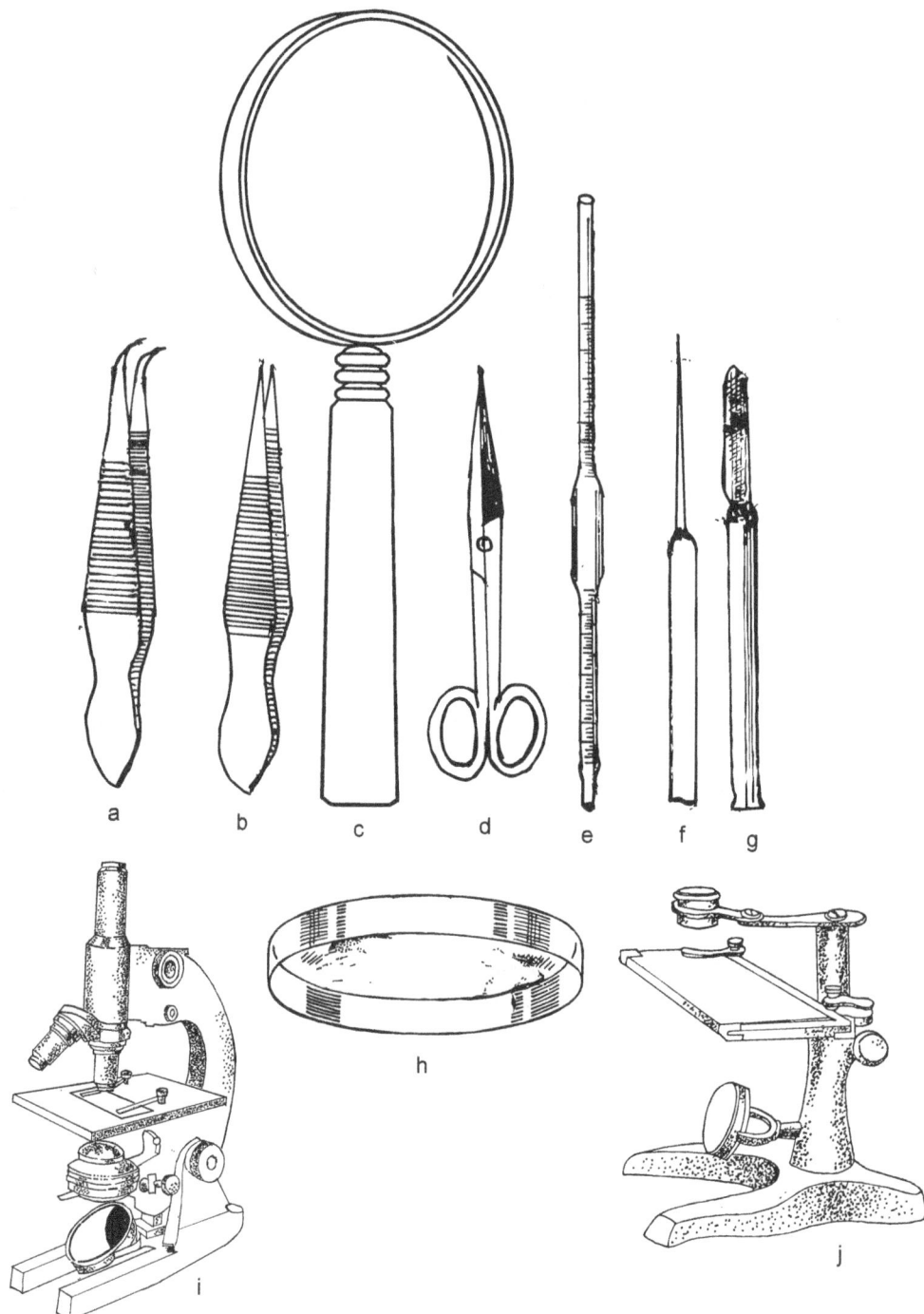

Fig. 17.1 : Instruments used for investigation of fish diseases. (a) Curved forcep, (b) Straight forcep, (c) Hand lens, (d) Scissor, (e) Pipette, (f) Needle, (g) Scalpel, (h) Petri-dish, (i) Ordinary Student's Microscope, (j) A simple microscope.

3. **Internal Examination of an Infected Fish:** When the study of external examination is over, the fish is sacrificed. In this process, fishes are cut-off their heads or necks and as a result the brain is cut at a point posterior to the eyes.

(i) *Preparation of Smears of Skins and Gills*: Fish gills are scraped and the material is kept onto a slide and then covered with a coverslip. Sometimes, dilute Indian ink is added to the slide to make the parasites more distinctly visible.

For smear preparation, the fish body is also scraped with the aid of a spatula along the skin from head to tail. This helps obtain any parasites from the skin, if present. In some skin diseases (such as Ichthyophthirius infection) and external ulceration, skin smear is enough to establish diagnosis. But in case of internal diseases, an autopsy is necessary.

(ii) *Autopsy*: In some diseases such as tuberculosis, ichthyoporidiosis, bacterial haemorrhagic septicemia, etc. smear and culture are taken from the internal organs of fish. Analysis of 1 to 2 per cent of the total fish stock in a culture pond is enough to ascertain the overall state of fish health.

For performing an autopsy, four incisions such as ventral, 1ateral, opercular, and cranial are made. To expose the opercular chamber and the gills, the operculum is lifted with the aid of a forcep in one hand while it is removed with a scissor in other hand. After four successive incisions, however, the internal organs such as brain, gills, and general viscera are exposed for further examination.

(iii) *Observation With the Naked Eye*: When the internal organs are completely exposed, they are examined either with a magnifying glass or with the naked eye and the following points are noted.

* Pale color of gills and kidney.

* Yellowish coloration of the liver.

* Congestion and redening of the liver.

* Transparency of the intestine.

* Hardening of the bile and gall bladder.

* Fluid in the peritoneal cavity.

* White nodules on the gills, liver, and kidney.

* Smell in the abdominal cavity and condition of the eye.

All these observations however, provide several important informations for further examination. Deeply-seated eyes in the socket or bulging eyes from the socket, for example, indicate a pathological symptom. Similarly, yellow-colored livers clearly denote a fatty degeneration of the organ and the initiation of a destruction of the hepatic cells.

(iv) *Observation With the Microscope*: For microscopical examinations, a very small piece of the organ to be observed is kept on a slide and then a drop of clean water is placed over the piece. The piece of organ is then macerated by means of two needles. A coverslip is placed on the macerated portion of the slide with light pressure. For examination of gill filaments, some filaments from the gill are removed, kept on the slide and then a coverslip is placed. By this method, only temporary mounted slides are prepared.

For clear diagnosis of the organ, the preparation of permanently-mounted slide is made through specialized methods called *histological techniques.* This is a more exhaustive study, difficult to follow and takes a lot of time and hence it is generally not recommended for fish farmers. This technique may, however, receive attention because of academic and research interests, but the preponderance of investigation over extensive areas concerning to high-value and attractive food fishes with large production biomass is *sine-qua-non.*

17.5 Histological Techniques of Investigation of Fish Disease

To perform a more exhaustive investigation for clear diagnosis, pieces of diseased tissue should be fixed with several fixatives such as alcohol, formalin, and Bouin's fluid. These fixatives can result in cessation of all the vital activities of the tissue and following specialized techniques (such as histological, histochemical, bacteriological, virological, haematological, and serological), the tissue is prepared for microscopical examination.

17.6 Diagnosis of Fish Disease

Disease problems have an important consideration in fish culture management protocols. Most of the disease outbreaks are caused as a result of environmental problems stemming from intensive farming systems and infection of fishes by exotic pathogens. Conventionally, successful fish culture operation focuses on specific diseases of attributable economic impact, and the application of standard disgnostic procedures along with preventive and control measures.

Diagnosis of fish diseases is the recognition of a disease by its symptoms (Table 17.1) that manifest due to infection caused by fungi, bacteria, viruses, protozoa, and helminths. If correct diagnosis is not made, medicament will not be possible. Fish biologists and aquaculturists should be fully acquainted with details as to the condition of cultured water and soil, nature and type of host, source of the fish, and development of the infection. But unfortunately, most of the farmers are not efficient to understand the symptoms of fish disease and the correct diagnosis. Commercial diagnostic kits and vaccines are now available that have encouraged the control of fish diseases. Most of the fish diseases are still less clearly understood. Many diseases are considered as syndromes that involves complex pathogen-environment-fish interactions. Moreover, in different fish culture systems where many species of fish are equally involved and where there is limited water exchange facilities, fish stocks could potentially be affected by an equally wide variety of pathogens. Even with the emergence of an increasing number of pathogens, especially the viruses, it is very difficult to justify the development of diagnostic kits, vaccines, and chemicals to provide for large number of species involved in culture operations.

Table 17.1 Some Common Symptoms and Diagnosis of Fish Diseases

	Symptoms	Diagnosis
I.	**Body Form**	
1.	Deformities of the body	Whirling disease, tuberculosis, .rickets, bacterial haemorrhagic septicemia
2.	Softness and flexibility	Pox disease of the vertebral column
3.	Fraying of the fins	Bacterial fin rot, costiasis, alkalosis
4.	Red coloration of the body due to haemorrhage	Acidosis
5.	Fin damage	Trauma, ectoparasites, bacterial fin rot
6.	Raised scales	Bacterial haemorrhagic septicemia (BHS)
7.	Emaciation	Tuberculosis, *Octomitus*, Ascarids
8.	Swollen belly and sunken eyes	BHS
9.	Reddened anus	Enteritis
II	**Skin**	
1.	Bluish-white turbidity of the epidermis	Alkalosis; Infection by *Costia, Trichodina* and *Gyrodactylus*
2.	Cotton wool-like formation	Mycosis
3.	Rounded and reddened areas on the skin	Bites of leeches and carp lice
4.	Milky turbidity and gelatinous infiltrations	Carp pox
5.	White spots (the size of a sand grain and up to 1 mm in diameter)	*Ichthyophthirius*
6.	Red, pink or black growth	Tumours
7.	Cauliflower-like papillomas	Cauliflower disease
8.	Spots and ulcers on the skin, scale raised	BHS, furunculosis, ulcer and kidney diseases
9.	Air-bubbles underneath the skin	Gas bubble disease
10.	Black coloration	Intestinal disturbances, whirling disease
III	**Gills**	
1.	Bluish-white turbidity	Infection by *Costia, Dactylogyrus* and *Trichodina*
2.	Necrosis and discoloration	Acidosis, alkalosis and columnaris
3.	Gasping at the water surface	Lack of oxygen
4.	Spotted discoloration	*Ichthyophthirius*
5.	Anaemia	Result of several diseases
6.	White pearl-like nodules	Sporozoans
7.	White spots on the gill filaments	Parasitic copepod (generally *Ergasilus*)
8.	Small worms of up to 1 mm long in smears	*Dactylogyrus*
9.	Necrosis and destruction of gill filaments	Gill rot disease
IV.	**Muscles**	
1.	Ulceration	*Ichthyosporidium*,, furunculosis

...contd

Table 17.1 – *contd...*

	Symptoms	Diagnosis
2.	Swellings with microscopical spores	Bubonic plague of fishes
3.	Coiled worms detected microscopically	Nematodes
V.	**Eyes**	
1.	Opacity of the lens and cornea	Infection by *Diplostomum* and tuberculosis
2.	Exophthalmus	Kidney disease, tuberculosis, *Ichthyosporidium*
VI.	**Internal Organs**	
1.	Variable nodules in the liver	Hepatoma
2.	White nodules in the liver, ovary, spleen, heart and brain	*Ichthyosporidium*
3.	Green-colored liver	Biliary stasis
4.	Yellowish or brownish liver and kidney swollen	BHS
5.	Inflamation of gas bladder	BHS
6.	Nodules in the intestine	Nodular coccidiosis
7.	Liquid intestinal content	BHS, *Octomitus*
VII.	**Blood**	
1.	Anaemia	Gill and blood parasites, BHS, and Intestinal coccidiosis
VIII.	**Nervous System**	
1.	Nodules in the brain and spinal cord	*Ichthyosporidium*
2.	Parasites in the blood of the brain	*Myxosoma* and *Myxobolus*
3.	Spasma and convulsions	Intoxication, *Octomitus*, whirling disease
IX.	**Eggs**	
1.	Threads in the egg surface	Mycosis produced by *Saprolegnia*
2.	Turbidity of the egg	Unfertilized eggs, egg damaged by environmental factors

17.7 Fish Disease

Fish are susceptible to hundreds of non-parasitic and parasitic diseases particularly when grown under controlled conditions. Adverse hydrological conditions often lead to parasitic attacks, as the resistance of fish decreased. Mechanical injuries sustained by a fish when handled as a slipshod manner during harvest and transport, may expedite parasitic infections. In addition to this, carelessness in stocking and feeding permits serious parasitic infections and mortality.

Seven principal types of fish disease are generally recognized in fisheries and aquaculture. These include fungal, protozoan, bacterial, viral, helminth, environmental diseases and crustacean parasites. Fish diseases, though most common in tropical and sub-tropical regions, occur to a large extent in other regions as well. The destructive effects of fish due to disease outbreaks are often very serious. Disease outbreaks not only rob much of the immune system of fish and survival efficiency but also drastically reduce the ecosystem productivity. In order to have a general idea on fish diseases, however, identification of disease-producing organisms, symptoms, prevention and control that follows are most important.

17.8 Fungal Disease

Fungal diseases of fish are termed as *fish mycoses* and this field of study is known as *fish mycopathology*. Within the true fungi, members of the phycomycetes (algoid fungi) are those which cause the more important mycotic diseases of fish. Some common fish fungi are shown in Figure 17.2. Fungi are known to attack eggs, fry, fingerlings and adults of fishes and as a general rule, fungal infection starts when the host gets injured either accidentally or as a result of infections other than fungus. Fungal infections in fishes may result in almost 100 per cent mortality.

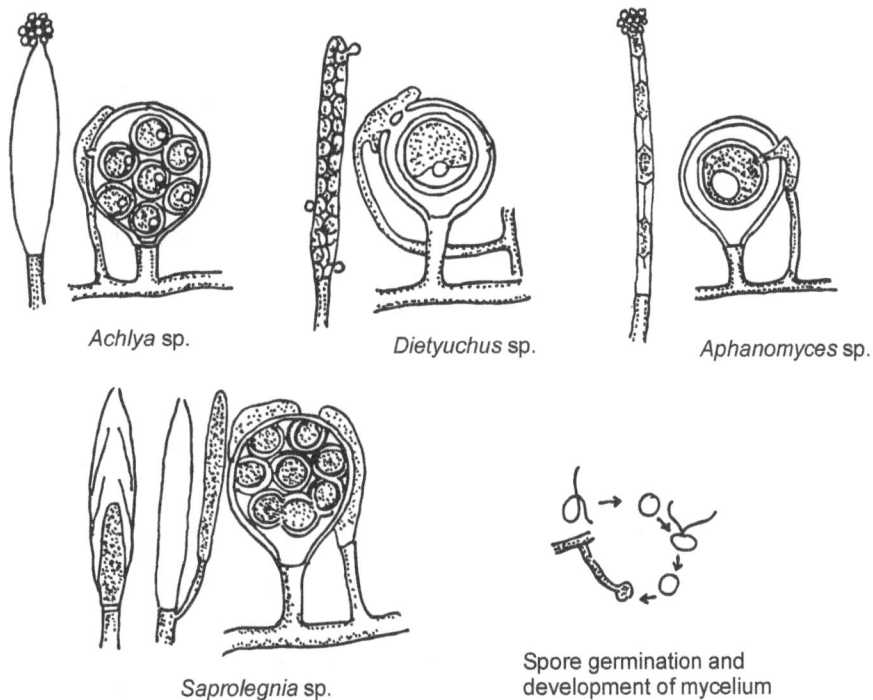

Achlya sp.   *Dietyuchus* sp.   *Aphanomyces* sp.

Saprolegnia sp.   Spore germination and development of mycelium

Fig. 17.2 : Some common fish fungi. Sporangia showing spore discharge and oogonia with oospores.

Types of Fungal Diseases

Fish mycoses may be divided into three main categories: (1) dermatomycoses or external mycoses, (2) systemic or internal mycoses, and (3) branchiomycoses or gill rot. Dermatomycoses include two types of dermal infection such as saprolegniasis and aphanomycosis.(Figure17.2a).

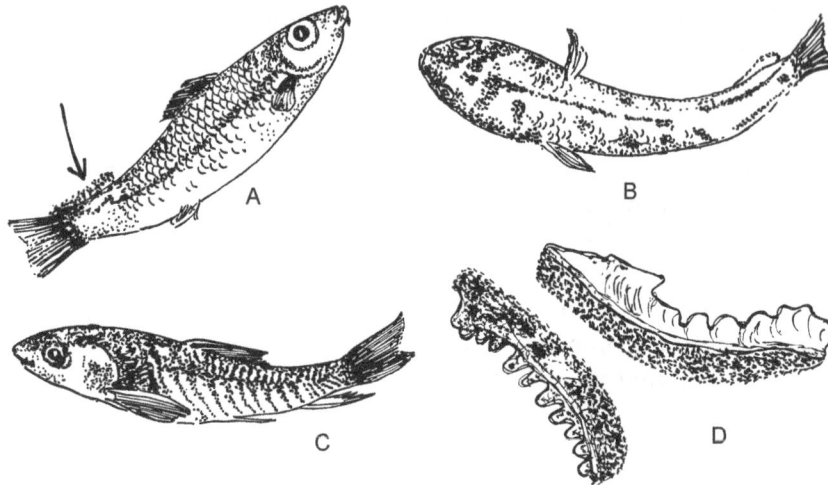

**Fig. 17.2a : Some common fungal diseases of fish. (a) Saprolegniasis; (b) Eye infection
(c) Aphanomycosis; (d) Gill infection, Branchiomycosis**

1. **Saprolegniasis:** Species of the genus *Saprolegnia* are usually implicated in fungal diseases of fish and fish eggs, although *Achyla*, *Aphanomyces*, *Leptomitus*, and *Pythium* have also been reported. These fungal diseases of fish are often considered secondary invaders following injury, but once they start growing on a fish the lessions continue to enlarge and may cause death unless medication is provided. Fungi often attack dead fish eggs and soon encompass adjacent live eggs, killing them, thus constituting one of the most important egg diseases. These fungi are widespread in natural freshwaters. Fish are more likely to be affected in temperate zone (early spring) following spawning activity. Under some conditions (such as crowding) fungus filaments could penetrate living eggs and fish.

(i) *Symptoms*: The disease is characterized by wooly, whitish blotches on the skin, fins, eyes, gills or mouth. The first indication of the infection is dullness of the body color. After sometime, the infected areas can be distinguished from the surrounding tissue by abnormal protrution of scales from the body surface. Later, the fungus appears as a small tuft growing in the centre of this region and the affected area increases in size as the fungus spreads. The activity of the fish is greatly reduced and several other spots appear on its body. In the last stage, fish becomes weak, emaciated, lethargic and gradually cease to exist. The infection has been observed in *Angiulla*, *Channa*, and *Osphronemus goramy*.

2. **Aphanomycosis:** It is quite different from saprolegniasis in that it causes blackening of the scales and subsequent exfoliation, exposing the body flesh of the host fish which results in its death. The fungus responsible for this disease is *Aphanomyces pisci*.

(i) *Symptoms*: At first a mycelium is developed, penetrates down to the cutis and even to the underlying muscles. Meanwhile, fungal hyphae also grow up on the skin. By their action, cells of skin and muscles begin to degenerate. In severe cases the fungus may penetrate even the skeleton.

Sametimes, eye infection is also observed which is characterized by the formation of a tuft of white or grey-white threads, resembling cotton wool, hanging out of the eyes. If not medicated, complete destruction of the eye occurs in a few days followed by penetration of mycelium into the brain leading to death.

The fungi infect fish eggs and rapidly destroy fish seeds. Genera of water moulds such as *Pythium, Leptolegnia,Protoachlya* and *Leptomitus* are involved in egg infection.

3. **Branchiomycosis:** Branchiomycosis is caused by two species of phycomycetous fungi: *Branchiomyces dermigrans* and *B. sanguinis*. The fungus forms long tubules in the gill epithelium. 'The fungal hyphae displace host tissues and the supply of blood to the gills are inhibited by causing swellings and destruction of the blood vessels. This results in yellowish-brown discoloration and eventual disorganization of the gill tissues. The affected fish gasp for air and ultimately die.

Fungal infection may sometimes results in systemic mycosis due to infection of the genus *Ichthyophonous hoferi*. The infected fish exhibits curious swinging movement while swimming and hence this disease is sometimes calles as *swinging disease*.

The disease appears in summer following a period of heat. Losses among 2 and 3-year old fish may reach from 30 to 60 per cent.

Water rich in organic substances tends to favor the appearance of branchiomycosis. Organically-fertilized ponds and tanks are susceptible to outbreaks of the disease.

The fungus predominantly attacks the oxygen-rich efferent branchial arteries, the fungus returns to the gill filaments. The infected gills show epithelial proliferations in the areas of the respiratory surface and adhesions on the neighbouring branchial laminae. The blood capillaries are displaced by the fungus.

Control of Fungal Diseases

There are two methods of control of fungus: (1) mechanical and (2) chemical. The former method is effected by removing dead and infected eggs 2 or 3 times a week. This is a time-consuming procedure, and some healthy eggs are injured in this operation. For this reason, chemical control is effective.

Formalin at 1 : 500 to 1 : 1,000 for 10-15 minutes or copper sulfate at 1 :200,000 for 1 hour may also be used for egg treatment with good results.

The chemical method for controlling fingerlings and adult fish is simple, cheaper and more efficient. Generally there are two methods of applying chemicals for fungal treatment such as (i) dipping fish and (ii) long-term treatment (for 1 Hour). However, research on control of fish mycoses has revealed the potential efficacy of some chemicals such as formalin (0.1 per cent dip), potassium dichromate (0.1 per cent dip), potassium chromate (0.25 per cent dip), potassium permanganate (0.02 per cent dip), silver nitrate (1-2 per cent topical application), phenoxitol (0.02 per cent dip), crystal violet (0.002 per cent dip), hydroquinone (0.04 per cent dip), neutral red (0.01 per cent dip), and teepol (0.04 per cent dip).

For one-hour treatment, aqueous solution of chemical is added to ponds or aquaria, allow fishes to remain for 1 hour and replaced with freshwater. The degree of success encountered with the use of any chemical is related to several factors. Water temperature, pH, presence of interfering substances, salinity, and the method of application may reduce or enhance the activity of a chemical.

17.9 Protozoan Diseases

Three protozoan groups such as flagellates, sporozoa, and ciliates are the best known and most serious contagious pathogens of freshwater fishes. Coccidia and Myxosporidae have varied and often severe effects on fishes ranging from castration to muscle degeneration. Most parasitic forms have characteristic and readily-recognizable spore stages which facilitate identification and encourage study.

Protozoa occur abundantly in all natural water. When attack in large numbers, they make the fish uncomfortable and the fish responds to by copious secretion of mucus. Parasites of this group are organisms which must be studied with a microscope to verify their indentification. Some common fish protozoa are shown in Figure 17.3a.

Types of Protozoan Diseases

1. **Ichthyophthriasis :** It is better known as *white spot disease* and is the most serious disease of carp and catfishes and almost all cultured fishes are susceptible to this disease. The disease can be lethal to fish of all sizes.

    The most important and simple species is *Ichthyophthirius multifilis*. The diseased fish appear to be covered with small white spots, and as the disease progresses the spots may join together and form areas of a dirty white color which later fall away as spots on skin. The fish swim violently in an attempt to get rid themselves of annoying the skin irritation. The progress of the disease is rapid when the water temperature is high. The disease is very common in intensive fish culture systems and in certain types of breeding tanks.

2. **Trichodina:** Many species of *Trichodina* have been reported in fishes. The parasites are small (25-50 microns), having a macro-nucleus, micro-nucleus, and food vacuole, circular with a conspicuous toothed disk within their bodies; dart erratically about over the body surface, gills, fins, and urinary bladder. Cilia around margin may or may not be obvious. They cause serious problems to fish culture. Different fishes such as gold fish, catfish, common carp, snakehead, etc. serve as host for the parasite.

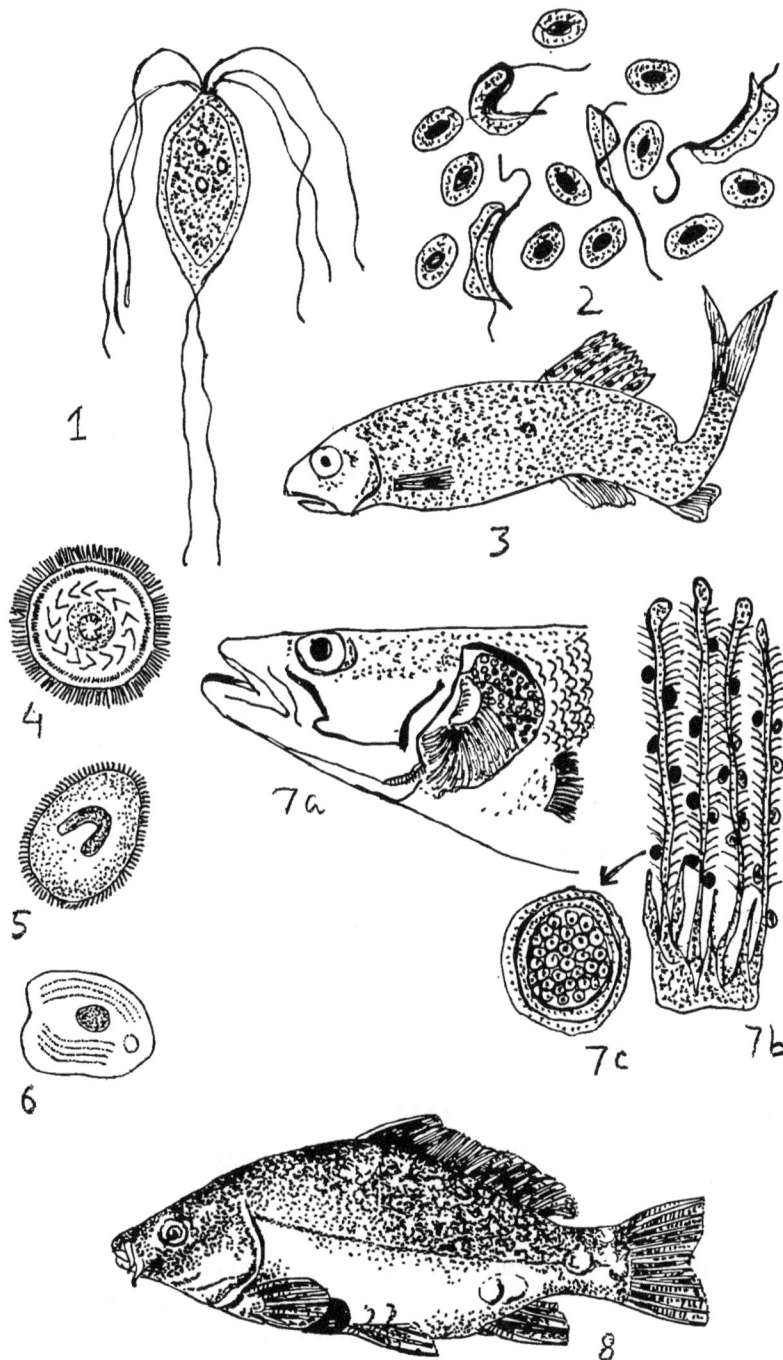

Fig. 17.3a : Some common parasitic protozoans in fish. (1) *Octomitus* **sp., (2)** *Trypanoplasma* **sp. in a blood smear; (3) Deformity of the vertebral column of a fish (Rainbow trout) suffering from whirling disease; (4)** *Trichodina* **sp., (5)** *Ichthyophthirius multifilis*; **(6)** *Chilodonella* **sp., (7a)** *Myxobolus* **sp. cysts grouped in the upper gill region of a pike-perch; (7b) Sporozoan cysts in gill filaments; (7c) Internal spores shown under high magnification; (8) Carp infected with** *Ichthyophthirius multifilis.*

3. **Chilodonella:** A number of species of this genus have been reported. They are oval-shaped and flattened in structure. The posterior part of the parasite appears as a heart and hence called *heart-shaped skin Parasite*. The micro-nucleus is round with variable positions. The macro-nucleus is oval and its size is about one-third that of the body. These highly destructive organisms creep rapidly about over the fins, body, and gills. Under high magnification, faint bands of cilia can be observed. The skin of the infected fish exhibits a bluish-white opaqueness. The affected skin may show a stony relief which may be so pronounced as to resemble small pox. The fish scraps themselves against the bottom and the sides of the pond/tank and swim listlessly and with considerable fatigue. The fish breath with difficulty. Chilodonella parasitizes the skin and gills in a manner which makes it an obligatory ecto-parasite. The parasite increases in numbers by continual division, and it exercises a destructive effect on the gills and skin. It is insensitive to temperature fluctuations and is found in open waters.

4. **Myxosporidiosis:** It is an all-season menace of stagnant water fish cultivation in the tropics. It is more virulent in intensive fish culture systems.

   It is called *nodular disease*. The nodules can be more than a millimetre in size. The nodules are also called *cysts*, in which numerous spores are found. The spores set free into the water by the eventual rupture of cyst wall. The most common genus of the myxosporidiosis is *Myxobolus*. Other species such as *Maxidium, Henneguya Chloromyxum*, etc. sometimes create difficult situations to fish farmers.

   Myxobolus infection can be easily recognized with naked eyes by the presence of round cysts on fins and gills of carps, pike, pike-perch and catfishes The spores are quite resistant to unfavorable conditions and are the infective stage of this parasite. When swallowing by the fish, each spore generates into a trophozoite in the intestine. It ultimately makes its way into the blood vessels and when carried to gills may complete the life cycle and form a cyst there. Epidermis of gills and fins become atrophied.

   The presence of cysts cause irritation and the fish robs the body against solid objects leading to haemorrhage. Retardation of the growth of fish due to heavy infection of this parasite is a good clinical sign.

   The pathological effects of the disease in fish are characterized by weakness, emaciation, raising of the scales, falling of scales, loss of chromatophores, and perforation of the scales.

5. **Whirling Disease:** One of the most important sporozoan diseases is the tumbling sickness of different species of trout. The disease is caused by *Myxosomsa cerebralis* spores. Whirling disease sometimes referred to as *blacktail*. It is a chronic and non-contagious parasitic infection of trout. This parasite occurs in the nervous system and they penetrate into the cartilage thus preventing normal ossification and producing skeletal deformities.

The disease can be diagnosed on the basis of changes in fish behavior and appearance. When feeding schedules are followed, some infected individuals exhibit an abnormal tail-chasing (whirling) behavior while swimming. The caudal peduncle and tail sometimes become dark or even black. Affected fish lose their sense of balance and turn around in an awkward manner. The body is turned around the longitudinal axis over 180 or even 360 degrees some 10 or 20 times before the fish is exhausted and takes some rest at the bottom of the pond. Diseased fish either die or heal spontaneously and may remain crippled for life. Individually, these symptoms are less than convincing, but if these are evident within a population, they together constitute a sound clinical diagnosis.

Spores of *M. cerebralis* are released into the ecosystem when the fish dies and decomposes or is consumed by scavengers. Spores are also released in the feces of predators that have consumed infected fish. The liberated spores are then ingested by tubificid oligochaetes (such as *Tubifex tubifex*) and a new phase begins and ends several months later with maturation of new forms known as *actinosporeans*. The transformed organism, provisionally named *Triactinomyxon gyrosalmo*, infects trout and initiates the alternate phase of whirling disease. The route of entry of that infection is assumed to be through gills. Once in the fish, spores of the parasite mature after about 3-months.

6. **Soft-Egg Disease:** The disease is an abnormal condition encountered in incubating fish eggs. It is also called as *soft-shell*. This disease is, however, identified by a flaccid, easily deformed character in incubating eggs. The affected eggs lack entirely the normal firm rebounding quality.

   The disease is considered to be the most likely due to an amoeba and reports on the occurrence of soft-egg disease are frequent and several hatcheries have had losses from it for period of several years.

7. **Mastigophora:** Different species of *Hexamita* occur as parasites in the intestine and the stomach of many species of freshwater fish. Symptoms are characterized by the anomalous swimming behavior (staggering movement and spasmodic dipping of the anal fin) and emaciation. The parasites have an oval-shaped body with two groups of three short swimming flagella at the obtuse end and two longer flagella at the opposite end.

   Sleeping sickness in fish is caused by several species of *Tripanosoma* and *Trypanoplasma*, the later being more common causative agents. *Trypanosoma* differs from *Trypanoplama* in having one flagella at the anterior end of the body.

   Different species of *Costia* cause sliminess of the skin of various species of fish such as *Labeo* spp., *Catla catla, Cirrhinus mrigala, Clarias batrachus, Heteropneustes fossilis*, snakeheads, etc. The disease produced by costia is termed as *costiasis* and the most common mastigophoran infection observed in pond fishes of Indo-Pacific regions.

Control of Protozoan Diseases

1. **Ichthyophthiriasis:** Treatment of Ichthyophthiriasis is difficult because of the extreme variability in the time required for completion of the life cycle. Prolonged and multiple treatments are necessary to effect a complete core.

    One of the most effective measures is to increase in the temperature of water upto 32°C daily for 3 to 5 days since this temperature level has been found to be lethal to the parasite and will speed the treatment. This method is highly effective for tropical pond fish that can resist high temperatures.

    Chemical applications are greatly affected by water quality. Chemicals which are effective in soft waters are frequently ineffective in waters with a high content of dissolved solids. Some of the suggested treatments for the disease are listed in Table 17.2. It should be noted that some of these chemicals are satisfactoty. Conditions of the fish and hatcheries greatly affect the success of treatments.

Table 17.2 : Suggested Treatments for Ichthyophthiriasis

Chemical	Application rate (mg/l)	Treatment Schedule
Formalin	200	Daily for 1 hour if 10ºC
Formalin	250	Daily for 1 hour if 15ºC
Malachite green	0.1	At 3-day intervals
Malachite green plus formalin	0.1 + 25	At 3-day intervals
Copper sulfate	0.5	Weekly in ponds
Methylene blue	2.0	Daily
Sodium chloride	7,000	For several days
Acriflavine	10	3 to 20 days
Quinine hydrochloride	20	3 to 20 days

Compiled from data published by the United States Department of the Interior, Fish and Wildlife Service, Washington, D.C., Fish Disease Leaflet No.2, August, 1974.

It should be kept in mind that the best control for the disease is prevention. New fish should be quarantined at temperatures near 22°C for at least 14 days to ascertain if fish carry the parasite.

Since this disease is highly infective, the infected fish must be destroyed, disinfected the hatchery, and make certain that the water supply does not contain pathogenic microorganisms. Sanitation of infected ponds is performed by applying about 1 kg of calcium cynamide or another suitable disinfectant per square metre bottom area after draining off the water, to be added in two portions — the first in winter and the second half at the summer season when the pond becomes dry. The method is strongly recommended and is also effective for disinfection in case of other protozoan diseases.

2. **Trichodina:**

(i) Treatment of 25 mg/l of formalin in ponds or 250 mg/l formalin for 1 hour in tanks.

(ii) Treatment of 3 per cent common salt solution or 1 : 1,000 acetic acid solution for 5-10 minutes.

3. **Costiasis:** Control methods of costiasis consist of bath for about 10 minutes in 1 : 2,500 formalin solution or 1 : 500 glacial acetic acid and 3 per cent common salt solution.

4. **Myxosporidiosis:** No chemical treatment has been found to be effective on the Myxosporidian cysts. The infections have been observed in Indian major carps in water having less than 400 mg/l chloride content. Weak solution of common salt can be used to disinfect the infected fish.

5 **Chilodonella:** This parasite can be controlled by treating infected fish with 25 mg/l formalin in ,ponds or 250 mg/l formalin for 1 hour in tanks.

6. **Soft-egg disease :**

(i) Sanitary measures and frequent disinfection of farm equipments.

(ii) Use of 1 : 18,000 solution of gentian violet for 5 minutes every 48 hours.

(iii) Filtration of water used for egg incubation has been found to be economic and efficient method of prevention.

7. **Whirling disease:**

(i) Avoidance of fish stocks from geographic region where the disease is enzootic.

(ii) The use of best drugs such as furoxone, benomyl and fumagillin drastically inhibit spore development.

(iii) Where the disease cannot be avoided and where the potential for spread of the infection is not a consideration, the propagation of resistant species of fish is highly effective.

(iv) Chemicals such as calcium hydroxide or calcium cyanamide are recommended and used for decontamination or practical eradication of tubificid worms.

(v) Decontamination process is carried out by combining filtration to remove or reduce suspended materials with ultraviolet irradiation.

(vi) Hot smoking of fish feed (at 66°C for 40 minutes) kills spores and consequently does not create any threat to the spread of the disease.

Therapy can effectively reduce losses, but inevitably results in an attendant development of pathogen carriers among the survivors. Frequent administration of drugs which are termed as *chemoprophylaxis* though inhibit spore development, drastically retard fish growth by 40-50 per cent.

17.10 Bacterial Disease

Most of the bacteria infecting fish can harm only if fishes are injured or weakened. Occasionally, bacteria possessing a higher degree of pathogenecity, may completely attack healthy fishes. The healthy fish has a wide range of defense mechanisms and the immune system which protect itself from microbes are capable of causing damage. Pathogenic bacteria are, however, capable of causing diseases, that is, a change in the normal structure or function of the fish's body. This changes to the host and manifested as a set of symptoms, which may be due to the effect of bacterial products such as toxins. Ulcers on the body of the fish, fluid accumulation inside the body cavity, scale protrusion, intestinal inflamation, opaque eyes, tumours, etc. are very common symptoms of bacterial disease in fishes. A number of bacteria have been noted in farmed fish which are highly infectious.

1. **Furunculosis:** This disease which is very common in salmonoids, is caused by *Aeromonas salmonicida*.

(a) *Symptoms :*

(i) Heavy infection of the intestine followed by haemorrhages in the muscles and the production of swellings or boils (furuncles) at the surface, generally on the sides of the body (Figure 17.3b)

(ii) Appearance of small spots in the liver, eyes and on the fins as well as opercula.

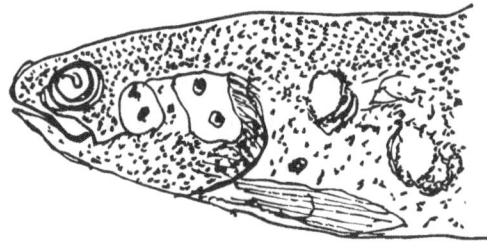

Fig. 17.3b : Furunculosis in trout showing the formation of boils.

(b) *Treatment*: The best treatment of furunculosis is the (i) removal and immediate destruction of dead and heavily-infected fish. The infected fish should be isolated in special ponds with utmost care, (ii) farm utensils should be disinfected with aqueous solution of potassium permanganate (1.0 g in 50 litres of water), (iii) the empty ponds should also be disinfected with quick lime, (iv) balanced and vitamin-rich food must be provided to fish stocks at regular intervals, (v) quality water should be supplied to fish farms during the entire culture period and must be free of organic impurities, (vi) several drugs such as sulfamerazine, sulfadiazine, sulfa-methazine and the like are mixed with the food at the rate of 10 g /45 kg of fish/day, and (vii) the use of chloramphenicol and oxytetracycline as antibiotics at the rate varying between 2.5 and 3.5 g/45 kg of fish per day.

2. **Red Pest Disease:** The disease has been noted in many species of freshwater fish (especially in pike, salmon, eel, cyprinids, Pike-perch and perch) and is caused by different strains of *Pseudomonas*, *Aeromonas* and *Chromobacterium*.

(a) *Symptoms :*

(i) red coloration on the skin of the belly and fins.

(ii) necrosis and haemorrhages in the gills.

(iii) infected fish exhibits slow and irregular movements and come to the surface shortly before death.

(b) *Treatment*: It is difficult to successfully combat this disease. Affected fish should be removed as quick as possible, since they die off very rapidly in ponds.

3. Red Spot of Fish: Different strains of bacteria such as *Aeromonas Pseudomonas:,Proteus vulgaris* and physical factors (such as cold) are responsible for red spot disease.

(a) *Symptoms* :

(i) appearance of muscular necrosis in the head region and at the base of the fins.

(ii) appearance of cutaneous blotches which may be round, oval or elongated in size and in some places, loss of scales may be observed.

(iii) the formation of blotches gradually extend to cover the entire body and these may be subject to a secondary mycotic infection.

The disease generally occurs in spring or summer. The losses may drastically affect the yield of fish in ponds and lakes and fish mortalities of about 60 per cent have been registered in several cases. Similar control measure may be adopted to this disease as for the red pest disease.

4. Infectious Abdominal Dropsy (Bacterial Haemorrhagic Septicemia) : The infectious abdominal dropsy or sometimes referred to as *swelling of the belly* (Figure 17.4 a) is the most serious disease affecting carp cultivation in many countries of the world and constantly causing substantial losses. However, various sub-types of bacteria have been described, but generally consider all of them induced by *Achromobactor punctata*.

(a) *Symptoms*: Two main types of external symptoms have been reported such as typical dropsy and the ulcerative type (Figure 17.4b) Black-white or red-colored ulcers appear in the skin and muscles. Ulcers cause loss of the scales and fins and skeletal deformities. In case of typical dropsy, the anus is inflamed and prolapsed. A foul-smelling yellowish or bloody watery fluid appears in the abdominal cavity. The intestine may be inflamed. Internal symptoms are characterized by the yellow, dark-yellow, gray-yellow or pea-green coloration of the liver with, at times, white or yellowish-colored blotches.

The infected fish can easily be recognized by their languid and listless swimming and they tend to keep themselves near to the sides of the pond.

(b) *Treatment*:

(i) use of chloromycetin or streptomycin in feed.

(ii) attention must be given to the hygienic aspects of fish farming such as feeding and water quality.

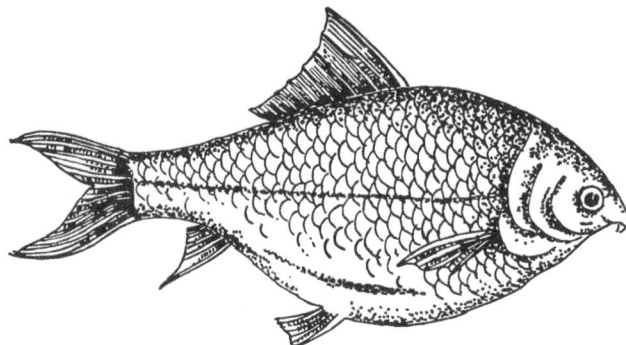

Fig. 17.4a : Abdominal dropsy of carp.

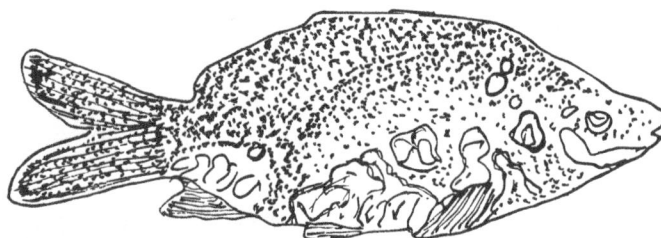

Fig. 17.4b : Carp with bacterial hemorrhagic septicemia (ulcerative form).

5. **Fish Tuberculosis:** The bacterium *Mycobacterium piscium* has been isolated from carp, goldfish, catfishes, air-breathing fish and eel. However, fish tuberculosis may occur as epidemics and cause mass mortalities particularly in ponds and large commercial installations.

(a) *Symptoms* :

  (i) loss of weight and appetite,

 (ii) loss of body color and paleness.

(iii) scale defects and desquamation.

(iv) ulcers and fin rot.

 (v) deformities of the vertebral column and mandibles.

(vi) exophthalmos, falling out of the eyes.

(vii) disturbance of swimming.

(viii) formation of nodules in the internal organs.

(b) *Treatment*:

 (i) Stocking the pond with fish at high rates should be avoided.

(ii) general cleanliness of ponds.

(iii) contact between separate ponds must be avoided on all occasions since mixed ponds constitute permanent foci of infection of this disease.

6. **Bacterial Tail Rot:** Bacteria of the genus *Aeromonus* and *Pseudomons* have been found to be the main causative organisms for tail/fin rot disease (Figure 17.5).

(a) *Symptoms* : The disease is characterized by a slight cloudiness of the outer edges of the fins that results in gradual reduction in the size of the fins. The disease becomes apparent in breeding ponds.

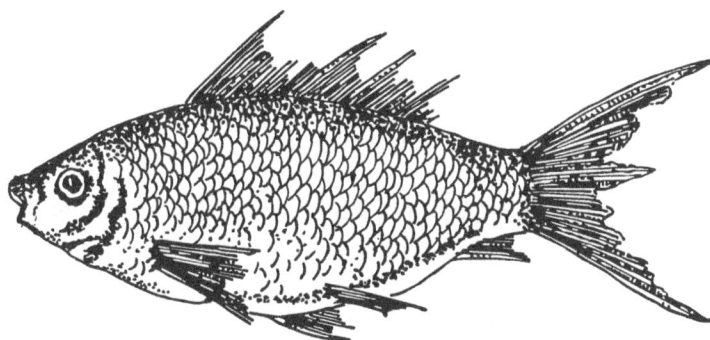

Fig. 17.5 : Fin rot disease of carp.

(b) *Treatment*:

(i) long bath in acriflavine (1 g/100 litre of water) or in sulfonamide (1 g/10 litre of water) is very effective.

(ii) use of malachite green dips and disinfection of the tanks with chlorine.

(iii) use of copper sulfate dips (1 g/2,000 litre of water).

7. **Bacterial Gill Disease:** The first indication of this disease is the loss of appetite and lethargic condition. Diseased fish tend to orient themselves into the flow of water. If fish are not treated after the first symptoms are noted, the mortality increases rapidly and more than 50 per cent of the fish may die within 24 to 48 hrs.

The disease is associated with the presence of myxobacteria in large numbers on the gill epithelium and over-crowding in ponds. The environment has, however, a great influence on the occurrence of the disease. It has been noted that fish receiving a nutritionally poor diet are very susceptible to bacterial gill disease.

(a) *Symptoms*: With bacterial gill disease, the hyperplasia starts at the distal end of the gill lamellae. Fusion of the ends may result in holes. Lamellar fusion and hyperplasia are greatest at the distal ends of the gill filaments.

(b) *Treatments*: The maintenance of water quality and avoidance of over-crowding helps prevent fish stocks from bacterial gill disease. Water sources with high turbidity and wild fish are dangerous. If recirculation is inevitable, the water must be reconditioned to rid the pond of waste products and to reduce any bacterial build-up.

Organic mercurial and quaternary ammonium compounds have been the most effective chemicals for controlling the disease. Ethyl mercury phosphate, for example, is used as a bath at 1-2 mg/litre. However, the high toxicity of mercurials to both man and fish and the high retention in flesh makes the use of mercurials extremely questionable. For control of an outbreak, treatments are continued daily or 7-10 day intervals until the fish stocks exhibit normal behavior.

8. **Columnaris Disease:** Columnaris disease popularly known as *cotton wool disease* is very common and widespread disease of freshwater fish. The causative organism of this infection is a myxobacterium, *Chondrococcus columnaris*.

(a) *Symptoms*: The disease is characterized by the presence of gray-white spots on the head, fins, gills, and skin. These lesions gradually become ulcerated, and the fin rays are frayed and exposed. The loss of the scales is very common when destruction of the skin is completed. In scaleless fish, a reddish-colored periphery is observed around the lesion.

(b) *Treatment*: The disease may be treated by dipping the fish stock in (i) 1 : 2,000 copper sulfate solution for 2 minutes, (ii) chloromycetin (@ 5-10 mg/litre), (iii) 1 : 15,000 malachite green solution for 30 seconds and (iv) acriflavin (2 mg/gallon of water). The use of sulfa drugs to the food has been very successful in the treatment of Columnaris disease.

9. **Peduncle Disease:** Peduncle disease, also termed as *coldwater disease*, is very common in temperate regions where the water temperature drops below 10° C. The disease is characterized by the typical lesion which starts as a small whitish or bluish spot behind the dorsal fin. In several cases, the entire caudal area is lost leaving exposed vertebrae. The causative agent is a myxobacterium, *Cytophaga psychrophila*.

(a) *Treatment* : (i) use of sulfa drugs at the rate of 100 mg/kg/day for 10-20 days. (ii) disinfection of farm equipments and avoidance of contact between wild and uninfected fish stocks are the two guiding principles to control peduncle disease. Wild fish are generally suspected to carry any bacterial disease.

17.11 Epizootic Ulcerative Syndrome*

Since the middle of the 1970s, the dreaded fish disease, popularly known as the *Epizootic Ulcerative Syndrome* (EUS) (Figure 17.6) has been causing considerable damage to many freshwater fish species of tropical and sub-tropical regions. Though the disease was previously referred to as *Ulcerative Fish Disease* (UFD) or Ulcerative Disease *Syndrome* (UDS), EUS was the name adopted in 1986 by experts of the FAO (Food and Agricultural Organizations) to describe the dangerous disorder of fishes in Asia-Pacific region.

---

* For further detail, see FAO (1986), Lilley *et. al.* (1999), Chinabut and Roberts (1999).

Fig. 17.6 : Epizootic Ulcerative Syndrome of Catfish

Simply defined, a group of symptoms which consistently occur together, are termed as *Sydrome*. However, the EUS has been defined as a seasonal epizootic condition of fresh warmwater fish of multiple and complex assignment to cause disease (etiology) which is characterized by the presence of fungi, Several species of pathogenic micro-organisms such as *Vibrio anguillarum, Aeromonas hydrophilla, Micrococcus, Escherichia coli, Pseuudomonas aeruginosa* (Bacteria); *Rhabdovirus* (Virus); and *Aphanomyces invadans* (Fungi) are usually implicated in EUS of fish. One or more invasive fungi and a range of opportunistic bacteria are involved in this disease. These species are, however, considered as secondary invaders following injury but once they start growing, the lesion usually continue to enlarge and may cause death unless suitable drugs are provided.

(a) *Symptoms* :

(i) presence. of lesions and ulcerations of varied appearance and size ranging from shallow haemorrhagic lesions during the early stages to large deep necrotic ulcers entirely damage the muscles. Generally, lesions are developed dorsal to the body and head and on the mouth as well as tail regions.

(ii) affected fishes become lethargic and have the tendency to float on water, sometimes with head projected out.

(b) *Treatment*: No therapeutic treatment of EUS is possible. However, treatment of diseased fish stocks may be accomplished by the application of lime, common salt, potassium permanganate, etc. Application of these compounds should be continued until the outbreak has ended.

Use of certified fish stocks, though effective in combating the EUS, is not always feasible; but much can be achieved by controlling the original stock and by quarantine. Practices such as fallowing and avoiding contact between fresh and infected fish are now becoming standard control procedures.

Attention to hygiene and environmental quality always pay dividends. Being aware of the critical times of fish together with a rigorous monitoring program prevents outbreaks of such disease which would otherwise results in high mortality.

17.12 Viral Disease

Only a limited number of fish disease could be attributed to virus agents. The importance of carp and trout farming, greatly favors the increasingly obvious spread of viral infections. Fish pox and viral haemorrhagic septicemia (VHS) are the two common viral infections generally found in some carps such as common carp and Prussian carp and some other fish such as European pike perch.

(a) *Symptoms*: Viral haemorrhagic septicemia diesease is characterized by listlessness, slow movement at the surface of the water, grey-brown pale color of the gills, swelling of the belly, and yellowish-colored anus. Pox disease is also characterized by the occurrence of white spots on the thickened epidermis.

(b) *Treatment* :

(i) maintenance of hygienic condition of ponds.

(ii) quick removal and destruction of affected fish.

(iii) infected ponds and hatchery utensils should be disinfected with quicklime or with calcium cyanamide.

(iv) utmost care should be taken to ensure that the oxygen supply is adequate and the number of fish in the pond should be within judicious limits.

17.13 Worm Disease

Previous sections remind us that fungal, protozoa, bacterial, and viral etiology are considered in the category of microbial diseases are referred to as *microbial parasites*. They are capable of destroying host tissues and multiplication within the fish. Resultant pathology and the course of the disease may depend very much on such factors as infective dose, virulence, and resistance to the individual host as well as host nutrition and other environmentally-influenced variables. However, there are some other parasites which are more larger than the microbial parasites that also cause the invassive diseases, are non-multiplicative in fish body once invasion has occurred. The various parasitic worms, copepods, nematodes, and annelids, which although of occasional concern, possibly do not exert serious effects on fish populations.

Some worms and their larval forms are skin and gill parasites which are generally termed as *ectoparasites*, whereas others live in the visceral cavity, the internal organs, and the musculature which are termed as *endoparasites*. The effects of worms and their larvae on fish include growth retardation, tissue disruption, metabolic disturbances, and even death in heavy infections. Added economic effects include discarding of otherwise edible fish products and delay in processing operations.

Fish ectoparasites are monogenetic trematodes which parasitize the skin, fin, and gills have characteristic hooks and one or more acetabuli at the caudal end of the body, by means of which they fix themselves onto the fish body. Development of ectoparasites occurs directly within the host, or conversely by passing through free-swimming larval forms which later seek out a suitable host. Since these parasites can multiply on the fish without additional hosts, they are called *monogenetic flukes*. Endoparasites are, on the other hand, digenetic trematodes and the fish may be either a definitive or an intermediate host. Some common fish worms are shown in Figure 17.7.

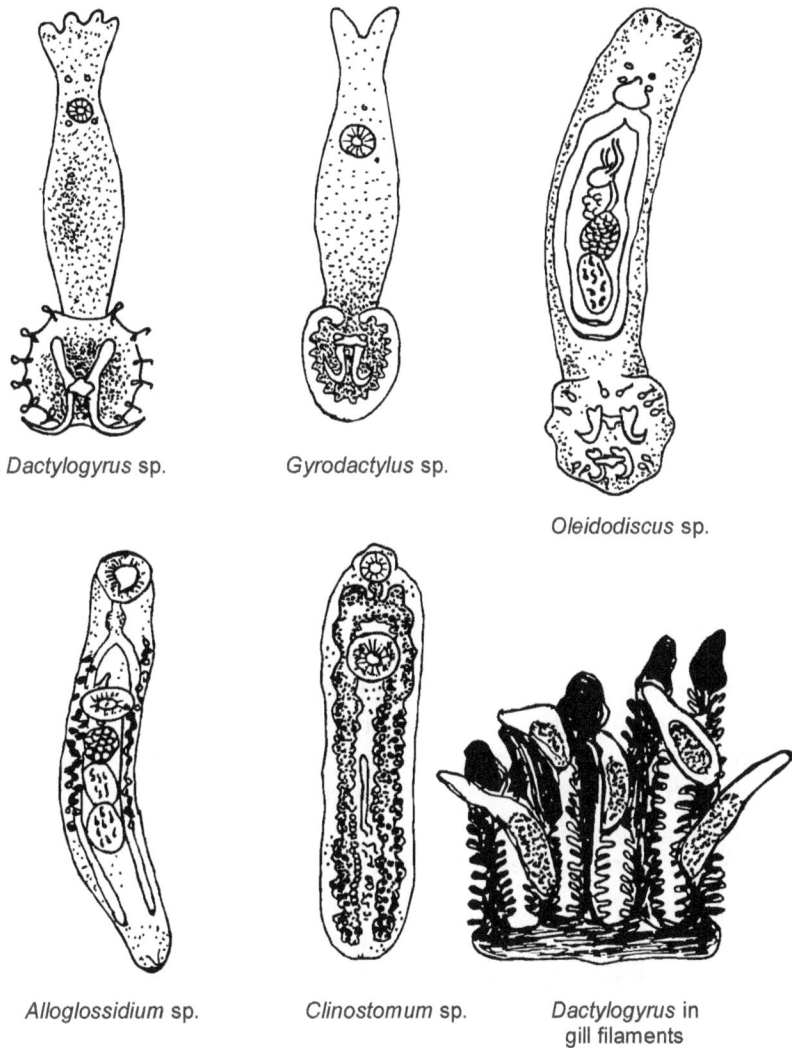

Dactylogyrus sp.   *Gyrodactylus* sp.

Oleidodiscus sp.

Alloglossidium sp.   *Clinostomum* sp.   *Dactylogyrus* in
gill filaments

Fig. 17.7: Some common fish worms.

1. **Dactylogyrus:**

(a) *Symptoms*: The external symptoms are not generally noted, except in cases where the infestation is severe when the gill edges become thickened and the opercula appear to be opened. The worms are localized on the gill filament; but when present in large numbers, they become distributed all over the body.

(b) *Treatment* :

(i) short bath in formalin (25 cc formalin solution in 100 litres of water) for 30 minutes.

(ii) filtration of water supplies to fish farms removes cercarian larvae of fish trematodes.

(iii) treatment of fish with 25 mg/l formalin in ponds or 250 mg/l formalin for one hour in tanks completely eliminate these parasites.

(iv) use of potassium permanganate (5 mg/l) or potassium dichromate (20 mg/l).

2. **Gyrodactylus:** This worm lives on the skin of fish and seriously infected fish exhibit an opacity of the skin often with inflamed and reddened areas. severely infested fish with this parasite may cause turbidity of the cornea followed by blindness in many fish species. In extensive fish culture systems where metabolic by-products are accumulated, *Gyrodactylus* infections are very common.

Short bath in formalin is effective for combating this parasite. Other methods that should be adopted for preventing this species are similar to that of *Dactylogyrus*.

3. **Disease Produced by Digenetic Trematodes :** Fish are hosts for adult parasites and frequently they act as intermediate hosts in their life cycles. Many flukes attain the adult-stage in the intestine, swim bladder, or urinaly bladder of fish. Such flukes have a flattened body, worm-like form, with an oral sucker at the anterior end and a ventral sucker near the mid-third of the body. A typical fluke of this type is *Alloglossidium corti*.

Flukes which use fish as intermediate hosts penetrate the body and encystation occurs in the flesh or mesenteries. Most of the flukes are larval stages of worms which are adult parasites in birds. Encysted worms give a grubby appearance to fish flesh. One of the most important parasites of this type is *Clinostomum marginatum*, which has a characteristic yellow color. Therefore, the encysted worms are often referred to as *yellow grub of fishes*. Severe infections due to digenetic flukes in most instances, cause major concern to fish culturists.

Another important species of fluke *Diplostomum spathaceum* is occasionally found in trout hatcheries. This fluke invades the eye, swells the cornea through accumulation of liquid in the anterior chamber of the eye and a milky turbidity of the lens is developed.

Since aquatic birds and snails are involved in the life cycle of these parasites, their removal from the vicinity of the ponds will drastically reduce the risk of disease infection considerably.

4. Tapeworms: Tapeworms have an intricate life cycle with change of hosts. They are common in intestine of fishes. They are long, flattened, segmented which taper toward one end. Closer examination at the smaller end reveals the presence of four small suckers.

Tapeworms may be divided into three groups: (i) the larval stage occurs in small aquatic animals which are consumed by the fish that is the final host, (ii) the larvae live in the bellies of one species of fish and the adults in another one, which feeds on the former and (iii) this group includes worms with a multiple change of hosts.

Common examples of the first group are the tapeworms *Ligula simplicissima*, *Proteocephalus*, and *Caryophyllaeus laticeps*. While larval stages of the first two genera live in *Daphnia*, the larval stage of the third genera live in *Tubifex*. The common species of the second group includes *Triaenophorus nodulosus*. Whilst the first larval stage lives in *Cyclops*, the adult worms have been noted in pike.

The tapeworm that belongs to the third group includes the dangerous species *Diphyllobothrium latum*. From a hygienic point of view, the group is the most important. The adult worm may live in the human intestine. Fertilization is internal and fertilized eggs are developed into embryos or capsules, leaves the fluke's body through its gonopore into host's intestine, and finally ejected out with its faeces. Under suitable conditions of the environment, encapsulated embryos differentiate into free-swimming, ciliated miracidium larvae. If these larvae are consumed by a Cyclops, they develop into pro-cercoid larvae. If the cyclops is devoured by a fish, the pro-cercoid develops into the last larval stage — the plero-cercoid larvae which are situated in the liver and muscles. The most common hosts are eel, Northern pike, lake perch, burbot, lake trout, yellow perch, walleye and numerous cyprinids.

The fish as such does not seem to suffer much from these tapeworms, which are dormant until the fish is consumed by humans. The latter is infected in case he fails to apply adequate heat treatment to the fish prior to consumption. In many countries, however, these tapeworms seem to be nearly extinct now, owing to good hygiene which prevents disposal of faeces in fish culture ecosystems on the one hand and the use of better cooking methods on the other. Tapeworms cause acute anemia and many other typical symptoms of helminthiasis.

Many cyprinids are sometimes infested with the tapeworm *Caryophyllaeus laticeps* which lives in fish intestine and hence fish are the final host of this parasite. The oncosphere is ingested by *Tubifex* where it leaves the egg. When a fish consumes such tubificid worm, the parasite continues its growth and develops into a mature worm.

Another tapeworm, *Triaenophorus modulosus*, has the fish either as a definitive host or as an intermediate host. This parasite is frequently found in pike, burbot and other fish. Fertilized eggs are laid by the worm and ultimately develops into a miracidium larva which is consumed by a copepod — the first intermediate host.

Inside the abdominal cavity of the copepod, the miracidium becomes a pro-cercoid larva. The copepods are eaten by the fish where the pro-cercoid larva is transformed into plero-cercoid larva. Typical symptoms of the disease are developed by the invasion of the liver by the plero-cercoid which are characterized by the cirrhosis of the liver, after which the fish immediately dies. Resistant of the fish to infectious diseases becomes diminished and they are very much sensitive to hunger and lack of oxygen.

(i) *Control of Tapeworm*: Tapeworm infection may be controlled by several methods that follows:

* by removing sick/infected fish.
* exhaustive fishing from ponds and lakes where infections are suspected.
* by preventing the free entry of fish-eating birds.
* by removing infected copepods from the water supply by sand- gravel filtration and treating the infected fish with Di-n- butyl tin oxide at the rate of 250 mg/kg of fish or 0.3 per cent of food for 5 days.
* destruction of infected copepods and tubificid worms have been successful in preventing the spread of these helminths.

5. **Nematodes:** Nematodes (or roundworms) are occasionally found in several species of freshwater fish but are not as common as tapewoms or flukes. They can be recognized by their elongated, cylindrical shape, lack of undifferentiation, and smooth body surface. Fish may be the intermediate or the final hosts of the nematodes, or even both at the same time.

Several nanatodes such as *Contracaecum* sp., *Cucullans elegans*, *Filaria sanguinae* and *Cystoopsis acipenseris* have been noted in catfish, flatfish, crucian carp, and sturgeon, respectively. Nematodes, however, live as internal parasites in different organs of fish such as intestine, eyes, stomach, muscles, peritoneal cavity, and swim bladder.

In large fish, nematode infections generally do not cause much harm unless they become numerous. Massive infections, however, result in an anemia and inflamation of the internal organs of fish. Nematodes can be removed by using santonin at the rate of 0.04 g/fish for sturgeon. Moreover, for controlling intestinal nematodes, the use of piperazine citrate has been very successful.

6. **Acanthocephala:** The most important genera of acanthocephala are *Echinorhynchus*, *Acanthocephalus*, *Corynosoma*, *Pomphorhynchus* and *Rhadinorhynchus*. All are dangerous intestinal parasites, producing perforation of the intestine. Generally, the larvae of these parasites live in freshwater shrimp (*Gammarus pulex*), the water louse (*Asellus aquaticus*), or in insect larvae.

Acanthocephala are hard to remove from the body of fish since their protrusible proboscis bear a large number of hooks hold firmly to the intestinal wall. Mild infections with these parasites do not appear to cause serious injury to the fish, but inflammation occurs in the area where the parasites adhere and the lesions they cause may serve as foci of secondary bacterial infections.

Treatment of acanthocephalan infections with Di-n-butyl tin oxide at the rates varying between 100 and 200 mg/kg of fish has shown encouraging results. In farmed ponds, both the parasites and their hosts may be destroyed by means of quicklime and drying out of ponds.

7. **Leeches:** Leeches are large external parasites with membered bodies and a large sucking disc at each end. Of all the leeches of fish, the most important and ordinary fish leech *Piscicola geomtra* reaches a length varying between 2 and 4 cm. Some other genera includes *Pontobdella, Branchelion, Callobdella, Cystobranchus*, and *Myzobdella* (Fig. 17.8) The infected fish are restless and are more sensitive to *Trypanoplasma* and *Aeromonas liquefaciens*. Fish leeches feed on the blood of the fish (Fig. 17.9) and may cause considerable harm. They remain at the pond bottom, held fast by their posterior sucker. As soon as any fish moves, they immediately attach themselves to it by means of the anterior sucker.

Removal of leeches is done mechanically by means of forceps (with blunt ends) after the leeches have been paralyzed by 15 minute treatment with 2.5 per cent sodium chloride solution. Ponds in which leeches have been seen should be emptied and disinfected with quicklime.

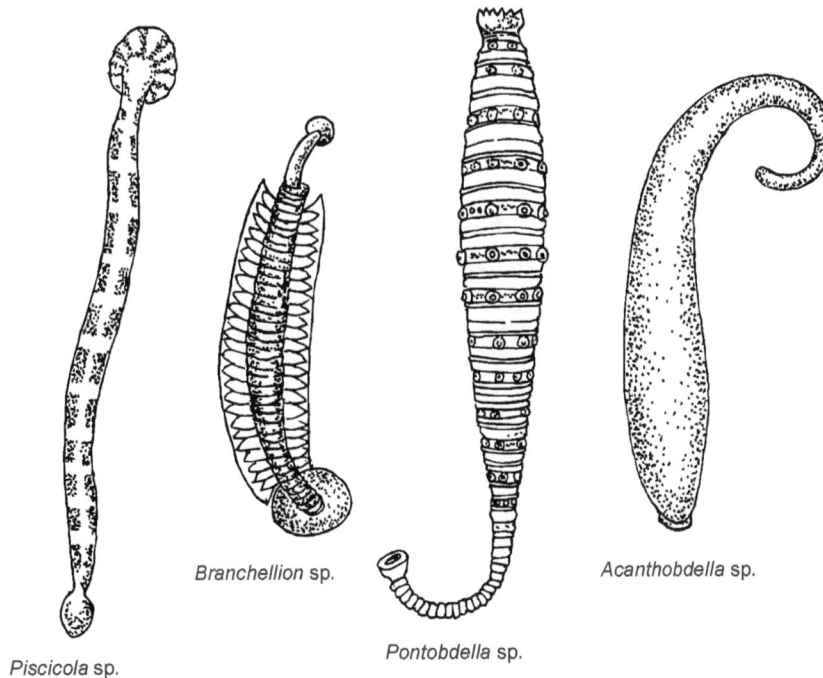

Branchellion sp.

Acanthobdella sp.

Pontobdella sp.

Piscicola sp.

Fig. 17.8 : Some common fish leeches.

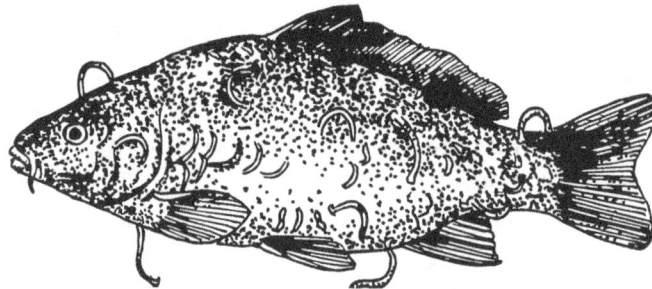

Fig. 17.9 : Carp infected with *Psicicola* sp.

17.14 Crustacean Parasites of Fish

Crustacean parasites or so-called *parasitic copepods* (Fig. 17.10) not often develop epizootic populations in natural waters but they may cause serious problems when fish are cultured, and epizootics occur in hatcheries. Of the 10 genera of parasitic copepods found in tropical and sub-tropical freshwater fishes, *Argulus* and *Lernaea* are the most important because of their world-wide distribution and ability to infect many species of fishes of the families Anabantidae, Catostomidae, Cyprinidae, Lepisosteidae, Percidae, Salmonidae, Sciaenidae, etc.

1. ***Argulus* sp. :** They are commonly referred to as *fish lice*. They have a flattened, saucer-shaped body, and can be observed creeping rapidly over the body of a fish. There is presence of jointed legs and two large sucking disks for attachment. It is the only large (1-22 mm long) external fish parasite and can be seen with the naked eye.

   Argulids puncture the skin and feed on blood through the oral sting. Feeding sites often become ulcerated and haemorrhagic, providing ready access to secondary infections by other parasites, viruses, fungi, and bacteria.

   Advanced infections by Argulids are characterized by weakness, erratic swimming, considerable flashing, and lack of growth. Moribund fish appear exhausted and cannot maintain equilibrium.

(a) *Control*:

(i) Removal of all dead and moribund fish, hard substrate, and terring of concrete structures prevents their use for egg deposition.

(ii) Wooden lattices are sometimes placed in the pond to serve as artificial substrates for removal of Argulids at intervals and to destroy eggs laid on fish body.

(iii) Pond inlets should be covered with 1/8 - inch mesh screen which prevents adult parasites from entering the ponds.

(iv) Complete drying of ponds kills eggs, larvae and adults within 24 hours.

Carapace

First maxilliped

Trunk

Egg pouch

Achtheres ambloplitis

Eye

Suctorial cup

Probosis

Thoracic appendages

Testes

Abdomen

An optically cleared *Argulus* sp.

Antennule

Hook-like antenna

Cephalothorax

Thoracic legs

Egg pouch

Egg sac

Ergasilus sp.

Larnea sp.

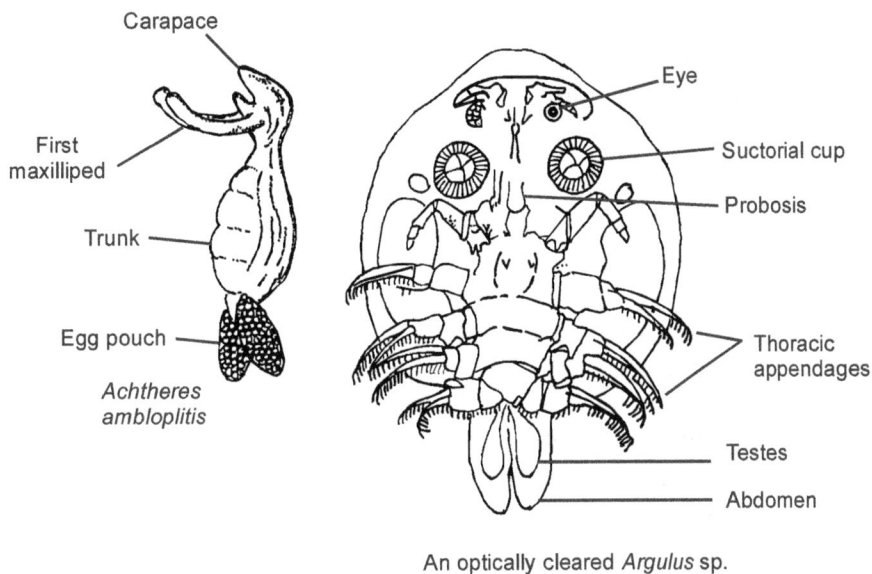

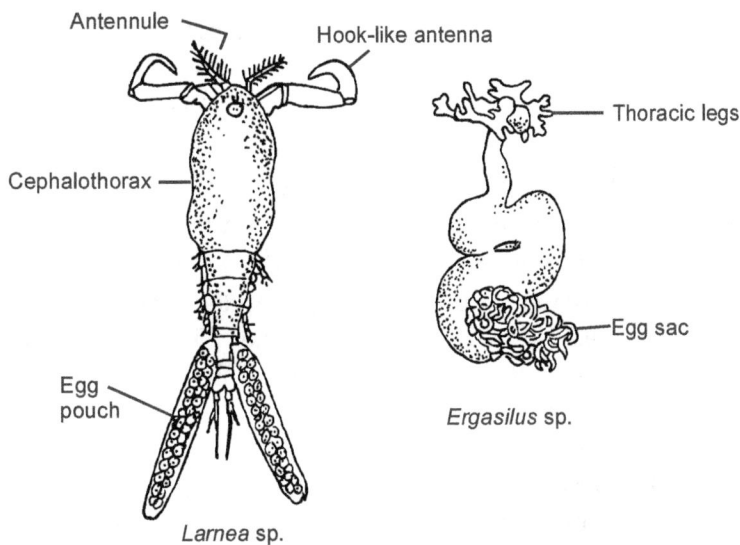

Fig. 17.10 : A few representatives of crustacean parasites of fish.

(v) Fish are bathed in potassium permanganate (at the rate of 10 mg/L) or an insecticide. Malathion (@ 0.25 mg/L), methyl parathion (@ 0.125 mg/L), benzine hexachloride and ammonium chloride (@ 500 mg/L) may be used with great caution and except potassium permanganate and ammonium chloride, other insecticides are not registered for use on fish as they are highly toxic to man.

2. *Lernaea* sp.: The anchor parasite or so-called *anchor worm* of the genus *Lernaea* is a copepod parasite on freshwater fish. They are so firmly attached to their hosts that care must be exercised in removing them. The mature female parasite is 5 to

22 mm long and is composed of a head and an elongated body with degenerate swimming legs. Egg sacs contain about 200 eggs and are attached to the posterior end. The head expands into large cephalic horns. The male parasite resembles the female at the 6th larval stage, but is about 25 per cent smaller.

(1) *Symptoms*

* Massive mortalities of significant economic importance have occurred in carp, gold fish, catfish, and trout. The parasite causes significant retardation of fish growth and loss of weight.

* The parasite destroys scales and cause ulcerated areas at the point of penetration. The major injury to the host results from the consumption of blood and from the presence of an open wound at the attachmemt site, which allows the establishment of secondary infections by bacteria, fungi, and viruses.

* Light infections may cause the fish to swim erratically, flashing against the sides and bottom of ponds. Heavily infested fish become convulsive or appear moribund, sometimes swimming upside down or hanging vertically in the water.

(ii) *Control*: The following treatments, which give only partial or limited control, have been used; but results differ because of differences in water chemistry, temperature, and other factors. Several treatment chemicals are toxic to fish and all should be tested with small numbers of fish before their extensive use is permitted.

* Use of common salt at the rate of 0.8 to 1.8 per cent for 3 days as a dip.

* Use of 250 mg/L solution of formalin for 30-60 minutes every 3 weeks has been found to destroy larval foms of the parasite with varying degree of effectiveness.

* A stock solution of potassium pemanganate is added slowly until a concentration of 25 mg/L is attained. After 90 minutes, another 25 mg/L is added to the same water, which is then drained 60 minutes later. Fish are kept in freshwater for at least 40 hours before marketing. This chemical is highly effective against adult copepods and is inexpensive.

3. *Achtheres* **sp.:** This parasitic copepod having total length of 2-7 mm is very common on the gills of catfishes where its light-colored body stands out in sharp contrast to the deep-red gill filaments. Legs have disappeared from the body and two mouth parts have become modified to form curved appendages which helps attach the parasite to the gills. Egg sacs protrude from the posterior end of the body. These parasites consume blood from the gills and are serious menace to catfish and many other freshwater fish.

4. *Ergasilus* **sp.:** This is another parasitic copepod which attacks the gills of fishes (particularly catfish). This form resembles the free-living copepod, *Cyclops*, but the second antennae are enlarged, terminating in large claws which serve as a means of attachment. The parasite feeds on body fluids and blood.

(i) *Prevention of Achtheres and Ergasilus*: Prevention of copepod parasites, the first order of control, is attained by the introduction of parasite-free fish and the use of fish-free water supplies. Pre-filling of ponds at least 1 week before use and filtration of water supplies are possible methods of excluding the free-swimming stages from the hatchery. Where definite host specificity exists, removal of host fish from the water supply is effective.

(ii) *Control of Achtheres* sp *and Ergasilus* sp. :

* Calcium chloride (0.85 per cent), copper sulfate (2 per cent), potassium chlorate (0.2 per cent), and sodium chloride (1.5 per cent) are generally recommended. The treatment duration should be 3-months, but 3- or 4-day treatment intervals would be more economical and less harmful to the fish.

* Destruction of infected fish and reduction of the density of fish in brood stock ponds.

* Use of younger and less heavily infested brood fish.

* Complete drying of ponds before restocking, and prophylactic treatment of young fish.

17.15 Environmental Diseases Produced by Physical and Chemical Factors

Several environmental factors such as water temperature, pH, dissolved gases and pollutants come into play in the state of proliferation of disease-causing organisms and their role in causing disease among fishes. Whether the limits of these factors are, however, crossed and wide fluctuations in these limits occur, different biochemical and physiological changes occur within the fish. Consequently, fish are either killed or their systems are damaged to the point where they become susceptible to other diseases.

The most common diseases of fishes associated with the physical and chemical factors include osmotic pressure, temperature, pH, dissolved air, algae, carbon di-oxide, acidosis and alkalosis, poisoning with heavy metals, nutritional ailments, and metabolic disturbances. We shall consider these factors one by one.

Osmotic Pressure

Osmotic pressure is the hydrostatic pressure which is applied to the solution in order to make the chemical potential of the water in the solution equal to that of pure free water at the same temperature, or to prevent the movement of osmotic water through a semipermeable membrane between the pure water and solution. When a living cell is in a quality water environment, a higher osmotic pressure is observed inside the cell. The semi-permeability of the cell membrane permits some substances to diffuse out of the cell, but the greatest effect is, in respect of diffusion of water into the cell, to equalize the osmotic pressure. This causes flooding of the cell until it brusts. A relatively small range of osmotic pressure in the cell structure of fishes maintain the level of water and salt in a way that would prevent cells from collapsing.

Diagnosis of adverse effects of osmotic pressure on fishes is difficult. Freshwater fishes subject to high salinity concentrations become lethargic, seek the shore areas of the pond or may attempt to leave the water. Estimation of total dissolved solids, salinity, etc. generally indicate ranges tolerated by the fish species involved.

Temperature Disturbances

Each species of fish has an inharent temperature range within which it can survive, grow and reproduce. For example, a suitable water temperature for pike, salmon and trout is 10°C; for perch, it is 12°C; for major carps, the range is 20-37°C; and for tilapia, the temperature varies between 10 and 32°C. However, natural water has a range of temperature similar to requirements of the fish species inhabiting there. Wide fluctuations in environmental temperatures beyond the normal survival range of fishes may cause thermal trauma. Disorder in fishes due to thermal trauma include effects on the cardio-vascular system, the nervous system,cessation or reduction of enzymatic activities, and permanent deterioration of fish health. However, marked temperature fluctuations lead to abnormalities such as malformation of operculum, fins, scales, and vertebral column.

Very high temperatures (particularly in tropical regions where temperature of pond water exceeds 35°C in summer) drastically affect the fish particularly when dissolved oxygen concentrations are not adequate. Under such conditions, fishes often refuge to consume feed and are prone to intestinal inflamation. On the other hand, a cold shock brings about disturbances in fish equilibrium, skin damage, edema, and haemolysis. Sudden drop in temperature (from 35 to 22°C) may lead to low temperature rigor and ultimately death occurs. In general, temperature fluctuations of up to 12°C do not have serious consequences. Of course, unless adverse situations arise, such fluctuations in temperature do not occur generally in fish culture ecosystems.

Stress Due to Oxygen Deficiency

Under favorable environmental conditions, fish utilize an appreciable amount of oxygen from water. Cyprinids, for example, can utilize about 60 per cent of the oxygen supply and only 20-30 per cent under unfavorable conditions. A sudden fall in the oxygen level acts as a stress and fish exhibits several symptoms such as increased intensity of respiration, increase in tendency to absorb oxygen from water, reduction in food consumption, decline in weight, poor state of health, increase in possibility of parasitic infection, and surfacing behavior.

Excess Carbon Di-oxide

Carbon di-oxide can act as a limiting factor during transport of fish and at the same time also limit fish production where intensive fish culture in closed systems with recycled freshwater is maintained. The process that will encourage the increase of carbon di-oxide level in water will contribute toward an increase in the asphyxiation threshold of the fish since carbon di-oxide decreases the affinity of blood for oxygen and reduces its oxygen-carrying capacity.

Gas Bubble Disease

Any process that will encourage the generation of gases in ponds at high levels (such as oxygen, nitrogen, sulfur di-oxide and acetylene) will contribute to the appearance of bubble disease in fishes. Increase in oxygen content of water and the use of high nitrogen fertilizers lead to the formation of gas bubbles in water. Conditions that permit the accumulation of bubbles encourage apathetic state of fish and gas emboly (blockage of the blood vessels by gas bubbles).

In the pond bottom where anaerobic conditions prevail, the use of nitrogen fertilizers undergo partial decomposition and small bubbles of nitrogen, sulfur di-oxide and acetylene are generated. Dissolved nitrogen gas is absorbed by the fish during normal respiration and as soon as nitrogen reaches a critical blood level, it forms bubbles which is fatal to fish. In general, these conditions are favorable for the cause of bubble disease. When the bubbles are comparatively large in size that are developed in the intestine, upset of equilibrium, surfacing behavior and disturbance in normal swimming movements occur.

Gas bubbles are made up principally of oxygen and nitrogen. They accumulate in the blood, under the skin and in the eye region and as a corollary, this accumulation generally gives rise to the so-called *gas bubble disease*. The bubbles in the blood may cause death.

It should be kept in mind that under pressure, the solubility of all gases in water is increased. Sudden release of pressure on the air-water mixture permits gases to slowly release back to the atmosphere until the gas solubility and pressure are equalized. When fishes live in water under such condition, may develop gas bubble disease.

A sudden reduction of the gas pressure develops as a result of supersaturation of both oxygen and nitrogen gas in water following decrease in the pressure of gases. Intense solar radiation results in the generation of oxygen by aquatic plants and cause a supersaturation of oxygen in the water. Since the blood of fish is at higher pressure, supersaturation results when there is a rapid drop in the gas pressure in water. When the gas tension of the blood is higher than in the water, gases are released in the form of bubbles. The composition of the gas bubbles largely resembles to that of the atmospheric air, however,

1. **Control of Gas Bubble Disease:** Avoidance of (i) sudden differences in gas tension in water, (ii) restraining from the use of nitrogen fertilizers in high amounts, and (iii) the addition of new freshwater to the pond are possibly the most important methods recommended for controlling gas bubble disease.

Acidosis and Alkalosis

Similar to oscillations and variations in the oxygen level which drastically affect fish biomass in farmed ponds, slight or wide fluctuations of pH in water are likewise of prime importance. Therefore, periodical recording of pH data of pond water must be made as a control measure.

The majority of fish live in pH range that varies between 5.5 and 8.0, others between 6 and 7, 7 and 8 or only 7.0 and 7.5. However, strongly acid water (pH below 5.5) generates the same results as strongly alkaline water (pH above 9), although the lethal limit varies for each species.

Two main terms such as acidosis and alkalosis are generally used while acidity and alkalinity of water are considered.

1. **Acidosis:** More specifically, it can be said that acidosis is an actual or relative increase in the acidity of pond soil and water to an accumulation of acids or an excessive loss of bicarbonate. The hydrogen ion concentration of the pond water significantly increased and thus lowering the pH value. When pond waters become too acid, it is termed as *acidosis*. Acidosis results in ponds where there is (i) deficient in calcium salts, (ii) liberation of humic acid from the soil, and (iii) drainage of acid mine. Under such conditions, fishes come to exhibit very rapid increase in respiratory movements, grasping and a tendency to jump out of the water. Death occurs rapidly or slowly and often hidden between the aquatic plants.

(i) *Symptoms* : Darkening of the edges of gills, indolent swimming, dark-grayish deposits, milky turbidity of the skin and mucus secretion from the gills of fish are some of the main characteristic features due to acidosis in pond waters. In the event of mortalities due to acidosis, the pH of water must be rectified.

(ii) *Control*: In fish culture ecosystems where acidification occurs, the use of lime in the form of powdered calcium carbonate should be recommended so as to keep the pH value within normal limits.

2. **Alkalosis:** The term alkolosis refers to an actual or relative increase in the alkalinity of pond waters due to an accumulation of alkalies or reduction of acids. Depending on the fish species, alkalosis occurs at pH 8-9. In soft water, growth of macrophytes and abundant solar radiation generate oxygen which ultimately dissolves in water and consequently, insoluble calcium carbonate is formed from calcium bicarbonate. This process is termed as *biogenic decalcification*. From the remains of the calcium carbonate, calcium oxide is generated by removal of carbon di-oxide, and the oxide corrodes the gills and fins, resulting in a typical fraying of the latter. The lethal pH values of some fish are noted below.

Fish	Lethal pH value
Trout and Perch	9.2
Catfish and Tilapia	10.0
Pike	10.7
Carp and Tench	10.8
Snakeheads	9.8

To prevent alkalosis, the water should be buffered by means of a suitable calcium-containing substance. This substance helps preserve original hydrogen ion concentration of the pond water upon adding a base. Excessive growth of aquatic plants should be avoided. The occurrence of ammonium compounds in alkaline waters may, however, results in the generation of ammonia in lethal amounts.

Algae

Algal colonies form hair-like matted filaments that are suspended in water. An over-abundance of algae definitely results in clogging of gills, deteriorate the process of respiration and eventually lead to distress and even death. Algal toxicosis and disturbance in the oxygen content resulting from metabolic products developed by the formation of massive algal blooms which are more important and apparently common in ponds.

Algal blooms may appear in ponds due to either application of chemical fertilizers or organic matter. Toxins released by some species of blue-green algae (such as *Anabaena*, *Microcystis* etc.) destroy other phyto-plankton and cause surfacing of fish stocks. Persistance of the bloom will cause toxicosis for the fish stock, exhibiting symptoms like convulsions leading to death.

Heavy Metal Poisoning

The domestic and industrial sludges and pesticides are the major source of some potentially toxic heavy metals such as arsenic, zinc, mercury, iron, and cadmium. Sludges and pesticides carry significant quantities of inorganic as well as organic chemicals that can have harmful effects on fish. Mercury, arsenic, iron and zinc are the heavy metals most often involved. Heavy metals have adverse effects on fish culture ecosystems, but the effect on fish is even more serious and are considered as fish health hazard. Zinc impairs the health status, growth and survival rates of fish. Likewise, high levels of iron or mercury salts are deadly poisonous to fish. A thick layer of brown ferric oxide is formed on the gills that results in suppression of the absorption of oxygen, irritation, and reddening of the epithelium. In general, the effects of heavy metal poisoning in fish are characterized by restlessness, decrease or increase in frequency of the opercular beatings, secretion of slime, reduced sensitivity to stimulation, exhaustion, and death due to asphyxiation. The toxicity of the salts of heavy metals to fish is highly correlated with environmental factors especially pH, temperature and hardness of water, however.

Some management practices considered as essential for the maintenance of fish culture ecosystem are relevant that has great practical significance in relation to fish production. Contamination of fish culture ecosystems by heavy metals and pesticides, for example, should be avoided as far as possible. Likewise, the use of chemicals must be reduced drastically. This is a remedy already suggested but increase in tendency of agricultural production may always not be possible to adopt such remedy. The effects of some heavy metal poisoning in fishes can be overcome by adding lime. However, these potential solutions must receive attention.

Poisonous Substances

Fish culture ecosystems are frequently contaminated with a number of poisonous substances that, to a greater or less degree, are toxic to fish. The quantities of most of the substances in which contaminants are used have increased in recent years, enhancing the opportunity for contamination. They are present in waters in increasing amounts and are daily ingested by fish resulting in intoxication with poisonous compounds which are the most potent blockers of cellular oxygenation. They inhibit respiration by blocking oxidative phosphorylation at the cellular level.

The degree of effectiveness of poisonous substances depends largely on their concentrations, duration of action, possibility of elimination by the fish, and condition of the fish. However, the poisonous action of fish can be reduced if they are transferred from polluted ecosystems to unpolluted ones and their ability to recover the normal condition depends on the concentration of the poison and its duration of action.

Physiologically, several poisonous substances have many attributes. Poisons infiltrate either through the gills, skin or the alimentary canal. Poisons change the position of fish (leaning sideward, lying laterally, upside down position), abnormal movements (remaining still, whirling, increased/decreased respiratory frequency), necrosis, cramps, haemolysis, changes in body structure (discoloration of the skin, fins and gills), secretion of copious amounts of mucus, and increased susceptibility to other disease-causing agents. Other harmful effects on fish associated with the poisonous substances are noted below.

	Poisonous substance	Disorder
1.	Insecticides	Liver cells and blood vessels are damaged
2.	Ammonia (more than 1 mg/L)	Haemorrhage in internal organs and extenal bleeding
3.	Chlorine (less than 4 mg/ L)	Discoloration and damage of gills
4.	Phenols, synthetic detergents and alkalies	Discoloration of gills, decrease in red blood corpuscles and haemorrhage on skin, head as well as the base of fins
5.	Arsenic salts and acids	Blindness
6.	Hydrogen sulfide (more than 6 mg/L)	Reddish-violet color of gill filaments
7.	Iron and nitrate	Increase in red blood cells

Whatever their symptoms observed during the course of fish poisoning, poisons can reach the pond soil, where they become part of the life cycle of soil → water → plankton → fish → human. Unfortunately, once the poisons become part of this cycle, they may accumulate in fish flesh to toxic levels. This situation has already resulted in restrictions on the use for human food of many fish. Also, it has become necessary to curtail the release of poisonous substances.

1. **Some Common Poisonous Substances:** The most common poisonous substances that are closely associated with the fish disturbance include hydrocyanic acid, phenols, nitrogen compounds, free chlorine, mercury compounds, arsenic, organo-

chlorine and organo-phosphorus compounds. These are the most violent toxins found in many fresh-water fish culture ecosystems.

(i)   *Cyanide:*. Cyanide is highly toxic to the blood and acts even more violently when in association with ammonia.

(ii) *Chlorine:* Free chlorine at the rates varying between 0.8 and 1.5 mg per litre attacks the gills and destroy the gill epithelium and its toxic action affects the body to a large extent. Chlorine is a toxic substance of slow action. Chlorinated water may be neutralized mainly with sodium thiosulfate @ 1 g/10 litre of water.

(iii) *Phenol:* Phenol is one of the most violent neurotoxins found in the residual waters of the coal industry. Since phenols are soluble in lipids, they are deposited mainly in the fatty tissues of the fish. Therefore, fatty fish such as eel, carp, salmon, etc. are more vulnerable than others to such deposition. Though absorption of phenols is very rapid, several weeks are required before they can be eliminated from the body even when fishes are kept in clean freshwater.

(iv) *Nitrogen Compounds:* Brown blood disease which occurs in fish culture systems where there is possibility of nitrite accumulation with high nitrite concentrations in combination with low concentrations of other monovalent anions. The disease is observed in fish ponds with nitrite levels as low as 0.8 mg/l.

The disease has been reported especially in intensively cultured salmonoids and channel catfish ponds when haemoglobin is oxidized by nitrite to form methemoglobin and hence the brown blood disease is referred to as *methemoglobinemia*.

In general, nitrites may accumulate under three conditions. First, excretion by fish of ammonia that is converted to nitrite in excess of the rate of conversion of nitrite to nitrate. At lower water temperatures, the conversion of ammonia to nitrite by bacteria of the genus *Nitrosomonas* is facilitated over the conversion of nitrite to nitrate by *Nitrobactor*. This possible temperature effect may contribute to the higher incidence of this disease during the cooler months of the year. Second, aerobic decomposition of excess organic substances in a pond and third, anaerobic decomposition of organic matter in the pond mud.

Water quality is closely tied to the severity of this disease. Disease-related; mortality increases with increasing nitrite concentration, decreasing chloride concentration, pH, and alkalinity. Since the respiratory efficiency of fish reduced by this disease, the severity of losses increases when dissolved oxygen levels in water momentously decreased.

Clinical symptoms include lethargy, crowding near water inlets or aeration equipments, a characteristic chocolate brown color of the blood and a sharp rise in mortality.

Metabolites of nitrogen compounds are highly toxic to fish and limits of the toxicity are extremely variable among different species of fish. The degree of toxicity also depends upon environmental factors especially temperature and pH of water. Ammonia is a neurotoxin and gill hyperplasia has been noted in many freshwater fishes exposed to as little as 0.009 mg/ L non-dissociated ammonia for several weeks.

Symptoms of ammonia toxicity include impairment of growth, reduction in resistance to infection and gill haemorrhage. Cases of bacterial infection to gills, skin and fins are attributed to ammonia.

* *Control of Brown Blood Disease*: Brown blood disease can be treated through application of either calcium chloride or sodium chloride to the culture system. Calcium chloride has been demonstrated to be more effective than sodium chloride in treating nitrite toxicity in fish. However, the monovalent anion (Cl⁻) competes with nitrite for uptake sites on the gills and elsewhere. The presence of chlorine in the culture system drastically reduces the amount of nitrite entering the fish which ultimately reduces the conversion of haemoglobin to methemoglobin. Treatment levels are expressed as, a ratio of chloride to nitrite. A chloride to nitrite ratio of 16 : 1, for example, has been shown to completely suppress methemoglobin formation in channel catfish ponds under conditions of low hardness and alkalinity (less than 100 mg/l), rainbow trout can be protected against brown blood disease by the intra-peritoneal injection of methylene blue at the rate of 10 mg/ kg of fish or intravenous injection of ascorbic acid at the rate of 5 mg/day. The disease is generally not restricted geographically; it can occur wherever fish are intensively cultured.

Nutritional Ailments

A diet which is not in balanced condition or there is deficient/excess in any essential components such as protein, carbohydrate and fat may cause disorder of fish health. Vitamins and mineral requirements are also inevitable for better production. Nutritional disease of many species of fish is principally associated with vitamins and mineral deficiencies and the signs of which are shown in Chapter 9 of Volume 1. Excessive fat in the diet and over-feeding, in general, may lead to degeneration and disease of the liver as well as increased susceptibility to infectious diseases. Use of low quality diet may also results in problems such as emaciation, chocking, and constipation.

Metabolic Disturbances

One of the greatest problems in fish culture strategy is the availability of physiologically balanced food. Fish survival, growth, and reproduction principally depends to a large scale on a suitable diet.

The term metabolism refers to understand the absorption of food, its decomposition and rebuilding with the body, and the elimination of certain end-products. During metabolism, energy is liberated which is utilized for locomotion. However, the following varieties of metabolic disorders are commonly encountered in fish pathology.

1. Bone degeneration
2. Gastritis and enteritis
3. Vitamin deficiency
4. Enzymatic disorders

Degeneration of fish bones is caused by abnormalities in enzymatic activity. The bones become softened due to loss of calcium. Causes of this condition include hormonal disturbances and bacterial infections.

Gastritis and enteritis are very common in tropical fish. Vitamin deficiencies and lack of one of the three basic foodstuffs (such as proteins, fats, and carbohydrates) cause inflamation of the alimentary canal. A serious disease is found following the feeding of tubifex or chironomid larvae grown on mud waters which are severely contaminated with pollutants. The disease is characterized by swelling and reddish inflamation of liver and other internal organs and scale protrusion. Feeding of live food with care is, perhaps, the best way of avoiding enteric and gastric disorders.

The most common metabolic disorders of alimentary origin is fatty infiltration of the intestine, liver, and ovary. Older fish are more succumbed to fatty infiltration than the young ones. Fatty infiltration definitely instigates the death of the fish. A more balanced feed must be provided to avoid such metabolic disorders.

17.16 Effects of Parasites on Human Health

One of the most important aquaculture products is the food fish and consequently, farming systems need to guard against any infections which may affect consumers. Though diseased fish have the potential to adversely affect their market price, human infections from farmed fish are rare and harvested fish from natural resources are much more likely to be a major source of human infection. With fish as food, however, the only parasites that are dangerous to man are the tapeworms of the cestode genus *Diphyllobothrium* and the liver flukes *Haplorchis, Opistorchis, Metorchis* and related genera. Whilst the cestode genus *Diphyllobothrium* is present in continental Europe, several species of digenean trematodes are found as metacercariae and freshwater fish in tropical and sub-tropical regions. Farmed pond fish such as carps and tilapias are particularly vulnerable when grown in semi-intensive systems. Where there is little attempt to curb molluscan intermediate and piscivorous final hosts, parasites generally remain undetected owing to their very small size. Whatever the parasitic species, all potential for human infection is prevented by cooking above 65 °C and deep freezing at –20°C for more than 5 days.

Several species of bacteria cause serious skin infection in man. They enter through wounds and induce inflamations. Consequently, fishermen frequently suffer from this skin trouble. Other infectious diseases of fish are of minor importance to human beings because they cannot be transmitted. Of course, some may affect human health in a general way by rendering fish unpalatable by enhancing putrefaction and other breakdown processes.

17.17 Conclusion

Diseases of freshwater fish play a critical role in fisheries and aquaculture sectors across the world as diseased fish drastically affect the market price and human health. Contamination of water with toxic substances, pathogenic micro-organisms parasites, and disease-related physico-chemical factors have a profound impact on fish stocks. Helminths, bacteria, and protozoa are potentially serious pathogens for all freshwater fishes.Therefore, an all round effort has to be made by fish farmers to maintain a healthy aquatic ecosystem for ensuring adequate health conditions for the farmed fish. And at the same time, thousands of rupees must be spent to eradicate diseases from farmed ponds and to prevent fish ponds from potential introduction of pathogens considerably.

Farmers have found that in almost every cases where therapeutic measures have been used, there has been a degree of success ranging from complete alleviation to some improvements of symptoms. Farmers have also had the opportunity to evaluate the use of some common and widely recognized chemicals especially potassium permanganate, sodium chloride, copper sulfate, malachite green,formalin solution, and ammonia. These measures are of considerable value in the treatment of many edible freshwater fish species.

Prevention and control aspects have attracted the attention to fish culturists. Efforts will have to be made to integrate fish culture with disease-related hassles in the areas concerned. If fish culturists are serious about raising their farm productivity in general and production in particular, due attention must be given that would pave the way for producing robust fish.

Prevention of fish diseases requires patience and considerable skill on the part of the farmers. When translocation of fishes carries a risk of disease introduction, the strategy stances to detect and exclude the pathogens in question must be developed and applied before such movement of fishes should be permitted. When environment changes are contemplated and its effects on fish biomass have to be predicted or assessed, the biology and ecology of pathogens and parasites present in the ecosystem must be carefully considered in this context.

References

Almacher, E. 1997. Text Book of Fish Diseases. Narendra Publishing House, Delhi, India.

Biswas, K. P. 2000. Prevention and Control of Fish and Prawn Diseases. Narendra Publishing House, Delhi, India.

Chinabut, S. and Roberts, R. J. 1999. Pathology and histopathology of Epizootic Ulcerative Syndrome (EUS), AAHRI, Bangkok, 33 PP.

FAO. 1986. Report of the expert consultation on Ulcerative fish disease in the Asia-Pacific region. FAO regional office for Asia and Pacific, Bangkok, Thailand.

Gilda, L. 1991. Guide in submitting specimens for disease diagnosis Agriculture Deptt. Southeast Asian Fisheries Div. Centre, Philippines.

Lilley, J. H., Callinan, R. B. et. al. 1999. Epizootic Ulcerative Syndrome. Technical Handbook, Aquatic Animal Health Research Institute (AAHRI), Bangkok, Thailand, 88 PP.

Questions

1. What is disease? Why the study of fish disease is so important in fish culture industry?

2. What are the reasons for occurrence of fish disease?

3. Explain how aquatic ecosystem is closely related to fish disease.

4. What are the techniques that are essential for investigation of fish disease? Explain.

5. What is diagnosis? Why it is so necessary in fish culture?

6. Mention different types of fish mycoses. What are the symptoms of fish mycoses? How fungal infections in fish are controlled?

7. Name some common protozoa which cause fish disease. What are the symptoms of protozoan diseases in fish? How protozoan infections are controlled ?

8. What are the common symptoms of bacterial diseases in fish? Name some common bacterial diseases found in fish. How bacterial infections are controlled?

9. What is syndrome ? What do you mean by the term EUS ? Why EUS is fatal to fish.

10. Mention some helminths that cause diseases of fish. What are the difference between monogenetic and digenetic trematodes?

11. What are the main symptoms of helminth infection? How helminth infections are controlled?

12. Name some common crustacean parasites found in fish. What are the symptoms of crustacean infection? How these parasites are controlled ?

13. What do you mean by the environmental stress-induced fish disease? Discuss some common diseases of fish associated with the physical and chemical factors of fish culture ecosystems.

14. Discuss the role of microbial diseases in fish culture industry.

15. Suggest some definite steps which should be taken to control fish diseases.

16. Write notes on the following:

(a) Fish lice, (b) Pathogenic and non-pathogenic diseases, (c) Fish mycoses, (d) Branchiomycoses, (e) Saprolegniasis, (f) Ichthyopthiriasis, (g) Myxoboliasis, (h) Whirling disease, (I) Furunculosis, (j) Red pest disease, (k) Abdominal dropsy, (L) Tail and gill rot disease, (m) Peduncle disease, (n) EUS, (0) Dactylogyrus and Gyrodactylus, (p) Anchor parasites, (q) Oxygen deficiency disease, (r) Gas bubble disease, (s) Acidosis and alkalosis, (t) Biogenic decalcification.

18

Conservation of Commercially
Important Freshwater Fish Biodiversity

In this chapter we shall provide a comprehensive picture of fish biodiversity and its conservation strategy. Fish biodiversity conservation is undoubtedly one of the most important aspects of aquaculture production. Conservation of threatened and endangered fish and fish biodiversity together determine the production potential in many countries of the world. At the same time, conservation of fish genetic diversity becomes a vital aspect on account of certain criteria besides its crucial role in producing more fish and profit to farmers. It is essential that we get a good picture of the general considerations, scope and value, importance, significance, reasons for the loss of biodiversity and factors affecting fish biodiversity before we proceed with a discussion on the various issues relating to fish biodiversity conservation.

18.1 General Considerations

Increased activity of aquaculture/fishery in general and fish culture in particular and human interference have affected different aquatic resources on a variety of ways. One of these effects has led to degradation of aquatic habitats including commercial fishing waters. Most of the freshwater areas are arid with problems of salinity, pollution, acidity, metal and pesticide toxicity. All these factors are definitely responsible for drastic reduction in the diversity of aquatic life including fish. Restoration of these degraded waters along with fish diversity is, therefore, a major concern both in developed and developing countries of the world. This is also necessary due to rapid development of aquaculture industries leading to decreasing areas of available water for sustained yield of fish.

Though available conventional methods of conservation of fish biodiversity are relatively insufficient, use of biotechnology, genetics, and other latest techniques that follows have become major forces for the existence of indigenous fish which are considered for culture. The products of biotechnology including prevention and care of fish diseases, devices for ecosystem protection, conservation of fish fauna, and fish germplasm resources are playing a very important role in quality fish seed production, stocking of broodfish, trade, profit, and the quality of farmers' life throughout the world.

Conservation strategy of aquatic resources involves the use of kaliedoscope variety of fish in particular for mass consumption and export. Therefore, extinction of commercially-important indigenous species and destruction of ecosystems have become a major concern to fishery scientists. One of the major efforts has been to conduct a survey and conserve biodiversity of the nation so as to protect world fish from extinction. Fish sanctuaries have been established and other efforts are being made for conservation of germplasm at the global arena. Biodiversity studies on fish therefore include (i) An examination of the full array of fish over the face of this planet according to a system and (2) A study on the methods by which diversity can be maintained and used for the benefit of mankind. A coherent discussion on biodiversity in a book on fish culture is pertinent because biodiversity is being utilized to purvey genes from wild species for study of biodiversity. A discussion on biodiversity has become important also because developed countries have been utilized biodiversity resources available in the developing world without paying any compensation. During nineties, several biodiversity conservation strategies were held for discussions on steps required to be taken by many countries to preserve the biodiversity at the global level.

It is fortunate that tropical countries are blessed with vast and varied fish germplasm resources widely distributed in different aquatic ecosystems. A typical example on the aspect concerned will realize this importance. The freshwater resources of India include about 30,000 kilometres length of rivers, 1,13,000 kilometres of canals, 1.8 million hectare of reservoirs, 1.5 hectare of tanks and ponds and 0.8 million hectare water in the form of derelict, stagnant and swamp. About 28,500 finfish species exist throughout the world that represent more than half of the vertebrate biodiversity. However, about 2,200 species of finfish have been recorded from different ecosystems of the country. Of then, about 500 species (22 per cent of the total finfish) are important from commercial point of view. These fishes have acquired a wide variety of forms and habitats as reflected in their adaptation to live in varying biotypes that ranged from cold mountain ecosystems to the deepest depth of the ocean. The approximate ecosystem-wise distribution of fish germplasm resources of the country is noted below.

Ecosystem	Number	Per cent of Total
Coldwater	73	3.32
Warmwater	544	24.73
Coastal brackishwater	143	6.50
Marine	11,440	65.45

Source: Das and Pandey (1999)

Since most of the countries profoundly realize the need for preserving fish biodiversity along with other aquatic biotic resources and the strategy concerned is being looked upon as pre-requisite for the well-being, it is inevitable to have a general idea on different causative factors which are responsible for the extinction/loss of fish populations and to suggest/adopt possible strategies for conservation and maintenance of fish biodiversity for sustainable fish culture.

18.2 Fish Biodiversity as a Resource of Mankind

From time immemorial, man has been dependent on natural resources for his livelihood. In course of time, man has realized that these natural gifts available around him have some usefulness with regard to nutrition in particular. Although fish and other aquatic organisms have existed over the earth's surface much before the appearance of man, their utilization in full potential was not possible due to unavailability of adequate technology. With time, man could learn to capture and culture fish by adopting different methods and acquired knowledge to use them for nutrition.

Any material which can be converted in a way that it becomes more precious and profitable, can be regarded as a *resource*. In other words, it is possible to achieve precious items from any resource. Thus different forms of water, land, forest, as well as human beings are resources. Fish being an important aquatic item, is termed as a resource provided that an adequate state-of-the-art technology is available and adopted to convert them into more useful items to fulfil the protein requirements of mankind.

Value of Diversity

The value of fish biodiversity includes culture of edible fish and fishing from different forms of water and other aquatic species for human need. But in the light of extensive use of fish for nutrition, the absolute value of fish biodiversity is far-reaching. There are three distinct characteristic values of biodiversity that follows.

1. **International Export Value:** Several species of high-priced edible fish if well preserved and cultured, frozen and processed products may be exported for foreign currency earnings. Due to continuous effort towards post-harvesting technology of fish and fish products, the demand for Indian aqua-products in Western countries has substantially increased and there is immence scope in this respect.

2. **Regional Market Value:** The human society is entirely dependent on different agricultural products and their market value is high in tropical countries than in temperate zone counterpart. In the regional economics, the accessed products are considered as regional market value. The price of these products increased several folds before it is being transferred to consumers.

3. **Culture Value:** Conservation of endangered table fish species of high commercial value following the adoption of technology is one of the most important characteristic values that should be kept in mind. This would not only provide adequate fish for both rural and urban people but to prevent them from malnutrition to a large extent.

18.3 Definition of Biodiversity

Simply defined, it is the association of a kaliedoscope variety of species that have occurred in different environments. A general definition of biodiversity has been furnished in the conference on Biological Diversity which was held in Reo-degenerio in 1992. The definition states that it is the variability among living organisms from all resources including terrestrial, marine, and other aquatic ecosystems and the ecological complexes of which they are part; this includes diversity within species, between species, and of ecosystems. However, biodiversity can be defined in different ways. Two simple definitions which convey the essence of the term are noted below.

1. It is a term encompassing the variety of organisms at all levels, from genetic variants belonging to the same species, to species diversity and including the variety of ecosystems. *– Mackenzie (1999)*

2. It is the variety of types of organisms, habitats, and ecosysterms on earth or in a particular place. *– Rickleft and Miller (2000)*

18.4 Why Fish Biodiversity Should be Preserved?

Though the need to conserve fish biodiversity and to avoid ecosystem degradation cannot be denied by many people, others have questioned about the value of preservation. The vindication for financial investment in fish species conservation has been challenged. Although the policy decisions are the matter of economics, model economics has failed to allot more than ostensible value to ecological resources.

Vindication for fish conservation centres on the human benefits that will accrue from maintaining a full range of biodiversity and avoiding ecosystem degradation. In fish culture strategy front, these can be summarized as utilitarian, scientific, and aesthetic reasons.

Utilitarian Reasons

In terms of food value, fish are superior to other forms of meat. They are tasty, tender and easily digestible. Different kinds of fish vary in their palatability and individual taste. Fish and other aquatic animals of commercial importance are rich in proteins, fats, essential minerals, and vitamins. In fact, almost three-quarters of the total fish catch are used for food.

Besides providing human food, fish and their by-products are equally useful in many other ways. Fish wastes from processing plants are made into fish meals, glues, oils and fertilizers. Many industrial processes depend very much on fish catch and the construction of crafts, gears, tin cans and making of salt and ice, for examples, should be remembered.

Scientific Reasons

Human society depends on the management of a functional aquatic ecosystem. Management of aquatic ecosystems is inevitable to the avoidance of pollution as discussed in Chapters 10, 11, and 12 of Volume 1. Loss of balanced aquatic ecosystems can have profound effects on species survival and may result in drastic changes in habitat conditions.

Aesthetic Reasons

Human society receives pleasure from different aquatic environments and the presence of various species of fish. Our valuation of nature passes through the pores of design, literature, and music and influences our recreational prosecutions. Some people remember that contact with ecosystems is necessary for human nutrition.

Other values of fish biodiversity include economic and commercial uses, education, and philosophical.

18.5 Types of Biodiversity

Though biodiversity may be expressed in different ways, environmentalists have classified all types of biodiversity into three classes that follows.

Diversity at the Level of Ecosystem

On the basis of ecosystem level, different types of ecosystem such as terrestrial, freshwater and marine can be recognized that have their own characteristic features. Freshwater ecosystems have been further divided into ponds, lakes, rivers, bogs, wetlands and the like. All these ecosystems have their own specific characters in which a distinct biodiversity gradually developed. Though different species form a community in different ecosystems, the community of an ecosystem differs collectively from that of the community of other ecosystems. For example, formation of a community in a pond ecosystem and its characteristic feature is entirely different from that of a lake or river.

Diversity at Species Level

Each community is composed of a variety of populations. Again, a population is formed by a particular species. Therefore, a number of species aggregate in a community that constitutes biodiversity. But one thing should be kept in mind that the number of species in a particular population must be taken into consideration. If we consider a lake ecosystem, it would be clear that it indicates the variety of fish species in that ecosystem. The picture of species diversity may be distinct from that of the same type of ecosystem in distant habitats. A strong interspecific interaction gradually developed and as a result a balanced ecosystem is established. For this reason, the presence of an unit member for each particular species in a species diversity is highly significant.

Diversity at Genetic Level*

Biodiversity is, in fact, a genetic diversity and production of hybrids which are much improved than either parent is a major innovation to improve fish production. As a generalization, a variety of fish species exist in a species population and they are classified as sub-species or race. These divisions distinctly provide a biodiversity signal to the species concerned. In reality, fish biodiversity determines the basis of genetic structure.

---

* For further detail, see Reisenbichler *et. al.* (1992).

18.6 Scientific Planning is Required for Fish Biodiversity Conservation

Biodiversity history has an abundance of instances of the loss of biodiversity on account of the lack of adequate planning to conserve biodiversity. It also abounds in success stories emanating from good planning on biodiversity. This is quite natural since it is the planning of biodiversity which provides the framework for all decisions on conservation of fish biodiversity. Decisions on various aspects include (1) fish germplasm resources, (2) broodstock conservation and management, (3) financial assistance, and (4) growing awareness among fish culturists.

Stated in a simple term, the meaning of biodiversity is the complete, precise and unbeatable plan designed specifically for attaining the biodiversity objectives of the world. The biodiversity objectives indicate what the nation wants to achieve; the strategy of biodiversity conservation provides the design for achieving them.

It is the fish conservation strategy that decides the success at the production level which in turn decides the total success of the nation. The linkage between conservation strategy and over-success is direct and vital. And in this linkage lies the significance of biodiversity conservation strategy.

Since realization of the biodiversity objectives is the purpose of conservation strategy, it is only logical that the conservation strategy takes its direction and cue from the biodiversity objectives of the nation.

18.7 Extinction of Fish Species

Fossil records indicate that extinction has always been part of the story of life on earth. Since the appearance of humans, new causes of extinction have arisen. Extinction, however, due to over-hunting of fish possibly began this incline, whereas pollution of water, widespread use of pesticides, loss of habitat, and agricultural intensification are among the principal causes of the rarity and extinction of fish species during the last 200 years or so. Most of the fish species are trembled in the balance. There is strong evidence that rates of extinction and variety of many valuable food fish have accelerated since industrial and agricultural revolutions.

A growing body of research has shown that extinction rates for fish are between 20 and 80 times greater than the natural background rate. Numerous species and sub-species of fish (about 40 taxa) that have become extinct in North America, for example, during the last 100 years. The lack of extinct species reported from other parts of the world may be a matter of maintenance of records than management of the faunas.

Many species of fish are rare and their long-term survival is also in doubt. Rare species are genetically destroyed which means that their gene pool drastically reduced and they have lost their ability to adapt to change in their ecosystem. Drastic environmental changes can bring about the extinction of a rare species of fish.

Types of Extinction

Extinction may be of two types such as (1) Natural extinction and (2) Man-made extinction.

1. **Natural Extinction:** The species which are vulnerable to extinction are characterized by small-sized breeding population associated with high coefficient of variation. Small-sized breeding population of fish promotes inbreeding. When the size of breeding population drops to a significant minimum, the likelihood of extinction is accelerated still further due to depression of inbreeding. The longevity of a species, however, generally varies from 0.5 to 1.0 million years to as much as 3 million years.

2. **Man-Made Extinction:** Destruction of fish habitats by human interference has resulted in drastic loss of fish biodiversity across the world. The rate of extinction is more severe particularly in tropical regions where population pressure is high for years. Some species of fish in India have been reduced to a significant degree (Table 18.1) and the rate of loss of fish species is increasing at an alarming rate. At the same time, over-exploitation of aquatic resources results in elimination of commercially important species of fish. It is, however, very difficult for fish population to attain salvation from human pressure. Increase in human population in future is supposed to be accompanied with several-folds increase in economic activity. If this activity is observed in future without remembering the harmful effects on fish populations, almost all edible fish biodiversity is destined to be completely wiped out from aquatic resources. This vividly demonstrates that steps must be taken to meet the exigencies of the difficult situation arises as a result of the loss of fish biodiversity and hence there is need to conserve them at any cost.

Table 18.1 : Fish Germplasm Resources of India and their Conservation Status

Type of Ecosystem	Total	Endangered	Vulnerable	Rare	Intermediate	Total
Coldwater	73	1	4	–	12	17
Warmwater	544	3	13	2	28	46
Brackishwater	143	–	2	–	4	6
Marine	1,140	–	2	–	8	10

Source : Pandey and Das (2002)

18.8 Threatened and Endangered Fish*

In many places of the world, numerous fish species are in danger of extinction and unfortunately, many important species have become extinct. The International Union for the Conservation of Nature (IUCN, 1988) have listed 77 species as endangered globally. According to the Endangered Species Committee of the American Fisheries society, there are over 70 endangered North American fish species and over 80 that are threatened with

---

* For further detail, see Baillie and Groombridge (1996).

endangerment. Similarly, out of 800 freshwater fish species which are distributed in European waters, about 100 species have been declared as threatened. On the basis of IUCN, a list has been prepared by the National Bureau of Fish Genetic Resources and has identified 21 vulnerable, 4 endangered, and 22 rare fish species from different waters of India. These data clearly indicate the records of threatened and endangered fish species from other countries of the world are highly alarming.

Many species and sub-species of fishes have severely restricted in a particular ecosystem. Involved reasons may be narrow temperature limits, high oxygen demand, necessary migration routes, limited types of spawning substrate, specific food for different life stages, desiccation of large bodies of water, and glacial relics. For whatever reason the ranges are limited, some species of fish in glacial periods have little amount of freedom in their fight to survive if slow natural processes are disturbed by human activities.

Manipulation of streams for industrial and domestic purposes or construction of dams, inter-connection of waterways, organic and chemical pollutants, siltation, channelization of rivers and the like can upset the normal ecology of fishes and at the same time can bring drastic change in habitats from bad to worse condition. These situations have been unkind to fishes.

If we are to prevent the rapid extinction of fish species due to human activities, we must (1) identify the species at risk, (2) investigate why they have become vulnerable, and (3) attempt to solve the problem.

The IUCN has published detailed lists of 762 freshwater fish species at risk of extinction in a series known as the *Red Data Book* and four categories of risk have been noted.

1. **Rare:** Fish with small population either restricted geographically which has localized habitats or with widely scattered individuals. These species are at risk of becoming more rare but they are not in a danger position of extinction.

2. **Vulnerable:** These fishes are under threat of or actually declining in number, or fishes which have been seriously depleted in the past and have yet to be recovered.

3. **Endangered:** Fish with low population numbers that are in danger of becoming extinct.

4. **Extinct:** Fish which cannot be found in areas they inhabited nor in other habitats.

   Section 2.28 reminds us that the scarcity of different species of fish both for commercial and academic importance in freshwater ecosystems as a result of changes in natural communities due specially to pollution, habitat annihilation, and 'global warming' are on the increase. Fluctuations in one system grow into larger pattern of changes affecting the structure of ecosystem communities. Large-scale conservations in the structure of freshwater ecosystems will have an impact on the food chain. Progressively shorter food chains will give rise to painstaking species with rapid reproductive rates such as algae in water. Harvestable fish stocks will decrease considerably and their gradual reduction will result in casualties among other edible aquatic species which feed on them. The disastrous pattern of alteration is being accelerated by continuous accumulation of persistent pesticides, herbicides, heavy metals and their residues and the overall changes in water/soil quality variables in the breeding grounds. Reproductive failure in fish is one of the palpable consequences of toxicant residues. Already many species of freshwater fish have been listed as endangered/vulnerable/rare/ extinct. It is the alarming signal to extinction. If this trend continues, it is expected that in the beginning of the 22nd century, more than 80 per cent of the total freshwater fish species will be effaced from extensive unmanageable water areas and the entire fish culture system will be disrupted. This is the most significant event in tropical and sub-tropical countries compared to that of the temperate zone counterparts where legal actions have been taken beforehand in many areas to ensure good results.

18.9 Warmwater and Coldwater Fish Biodiversity: Indian Context

India is furnished with a luxuriant fish genetic biodiversity and has occupied the 9th position among countries with freshwater 'mega-biodiversity'. This rank has depressed to a greater or less degree by anthropogenic interventions mostly due to urbanization, damming of rivers and abstraction of water causing ecosystem degradation. All these factors coupled with pollution and habitat destruction have aggravated fish genetic biodiversity.

Over the years, many species of finfish have been catalogued on the subject of biodiversity. In the last two decades in particular, the expansion in the catalogue on the subject concerned has been so tremendous that a vivid picture has come out. In the Indian context, however, a variety of warmwater and coldwater fish species acquire a species importance since most of the species are cultured in production ponds, tanks,and lakes. Some commercially important species of fish are given below.

Warmwater Fish Biodiversity

1. **Carp :** *Labeo rohita, L. calbasu, L.gonius, L.bata, L.kontius, L. fimbriatus, Catla catla, Cirrhinus mrigala, C. reba, C. cirrhosa, Cyprinus carpio* var *communis, C. carpio* var *nudus, C. carpio* var *specularis, Ctenopharyngodon idella, Hypophthalmichthys molitrix, Puntius javanicus,*

2. **Cichlid :** *Oreochromis mossambicus, O. niloticus.*

3. **Carnivorous Fish :** A number of fish species have been noted in section 6.3 of this volume.

Coldwater Fish Biodiversity

Coldwater ecosystems at higher altitudes have largely unexploited fish stocks since these water bodies are deficient in species and numbers. Many of such waters remain a welcome reserve of indigenous fish. Only a few introductions of exotics have been done with the purpose of increasing commercial catches. A number of coldwater fish diversity have, however, been listed in Chapter 10 of this Volume.

18.10 Freshwater Fish Resources in Several Asian Ecosystems

Since tropical countries are rich in fish fauna, a list of freshwater fish biodiversity has been prepared from a study made by the World Bank. Twenty countries have been brought under consideration and three countries such as India, China and Indonesia may be found to have the maximum diversity of fish species. Obviously, the value of freshwater ecosystems in these regions is immense owing to high biodiversity of fish species and further studies on this aspect would record much more species. The number of freshwater fish species in twenty countries are given below.

Country	Number of fish species	Country	Number of fish species
Afganistan	84	Korea	90
Bangladesh	260	Lao	262
Bhutan	40	Malaysia	600
Brunei	55	Nepal	130
Burma:	300	Pakistan	160
Cambodia	215	Singapore	45
China	1,010	Sri Lanka	90
India	748	Taiwan	95
Indonesia	1,300	Thailand	690
Japan	150	Vietnam	450

Source : World Bank (1996)

18.11 Factors Involved in Fish Biodiversity Loss

Though the decline of individual fish species is often related to more than one factors, the most important reasons involved in fish biodiversity loss have been discussed in the following sub-sections.

Habitat Destruction

Description on the declining incidence of fish populations must include habitat destruction which is characterized by the construction of roads and buildings coupled with over-grazing of cattle populations along the sloppy zones, construction of dams, deforestation, and excessive withdrawal of water from rivers/reservoirs for agricultural, domestic and industrial uses. Diversion of water greatly reduced the river's flow downstream where it fed a series of ponds/lakes. Consequently, these water bodies, formerly teemed with fish, are now largely desiccated. Damming of the Himalayan river in India, for example, has flooded many spawning grounds of schizothoracines and mahseers or made it impossible for them to reach spawning grounds.

Over-explotation

A state of chronic over-exploitation exists in the world's many freshwater ecosystems. Several freshwater fishing grounds show a steep decline in size of catch. In India, mahseer and schizothoracine fisheries have declined by more than 60 per cent since the 1970s. At the same time, some species of fish such as *Pangasius pangasius, Eutropichthys vacha, Osteobrama, Nandus,* etc. are gradually declining at a rapid rate and will never be recovered the pristine conditions. The featherback fishery in India has collapsed completely.

Excessive fishing produces a population of young fish because fish are caught as soon as they reach a catchable size. Young fish would grow rapidly if they are kept for prolonged period and would soon reach valuable size. This will give heavier landings of better quality fish. However, if fish are landed before spawning, the reproductive potential of the stock will be seriously diminished, a catastrophic decline in numbers will occur and extinction of local fish species may result.

Wanton Destruction

Poachings of broodfish and juveniles and use of destructive fishing devices (such as dynanite, ichthyotoxin and bleaching powder) in different river stretches in many tropical countries dramatically affect the population of several species of fish. Fish stocks are killed by fishermen during the breeding season using gears. The fish population migrates in shoals in shallow running streams for spawning. At that time fishes are in gravid condition and the females are heavily loaded with ripe eggs and hence fishes are vulnerable to all these destructive fishing methods.

Aquatic Pollution

Rivers, streams, lakes, and marshes are specialized habitats of fish and other aquatic animals. Their ecosystems are sensitive to changes induced by human activity in the water balance and in water chemistry. Industrial society not only makes radical physical changes in water flow by construction and engineering works, but also pollutes and contaminates waters with a large variety of wastes.

The sources of water pollutants are many and varied. Some industrial plants dispose of toxic metals and organic compounds by discharging into lakes, rivers and streams. Many communities still discharge untreated or partly-treated sewage/wastes into surface waters. In urban and suberban areas, pollutant matter entering streams and lakes includes salts and lawn conditioners (lime and fertilizers) which can also contaminate ground waters. In agricultural regions, important sources of pollutants are pesticides, fertilizers, and livestock wastes.

The most chemical pollutants are chloride, sulfate, sodium, nitrate, phosphate, and calcium ions. These pollutants are derived from various sources. For example, important sources of nitrateions are fertilizers and sewage effluents. Excessive concentrations of nitrate in freshwater are highly toxic and, at the same time, their removal is difficult and expensive. Phosphate ions are contributed in part by fertilizers and by detergents in sewage effluents. Chloride and sodium ions are contributed by fallout from polluted air. However, pollutants are introduced into streams and rivers in quantities that are locally damaging or lethal to fish communities. All these phenomena cause havoc to the fish genetic threshold which ultimately lead to permanent damage to genetic resources in addition to their direct toxic effects.

Acid Rain

Scientists have called attention to marked increases in the acidity of precipitation over recent decades, particularly in relation to forest decline. In the absence of contaminating atmosphere gases, rain water has a pH of about 5.6 owing to the presence of carbonic acid formed from carbon di-oxide in the atmosphere. In extreme cases of dense fog, the pH may drop to nearly 3.0, which is a serious potential hazard both to society and aquatic ecosystem.

Acid precipitation, popularly called *acid rain,* is apparently due to the oxidation of nitrogen- and sulfur-containing gases that dissolve in the water vapor of the atmosphere to form nitric and sulfuric acids. Oxides of nitrogen and sulfur move to the atmosphere, converted to inorganic acids, and return to the earth in rain and snow. Such cycling is responsible for lowering the pH of precipitation in many parts of the world.

Acid rain is assumed to be responsible for marked increase in acidity in many lakes, reservoirs and rivers across the world. Most of the fish species will not tolerate pH level below about 4. 5. As a consequence, the increased acidity of aquatic ecosystems is thought to have essentially eliminated almost all species of fish from acidified aquatic ecosystems.

In general, there is a tendency for all lakes to become more acidic with time as a result of natural aging processes, but studies have shown that observed rates of change in pH value since the middle of the nineteenth century have exceeded the expected natural rates in many lakes.

1. **Effects of Acid Rain on Aquatic Ecosystem:** Effects of acid rain are more pronounced on the acidity of water than on soil acidity. Soils are sufficiently buffered to accomodate acid rain with little or no increase in soil acidity on an annual basis. But continual inputs.of acid rain at pH varying between 3.0 and 4.5 would have significant effects on the pH of soils, especially those that are weakly buffered. However, most of the investigations on the impact of acid rain on aquatic communities have involved fish populations. Many fish species have completely disappeared with lake acidification. In addition to this effect, fish in acid lakes/ ponds also succumb to toxic concentrations of metals that are leached from the surrounding rocks by the acids. The problem of acid rain not only stems from the rain-fall itself but flow-off also affects surface and ground waters. As a consequence, acidification of a wide range of aquatic ecosystem can trigger off a range of impacts on the chemistry, ecology, and productivity. This condition has definitely serious consequences on the existence of fish biodiversity.

There are two obvious ways to alleviate the effects of acid rain on aquatic ecosystems. First, the emission of sulfur and nitrogen oxides must be reduced drastically. This is the redress already recommended, but economic and political decisions may delay its implementation. Second, the effects of the acid rain on water pH can be surmounted by applying lime. This potential solution may be less costly than the first but difficult to achieve, since it will require commitment of resources by farmers. Unfortunately, many adverse effects of acid rain on steep mountainous regions have been noted where waters may be shallow and where application of lime is difficult if not impossible. Therefore, both potential solutions must receive attention.

Natural Calamities

Natural calamities in the form of floods, volcanic eruptions, cyclones, landslides, and earthquakes have serious effects on fish biodiversity. These calamities are becoming increasingly significant both in terms of impact and number of .events. Evidences indicate that the frequency of natural disasters has increased during the last two centuries. Though these events cannot be prevented, their impact can be reduced if effective measures are adopted to reduce their severity and frequency. These calamities are, however, regarded as the most destructive force of nature.

Effective measures involve the (1) development of the computer network for disaster-related information, (2) World Meterological Organization, and (3) Satellite Measurements. These global database provide relevant data that helps realize the phenomena causing disasters. Widespread concern has been expressed that the global warming will drastically alter the climatic conditions in future with the consequence of the loss of vulnerable fish stocks.

Disease

Among different types of diseases caused by bacteria, fungi, and viruses, the most virulent and dreaded one is the Epizootic Ulcerative Syndrome. This syndrome has completely destroyed large populations of a number of both economical and non-economical species of fish. The loss of genetic variability is a serious threat to fish biodiversity to a large extent. It should be kept in mind that intensive fish farming systems have increased the risk of disease outbreaks significantly.

Introduction of Introduced Fish

Introduction of introduced fish into the native waters has serious consequences on the existing local fish fauna such as *Osteobrama belangeri*, Indian major carps and other commercial species of fish. A number of introduced fish such as *Salmo salar, S.. trutta fario, Oncorhynchus mykiss, Salveninus fontinalis, Cyprinus carpio, Carassius carassius, C. auratus, Oreochromis mossambicus, O. niloticus, Ctenopharyngodon idella, Hypophthalmichthys molitrix, Tinca tinca, Osphronemus gouramy, Gambusia affinis, Lebestes reticulatus, Clarias gariepenus*, etc. have adapted well to local environments, started feeding on the available natural food organisms and achieved good growth rates. Several years after their introduction, reproduction of the introduced fish continues and their number is increased due to abundant progeny. Consequently, the introduced fish become stabilized.

Composite fish culture systems have always been the backbone of fish culture industries in rural areas. This favorable developnent on polyculture scene has been largely the result of the remarkable aquacultural breakthrough achieved by most of the Asian countries in the early seventies through the technological route referred to as *fish revolution*. Progress on the composite fish culture front has resulted in the contamination of other ecosystems by introduced species and at the same time created menace among aquaculturists by changing fish biodiversity and genetic resources through inter-breeding, competition for food, and through transmission of diseases.

The result from the introduction of exotic fish in many inland water bodies exhibits the need for close management of the introduced stocks from the beginning. Local fish stocks can be saved from disappearance by harvesting the excess introduced fish to avoid the problem of cannibalism.

18.12 Genetic Problems in Endangered Fish

Although genetic engineering (See Panel 18.A) has enormous potential with regard to fish production and their gene pool conservation, concern over the possible hassles of the precious term genetic engineering for increased production and conservation strategies have stimulated research on the fate of endangered fish species in natural waters. And serious attention has been given to commercially important species of farmed fish which are cultured in different ecosystems.

It should always be remembered that genetic improvement of wild stocks has substantially increased the production of farmed fish. Genetic selection for fish phenotypic improvement has, however, exhibited dramatic results. Aquaculturists definitely prefer to use genetic selection to improve the growth rate, food conversion efficiency, disease resistance, and fecundity of fish to a large extent.

Over-fishing drastically reduce population size and because large-scale individuals are selectively eliminated, this practice is commensurated with the selection for small-sized fish. Due to rapid environmental changes resulting from pollution, adaptation of natural fish stocks significantly reduced. Inbreeding, negative selection and lack of adaptation are regarded as genetic causes for decline of natural fisheries and lack of redemption. It seems quite plausible that restocking strategies involving hatchery stocks are likely to be solved that problem since these stocks were selected for adjustment to hatchery conditions.

Habitat destruction coupled with over-exploitation results in shrinkage of fish biomass. Consequently, a number of fish species have degenerated rapidly from several common fishing premises and some have even become endangered too. Since genetic variations are the raw materials in fish population which enable them to adapt to the changed environment, any loss of genetic variation causes in erosion of developmental elasticity.

PANEL - 18.A
GENETIC ENGINEERING

Genetic engineering is the manipulation of an organisms genetic material (DNA) to alter the organism's characteristics and physical traits in a precise manner. Aquaculture production has observed great advances through use of genetic engineering, and has enabled fishes to be tailored to consumer requirements. The possible use of genetic engineering is staggering in their variety and scientists are thinking up new applications every day. In principle, any product or technique that depends in some way on a living organism is a candidate for genetic engineering. And the application of genetic technologies has the potential for profound effects on life.

Richness of fish community (and also other animals and plants) in a given habitat interact ceaselessly with the ecosystem. They cannot be replaced because the species becomes adapted in a given habitat after a long course of time. If a species extincts, it indicates that the entire gene pool of that particular species will be disappeared. The real value of fish species richness lies on the information that are encapsulated in the genes. There is, however, urgent need for future to protect the genes from destruction.

Genetic engineering essentially involves the detailed study of the nature, concept, and synthesis of genes; tools used for genetic engineering; techniques of genetic engineering, and the genetic engineering for aquaculture/agriculture production and conservation of valuable species. All these studies have significantly made it feasible to understand the subject of genetic engineering with regard to fish production and conservation of threatened fish species.

In genetic engineering programs, however, it has paved the way for mapping the whole genome of an organism to trace out the function of the genes, slice and transfer into another organism. Owing to the success achieved from gene cloning, many aquaculture species have been obtained through genetically- engineered cells, and it is hoped that fish production can not only be increased several folds during the current century, but also be conserved the gene pool of endangered fish species. Recombinant DNA technology has become the major thrust and has made it easier to detect several types of finfish/shellfish diseases and cure them accordingly. Gene bank and DNA clone bank have been constructed to make available different types of genes of its known function. Thus this technology has made it possible to protect endangered fish species from extinction.

This contribute to the probability of extermination of fish population from any aquatic environment. Such genetic problems in the small genetically-effective population take the form of genetic drift and inbreeding depression has great practical significance.

Genetic Drift[1]

Variations in gene frequencies which arise because of sampling errors in finite populations are termed as *genetic drift*. The smaller the population, the greater will be the importance of genetic drift (the bottleneck effect) in very large fish populations and its influence will be negligible. There are, however, some unforeseen changes that occur in gene frequencies from generation to generation in all populations. Changes in gene frequencies are observed as sampling variation in small populations. In some generations, the gene frequencies are changed by chance during reproduction; in others, it significanly decreasesd and in still others it may remain the same. These oscillations in gene frequencies occur at random in each family of fishes in the population. All the effects of random genetic drift have one feature in common: they involve a loss of genetic variability. As population lose their variability, they become more homozygous that results in less heterozygous. Drift mainly depends upon the number of breeding individuals who could produce the next generation. It is unlikely that random genetic drift will affect allelic frequencies at a gene locus over long periods of time. The effects of drift can be studied by monitoring the reduction in the frequency of heterozygotes.

Inbreeding Depression[2]

Inbreeding is defined as the mating of individual fishes related to common ancestry that share more genes to descent than those individuals randomly selected from the population. Inbreeding results in a predictable increase in homozygous genotypes which affect different traits.

Common inbreeding impacts may be readily apparent such as gross deformities, stunted body growth, and disease resistance of many aquaculture stocks such as Indian major carps, silver carp, and mahseer. Inbeeding depression, however, is probably the most serious problem of threatened fish with small population which is frequently associated with a general decline in fitness. In an effective population size (Ne)[3] of fish, inbreeding depression to the tune of 2-17 per cent has been reported in one year among the Indian major carps with the Ne ranging from 3 to 30.

18.13 Conservation of Fish Germplasm[4]

The evaluation and management of germplasm have contributed substantially to modern animal husbandry and agriculture. But aquaculture has so far benefitted very little from efficient genetic markers and genetic improvement programs. Among many reasons for this are the delay in the development of suitable techniques for the marked improvement

1. For further detail, see Nei and Chakraborty (1975).
2. For further study, see Eknath and Doyle (1990), Pandey *et. al.* (1975).
3. Ne = (4 Nm Nf) / (Nm + Nf), where Nm is the actual number of breeding males and Nf is the actual number of breeding females.
4. For further study, see Das (1992).

of genetic markers, chromosome banding, biochemical genetics, and genetic introgression of several farmed fish. These studies require long-term experimentation with a large number of individuals and generations and, therefore, considerable time may elapse before useful results become available. Moreover research on farming technologies has, in mnay cases not reached that level. The only way to improve production and to achieve as maximum profit as possible is by the improvement of fish germplasm conservation. The present technologies are, however, too insufficient to benefit from the use of endangered stocks.

The molecular genetic markers (see Panel 18.B) are necessary for identification of different stocks and marker-assisted selection and genetic improvement since biological information about different species of fish such as Tilapia, Channel catfish, Blue catfish, and Rainbow trout is not adequate for undertaking conservation of genetic upgradation programs. For successful programs, a number of genetic markers have been developed and determined genetic variations not only among different species but within different species as well chromosome banding has been detected in more than 20 different commercial and endangered fish which help to detect the species-specific pattern of fish. The particular qualities that make a fish species different from other species through application of biochemical genetic techniques using several enzyme systems and iso-electric focusing (See Panel 18.B/1) of eye lens and haemoglobin have been carried out in many species of fish. Studies on population genetics have shown the genetic variation in isozyme and iso-electric focusing markers for several species of carps such as *Labeo rohita* and golden mahseer. Detection of genetic introgression of farmed fish stocks with gold fish genome has also been made using isozyme genetic markers which help to maintain pure brood-stocks. The biochemical genetic marking of fish helps resolve taxonomic dubiousness.

18.14 Management of Fish Germplasm

It is important to note that resources and adequate facilities should be available for regeneration, multiplication, evaluation, characterization, and distribution of fish germplasm. During management of the germplasm, its identification and genetic stability must be maintained. Further more, for safety, samples must be duplicated at various locations. If these cautions are not taken and facilities for management of collections are not expanded, some of the germplasm storages may be transformed into germplasm mortuaries.

18.15 Necessity of Fish Biodiversity Conservation*

Generally, fish conservation strategy activates the production- marketing-consumption chain. Actually it makes production through conservation of fish germplasm resources; it also generates profits and when the profits are ploughed back into additional yield, the cycle continues with increased vigour. Conservation also takes the world to higher levels of production. By converting the entire fish farmers/fishermen into a production-related

---

* For further study, see Smith and Chesser (1981), Brown *et. al.* (1989), Frankel and Soule (1981), Minckley and Deacon (1991).

PANEL - 18.B
GENETIC MARKERS

Markers are the objects generally used to indicate a position or place. A number of biochemical genetic markers have been identified for stock discrimination to the expression of growth factors during maturation. Molecular genetic markers include isozymes (separable forms of enzymes encoded by one or more loci), allozymes (products of different alleles at the same locus), nuclear DNA, mitochondrial DNA, non-repetitive and repetitive DNAs, coding and non-coding DNAs, etc. However, for aquacultural applications, the wealth of isozyme analysis technique gives a considerable advantage to conserve germplasm resources for many fishes because it is inexpensive, and the isozyme data constitute the largest existing genetic data set for many species of fish.

PANEL - 18.B/1
ISO-ELECTRIC FOCUSING

Separation of proteins having different molecular sizes or molecular masses can be done through electrophoresis. Although Sodium Dodecyl sulfate Polyacrylamide Gel Electrophoresis (SDS-PAGE) is most widely used method for analysis of protein mixture qualitatively, this method is of no use in the separation of enzymes since enzymes (proteins) are denatured in SDS-PAGE procedure. These polyacrylamide gels are again used. The enzyme of interest is identified by incubating the gel in the appropriate substrate solution so that a colored product is developed at the site of the enzyme.

Iso-electric focusing technique is particularly useful in separating isoenzymes (also called *isozymes*). The technique essentially involves electrophoresis which depends upon the charge carried by a molecule but at its iso-electric pH, a molecule will not migrate in an electric field. If electrophoresis is carried out in an electrolyte which has a pH gradient, molecules will migrate to the point where the pH of the solution is same as their iso-electric pH (the pH at which a molecule has no net charge) and remain there as long as the gradient and potential differences are maintained.

The method is based on the separation of isoenzyme molecules on the basis of their iso-electric points. At these points, molecules carry no net charges and are, therefore, electrophoretically immobile. The method has a high resolution being able to seperate enzymes that differ in their iso-electric points by as little as 0.01 of a pH unit. Separation is performed by applying a potential difference across the gel that contains a pH gradient (For further detail, see Rana, 2005).

community, conservation of fish germplasm resources steadily alienates the world away from the vicious circle of low production and low development. It is this feature that enables conservation to enhance the production strategy stances and consequently the development of native fish populations proceeds perfectly. Adoption of conservation steps/ strategies can offer a new hope to the aquaculturists. In fact, it is the recognition that fish biodiversity conservation can play a catalytic role in her production plans that should be forced the developing nations in general, to adopt conservation and rehabilitation programs as understood and practised in the developed nations.

The above discussion dictates us that the maintenance of fish biodiversity can be regarded as an essential prerequisite for the current and future generations. There are three basic reasons that follow for the preservation of biotic resources around the world.

1. Diversity seems appropriately delightful in most ecosystems. Though it is not applicable in general, it often applies to certain species regularly confronted with man.

2. Species diversity and genetic variability are essential for long-term subsistence of stable and complex ecosystems and species. Some genetic traits from diverse germplasm resources may be useful to increase the aquaculture production through gene technology and hybridization. These areas of research have revolutionized the potential of germplasm resources.

3. During the course of evolution, different species of an ecosystem co-evolve for their mutual benefits. If any species is abolished, the ecological balance is also upset and consequent to the detriment of each species in particular and to the community in general. Conservation of fish germplasm through a novel approach termed as *cryopreservation* (to be discussed later on) in liquid nitrogen on a long term basis is highly significant. Suggestions have been made that germplasm bank should be attached to some of the International Research Institutes that would hold responsibility for the storage, maintenance, distribution, and exchange of these disease-free germplasm of the important fish species.

Facilities for storage of genetic stock of fish species can be developed in large-sized cylinders (20-40 litres capacity) where liquid nitrogen does not require refilling for 6 -1 0 months. Germplasm bank in such a device where facilities of cryopreservation of genetic resources of a variety of fish species are thus available and on the demand, the germplasm can be supplied both nationally and internationally. The potential and prospects of cryopreservation of fish gametes (sperm and egg), embryos, tissues, and establishment of germplasm bank are shown in Figure 18.1.

18.16 Storage of Fish Germplasm

Conservation of fish genetic resources or germplasm by several national such as National Bureau of Fish Genetic Resources (NBFGR) and international organizations such as International Board of Animal Genetic Resources (IBAGR) have become thrust areas of fish biotechnology. For this purpose, various strategies for storage of gemplasm without

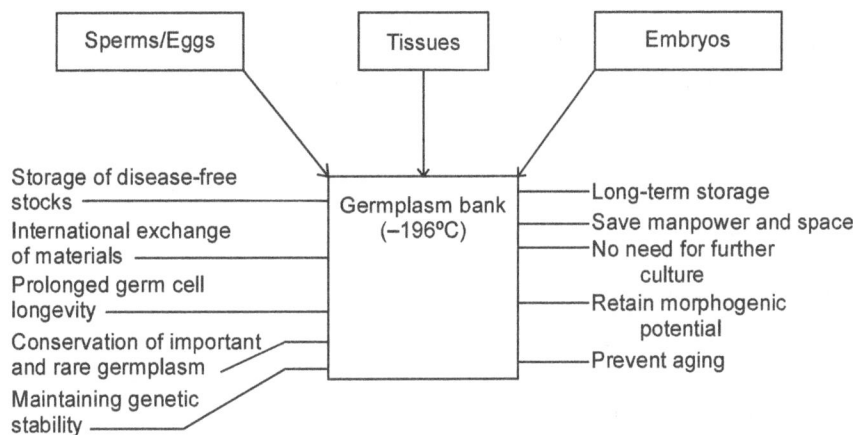

Fig. 18.1 : Potentials and prospects of fish germplasm conservation by Cryopreservation.

causing loss of its viability have been developed and are being utilized more and more widely. Fish germplasm can be stored for prolonged periods in a variety of forms such as fish milt (sperms and eggs), embryos, cells, and tissues. Successful freezing of these structures and to preserve them for long periods without any loss of their potentiality is of considerable importance in the improvement of fish culture/aquaculture/fisheries. It has a great practical significance and would definitely open new perspectives in establishing genetic material reserves for selective breeding, provide a means of conserving endangered species by gene banking, executing cross-breeding of desirable stocks and by improving the existing induced breeding technology.

Technique for Freezing Fish Gametes

In order to freeze fish gametes or embryos, they are dehydrated by using a chemical called as *cryoprotectant*. Ethylene glycol, glycerol, and propylene glycol are used as cryoprotectants for fish sperms, but the most widely used compound is dimethyl sulphoxide (DMSO) which has good potential as a cryopreservative agent. When equilibration is achieved, fish milt and embryos are cooled slowly in a controlled manner from 25 to –38°C. The cooling is particularly done in a programable controlled rate freezer. The milt and embryos at –38°C are submerged into liquid nitrogen (–196°C) and stored for the posterity and used whenever required.

Fish milt is stripped into dry tubes without contamination by faeces, urine, or toxic substances. The milt is diluted with an extender solution before freezing. The extender solution is a mixture of salts at suitable pH which helps maintain the viability of sperms at the time of refrigeration without activating them. The solution also contains a cryoprotectant which binds with electrolytes during the freezing process, thus preventing the solution from developing lethal concentrations. They help to reduce the freezing point of the intra-cellular fluid.

The techniques involving storage of tissues include the use of germplasm by anyone of the following methods: (1) reduction in oxygen concentration, (2) addition of chemical retardants, and (3) lowering the temperature. Of them, the last method is highly effective for long-term storage that involves storage at a very low temperature using liquid nitrogen. This is explained as *cryopreservation* (see Panel 18.C) which allows storage for virtually indefinite periods. DMSO (0.5 M or 10 per cent) or some other media (such as PMSO + glycerol + sucrose) are used for step-wise freezing to preserve in liquid nitrogen. The milt can be used whenever required by rapid thawing.

Methods for Freshwater Fish Milt Cryopreservation

Unlike sperm of higher vertebrates, fish spermatozoan remain quiscent and become activated the moment they come in contact with water. When spawning of fish occurs in freshwater ecosystems, spermatozoa remain active for 1-4 minutes during which they are able to fertilize the eggs. In general, fish sperm cryopreservation technology involves proper collection of fish milt, addition of extender to prevent depletion of sperm energy reserve and to maintain sperm in quiscent condition but alive, use of cryoprotectants to reduce thermal shock, and use of liquid nitrogen as a freezing and storage agent.

Milt collection generally involves (1) the withdrawal of feeding for a day or two before collection so as to reduce void of faecal matter, (2) anaesthetising the donor, (3) keeping genital aperture and collecting glasswares in dry condition during stripping, and subsequent keeping extruded milt in ice or refregerator. Dimethyl sulfoxide rapidly penetrates into the cellular membranes and is used at 10 per cent concentration. Sperms have been cryopreserved in glass ampules either in sealed or in unsealed condition. The storage period of sperm varies from 34 days (*Labeo rohita*) to 5 years (Pink salmon). The use of sperm is generally carried out after thawing it in water at temperature ranging from 10 to 60°C.

18.17 Developnent of Fish Sperm Cryopreservation Technology

Developmental strategy of commercially important food fish (and also endangered species) will provide an effective experience to conservationists and aquaculturists on the formulation and implementation of cryopreservation technology. It would serve as a good example of the Indian aquaculture to attain the success of cryopreservation following the adoption of a suitable technology. However, development of cryopreservation technology involves the following subjects.

Evaluation of Fish Semen

This is a very important criterion to judge the efficacy of sperm before cryopreservation and subsequent fertilization. Good quality semen is characterized by the color, volumes, density, pH, morphological abnormalities, and mortality of spermatozoa. A thorough knowledge on the traits of sperms will help to observe the sperm consistently, fertilizing capacity, maturlty, environmental condition, and to formulate adequate extender solution.

PANEL - 18.C
CRYOPRESERVATION

The study of the effects of extremely low temperatures on living tissues of plants and animals are termed as *cryobiology*. It has several applications in scientific research, medicine, and surgery. It is also used to preserve sperm, eggs and embryos.

Water freezes at O°C. The temperatures used in cryobiology generally vary from about -100°C to -273 °C which are much lower than those observed in the coldest natural environments. Since these temperatures cannot be reached under normal circumstances, cryobiologists use special substances which are termed as *refrigerants*. Liquid nitrogen is widely used as one of the most common refrigerants because it is cheap, plentiful, and unreactive. Liquid nitrogen has a boiling point of -96°C and therefore, it remains a liquid state at temperatures lower than this.

Cryopreservation of Eggs and Sperm : Cryopreservation refers to the preservation of living tissues by freezing. As temperature decreases during the process of cryopreservation considerably, the water within tissues freezes. Under normal conditions, this freezing would damage the tissues and consequentially kill the organism. To solve this problem, cryobiologists have been forced to develop techniques to reduce the temperature without deteriorating the tissues. The development of techniques for freezing eggs and sperm using liquid nitrogen significantly increased the use of artificial propagation and hybridization for animal stocks. Frozen materials can be transported to long distances and are preserved for many years.

Cryopreservation of Embryos: Cryopreservation techniques are used for the preservation of embryos. Fish embryo, for example, can be preserved in liquid nitrogen after slow pre-cooling in a suitable chemical at –196°C. These embryos remain at a very early stage of development and consist of a very few cells which are not specialized to undertake any specific functions. It should be pointed out, however, that the freezing technique is labor-intensive and highly expensive and still remains as developmental stage. Unless it is ascertained that the chemical substances used in freezing embryos do not cause any long-term genetic effects, cryopreservation of embryos will be considered as an important device for long-term preservation.

Cryopreservation and Conservation: Conservationists use this technique to conserve endangered animals. Conservationists collect endangered animals from the wild and try to breed them under controlled condition. Many such breeding programs in sanctuaries and zoos are being carried out across the world. The population of the endangered species in each region is very small. To avoid inbreeding, conservationists use cryopreservation techniques to introduce new genetic materials.

Milt Collection

Development of suitable fish collection methodology is the real starting point of conducting cryopreservation experiments. Suitable techniques help to reduce the damage of the milt and injuries to the donor fish. Instead of stripping, it is perhaps better to draw the milt directly with the aid of a syringe or cathetor. Several aquaculture farms generally take effective suggestions in their strategic milt collection efforts into consideration. They are active in producing good quality of milt from the donor fish. The recent development of the state-of-the-art technology undertaken by NBFGR on this line is worth citing.

Testing of Extenders

For a cryopreservation enterprise, its technology includes the selection of suitable extender solutions. Through selection of extender solution, aquaculturists try to cryopreserve fish milt that helps formulate the future development programs. The use of extender 189 M (NaCl 730 mg, $NaHCO_3$ 500 mg, fructose 500 mg, vegetable lecithin 750 mg, mannitol 500 mg, and 100 ml of distilled water) has, however, been proved to be effective. In cases where liquid nitrogen preserves carp milt, the use of 189 M extender is effective to fertilize fresh ova and 60 per cent of the ova may produce healthy larvae.

Testing of Fertilizing Capacity of the Cryopreserved Sperms

The capacity of well-preserved sperms to fertilize the eggs and produce robust larvae principally depends on the success of the sperm cryopreservation technology. To achieve optimum utilization of cryopreserved sperm to fertilize large number of eggs, suitable ova-sperm ratio must be taken into consideration. Fertilization capacity of cryopreserved sperms should be tested before their use in fertilization and hatching programs is permitted. Testing of cryopreserved fish requires sophisticated techniques that deserve serious consideration.

18.18 Need For Integration of Fish Biodiversity with Technology

Though most of the tropical countries are rich in fish biodiversity, little attention has so far been given to preserve this kaliedoscope variety of fish species. Administrators and policy-makers are restricted mostly to reptiles-birds-mammals syndrome. In many developing countries, fish biodiversitp is generally considered as no more than a ruffle of ecosystem agencies which have no expertise to conserve biodiversity at production levels. Even these countries have yet to perceive the extensive economic, scientific, political, technological and ecological potentialities of fish biodiversity. If this continues, these countries are likely to be limited in the race to preserve and sustainable utilization of their biodiversity for the benefit of the society and the world at large. Scientists, planners, and policy makers have to realize that conservation and sustainable utilization of fish biodiversity has to be central to all development planning in these countries, since these countries are rich in fish biodiversity. Hence, there is urgent need for effective coordination among different institutes, organizations and countries.

18.19 Strategy For Fish Biodiversity Conservation

Several strategies have been suggested for preserving fish biodiversity. These include (1) No undisturbed water bodies should be used for human interest and since many of these water bodies are the main repository of fish, the use of these waters will lead to the loss of biodiversity, (2) Catalogues of fish genetic resources and fish biological inventories should be prepared so that the threatened and endangered fish may be prevented from extinction, (3) steps should be taken to reduce acid precipitation, (4) In developing and under-developed countries, birth rates should be lowered and high production systems should also be developed so that preservation and sustainable exploitation of fish biodiversity may proceed smoothly. Developed countries should provide massive financial assistance to fulfil this purpose without any lapse of time, (5) to avoid destruction of value-added fish species, National Fish-Security Principles should be formulated, and (6) Effective measures for conservation of fish biodiversity should be developed and enforced in all nations.

Since conservation and rehabilitation of endangered and threatened fish is a problem of global importance, (1) the production of stocking materials through artificial propagation and their distribution in natural waters, (2) collection, classification and evaluation of information of fish germplasm resources, (3) cataloguing of fish genotypes, (4) monitoring of the introduced species in natural ecosystems, (5) maintenance and preservation of fish germplasm, (6) protection of fish biodiversity by enforcing laws and regulations, and (7) establishment of fish sanctuaries are some of the solutions which must be considered to save the endangered species from extinction. And a comprehensive account of fish conservation methods makes a separate discipline of biodiversity studies generally termed as *conservation biology* which is beyond the scope to discuss in this chapter.

GERMPLASM BANK

The first decade of the 21st century was the dawn of expansion of fish germplasm bank development around the world. Germplasm bank (See Chart 18.1) is generally considered as an important plank for fish biodiversity conservation. Gene banks are established for *in-situ* and *ex-situ* conservation areas of endangered fish species where they can find their safe habitats. One of the objectives of these proposed areas is to establish gene banks and to improve the reproductive capacity of endangered fish following different reproductive technologies. Gene banks should be established in selected areas in collaboration with the top ranking institutes and aquatic sanctuaries (See Panel 18.D). And approach must be standardized with the molecular biology and reproductive strategies in non-endangered fish and then assemble to endangered ones.

1. *In-situ Conservation Areas*: This type of conservation applies only to wild fish fauna but not to farmed/ cultured fish because *in-situ* conservation essentially involves adequate management of wild fish in their natural ecosystems. Many aquatic ecosystems are self-contained and self-sustained units and are home to a spectacular diversity of aquatic fauna including fish. *In-situ* conservation areas include Sanctuaries, National Reserves, Cultural Landscapes, and Fish Parks, however.

CHART - 18.1
GERMAPLASM BANK OR GENE BANK OR GENOMIC LIBRARY

Gene bank, germplasm bank or genomic library is a complete collection of cloned DNA fragments which comprises the entire genome of an organism. Gene bank is constructed by a shot-gun experiment where the entire genome of a cell is cloned in the form of random and unidentical clones. The clones of DNAs are produced by (a) isolation of DNA fragments to be cloned, (b) joining the fragments to a suitable vector (generally Phage I), (c) introduction of recombinant DNA into host cells at high efficiency to get a large number of independent clones, (d) selection of the desired clones, and (e) use of clones for the construction of gene bank (See Figure 18.2). Once such a library or a bank is available, clones can be perpetuated for indefinite period of time in a plasmid vector and retrieved as and when needed for a variety of purposes.

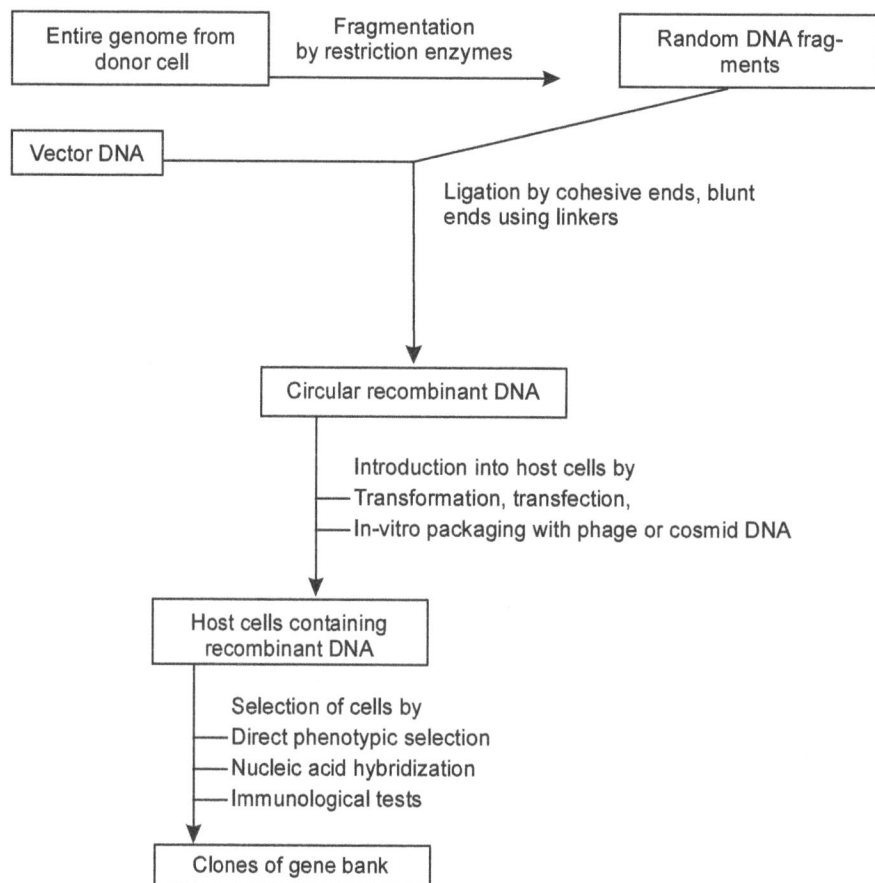

```
┌─────────────────────┐   Fragmentation        ┌─────────────────────┐
│ Entire genome from  │   by restriction enzymes│ Random DNA frag-    │
│ donor cell          │ ──────────────────────→ │ ments               │
└─────────────────────┘                          └─────────────────────┘

┌─────────────┐
│ Vector DNA  │─────────────────────────┐
└─────────────┘                         │
                    Ligation by cohesive ends, blunt
                    ends using linkers
                                        │
                          ┌─────────────────────────┐
                          │ Circular recombinant DNA │
                          └─────────────────────────┘
                                        │
                    Introduction into host cells by
                    ── Transformation, transfection,
                    ── In-vitro packaging with phage or cosmid DNA
                                        │
                          ┌─────────────────────────┐
                          │ Host cells containing   │
                          │ recombinant DNA         │
                          └─────────────────────────┘
                                        │
                    Selection of cells by
                    ── Direct phenotypic selection
                    ── Nucleic acid hybridization
                    ── Immunological tests
                                        │
                          ┌─────────────────────────┐
                          │ Clones of gene bank     │
                          └─────────────────────────┘
```

Fig. 18.2 : Methods of Construction of a Gene Bank

Due to disappearance of natural fish stocks at rapid rate, genetic variabilities are being lost for ever. Therefore, to save the threatened and gradually disappearing fish species and to meet the world demand of fish protein, it is cardinal to conserve them in full potential.

The gene bank has taken the challenge of conserving the gene pool of ecologically and economically important fish species.

PANEL 18.D
AQUATIC SANCTUARIES FOR FISH CONSERVATION

Though protected areas can be established in different ways, the two common methods are government action (at national, regional or local levels) and lease/purchase of water bodies carried out by private conservation organizations. The government of a developing nation in the tropics in collaboration with the international conservation organizations, banks, and governments of developed countries play a significant role on fish biodiversity conservation. Such protected areas developed in many tropical regions commonly have high conservation potential. Developed nations provide adequate fund, training and scientific as well as management expertise to set up protected areas in the region concerned. In contrast to temperate regions, the high conservation potential of tropical productive freshwater ecosystems helps illustrate why waters of tropical regions have not been extensively utilized. These waters are high in nutrients, close to sources of fertilizers and manures and high manpower, making it easy to develop/implement conservation policies.

One of the primary methods of conserving freshwater endangered fishes is the establishment of fish sanctuaries that has great conservation potential. They are similar to strict nature reserve but some human manipulations may be inevitable to maintain the characteristics of the ecosystem and some controlled harvesting may be permitted. They also permit non-destructive uses of the ecosystem by resident people and provide opportunities for recreation and tourism.

In India, two aquatic sanctuaries have been set up in two different wildlife protected areas of Uttar Pradesh: (1) Samaspur Bird Sanctuary (SBS) and (2) Katraniaghat wildlife Sanctuary (KWS). Monthly fishing data clearly dictates the availability of endangered fish. However, a total number of 39 species belonging to 9 orders, 12 families and 26 genera have been recorded from SBS. On the other hand, KWS is inhabited by 57 species representing 7 orders, 18 families and 40 genera. Aquaculturists must recognize the aquatic sanctuaries as important resources of endangered fish that should not be severely damaged.

Problems of Aquatic Sanctuaries : Establishment of aquatic sanctuaries are generally considered as the best way of protecting wild fish stocks. Since the population pressure is high for years, the demand for fish and waters has increased considerably. Most of the sanctuaries are likely to occur as small pockets of wild fish which are circumscribed by house, holly places, roads, industries, and agricultural farm lands.

The isolations cause several problems. *First*, the sanctuaries within the protected areas are very attractive to tourists and people living around the sanctuary. In addition to this, protected areas are often invaded by domestic animals. These encroachments are illegal; but for many people, they provide the only opportunity for short-term survival. *Second*, since nature is always dynamic, aquatic populations generally increase and decrease cyclically. These cycles may be the result of fluctuations in climate, rainfall and disease, cold winter or warm climate. These environmental fluctuations may affect fish populations considerably, or they may have an indirect effect by decreasing fish community that the human populations consume fish as food. Therefore, sanctuaries afford only limited protection of fish stocks.

Suggestions to Improve Fish Stocks : Though a number of ways have been suggested to improve fish stocks and to prevent them from disappearance or possible extinction of several species of fish in different freshwater areas of the world, there is general agreement that protection of catchments should be considered first in the list of priorities and protection of water through prevention of pollution from both non-point and point sources is also necessary. Another priority is the need for enforcement of fishery laws and regulations, which should assist in the conservation of the most threatened fish species. Stocks of native fish can be saved with the help of stocking at regular intervals of hatchery-produced fingerlings. Shortage of experts and repugnance to implement the strategy is often an impediment to expedite the development of aquatic sanctuaries.

Major efforts are being planned across the world by a number of government and voluntary organizations to conduct research on *in-situ* conservation of fish. In India, for example, a number of associations/clubs such as Himachal Angling Association, Palamber; Assam Anglers Association, Tezpur and many religious places as well as sanctuaries (See Panel 18.D) (such as Andhreta Machhal on Awakhal, Nagachula in Mandi; Owakhud, Mamta and Samtole sanctuaries in the District of Kangra, Himachal Pradesh; New Damyanti Tal in the District of Nainital, Corbet Park in Uttar Pradesh, Rishikesh and Hardwar Tamples in the Central Himalaya have been used for protection of endangered fish because these water bodies are very productive. Fishing on these water bodies has banned due to religious sanctity.

On the government front, the Karnataka Tourism Department Corporation in collaboration with Wildlife Association of South India has been granted lease on the long stretches of the river Kauvery which helps allure the anglers even from UK, Japan and Europe. With efforts and constant monitoring, they have done encouraging jobs by (i) creating a sense of awareness among farmers and anglers, (ii) protecting fish habitats, and (iii) controlling fish harvest.

A number of voluntary and local organizations also play a pivotal role in the conservation of depleting fish stocks of commercial importance. They can serve (i) to aid and advice the government, (ii) to act eyes and ears of the government, (iii) to educate the people at large, (iv) to create awareness among people in favor of conservation, (v) as valuable sources of information, (vi) research and development programs, and (vii) to disseminate the biodiversity conservation policies in a poular level through publications, seminars, farmers' day, and audio-visual programs.

In-situ conservation of fish also involves regular fishing. Indiscriminate killing of fish by destructive methods should be banned and made cognizable offence through legislation.

Non-availability of adequate watchmen has encouraged such fishing activities. Even regulation of adequate mesh size to protect juveniles from mass killing are not being observed. Attempts have, however, been made by developing nations to formulate their fishing regulations to conserve fish stocks in sport and commercial fishing premises.

(i) *Appraisal of In-situ Conservation:* Conservation of individual species is not a perfect strategy . Many endangered fish are found in a very small and remote area. To conserve these species one by one would take many years. With thousands of endangered species facing extinction, we may not have enough time to fight individual battle for each of these species.

As human populations have expanded over the past several centuries, protected areas have become unfit for the survival of many species of fish. In India, for example, the population of world's famous Himalayan mahseer, *Ompak pabo*, and *Notopterus chitala* has drastically reduced and it is expected that they will soon appear in the Red Data Book in future unless suitable steps are taken to conserve them from extinction. Similarly, Sevan trout, Sevan barbel and Sevan Khramulya in Armenia are either extremely rare or extinct.

2. *Ex-Situ Conservation Areas** : Though *in-situ* is the best strategy for long-term conservation of fish biodiversity in the preservation of natural communities and population in the wild for many rare species, *in-situ* conservation is not generally considered for the purpose due to increase in human interventions. 'The number of species may decline and extinct due to genetic drift and inbreeding, environmental variation, loss of habitat quality, competition among introduced species, over-exploitation, and disease. If all individuals are observed outside the protected areas or if a population is too small to remain, *in-situ* conservation may not be effective. Therefore, it is likely that individual species should be maintained in artificial environments under human supervision. This method is termed as *ex-situ conservation* which includes fish farms, large aquaria, and captive breeding programs.

Ex-situ conservation involves seed banks, embryo banks, gene banks, and genetic engineering. These are very important for fish crop improvement in future.

---

* For further study, see Minckley and Deacon (1991), Ponniah *et. al.* (1996).

In a gene bank, endangered fish are reared in captivity, breed therein and genetically managed to avoid inbreeding depression, domestication and unintended selection. The gene sequence of endangered or threatened fish species is, however, stored in the gene bank with their account number for each species. On the other hand, the embryo bank involves the keeping of adequate samples which are representatives of the natural genetic variations of endangered species in suspended state of animation at -196° C in liquid nitrogen. Establishment of gene bank by cryopreservation of eggs, milt and embryos helps provide the genetic materials of threatened categories of fish for undertaking intensive breeding programs of economically important species particularly *Salmo gairdneri gairdneri*, *Tor* spp, *Labeo rohita*, *Catla catla*, *Cyprinus carpio* and *Horabragrus brachysoma*, (yellow catfish).   However, gene banks have the advantage of being able to store large amount of genetic materials in a small space and will be available round the year for seasonal breeders.

(i)   *Ex-Situ Conservation Efforts at International Level*: Major efforts in *ex-situ* conservation of fish genetic resources have become possible due to technical support provided by the consultative group of international research centres. These centres have the capability to construct the world's largest *ex-situ* collection of fish gene pools. These germplasm collections are held in believe for use of present and future generations of researchers and farmers around the world. These research centres are committed to strengthen national research systems in genetic resource programs.

The main objectives of establishing fish genetic resource centres include several strategies that follow.

* To assist countries, particularly developing ones to evaluate and fulfil their needs for fish genetic resources.

* To strengthen and contribute to international collaboration in the conservation and use of fish genetic resources.

* To develop and encourage improved strategies and state-of-the-art technologies for fish genetic resource conservation.

Global interest in the conservation of fish biodiversity has, however, been stimulated several research organizations of high standards.

(ii) *Ex-situ Conservation Efforts in India*: In 1983, with the assistance of the Indian Council of Agricultural Research (ICAR), a fish genetic resource centre has been established in the name of the National Bureau of Fish Genetic Resources (NBFGR) located at Canal Ring Road, Lucknow. The bureau has appeared as one of the most pioneering research institutes of the country in the field of both *in-situ* and *ex-situ* conservation of fish biodiversity.

18.20 Conclusion

Among icthyologists and progressive fish-culturists, there is a growing awareness that the fish biodiversity harbored in this planet is an immensely valuable resource that must be preserved at any cost for the posterity. It should be kept in mind that the loss of fish biodiversity and genetic resources occurs worldwide which is not only jeopardize the fish culture ecosystem but cause poor production as well.

Fish biodiversity loss is most critical in tropical and sub-tropical regions. In these regions, fishes are most vulnerable to extinction. Many advanced fishing nations are judiciously exploiting aquatic resources to such an extent that the threatened and valuable food fish are on the brink of extinction. The overall management of freshwater ecosystems offers the most effective means of conserving fish biodiversity. Such practices, known collectively as *conservation*, have become increasingly popular in many parts of the world. Many organizations and research centres are actively engaged in conserving wild fish stocks at genetic level for sustained production of fish as food and recreation.Conservation strategies, particularly *in-situ* and *ex-situ*, are undoubtedly the most significant conservation practices developed in modern times.

Developing countries that are rich in fish biodiversity, have become a favored destination for developed countries seeking affordable bulk research and assistance. It should be added as a note that the global norms on protecting and commercially exploiting fish biodiversity are contradictory and are likely to remain so far a while. The need, therefore, for developing countries to come up with homegrown legislation is paramount. In addition to undertake extensive research and growing consciousness among people, stringent laws on the protection of aquatic biodiversity resources in general and fish biodiversity in particular must be implemented and enforced; otherwise, all efforts will be ended in smoke.

References

Baillie, J. and D. Groombridge. 1996. IUCN Red List of Threatened Animals. International Union for Conservation of Nature and Natural Resources. Gland, Switzerland.

Brown, A. H. D., O. H. Franel *et al.* 1989. The use of Plant Genetic Resources. Cambridge University Press, Cambridge.

Das, P. 1992. Conservation of Coldwater fish germplasm. Recent Researches in coldwater fisheries, 35-43. Ed. K. L. Sehgal. Todays and Tomorrows Printers and Publishers, New Delhi.

Das, P. and A. K. Pandey. 1999. Endangered fish species: Measures for rehabilitation and conservation. *Fishing Chimes*. 19 (6) : 31-34.

Eknath, A. E. and R. w. Doyle. 1990. Effective population size and state of inbreeding in aquaculture of Indian major carps. *Aquaculture*, 85 : 293-305.

Frankel, O. H. and M. Eo. Soule. 1981. Conservation Evolution. Cambridge University Press, Cambridge.

Minckley, W. L. and J. E. Deacon. 1991. Battle Against Extinction: Native Fish Management in America. Western University of Arizona Press, Tucson and London.

Nei, M. T. and R. Chakraborty. 1975. The bottleneck effect of genetic variability in populations. *Evolution.* 29 : 1-10.

Pandey, A. K., A. Mitra, and P. Das. 1998. Skeletal deformity and stunted body growth in the hatchery-bred progeny of *Cirrhinus mrigala J. Natcon.* 10 : 231-237.

Pandey, A. K. and P. Das. 2002. Fish biodiversity conservation: Theory and Practice. *Fishing Chimes.* 22 (5) : 14-24.

Ponnial, A. G., P. Das, and S. R. Verma. 1990. Fish Genetics and Biodiversity Conservation. *Nature Conservators*, Muzaffarnagar, 474 PP.

Reisenbichler, R. Re, J. D. Mcintryre and et al. 1992. Genetic variation in steelhead of Oregon and Northern California. *Trans. Amer, Fish. Soc.,* 121 : 158-169.

Smith, A. H. and R. K. Chesser. 1981. Rationale for conserving genetic variation of fish gene pools. *Fish Gene Pool Ecol. Bull.* (Stockholm) 34 : 13-90.

World Bank. 1996. Freshwater Fish Biodiversity in Asia. Tech Pap., 343.

Questions

1. Explain the term biodiversity and discuss why fish biodiversity should be conserved.

2. What are the reasons for biodiversity loss? What steps should be taken to conserve fish biodiversity?

3. What do you understand by the term germplasm? How the fish germplasm is preserved?

4. State what are the problems related to conservation of endangered fish species.

5. What are the differences between extinct and endangered fish species?

6. What are the importance, meaning, and significance of fish biodiversity?

7. Discuss the factors which affect fish biodiversity.

8. Give an account of fish biodiversity resources with reference to Asia.

9. What are the types of fish biodiversity? Explain the value of fish biodiversity.

10. What is cryopreservation? Explain how fish gametes (sperms and eggs) are preserved by cryopreservation method.

11. Discuss the technology developed in the preservation of carp sperm through cryopreservation.

12. What do you understand by the term *in-situ* and *ex-situ* conservation? Discuss the different approaches used for both *in-situ* and *ex-situ* conservation of endangered and commercial species of fish.

13. Explain how population growth, disease, acid rain, and pollution are related to the loss of fish biodiversity.

14. Distinguish the following terms:

(a) gene banks, (b) resource, (c) acid rain, (d) genetic engineering, (e) genetic drift, (f) inbreeding depression, (g) fish germplasm, (h) genetic markers, (i) NBFGR, (j) cryoprotectant, (k) fish sanctuaries, (1) Red Data Book, (m) effective population size.

19

Bioremediation in Fish Culture

The effects of toxicants on fish and aquatic ecosystems are evident everywhere. Onslaught of pollutants drastically upset the fish production potential by modifying physicochemical qualities of water and soil. Sustainable development of fish culture sector can be achieved by adopting eco-friendly fish culture practices through minimizing/eliminating adverse impacts on the activity of the surrounding environment. Maintenance of healthy ecosystem of fish culture ponds/tanks following the application of eco-friendly methods is termed as *bioremediation* which should be considered for achieving good production.

Bioremediation is a technology by which pollutants are eliminated from fish ponds and consequently contaminated waters are re-distributed and considered as best for fish culture. In this chapter, technologies being evolved for use of living organisms, prevention of pollution, and use of probiotics in fish culture have been briefly considered.

19.1 General Considerations*

The dogma of microbial reliability is that all natural compounds are biodegraded and consequently favorable environmental situations prevail. The term xenobiotics refers to artificial compounds that are foreign to biological systems and contains bonds and structures that do not occur in biological systems. Xenobiotics may be polychlorinated or polybrominated compounds or pesticides. Several species of algae, fungi, and bacteria are known to solubilize, transport, and deposit the toxicants and detoxify the complex chemicals. However, bioremediation is the application of some living micro-organisms to degrade toxicants or prevent pollution. The foundation of bioremediation is the tremendous natural ability of micro-organisms to deteriorate organic compounds. This ability could be improved by using genetically-modified micro-organisms. Provision of feed for cultured fish can be manipulated to enhance yield of fish biomass. Feed manipulation can be performed in two ways. *First*, feed should be used with a judicious ingradient mix for better utilization of nutrients and *second*, known and suitable additives are mixed in prescribed proportions to farm-made or commercial feeds for production of robust fish. If farmed fish have to be grown beyond their physiological limits, however, it is inevitable not only to manipulate fish stocks genetically but to administer some chemical substances as well. These chemical substances can be grouped as antibiotics, hormones, microbial cell-wall preparations, enzyme extracts, and the like. Use of additives to increase fish (and also shellfish) production has gained considerable importance in recent years.

---

* For further detail, see Alexander (1994), Tripathi (2005).

By applying the knowledge gained from fundamental research, bioremediation technology has been used to eliminate environmentally-hazard chemicals or to detoxify them into non-toxic forms.

Biodegradation referes to the process in bioremediation by which xenobiotics are transformed to less toxic states. Biodegradation of a compound does not mean that the breakdown products are less toxic or not desirable at all. Mineralization generally implies decomposition of a xenobiotic to carbon di-oxide and inorganic ions. This is the most enviable situation because the end products are non-toxic.

Recalcitrant molecules are polyaromatic compounds (such as tanin and lignin), persistent micro-organisms (melanin-rich algae), synthetic molecules (herbicides, insecticides, and fungicides), plastics and detergents. The persistence of xenobiotics in aquatic ecosystems ranges from day to years and minor alterations in biodegradable compounds can furnish them with recalcitrant molecules.

19.2 Approaches to Bioremediation

Two general approaches to bioremediation may be considered: (1) Betterment of potential activity of existing micro-organisms through manuring and fertilization, and (2) Use of suitable and selective micro-organisms.

The toxic materials remain in vapor, liquid or solid phases and therefore, bioremediation methods vary accordingly whether the waste materials involved are in their natural setting or these materials are eliminated and transported into a bioreactor. On the basis of removal and transportation of waste materials for treatment, two methods are followed: (1) *in-situ* bioremediation and (2) *ex-situ* bioremediation. The former method involves the contact between micro-organisms and dissolved toxicants. The micro-organisms act well only when the waste materials help them to generate energy and nutrients to develop more cells. When the native micro-organisms have no capacity to degrade toxicants, genetically-engineered micro-organisms may be applied during the process of *in-situ* bioremediation. *In-situ* bioremediation is of two types such as (1) Intrinsic bioremediation (when it occurs naturally) and (2) Engineered bioremediation (when skilfully arrange for the process to happen).

Conversion of pollutants into harmless forms through innate capabilities of naturally occurring microbial populations is called *intrinsic bioremediation*. Several conditions such as pH, temperature, carbonate, minerals, and nutrient availability favor the process of intrinsic bioremediation. When these conditions are unsuitable, bioremediation requires construction of engineered systems to supply substances that encourage micro-organisms to degrade pollutants and termed as *engineered bioremediation*. It expedites the desired biodegradation reactions by stimulating growth of more micro-organisms via optimizing physico-chemical conditions. Oxygen, electron acceptors (such as nitrate and sulfate) and nutrient carriers promote microbial growth.

19.3 What is Bioremediation

Bioremediation is the use of living micro-organisms capable of degrading environmental pollutants or to prevent pollution through waste treatment. Bioremediation is now being considered as an ideal technology which helps eliminate pollutants from the aquatic environment and rejuvenate contaminated waters for use in fish culture. The technology of bioremediation involves manipulation of micro-organisms in ponds to accelerate the process of mineralization of organic matters and to get rid of undesirable compounds. This environment-friendly technology is now being widely considered as top priority for mass aquatic food production around the world.

19.4 Necessary For Biorernediation

The use of this technology is necessary because fish culture ecosystems are being regularly contaminated with massive agricultural washes to concentrations that are detrimental to water quality and fish health. Some authorities have remarked that fish production potential may be doubled if these water areas can be made quite clean and kept unpolluted through bioremediation methods where the cost is a secondary factor. Scientists and government representatives of both developed and developing nations have recognized that this technology can have local, regional, national, and global applications and that both genetically-modified and indigenous microbes may play a significant role in this context.

19.5 Application of Bioremediation Strategy in Fish Culture

The most significant contribution to the aquatic micro-organisms is their ability to degrade pollutants.By degradation, toxicants are broken down, and convert aquatic ecosystems for use by fish culturists. Many naturally-occurring micro-organisms have better toxicant degradation kinetics and attack a wide range of toxicants. At the same time, some other micro-organisms can grow well under harsh environmental conditions such as high temperature and tolerance to organic solvents. This instruction has great practical significance in relation to bioremediation technology.

The application of genetic engineering has paved the way for the development of new strains of micro-organisms with high biodegradable capacity. Genetically-engineered micro-organisms, for example, may be developed through gene coding to encounter with the complex chlorinated hydrocarbons (such as dioxins) which are non-degradable by naturally-occurring micro-organisms. As the genetic changes of microbes occur, their survival and growth capabilities even under extreme environmental conditions are significant.

Two different central points should be taken into consideration regarding research and development on bioremediation programs throughout the world. These points include: (1) Development of innovative programs in bioremediation research for upgrading the traditional waste and waste-water treatment system adopted by many European countries to cope with the effects of specific chemical pollutants and (2) Methods developed by the Environmental Protection Agency, USA, to clean-up soil and water that have contaminated with xenobiotics and petroleum products.

Simple Decomposition Products and their Elimination

As changes in the soil organic matter occur during fish culture operations, simple inorganic products are developed. Along with carbon di-oxide, the most important but detrimental products are those containing essential nutrients for ecosystem productivity especially nitrogen, phosphorus and sulfur. For example, when proteins are attacked by microbes, amino acids are generated. These, in turn, are broken down and generate first ammonium compounds and sulfides and finally nitrates (through nitrification) and sulfates. Fortunately, most of the inorganic ions released by decomposition are readily available to water. Similar break down of other organic compounds helps release inorganic phosphates as well as cations.

The vast majority of fish culture systems accumulate nitrate as they do not contain a denitrifying filter. In aquatic ecosystems, nitrate is the major nutrient for the growth of phytoplankton and microbes. Hence micro-flora and fauna are robbed much of their growth potential and thus limits the process of denitrification.

The removal of nitrate from contaminated waters (or from waste-waters) helps prevent eutrophication. Nitrate concentration can be reduced from 75 mg/l to at least 0.1 mg/l when water is passed through a special device termed as *bioreactor* or *denitrifying filter*. A denitrifying filter creates an anaerobic region where anaerobic bacteria can grow well and reduce nitrate to nitrogen gas. The bioreactor is teemed with methylotrophic bacteria (*Methylophilus methylotrophus*) that carry out the process of denitrification. Methanol is first added to bioreactor to accelerate the growth of bacteria and later removed by trickling and activated carbon filters before the water is discharged for use in aquaculture or agriculture. The bacteria transform nitrate to nitrite and then to elemental nitrogen.

Elimination of Toxic Chemicals from Waste-water

Current waste-water treatment devices require sufficient quantities of desired specific micro-organisms that can degrade hydrocarbons and chlorinated solvents found in waste-waters. Certain species of bacteria such as *Pseudomonus capacia* are able to degrade chlorinated hydrocarbons. The waste-water can also be treated anaerobically in the bioreactor. Alkaline hydrolysis removes chloride. Effluent-containing unreacted toxicants are also treated with specific types of micro-organisms. Thus waste-waters containing huge quantities of toxic chemicals are degraded in the waste-treatment digester and the detoxified water is released safely into the ecosystem. Many developed countries have taken up bioremediation programs as cost-effective measure for restoring the quality of water.

Bioremediation is now considered as an alluring option to the physical removal and subsequent destruction of toxic substances. However, the cost of incineration of polluted substances, is at least, ten times than that of biological treatment if the latter method is carried out *in-situ* (*in-situ* bioremediation).

Though many international companies sale micro-organisms for bioremediation programs, most of their microbial products are of doubtful value. Indigenous microbes are able to degrade the pollutants and addition of introduced micro-organisms has been shown not to be involved in the process of degradation. In contrast to indigenous microbes, most of the introduced ones do not accelerate the rates of biodegradation. However, efforts have been made to alter the ecosystem by enhancing the activity of indigenous micro-organisms.

Bioremediation of Heavy Metals

The bioremeditation technology can also be used to treat waters that are contaminated with heavy metals. Bacteria, fungi and microbial-algae are able to extract these pollutants. After uptake, these pollutants are either accumulated or assimilated by these micro-organisms. A few examples will illustrate this situation. The genus *Zooglea ramigera* adsorbs cadmium and copper to the level of 100 and 300 mg/g dry weight, respectively. Bacteria belonging to the genera *Pseudomonas putida*, *Arthrobactor viscous*, and *Citrobactor* spp. remove several toxic heavy metals from effluents. Some fungi such as *Trichoderma*, *Aspergillus*, *Ophiostoma* and *Rhodotorula* are also able to absorb heavy metals in large amounts. Therefore, these micro-organisms seem to play an important role in detoxifying pollutants.

19.6 Mechanism of Bioremediation of Heavy Metals

Micro-organisms accumulate large amount of heavy metals in their cells. Several mechanisms operate in them for elimination of heavy metals from the ecosystem. Some important mechanisms are discussed below.

Metabolism-Dependent Accumulation

Metal ions are transported into the cells through cell membrane. As a result of metabolic processes, ions are precipitated around the cells, and synthesized intra-cellularly as metal-binding proteins.

Metabolism-Independent Accumulation

Although positively-charged ions are attracted to negatively-charged ligands in cell materials, the composition of biomass and other factors (such as pH and temperature) drastically affect biosorption.

Extra-Cellular Precipitation and Complexation

Fungi generate several extra-cellular products which can precipitate or form complex compounds with heavy metals. Many fungi, for example, release iron-binding compounds that chelate iron. The ferric (Fe^{3+}) chelates are taken up by the cell. In *Saccharomyces cerevisiae*, removal of metals is performed by their precipitation as sulfides.

Chromium (VI) (Cr^{6+}) is water-soluble and highly toxic and if reduced to Cr^{3+}, it becomes less soluble and less toxic. Arsenic (III) As^{3+}, on the other hand, is oxidized to $As5^{+}$, making it more easily precipitated by agricultural lime and phosphate. Likewise, Manganese (II) (Mn^{2+}), can be oxidized to Mn^{4+}, which precipitates as an insoluble manganese oxide. In a non-acidic aquatic ecosystem, oxidation of ferrous (Fe^{2+}) to ferric (Fe^{3+}) causes percipitation of ferric hydroxide. In anaerobic and SO^{2-}–rich environment, formation of reduced S^{2-} promotes formation of insoluble metal sulfides.

19.7 Bioremediation of Wastes

Pollutants are discharged into the ecosystem from a large number of industries. It has been reported that actinomycetes exhibit higher capacity to bind metal ions as compared to fungi and bacteria. The living microbes, however, accumulate metals intra-cellularly at higher concentrations, whereas dead cells precipitate metals in and around cell walls through several metabolic processes. The biomass of the fungus *Aspergillus niger* contains about 30 per cent of glucan and chitin. Chitin phosphate and chitosan phosphate present in fungi absorb substantial amount of cadmium, copper, mercury, cobalt, and calcium.

19.8 Bioremediation of Xenobiotics

Use of pesticides has helped, to a large extent, the modern society by improving the quality and quantity of the world's food production. Progressively, pesticide usage has become an integral part of the modern agriculture system. Aquatic animals are vulnerable to a vast variety of artificially-made foreign chemicals (termed as *Xenobiotics*) especially pollutants. These chemicals retain in the environment and do not undergo biological trarsformations. Use of xenobiotics with utmost care is important in learning how to cope with the pollutant onslaught.

Knowledge on the toxicity of xenobiotics is basic to a rational understanding of toxicology and management of aquatic environments. Such understanding is very important from both fisheries and aquaculture points of view.

Fishes Encounter with Thousands of Xenobiotics

A xenobiotic (Greek *Xenos means stranger*) is a compound that is foreign to the body of fish. The principal classes of xenobiotics which have been reported in fisheries and aquaculture sectors include various compounds such as polychlorinated biphenyls (PCBs), insecticides, and pesticides that have found their way into the aquatic ecosystem by one route or another. More than 2,00,000 manufactured chemicals exist in the environment. Most of the chemicals are subject to chemical alteration (termed as *metabolism*) in the ecosystem with different micro-organisms being the most important biotic factors essentially involved. Micro-organisms maintain a steady concentration of chemicals in the ecosystem concerned. The complete degradation of a pesticide molecule to its inorganic compounds that can be consequentially used in an oxidative cycle eliminates its potential toxicity from the ecosystem. The most important objective in relation to bioremediation of xenobiotics is to device bioremediation methods for detoxifying or eliminating high concentrations of harmful pesticide residues.

The characters which are responsible for pesticide biodegration are situated on plasmids (See Section 19.9) and transposons, and are grouped in clusters on the chromosomes. Understanding the characters affords some indications to the evolution of biodegradation pathways and fabricates the exercise of gene operation easier to create the genetically-engineered microbes which are able to degrade the toxicants.

Different enzymes catalyze the reactions involved in xenobiotic metabolism (for further detail, see Chapter 20 of this volume). The following section will, however, provide a brief idea of how enzymes are involved in detoxifying xenobiotic compounds.

19.9 Enzyme Technology in Bioremediation and Detoxification

Bioremediation and biodegradation can take place through use of exoenzymes. Enzymes are effective at a lower substrate concentration and are active in presence of toxins. Some examples of exoenzymes used in bioremediation are esterases, phosphatases, lipases, proteases, phenoloxidases, peroxidases, luccases, and tirosinases. The following brief examples will illustrate this point. Peroxidases catalyze phenolic aromatic or amine radical formation, and then react to form polymers. The polymers are insoluble in ecosystems. Laccases and tirosinases catalyze oxidative coupling reactions of organic compounds to themselves and to other organic materials. The toxicity of polymers and bound compounds in the ecosystem significantly decreased.

Gene Manipulation of Pesticide-Degrading Micro-organisms

Day after day, the number of xenobiotic-degrading micro-organisms in aquatic environments is on the increase. Pesticide-degrading genes of only a few micro-organisms have been characterized. Most of the genes which are responsible for catabolic degradation are located on the chromosomes; but in a limited number of cases, these genes are found on plasmids. The discovery of some pesticide-degrading genes (such as naphthalene degrading plasmid) has made it possible to construct a new genetically-engineered strains of *Pseudomons putida* that is potent to degrade other compounds such as xylene, toluene, camphor, hexanes, etc. Strains of *Glioeladium virens* (a fungus) produced by transferring a recombinant DNA plasmid are potential for use in the bioremediation of contaminated soil. Genes isolated from *Flavobacterium* sp. and *Pseudomonas diminuta* are highly associated with the degradation of pesticides such as parathion, methyl parathion, etc. These examples clearly depict how pesticide degradation occurs through plasmid-mediated genetically-engineered micro-organisms that holds great promise in the abatement of the pesticide pollution to a large extent.

19.10 Advantages and Disadvantages of Bioremediation

Though the process of bioremediation is very important to clean up fish culture ecosystems for sustained production, a number of advantages and disadvantages that follows have been identified.

Advantages

If complete mineralization occurs, the end products of toxicants are non-toxic. Other biological efficiency of the contaminated water is left relatively quiet. It is not costly when compared to physical method for clean up and very simple equipments are necessary for the purpose.

Disadvantages

A high waste concentration is sometimes needed to expedite the growth of micro-organisms. Bioremediation is limited by the environmental conditions that exist in the polluted sites. Under extreme conditions of the environment such as too hot, cold, dry, acidic and alkaline, bioremediation process is worked with reduced vigour or stopped completely. Also, bioremediation process is limited by the availability of time for treatment. In contrast to treat the contaminated sites by physical or chemical means, the bioremediation process takes longer time.

To sum up, the naturally selected strains of micro-organisms available in bioremedial products, due to their rapid biodegradation capabilities of organic wastes, would restore the health of the pond bottom for the next crop and consequently, would promote sustainable aquaculture development. By applying the knowledge gained from fundamental research, bioremediation technologies have been used to eliminate environmentally-hazard chemicals or to detoxify them into non-toxic forms. Bioremedial technology an be regarded as a perfect rejuvenator of aquaculture systems for high production. It is, however, hoped that farmers would attempt to apply these technologies and to increase production potential as well as their income.

19.11 Phytoremediation

Aquatic plants are important components of pond, lake, river, and reservoir ecosystems. Inspite of their nuisance features, the ecological and environmental significance of these plants, particularly their capability to improve water quality, has created substantial interest in utilizing them for beneficial purposes.

The recovery potential of a damaged ecosystem depends on the extent of damage caused and mode of operation of the stressor. A number of direct options such as diversion of wastewater input and its treatment, inactivation of nutrients in the input water, aeration, artificial bubbling, liming of acidified ponds/lakes, alum treatment, biomanipulation, etc. are widely used in many freshwater ecosystems. Several sophisticated ecotechnological models are successfully adopted to curb eutrophication mostly in large freshwater ponds and lakes in developed nations but these models are unfit for tropical small and shallow ponds utilized for fish culture in developing countries.

Simply defined, phytoremediation refers to the process by which hazardous substances are removed by vascular plants from contaminated water or wastewater. Some water-loving plants such as *Azolla, Pistia, Lemna, Eichornia,* etc. are known to tolerate, uptake and even accumulate heavy metals, metabolites of fertilizers and other toxicants. These plants have tremendous biotechnological potential to remove hazardous substances from water and even the use of genetically-engineered plants in fish culture ecosystems is on the increase.

Stepwise Lagoon System for Toxicant Removal

A simple design which is termed as *stepwise lagoon system* has been developed which helps eliminate toxic substances from the pond water. A lagoon is an artificial pool for the treatment or to accommodate an overspill from surface drains. However, about one metre cube asbestos containers are fixed in stepwise system (Figure 19.1) and various dilutions of the effluents from fish culture ponds are poured in the containers. Each container is loaded with *Eichornia* or *Pistia*. After a retention time of 2-3 days, the effluents are flushed out into the next container and again after 2-3 days into the third container. Finally, the water which is collected from the last container, is found to be free from toxic substances. In an experiment, a linear increase in the absorption of heavy metals, for example, has been noted with increasing their concentration in *Eichornia* and *Pistia*.

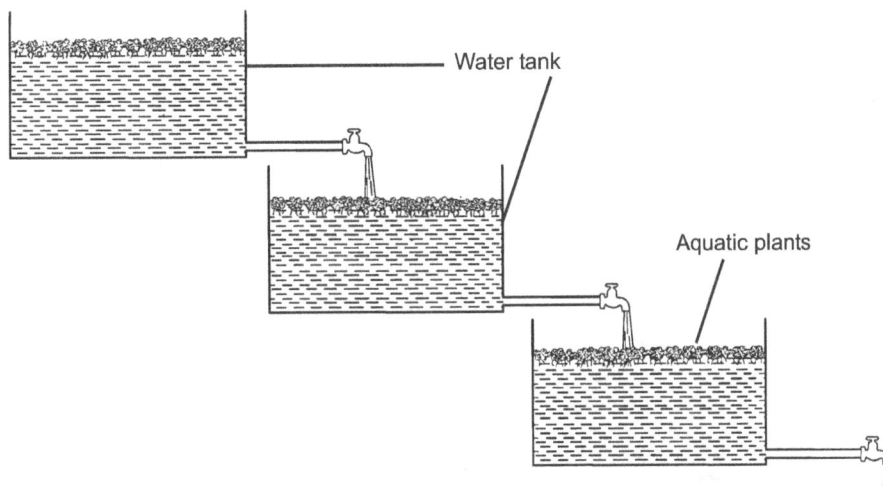

Fig. 19.1 : Diagramatic Design of Stepwise Lagoon System for Elimination of Toxic Substances

A Multi-looped Recycling Model of Fish and Duckweed Aquaculture

In this type of model, the water recycling system generally lasts for 2-6 hours, discarding the solid fish waste substances and avoiding ammonification as well as enriching the water with oxygen (Figure 19.2). The daily nutrient recycling systen via the cultivation of floating aquatic plant in aquaculture (especially termed as *duckweed aquaculture*) takes place after 10-14 days of thermophilic digestion of the solid wastes accumulated in intensive fish culture systems. The fermented dissolved organic and mineral compounds are used as nutrients for the growth of aquatic plants. Fresh aquatic plants regularly substitute up to 30 per cent of the total fish fodder.

To sum up, the use of biological agents in the reclamation of tropical freshwaters for fish culture is of great ecological significance. A vast number of tropical eutrophic water bodies are profitably used for fish culture after restoration through biological agents.

19.12 Probiotic Substances for Bioremediation*

The term probiotics was first coined by Parker in 1974. It was derived from two Greek words: *pro* and *bios* which denotes for life. The previous concept of disease treatment through biological means using bacterial supplements that consist of a single or mixed culture of selected non-pathogenic bacterial strains has now received considerable attention. Probiotics generally include certain kinds of micro-organisms such as bacteria, fungi, and actinomycetes. Fish and fish food organisms extract essential nutrients from the environment for their survival and growth. Nutrients are derived from microbial decomposition of organic wastes. It is important to note that a number of micro-organisms which are collectively termed as *effective microbiota* are known to improve the quality of water and inhibit the pathogens in fish culture ponds. Probiotics have a high fish production potential and hence they are widely recommended for commercial fish culture.

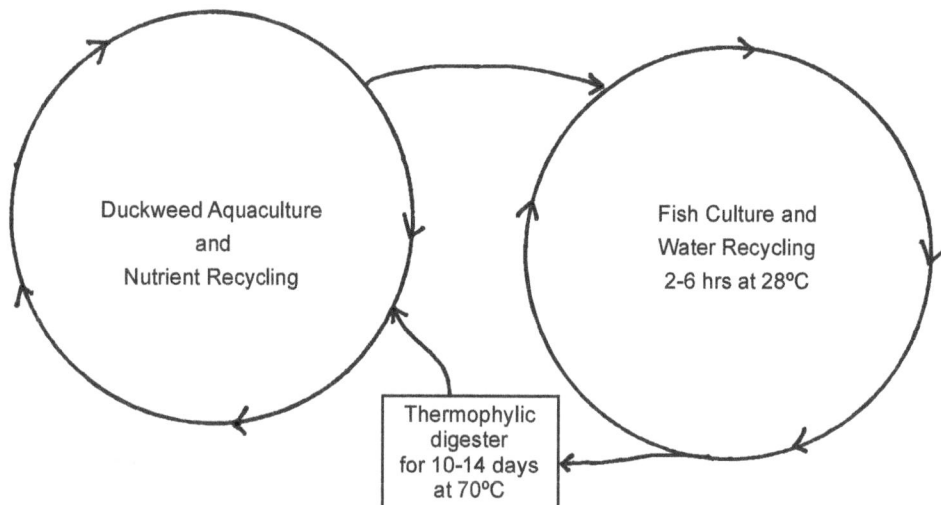

Fig. 19.2 : A simple multi-looped recycling model of fish and duckweed (*Lemna* sp.) aquaculture. The duckweed has unique palatable and crisp features as compared with other closely related species. While initial economic success resulting from marketing the duckweed product as a natural health food is welcomed at present, the principal goal has been to use duckweeds for recycling the pond water by generating a ready-to-serve leaf protein biomas. Though this remains a dreadful need in modern intensive fish farming and even though the recycled duckweed biomas was proven to be a useful available protein source, commercial acceptance of this technology should not be neglected.

Several water probiotic products have recently been developed and accordingly their potential use in fish culture industry is being recognized all over the world. These products are virtually a boon to fish farmers. They have the capacity to generate probiotic bacteria in large numbers when applied in grow-out ponds. These products help to improve the quality of water by balancing between the population of bacteria in water and reducing the pathogenic bacterial load.

---

* For further detail, see Fuller (1989).

Use of feed, nutrient carriers, and other production-related substances in intensive fish culture ponds result in accumulation of waste products in the pond mud. Degradation of these wastes also results in the production of ammonia, nitrite and hydrogen sulfide. These toxic substances require detoxification through nitrification. Several species of bacteria such as *Nitrosomonas* and *Nitrobactor* are involved in nitrification. Since pond waters are not always teemed with these bacteria, it is necessary to create conditions favorable for the multiplication of these bacteria. The water probiotic products contain selected strains of these nitrifying bacteria along with other natural probiotic strains of *Bacillus*, *Aerobactor*, and *Cellulomonas*. The products also contain biochemical accelerators and several enzymes such as amylase, protease, lipase, lactase, and hemicellulase. The scientific and technological advances have, however, resulted in enormous increases in the harvest of aquatic organisms of economic importance. An effective option which is being projected of late has been the efficient use of probiotic products in fish culture.

19.13 What is Probiotic*

A probiotic is a live microbial feed supplement which beneficially affect the host animal by improving its intestinal microbial balance. In fish culture, however, addition of selected strains of live bacteria to ponds or tanks is highly encouraging. It is also used in fish feed. Fish health is improved by eliminating pathogen load in pond water. At the same time, the water quality standard is also improved. Therefore, probiotics are used in fish culture not only as water additives but also as feed additives.

Probiotics contain different types of enzymes and bacteria. Biovet YC, for example, has the following compositions: *Lactobacillus acidophilus* (30,000 million Cfu/kg), *L.sporogenus* (30,000 million Cfu/kg), *Saccharomyces cerevisiae* (125,000 million Cfu/kg), alpha amylase (5 g/ kg), seaweed extract (50 g), and excipient (9.5 g). However,the terms probiotics, probiont, probiotic bacteria, friendly bacteria or beneficial bacteria are virtually synonymous and are used as a general term probiotic bacteria.

19.14 Types of Probiotics

Generally speaking, two forms of probiotics are available in the market such as (1) Dry spore form and (2) Liquid spore form. The former type is available in different brand names. Dry spores become active when they absorb water and in the presence of organic matter, spores germinate into vegetative forms. On the other hand, the liquid spore suspensions provide ready-to-use form in which the hydrated spores rapidly germinate, give rise to vegetative cells and multiply in large numbers in an activation medium. The activation medium contains formulated peptones, nutrients, and minerals which accelerates the growth and multiplication of bacteria. The ready-to-use probiotics in cultivation medium are by far the best than the powdered dry spore forms.

---

* For further detail, see Gibson and Roberfroid (1995).

19.15 Features of a Probiotic

A probiotic is characterized by the presencec of a strain (or more than one strain) efficient of generating beneficial influence on fish and aquatic ecosystems. Beneficial influence includes: neutralization of toxin, production of anti-bacterial compounds, alteration of microbial metabolism, stimulation of immune systems of farmed fish, production of enzymes and hydrogen sulfide, acceleration of the sediment decomposition, regulation of aquatic microflora, control of pathogenic bacteria, increase in population of fish food organisms, and reduction of toxic gases such as ammonia, methane, and hydrogen sulfide. In addition to these characteristic features, probiotics must have the following additional features: (1) They will be non-toxic, (2) They will contain viable cells in large numbers, (3) They will be capable of surviving and neutralizing the environrment of the alimentary canal of fish, (4) under field condition, the product would be stable and capable of remaining viable for prolonged period, (5) They will create better habitat for fish, and (6) They will be able to resist over-growth of algae by utilizing phosphate.

19.16 Mechanism of Action of Probiotic Bacteria

The probiotic bacteria have the capacity to generate antibiotics such as bactriacin and polymyxin that inhibit the spread of pathogenic bacteria. When the pathogenic bacteria such as *Vibrio* spp., for example, secrete mucus substances in large amounts to bind more organic matter as their food, probiotic bacteria produce exo-enzymes which breaks down the mucus substances, utilize the organic matter and make the nutrients unavailable to the growing vibrios. In other words, the probiotic bacteria compete with the food of vibrios resulting in their elimination. The enzymes are specific in digesting the organic waste matter in the pond water.

The production of mucus by pathogenic bacteria far in excess causes anoxic conditions, producing hydrogen sulfide and ammonia which are detrimental to fish. Thus, while different forms of bacillus bacteria remove mucus by generating exo-enzymes, they prevent the accumulation of ammonia and hydrogen sulfide and thus the quality of culture ecosystem is greatly improved.

The bacillus bacteria digest the excess dissolved organic carbon that eventually leads to algal growth, thereby controlling algal blooms.

19.17 Response of Fish and Aquatic Ecosystem to Probiotics

A number of probiotics such as Environ - AC, Biogreen, Biostart, BRF - 13A, Cling, PB - 32, PB2 - 44, Pond Pro - VC, livol, aquagram, Super-PS, Bioboost Forte, G- Probiotic, Biovet YC and the like are known to improve the water quality criteria and to stimulate the growth and survival of farmed fish. High quality formulated feeds if supplemented by small quantities of probiotics, are effective in enhancing the survival and growth of fish. Typical responses of fish and ecosystem to probiotic application are shown in Table 19.1. Increased response at optimal environmental factors of aquatic ecosystems suggests that probiotics exhibit synergistic effect on the potentiality of microbial activity in releasing

**Table 19.1 : Effects of Added Probiotic Biovet YC on Fish *Cirrhina mrigala*
and Aquatic Ecosystem under Indian Condition**

Treatment rate (per cent)	Parameter			
	Water			
	pH	Dissolved Oxygen (mg/l)	Total alkalinity (mg/l)	Ammonia nitrogen (mg/l)
Control	7.0 – 7.4	4.5 – 8.5	58.6 – 85.0	0.42 – 0.83
0.25	7.0 – 7.3	4.9 – 7.4	61.5 – 86.0	0.42 – 0.83
0.50	7.0 – 7.4	5.5 – 8.4	74.1 – 86.1	0.42 – 0.83
0.75	7.0 – 7.3	4.3 – 7.3	60.4 – 86.1	0.42 – 0.83

Fish					
Initial Weight (g)	Final Weight (g)	Net Gain (%)	Weight Gain (%)	Survival (%)	Food Conversion Ratio
1.83	4.34-	2.51	137.16	72.2	3.01
1.87	4.81	2.94	157.20	66.7	2.76
2.05	6.70	4.64	226.30	88.9	2.19
1.96	5.65	3.69	189.20	66.67	2.49

Source : Selected data from Murty ad Naik (2002)

active substances. The responses obtained from different rates of probiotics are generally higher. This is due to the fact that the probiotic substances have significant potential benefits like food conversion ratio, reduction in blue-green algae and nitrogen level, availability of dissolved oxygen, and control of obnoxious gases such as methane, ammonia and hydrogen sulfide.

The application of probiotics has some favorable effects on fish. Probiotics can serve as disease-resistant products, thereby protecting and conserving fish stocks. Unfortunately, certain amounts of probiotics may be lost through dissociation/dispersion in the ecosystem if proper dose is not evaluated. Therefore, encapsulated probiotic bacteria coated with suitable adhesives such as guargun, carragenan, etc. are used in feed. However, studies on the effects other probiotics on wide variety of fish species of economic importance under different farming conditions are badly needed.

19.18 Antibiotics Versus Probiotics

Antibiotics are chemicals prepared from micro-organisms that retard the growth of, or destroy, other micro-organisms. They are used in fish feed to treat infections caused by bacteria, fungus, protozoa, and worms. Antibiotics are produced only by micro-organisms, mostly fungi or bacteria. Compounds that are toxic to micro-organisms have been made synthetically.

For the last four decades, broad/narrow-spectrum antibiotics* are being widely used in all production systems of farmed animals including fish as therapeutic agents and growth promoters. However, the excessive use of antibiotics in fish culture systems has not only resulted in the development of resistant pathogenic bacteria but also makes the fish stocks vulnerable to diseases. For this reason, the use of antibiotics for high aquatic production is not considered in most cases as a permanent solution to avoid pathogenic bacteria from fish culture systems.

Recently, the presence of residual antibiotics in finfish and shellfish has led to rejection of imported products from the western markets (particularly from USA, UK, and Japan). Residual antibiotics have, however, been detected in several species of shellfish cultured in China and Thailand. Consequently, several restrictions have been imposed on the entire products. This is due to the fact that therapeutic use of antibiotics has resulted in intestinal disorders. This is perhaps due to the side effects of the antibiotics which tend to remove indigenous gut flora along with disease-causing agents.

In contrast to antibiotics, the use of some compounds derived from several species of bacteria such as *Bacillus subtilis, B. megateterium, B. polymyxa, B. pasteurii, B. laterorsporus B. circulans. B. macerans, B. licheniformis, Lactobacillus bulgaricus, L. acidophilus, L. casei, L. lactis, L. salivaricus, L. plantarum, Streptococcus thermophilus Enterococcus faccium, E. faccalis* and *Escherichia coli* have been found to promote significant yield of fish in productive water bodies.

Management of microbial populations in fish culture ecosystems to combat pathogenic bacteria and viruses are otherwise termed as *microbial biotechnology* and must take this strategy into consideration.

Probiotics are generally used through feed in the form of capsules, paste, granules, or powder before the occurrence of diseases whereas antibiotics are recommended at the onset of disease. Probiotics are not considered as drugs to combat diseases. Probiotics consist of single-cell organisms available in spore and vegetative forms. Unless activated, bacterial spores remain in dormant state and under favorable conditions, they become active and are transformed into vegetative forms.

19.19 Conclusion

Accumulation of toxic substances in fish culture ecosystems is becoming increasingly important to world fishery industry as the level of toxicants is on the increase. Studies on the application of biomemediation technology and probiotics confirm that the use of these technologies not only increase fish survival and growth but improve water quality as well. And this indicates that attention to the use of bioremediation products and probiotics in fish culture will likely to increase in the future.

---

* Broad-spectrum antibiotics are those which are effective against a range of micro-organisms and narrow-range spectrum antibiotics are effective against only a single species or one group of micro-organisms.

In most cases, ecosystem management practices in right directions throughout the culture period will reduce the risk of toxic accumulation considerably. But increasingly observed situations where severe problems emerge due to pollution can, perhaps, be solved only by the application of bioremediation technology. This technology is becoming a common strategy and will likely to become even more necessary in the future.

References

Alexander, M. 1994. Biodegradation and Bioremediation, Academic Press, San Diego.

Fuller, R. 1989. Probiotics in man and animals. *J. Appl. Bacteriology.* 66 : 265-278.

Gibson, G. R. and Roberfroid, M. 1995. Dietary modulation of the human colonic microbiota: Introducing the Concept of Probiotics. *J. Nutri,* 125 : 1401-1412.

Murty, H. S. and Naik, A. T. 2002. Growth performance of *Cirrhinus mrigala* in relation to a natural probiotic feed supplement. Biovet YC. *Fishing Chimes* 22 (4) : 54-56.

Queiroz, J. F. and Boyd, C. E. 1988. Effects of a bacterial inoculum in channel catfish ponds. *J. World Aquaculture. Soc,* 29 (1) : 67-73.

Tripathi, L. K. 2005. Water and Soil Management. ABD Publishers, Jaipur.

Questions

1. Describe how enzymes are involved in bioremediation for toxic compounds.
2. Identify two approaches to bioremediation.
3. What is biomemediation? In what ways it is good for ecosystem clean-up?
4. State how bioremediation strategy is applied in fish culture.
5. Mention how toxic substances are eliminated from aquatic ecosystems.
6. Discuss how probiotic substances act on aquatic ecosystems for high production.
7. What is probiotic ? Mention different types of probiotics. Mention the mechanism the action of probiotics.
8. Write notes on the following.

(a) Xenobiotics, (b) Phytoremediation, (c) Advantages and disadvantages of bioremediation, (d) Probiotics versus antibiotics, (e) Detoxification mechanism.

20

Aquatic Toxicology and Ecotoxicology in Relation to Fish Culture

In the past three decades, aquaculture industry has been growing in leaps and bounds. Though the global finfish and shellfish stocks in freshwater ecosystems have reached a plateau of production, their intensive culture had been frequently affected by a vast variety of toxicants resulting not only in huge loss of fish biomass but also destroy aquatic habitats to a large extent.

Toxicology is the investigation of unfavorable effects of any chemical on aquatic organisms. Potentially serious toxic effects of chemical substances (such as detergents, fertilizers, pesticides, insecticides, and heavy metals) in fish culture and fisheries include a variety of symptoms in fish that follow. Since the demand of these chemical agents as crop protection, production and disease control strategy stances has rapily increased in almost all the countries of the world after the introduction of high-yielding varieties of crops, it seems plausible that the contamination of water with toxicants and their accumulation in fish culture ecosystems present very real hazards to fish that has definitely robbed much of its fish production potential by disrupting the food chain and nutrient dynamics. Therefore, the toxic effects of different chemical agents on fish culture ecosystems must be evaluated accurately before their extensive use is permitted.

In this chapter, several issues pertaining to fish culture ecosystems such as biological effects of toxicants and their removal, biomonitoring of toxicants, safe disposal of toxicants, fish toxicity test, effects of toxicants on ecosystems, effects of acid deposition in aquatic ecosystems, ecotoxicology, etc. have been discussed in brief so that fish culturists may fully (i) recognize the long-term significance of water and soil and (ii) may become aware of the hazardous effects of toxic substances on many commercial species of fish when their culture operations are undertaken on economic basis.

20.1 General Considerations*

Pollution is defined as the disruption of a normal ecosystem (physico-chemical and biological properties of the ecosystem) as a result of human activities. Pollution can also be natural caused by hydrological processes in which the decomposed plant and animal substances and weathering products of minerals and soil ingredients are brought into the

---

* For further detail, see Abbasi (1989), Anna and Harding-Barlow (1987), Trivedy (1998, 2001), Yasue (1986), Handa (2000).

main water resources. These characteristics are, however, inter-dependent. Addition of an organic waste, for example, will not only influence the odor, color and biological properties but also influence the chemical characteristics simultaneously. In a fish cultute ecosystem, however, pollution generally occurs when sufficient quantities of harmful substances are intentionally or unintentionally released into the water resulting in severe damage to fish populations. Harmful substances which cause pollution are termed as *pollutants*. The use of chemical fertilizers, agricultural wastes, pesticides, and herbicides to increase crop production leads to water pollution and consequently, toxic substances are accumulated in the food chain. Toxic substances that are likely to accumulate in the organisms of food chain are generally termed as *toxicants*. Toxicants are transferred through food chain from one trophic level to other. As a generalization, four types of toxicants are encountered in fisheries sector such as (1) radio-active substances, (2) agricultural chemicals, (3) industrial wastes and (4) solid wastes. Among all these types of toxicants, the last three types are most common and important. Agricultural chemicals include nitrogen and phosphorus fertilizers, polychlorinated byphenyls, heavy metals, pesticides, and herbicides (Figure 20.1). Fish and fish food organisms are often contaminated with these substances. Consumption of contaminated fish results in serious human health hazards.

20.2 Ecological Aspects of Toxicants

Fish are always confronted with a vast variety of toxicants. In aquatic ecosystems, over a billion of cases of fish poisoning occur around the world each year, a large fraction are fatal. Fish deaths due to ingestion of chemicals have been markely increased in the past forty years as a result of serious exposure. Consequently, in many cases, total fish stocks from ponds/lakes have become vanished. Safety packaging and effective poisoning prevention training/education to farmers have not been taken into consideration.

From ecological point of view, two types of toxicants can be apprehended: (1) non-degradable toxicants and (2) degradable toxicants. In toxicology, however, the effects of chemical substances (or toxicants) on living forms are evaluated. Many toxicants such as trace metals, mercury and organic chemicals (such as DDT) are not susceptible to degradation and consequently accumulate in aquatic micro-organisms. Toxicants accumulate in phytoplankton cells which arethen consumed by zooplankton. When contaminated zooplanktons are consumed by fish, the toxicants persist within the fish flesh and ultimately entered into the human body. It is interesting to note that many chlorinated hydrocarbons such as DDT, aldrin, heptachlor, etc. are not soluble in water and hence contains several parts of pesticides per billion parts of water. When these chemicals are transported to different trophic levels, the concentration of toxicants in the highest trophic level greatly increased up to several thousand parts of pesticide in the body. The best known example is DDT. While this compound is generally present in water at the rate of 1-2 mg/l, it is further concentrated in fish flesh up to 3,000 mg/l. The increased accumulation of toxicants in the food chain is referred to as biological *magnification*. Many species of fish have shown adverse effects as a result of this accumulation. The ratio of a toxicant in any aquatic organism to that of the ecosystem is called *concentration factor* (CF).

Figure 20.1: Structural formulas of some agricultural chemicals: 1-3, Organo-chlorine compounds; 4-6, Organophosphorus compounds; 7, A carbamate insecticide; 8-12, Contact herbicides; 13-16, Systemic herbicides; 17-19, Contact insecticides; 20-21, Systemic insecticides; 22-23, Nitrogen fertilizers; 24, Single superphosphate; 25-26, Polychlorinated biphenyls (PCBs). At present, on a world-wide basis, more than 10,000 chemicals are in commercial use as pesticides, herbicides and a few members of nutrient carriers. It is estimated that the total number of pesticide formulations may exceed one lakh in the coming years. This report definitely reflects the overwhelming contribution to hazardous effects of these chemicals on aquatic life.

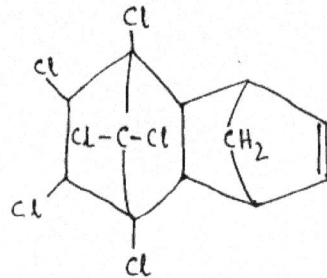

1. Aldrin

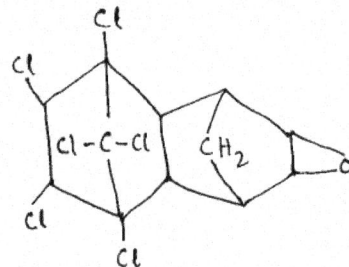

2. Dieldrin

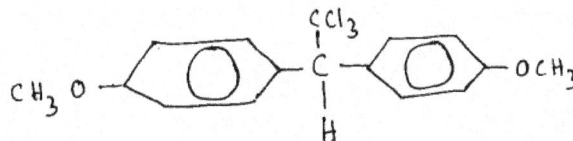

3. Methoxychlor

Organo-Chlorine Compounds

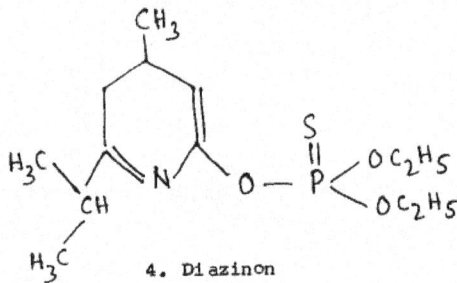

4. Diazinon

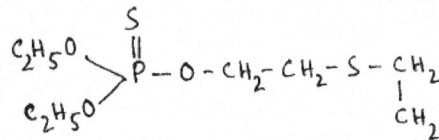

5. Dimeton

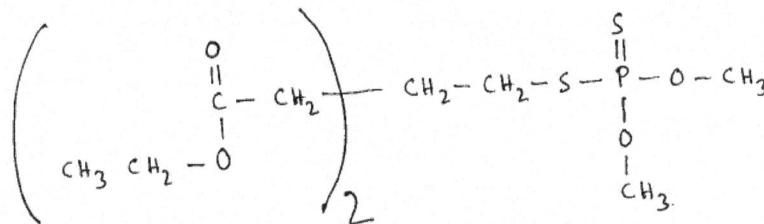

6. Malathion

Organo-Phosphorus Compounds

Figure 20.1–*Contd...*

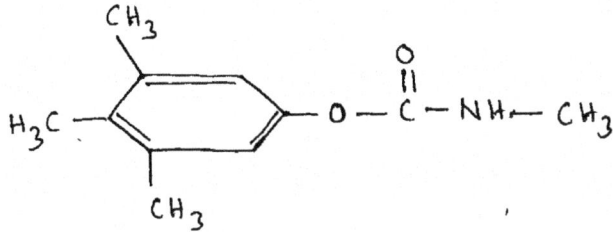

7. Methiocrab (A Carbamate insecticide)

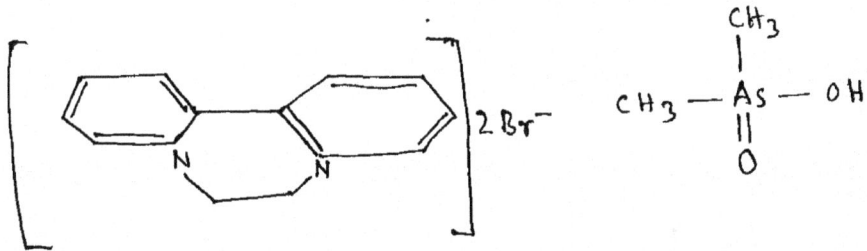

8. Diquat

9. Cacodylic acid

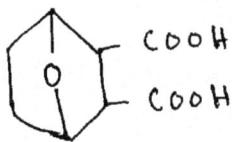

10. Endothal

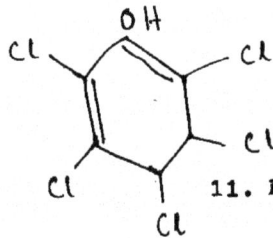

11. Pentachlorophenol

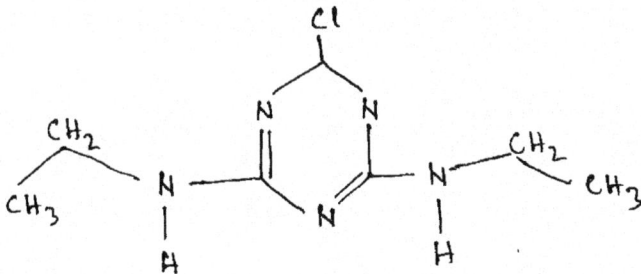

12. Simazine

Contact Herbicides

Figure 20.1—*Contd...*

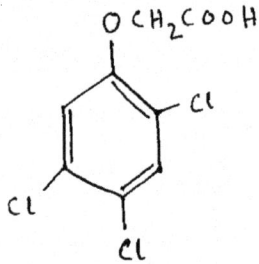

13. 2,4,5-T

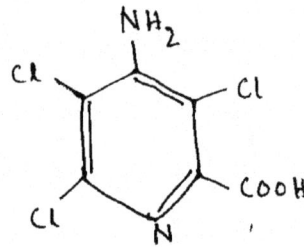

14. Picloran

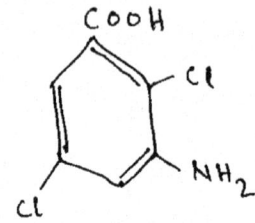

15. Chloramben

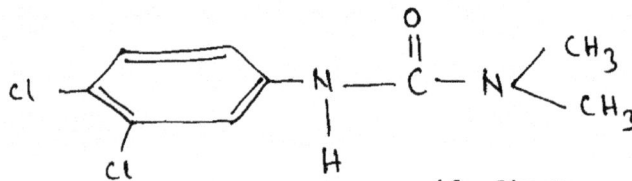

16. Diuron

Systemic Herbicides

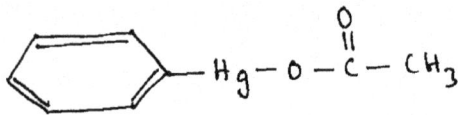

17. Phenyl mercury acetat

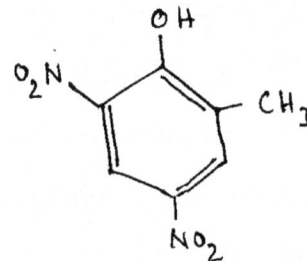

18. 4,6-Dinitro-O-Cresol

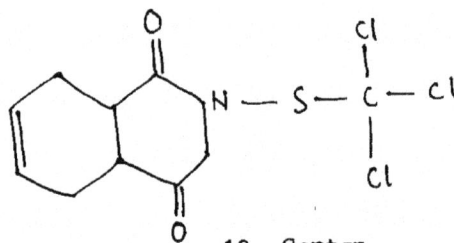

19. Captan

Contact Insecticides

Figure 20.1–_Contd..._

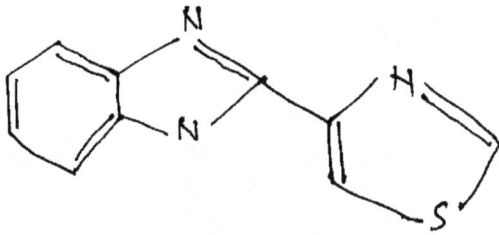

20. Thiabenzadole

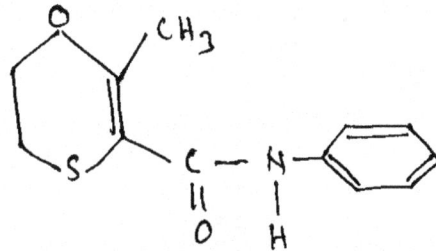

21. Carboxin

Systemic Insecticides

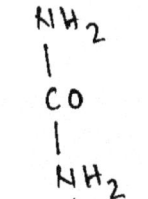

22. Urea

$(NH_4)_2 SO_4$

23. Ammonium Sulfate

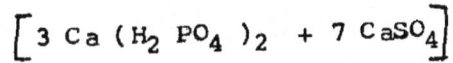

$[3 Ca (H_2 PO_4)_2 + 7 CaSO_4]$

24. Single Superphosphate

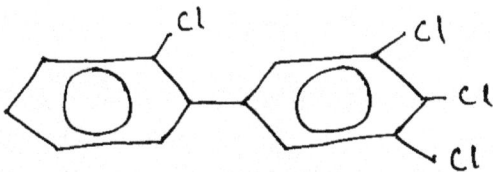

25. 2,3,4,5 – Tetra Chloro Biphenyl

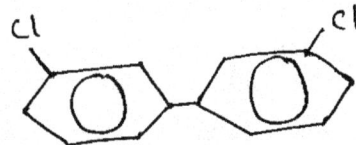

26. 3,3 – Dichloro biphenyl

Polychlorinated biphenyls

Biodegradable toxicants, such as domestic wastes, can be decomposed either by sewage treatment plants or by natural processes. Problems arise when the discharge of such toxicants into the aquatic ecosystem exceeds the capacity of ecosystem to decompose them. Due to accumulation of biodegradable pollutants and other wastes in freshwater ecosystems, they are not able to recycle and consequently their self-regulatory capacity is lost. The decomposition of toxicants by aerobic bacteria drastically reduced due to higher levels of pollution. Such water bodies become unfit for fish culture. Since decomposition of toxicants is an aerobic process, their accumulation in water results in decrease the content of its oxygen. Most of the toxicants are vulnerable to microbial decomposition although the rate of breakdown of some toxicants is slow, causing residual problems.

20.3 Contamination of Fish Culture Ecosystem With Toxicants

As noted earlier, aquatic ecosystems are contaminated by a number of toxic substances that, to a greater or less degree, are toxic to fish. Certain heavy metals such as arsenic, chromium, cadmium, and mercury are extremely toxic; lead and nickel are moderately so; and copper and zinc are relatively low in toxicity.

Sources and Accumulation

There are many sources for the chemical contaminants that can accumulate in waters. The sewage and other wastes, run-off fields, industrial effluents, agricultural discharges, thermal and nuclear power plants, and industrial wastes release into the aquatic ecosystems tonnes of these elements which is carried for kilometres and later deposited on the soil and plants.

A number of fungicides, herbicides, detergents, and fertilizer contain appreciable quantities of heavy metals. Limestone and superphosphate usually contain small quantities of cadmium, copper, manganese, zinc, and nickel. Arsenic, for many years, is used as an insecticide. Some of these heavy metals are found as constituents in specific organic pesticides and in industrial and domestic sewage sludge.

The amount of most of the products in which toxicants are used have increased notably in recent years, enhancing the opportunity for contamination. They are present in the ecosystem in increasing amounts and are consumed by fish and fish food organisms.

Phosphates are used in detergents. They favor excessive growth of algae which forms water blooms. These blooms consume most of the available oxygen from water and upon decay they produce foul odor. Some algae are known to produce toxins which kill animals including fish.

Concentration in Organisms

Whatever their sources, toxicants can reach the ecosystem where they become part of the life cycle of soil → water → plankton → fish → man (Figure 20.2). Unfortunately, once the toxicants become part of this cycle, they may accumulate in fish and human body to toxic levels. Residual accumulation of some pollutants such as sublethal concentrations each of benzene hexachloride (3.5 and 1.75 mg/l) and cadmium (1.5 and 0.75 mg/l) in *Cyprinus carpio* for example, (lethal concentration 50 or so called LC 50 for cadmium and benzene hexachloride = 3 and 7 mg/l, respectively) may reflect the possibility of toxicity to humans through food chain.

Human and Animal Wastes

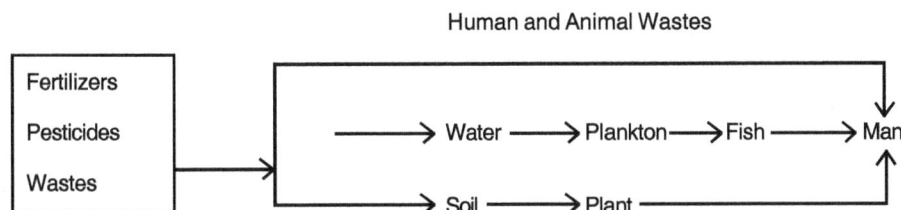

Figure 20.2: Source of toxicants and their cycling in the soil-water-organism-ecosystem. Note that the content of toxicants construct from sources to the target organisms indicating the vulnerability of humans to toxicants.

20.4 Biological Effects of Toxic Substances*

Modern agricultural practices generally use a large number of chemicals in the form of fertilizers, insecticides, pesticides, herbicides and heavy metals. At the residual forms of these chemicals along with other organic debris are trapped by the surface flow-off and thereby pollute the receiving water. The agricultural flow-off is considerably, rich in nitrogen, phosphorus, organic matter, pesticides, insecticides, herbicides, and heavy metals. While the first three forms create a lot of problems of eutrophication of aquatic ecosystems, the last four are responsible for causing toxicities to aquatic life.

Pesticides, herbicides, and heavy metals enter the human body as a result of consumption of contaminated aquatic food items particularly fish. These toxic substances accumulate in fish body and cause damage to kidney, gills, flesh, etc. or even result in death of fish stocks if taken in higher concentrations. Metabolites of fertilizers also cause similar effects. The presence of toxic substances in fish body is, however, widespread. For this reason, the over-use of fertilizers and pesticides has been severely criticized not only because of their effects on the aquatic ecosystem in general, but also because of their adverse chronic effects on fish and man in particular. The following sub-sections will provide an overall conception about the damaging effects of toxicants on fish communities.

Survival and Growth

Different agricultural chemicals at sub-lethal and lethal levels are toxic to eggs, fry, and fingerlings of fish and are known to reduce the survival and growth of many freshwater fish species. Toxicants when enter into the fish body, inhibit the activity of digestive enzymes and also cause severe damage to the buccal epithelium and taste buds

---

* For further study, see Murty (1986) and Motsumura (1985).

of fish and consequently, the palatable nature of food materials cannot be detected by the fish. As a result, the exposed fish lose their appetite.

Physiological Effects

The presence of toxicants in any fish culture ecosystem decrease in oxygen consumption; haemoglobin, total number of red blood cells, plasma protein in blood; glycogen and protein content in flesh and liver. At the same time, blood glucose and cholesterol levels also increased. Increase in mucus secretion from gill epithelium results in heavy precipitation of the mucus causing clogging of the respiratory sites leading to aphyxiation and death of fish.

The presence of toxicants in blood for prolonged periods may also exhibit several physiological abnormalities (such as alteration in blood chloride level) and the effects of toxicants on gill damage of freshwater fish have been well documented. Gills are responsible for chloride regulation and therefore, gill damage might be accompanied by an impairment of the salt balance. Fish when exposed to sub-lethal levels of urea (LC 50 = 1,340 mg/L) or calcium ammonium nitrate (LC 50 = 151.2 mg/L), for example, exhibited significant increase in blood chloride level at day 15. At 15 day onward, the blood chloride level decreased.

Histopathological Effects

A vast variety of toxic chemicals induce severe histopathological effects in fish. The liver, gills, kidney, stomach, intestine, and brain are the principal organs involved in damage. Tcxico-pathological manifestations are characterized by (1) necrosis and vacuolization of the epithelial cells of liver and kidney, (2) shedding off damaged epithelial cells in the lumen of the air bladder, (3) swelling and rupture of epithelial cells, and (4) changes in the epithelial lining of the branchial diverticulum.

Effects on Reproduction

Sub-lethal and lethal levels of toxicants inhibit spawning in fish. Drastic reduction in the fecundity rate and gonao-somatic indices of many freshwater fish have also been noted. Other effects on reproduction include arrested ovarian recrudescence, cessation of gametogenesis, inter-follicular oedema in ovary, and damage to the basement membrane in testes.

Behaviour of Fish

Exposed freshwater fish at lethal concentrations of toxicants generally exhibit high rate of opercular movement, and secretion of mucus all over the body and gills. Spontaneous, irregular and erratic movements, loss of equilibrium, and swimming upside-down are also common behavioral symptoms. Other symptomps include hyperexcitability, perpendicular movement towards the surface with jerks, violent swimming on the surface, los of balance, floating on the surface and drowning passively to the bottom. Fish mortality in fertilizer-contaminated waters may be attributed to the failure of cardiac muscles to contract due to increase in the levels of ammonium and potassium ions.

Disastrous Effects of Toxicants on Fish Biomass

Fish culture ecosystems that receive toxicants from neighbouring agricultural fields or from domestic sewage and industrial effluents may exhibit mass mortality of fish. One such episode was observed in a fish culture tank situated in Mysore, India in May 2001. Untreated sewage and industrial effluents accidentally entered into the tank were responsible for catastrophe. The tank has a water-spread area of about 100 hectare with an average depth of 6 m. The tank ecosystem was severely affected by the onslaught of toxicants and the fish kill continued for several days. More than 100 tonnes of fish valued at about Rs. 35 lakhs were damaged, more than two hundred fishermen were victimized and put out of work.

Development of Resistant Species

In areas where fertilizers and pesticides are extensively used, several resistant species of fish such as mosquito fish, bluegills, golden shiners, and black bullheads have evolved. The development of resistant fish may be beneficial with regard to their survival, but when a non-resistant predator consumes them, it can receive a lethal dose of toxic materials and get killed.

Antagonistic and Synergistic Effects*

Problems of pesticide and heavy metal toxicity to fish and aquatic organisms get further complicated by interactions when two or more chemical substances are present together. Several members of pesticide and heavy metal combinations have been found to be antagonistic in toxicity to fish. Mixture of two heavy metals such as copper sulfate and cadmium sulfate, for example, at different ratios (1:1, 2:2, 3:3, and 4:4) may exhibit less oxygen consumption than individual metals and this is likely to be due to variations in the mode of interactions of component metals to the target organisms. Therefore, the effects of mixture of two or more metals on the oxygen consumption rate of fish seems to be antagonistic. One metal acts as a chemical antagonistic of the other simply by ionic binding that makes the other metal unavailable for interactions with the haemoglobin of blood involved in oxygen consumption. As a generalization, the combined action of two or more chemical substances produces no or less effects and may thus exhibit antagonistic effects in the toxicity to aquatic organisms.

On the contrary, the combined action of two or more substances has been found to be synergistic in toxicity to aquatic organisms. Synergism is the phenomenon in which the combined action of two substances produces a greater effect than what would be expected by adding the individual effects of the two.

---

* For further detail, see Revera (1979).

The foregoing examples dictate us that all these effects will definitely help fish farmers to predict with accuracy the acceptable levels of toxicants in fish culture ecosystems. They can be used as early warning signals of potential pollution problems. At the same time, to protect fisheries and fish culture industries from contamination with toxicants the risk of toxicant exposures to aquatic systems needs to be assessed.

20.5 Types of Effects

Since most of the pesticides are non-specific in action, they not only affect the target organisms but also a number of other beneficial or neutral species (non-target organisms). In hundreds of cases, massive numbers of death in fishes have been attributed to pesticides; however, it is not known in how many of these cases gross negligence has played a significant role.

The most direct and obvious effects on organisms include the acute poisoning of affected individuals and, over an extended period of time, diminished reproduction as a result of sub-lethal effects. The interpretation or prognosis of direct biological effects is a very herculean task due to the following aspects.

1. Differences in the sensitivity of target organisms to chemicals.
2. The influence of physico-chemical conditions on the direct dose-response relationships.
3. The interactions with other chemical stressor in the target organisms including the possibility of additive, synergistic or antagonistic reactions.

There are few data available on the simultaneous effect of several chemicals with other stress factors (synergisms). Results from acute tests with fish have confirmed the presence of synergistic effects in pesticides (such as DDE and parathion).

The direct effects of the individual level can have a vast range of indirect effects on the level of fish poulation and ecosystem such as:

1. *Trophic effects*: Changes or interventions in the food webs and resulting shifts in the species composition.
2. Destructive impact on the habitat.
3. Ability of fish to withstand disease as a frequently observable consequence of selection by the pesticide as an environmental factor.

Similar to direct biological effects, indirect effects may be influenced to a critical degree by the presence of several chemicals and interactions with the physico-chemical factos of the environment. Moreover, indirect biological interactions and ecosystem processes may be subject to considerable time lag between exposure, the direct or primary biological reaction and finally, the indirect effect.

The direct and indirect biological reactions to a chemical are integrated at the ecosystem level. This combination of mechanisms leads to qualitatively new reactions to chemical stress as follows:

1. Direct or indirect effects on aquatic organisms which play a decisive role in the structure or function of the ecosystem.

2. Changes in the community structure.

3. Impact on ecosystem processes especially effects on primary production, decomposition, and the nutrient cycle.

20.6 Heavy Metal Toxicity*

The term heavy metal is a loose one. Heavy metals are, however, the metals that have a molecular weight greater than 55. While this includes manganese and iron, metals that are an ecosystem concern are regarded as chromium, nickel copper, zinc, selenium, arsenic, molybdenum, cadmium, and mercury. Some of these metals such as chromium, copper, and molybdenum are micro-nutrients which are necessary for plant and animal growth.

The most common features of metals are that (1) they are all relatively toxic even at fairly low concentrations, (2) they are readily concentrate by aquatic organisms, fish and plants, and (3) the seriousness of heavy metal contamination lies in the fact that they are generally water-soluble, non-degradable, vigorous oxidizing agents and are strongly bonded to many biochemical substances inhibiting their functions.

General Considerations

Freshwater ecosystems are often contaminated with certain heavy metals such as mercury, lead, cadmium, chromium, copper, and zinc through fertilizers and waste substances. Industrial waste disposal system is most often involved in contamination of aquatic ecosystems. These metals exist in nature and have their own background levels of concentrations. But when present in sufficient levels, they cause deleterious effects on ecosystems in general and fish in particular. Increase in the level of heavy metals in aquatic ecosystems in general cause concern about fish and fish food organisms because of the following reasons:

1. Synergistic effect of metals on fish.

2. Possible accumulation of metals in the body of fish through consumption of contaminated fish food organisms.

3. Persistence of metals in the ecosystem and their transformation into more toxic compounds (such as methyl mercury).

While certain metals are essential for normal growth and development of organisms, Some are highly or moderately toxic. A metal can be regarded as essential only if (1) the organism can neither grow nor complete its life cycle in absence of the element, (2) the

---

* For further detail, see Wittman (1983).

element cannot be replaced by any other element, and (3) the element exhibits a direct influence on organisms and its metabolism. Similarly, a metal is considered as toxic if it hampers growth and metabolism of fish when supplied with sub-lethal or lethal concentrations. Since certain metals are required in life processes, both finfish and shellfish have a capacity to accumulate metals within their body. The power of accumulation is enhanced by feeding and metabolic processes which can lead to enormously high concentrations. Freshwater finfish and shellfish apparently can accumulate metals in high quantities in their tissues (several hundred times greater than sea fish) either directly or indirectly from water through absorption and food chain respectively. Since metals form complexes with organic substances, they have a tendency to fix in the tissue and not to be excreted. This is, perhaps, one of the major problems that metals exhibit hazardous effects with respect to their concentrations on aquatic organisms.

Many species of finfish and shellfish have got the capacity to concentrate trace metals in their tissues at much higher levels than in other organisms. If concentrations of metals are not high enough to kill fish, but are high enough to destroy fish food organisms, there will be substantial damage to fishery and aquaculture. Therefore, the toxicity of a metal and its effects on organisms will depend upon (1) the chemical form of metals, (2) presence of other metals, (3) physiological status of organisms, and (4) physico-chemical factors such as temperature, salinity, pH, and dissolved oxygen of water. For examples, metals are less toxic at lower temperatures and higher salinities than at higher temperatures and lower salinities. Similarly, the toxicity of a metal is also dependent upon how long the metal will persist within the body of fish. Generally, most of the metals have a long residence time and hence exert their toxic effect over a long time.

Accumulation and Turnover of Metal in Aquatic Ecosystems

Most of the metallic elements occur naturally in aquatic ecosystems, and are classified as pollutants only when added by man in quantities sufficient to generate deleterious effects. In the natural environment, organisms are exposed chronically to sublethal concentrations of several contaminants simultaneously, and concentrations of metals present within the organisms result from the relative rate of metal accumulation and turnover. These rates vary among species and also with the concentration as well as physico-chemical forms of the metal. Assessment of potential effects of metal contaminations on aquatic organisms must consider organismic response not only to levels of metals in the ecosystem, but the contained metals as well. Furthermore, both the cycling of a particular metal and its toxic characteristics may depend upon the physico-chemical forms of the metal introduced into the system. An understanding of the potential physiological effects resulting from metal additions to aquatic ecosystems, therefore, requires knowledge of the complex physical and biological processes that control distribution and turnover of metals in the component organisms and in the entire ecosystem.

20.7 Protective Measures of Contaminated Fish Culture Ecosystems

Over the years, a number of laboratories and research institutes have developed toxicant removal devices. Under the All India Institute of Hygiene and Public Health and

India-Canada Environmental Facility, several hundreds arsenic ramoval devices, for exanple, have been installed in many areas of the country. Though some of the devices are well-known, there are others which are better known internationally and some indigenously developed on the basis of local understanding of the problems.

The sublethal and lethal effects of toxic compounds, their bio-accumulation, and their long-term effects on fish and aquatic ecosystems require regular monitoring and survellance before/during fish culture operations. Since the use of pesticides is not likely to be substituted by any other plant protection measure in near future, suitable devices should be adopted to protect fish culture ecosystems. Some of the common devices that are necessary for protecting fish and aquatic ecosystem from toxicants are noted below.

1. Construction of fish farms at a considerable distance from agricultural fields.

2. Construction of dykes to prevent the surface flow-off to fish culture ecosystems.

3. Selection of toxicants should be such that they should leave least amount of residues to the ecosystem.

4. Residues should be photodegradable and biodegradable.

5. Suitable devices for use of chamical substances to agricultural fields should be adopted by farmers to avoid contamination of water bodies.

20.8 Removal of Pollutants From Water*

To avoid the hassles of residual effects of toxicants on aquatic life, their complete or partial removal from fish culture ecosystems is inevitable so that fish culture may be undertaken in subsequent years. Since polluted water bodies are not fit for fish culture, suitable devices must be adopted to remove toxic substances from water. Suitable devices generally include absorption by aquatic plants, adsorption, activated carbon adsorption, electro-dialysis, ion-exchange, and reverse osmosis. But each of these devices has some drawbacks and limitations. Therefore, microbial populations have been used for removal of pesticides and heavy metals. The Council of Scientific and Industrial Research (CSIR), New Delhi, India, has recommended some important techniques that follow for potential removal of a variety of pollutants from water.

Heavy Metals

Certain heavy metals could be removed from contaminated water by microbes and aquatic plants. In the process of removal, several micro-organisms such as *Zooglea ramigera*, *Pseudomonas putida*, *Arthrobactor viscous*, and *Citrobactor* spp. are involved. These organisms have the capacity to remove heavy metals in appreciable amounts. The species *Zooglea ramigera*, for example, adsorb cadmium and copper up to 100 and 300 mg /g dry weight, respectively. Microbes degrade/detoxify the toxic substances and they play an important role to clean-up water and soil.

---

* For further study, see Alexander (1994), Tewari (2000), Bhart and Dkhar (1998).

Some species of aquatic plants are able to absorb appreciable quantities of heavy metals from polluted ecosystems. The process of absorption of toxicants and their subsequent detoxification/degradation is very rapid and does not require any additional investment. This aspect has, however, been described in Volume 1 (See Chapter 4).

Ammonia

Ammonia can be removed by ion-exchange method. This process involves the development of a weak acidic cation exchange, which helps remove ammonia as ammonium sulfate. This is widely used as a fertilizer.

Degradation of Chemicals*

Three main important mechanisms are generally involved in breakdown of chemicals in aquatic ecosystems such as auto-oxidation, microbial degradation, and degradation by sunlight.

1. **Oxidation:** Synthetic pesticides contain hydrocarbon skeletons with a variety of substituents (functional groups) such as amino, nitro, hydroxyl, halogens, sulfonate and other functional groups. Aliphatic hydrocarbons are oxidized to fatty acids which are degraded through beta-oxidation and the resulting fragments are finally metabolized via the tri-carboxylic cycle. Aromatic ring structures are metabolized by dihydroxylation and ring cleavage mechanisms (Fig. 20.3). Prior to these transformations, different substituents may partially or completely be eliminated. Substituents interrupt oxygenation and as a result compounds which are resistant to biodegradation are produced. such compounds are termed as *recalcitrant* (Fig. 20.4).

    A simple alteration in the substituents often exhibit the difference between recalcitrance and biodegradability. While the herbicide 2,4-D undergoes biodegradation within days, the other herbicide 2,4,5-T persists for more than 190 days. This is due to the structural differences between two herbicides. Similarly, propachlor with a tertiary amine group is not attacked by microbial amylase and persists for prolonged period, whereas prophan is rapidly cleaved by this enzyme and hence does not exhibit any residual toxicity. Methoxychlor is less persistent than DDT because the para-chloro substitution renders DDT greater chemical and biological stability, and the para-methoxy groups of methoxychlor are subject to dealkylation. In some pesticides, one portion of the pesticide is not degraded and another is susceptible to degradation. Some acylamide herbicides (such as propanyl) are cleaved by amidases and the aliphatic moiety is converted to oxygen and water (Figure 20.5). The aromatic moiety is stabilized by chlorine substitution and hence resists the process of mineralization.

2. **Microbes Involved in Degradation*:** Certain metabolic processes (such as fermentation, anaerobic metabolism, metabolism through induction and mutation)

---

* For further study, see Alexander (1994), Tewari (2000), Bhart and Dkhar (1998).

$$R-CH_2-CH_3 \xrightarrow[H_2O]{O_2+2H} R-CH_2-CH_2OH \xrightarrow{-2H} R-CH_2-CHO$$

Alkene   Primary alcohol   Aldehyde

Aliphatic hydrocarbon

$$R-CH_2-COOH$$

Fatty acid

Beta oxidation

Benzene   Catechol   Cis-cis muconic acid

Aromatic
hydrocarbon

2-Hydroxy-Cis,Cis
muconic semialhedide

Beta-Ketoadipic acid

$$HOOC-CH_2-CH_2-COOH$$

Succinic acid

$$H_3C-C-COOH \quad \text{(||O)}$$

Pyruvic acid   2-Keto-4-Pentenoic
acid

$$H_3C-CH \quad \text{(O)}$$

Acetaldehyde

$$H_3C-C-SCoA$$

Acetyl CoA

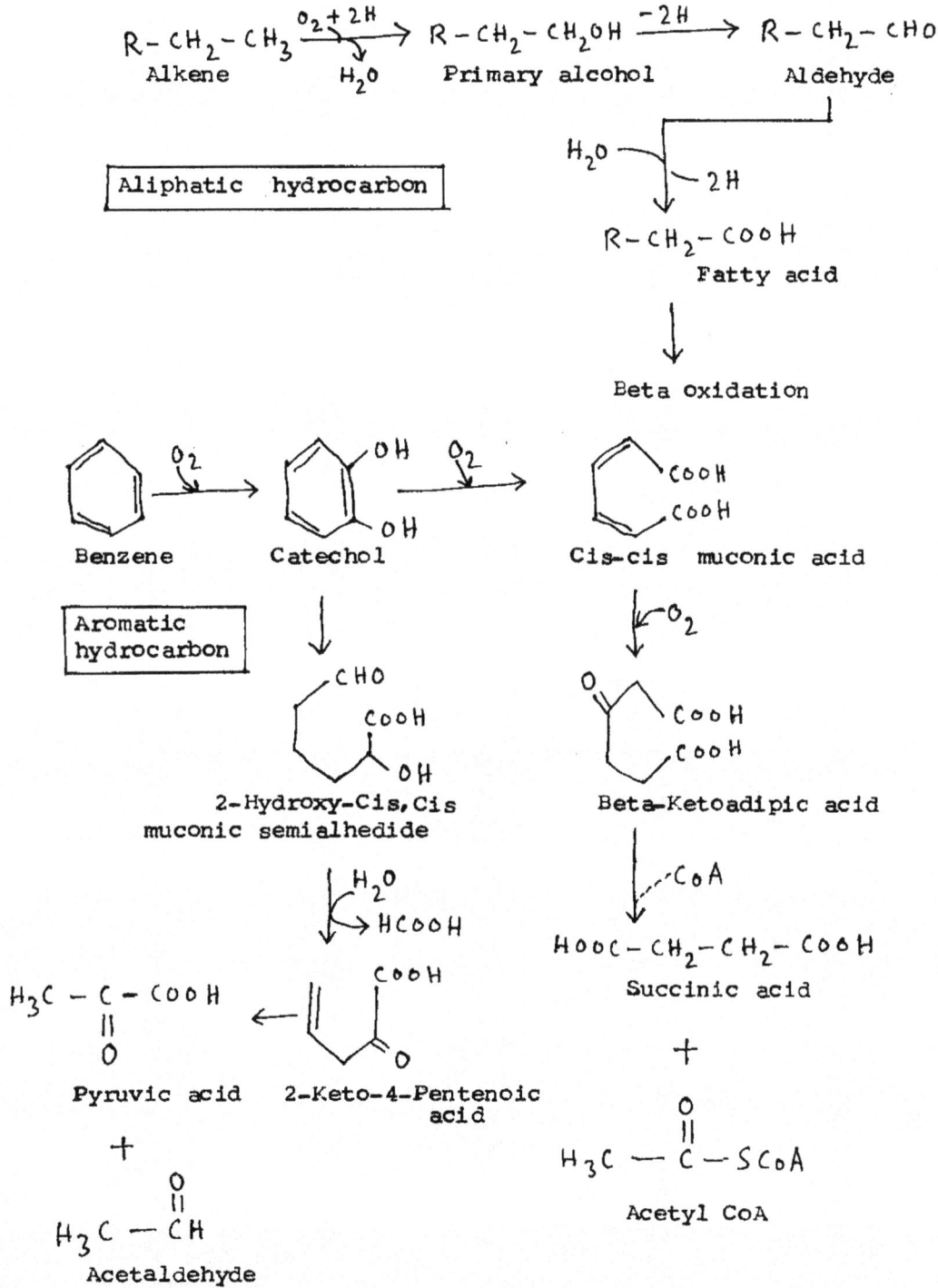

Fig. 20.3 : Degradation pathways for aliphatic and aromatic hydrocarbon pesticide moieties.

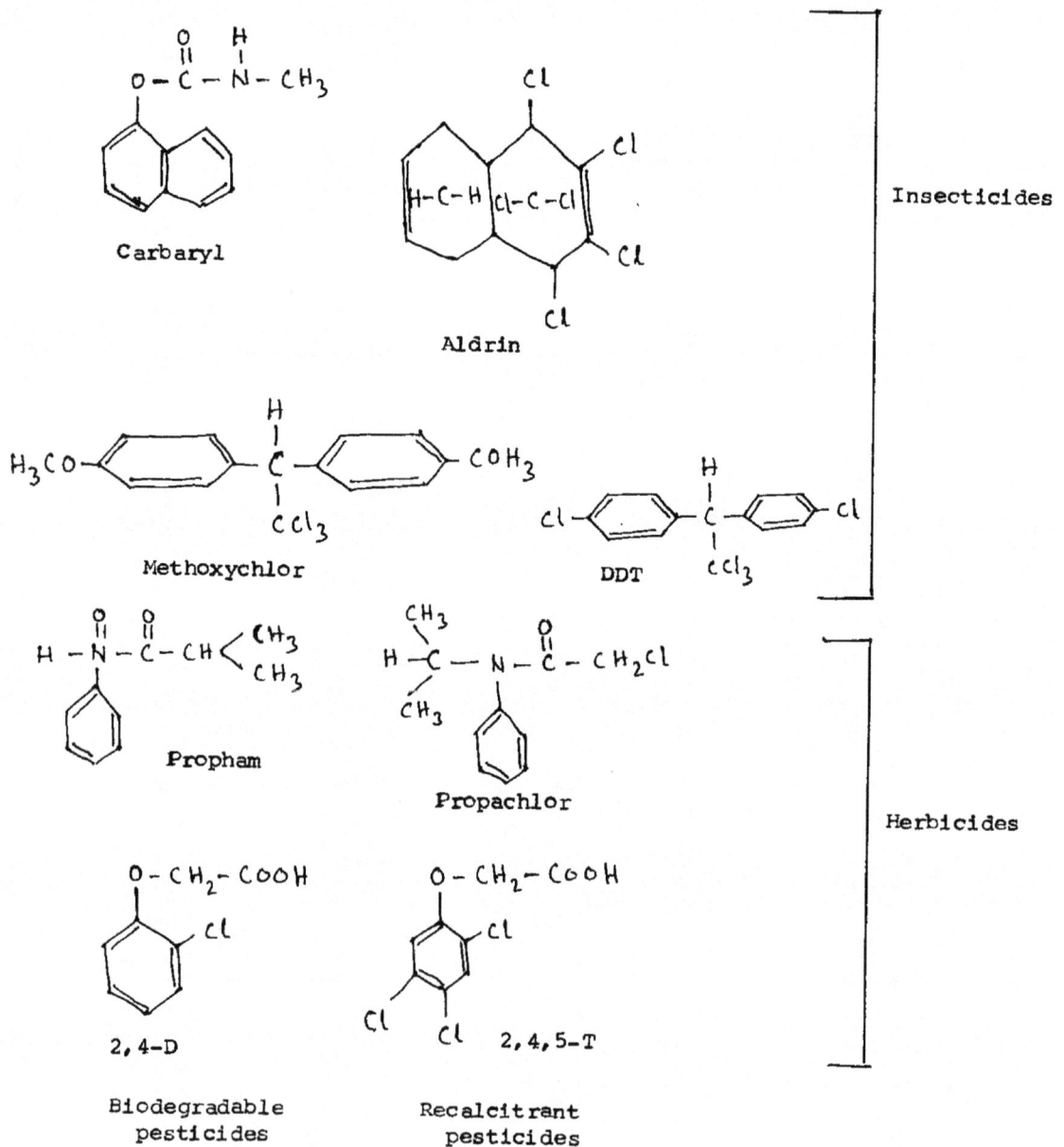

Fig. 20.4 : Structural formulas of some biodegradable and recalcitrant pesticides. The molecular features furnishing the compounds recalcitrant are the 5-chloro-substitution in 2,4,5-T, the N-alkyl substitution in propachlor, the multiple chloro-substitution in aldrin, and the two para-chloro-substitution in DDT.

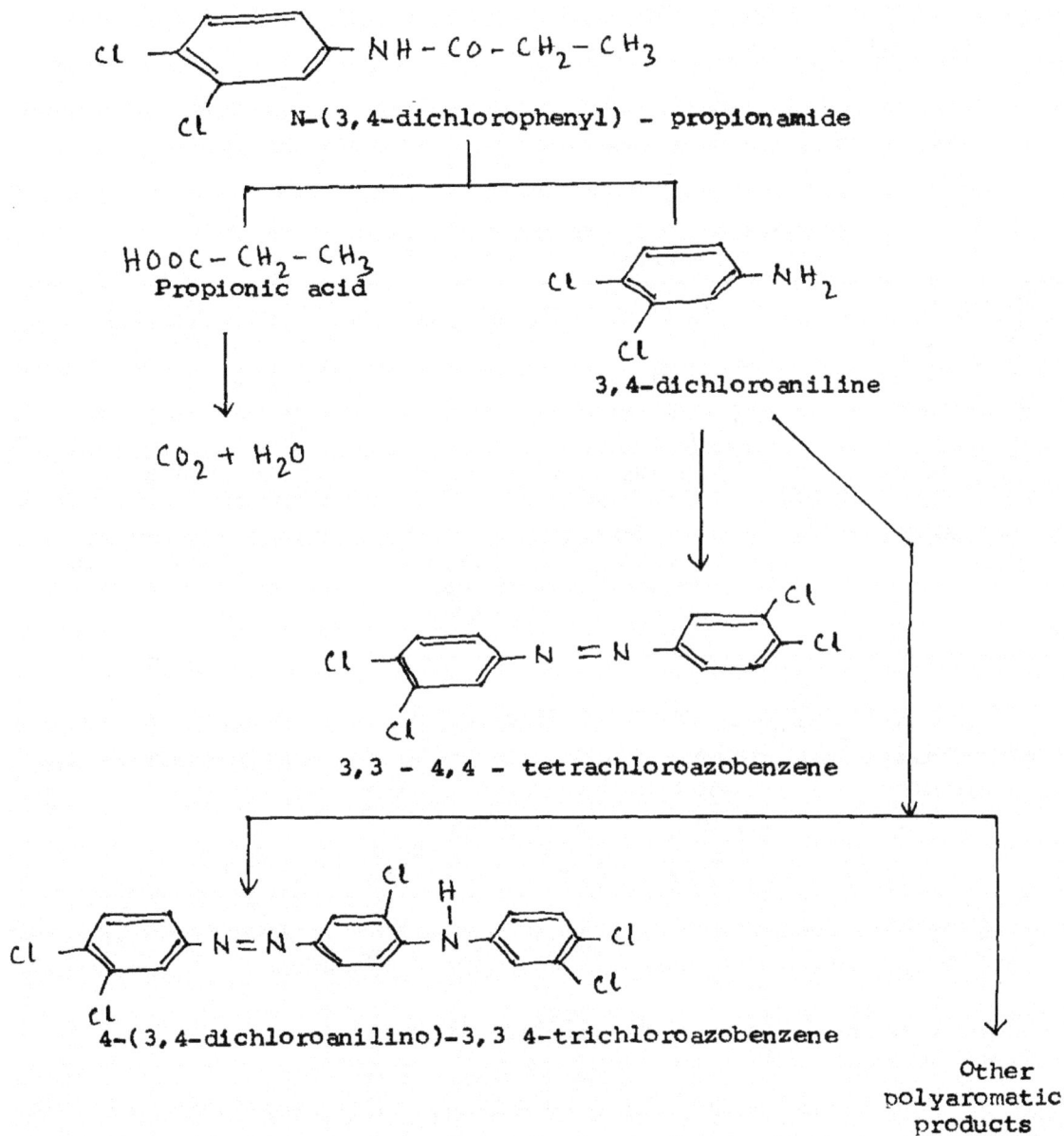

Fig. 20.5 : Biodegradation pathway for the herbicide propanil, N-(3, 4-dichlorophenyl)-propionamide. The aliphatic part of the molecule is degraded but the aromatic one is dimerized and polymerized to residues that are persistent in ecosystems. It should be noted that the transformation involves microbial synergism.

take place in micro-organisms that makes them unique tools for biodegradation or detoxification. A number of micro-organisms encourage degradation of different pesticides (Table 20.1). The micro-organisms partially or completely degraded a vast variety of toxicants by enzymatic or non-enzymatic processes. Degraded products are then accumulated in aquatic ecosystems. The toxicants are biologically transformed through reduction, oxidation, hydrolysis, and condensation. Parathion hydrolase, for example, extracted from *Pseudomonas diminuta* and *Flavobacterium* spp.can hydrolyze several organo-phosphorus pesticides such as cyanophos, diazinon, fenitrothion, and methyl parathion.

Table 20.1: Some Examples of Environmental Use of Micro-organisms in Detoxification/Biodegradation of Pesticides Used in Agriculture

Pesticide	Biodegraded micro-organism	Pathways/Enzymes used
Parathion, Diazinon	*Streptomyces, Arthrobacter*	Hydrolysis of alkyl and aryl bonds
Paraoxon, Chloropyrifan	*Pseudomonas, Flavobacterium*	Reductive transformation
Malathion, Fenitrothion	Corynebacterium, *Nocardia Trichoderma viride*	Phosphotriesterase
Carbaryl, Carbofuran	*Pseudomonas , Achromobactor*	Reductive demethylation, Hydrolases
Lindane, DDT, Heptachlor, Chlordane	*Aerobactor aerogenes, E. coli, Proteus vulgaris, Clostridium, Pseudomonas fluorescens, P. putida, Nocardia, Streptomyces, Saccharomyces, Aspergillus*	Reductive dechlorination (anaerobic), Dehydrochlorination (aerobic)

3. **Sunlight and Catalyst:** The use of solar power is also an important and simple technique for cleaning up polluted waters. Experiments have shown that a combination of a catalyst (such as titanium dioxide) and sunlight can destroy a number of toxic substances present in water.

To sum up, the overwhelming contribution of micro-organisms and aquatic plants to several processes ponders the omnipresence of micro-organisms, their significant contribution to the total bulk of living material, their collective ability to degrade/detoxify a vast variety of organic materials and hence deserve special attention.

20.9 Monitoring of Pollutants in Aquatic Ecosystems*

A fish culture ecosystem is said to be polluted when the quality of water is degenerated due to sewage and industrial discharges, degradation of proteinaceous subtances and surface flow-off that contain agriculture chemicals. In other words, aquatic ecosystem is damaged beyond that reasonable limit which is termed as *tolerance limit*. Many instrumental and non-instrumental methods have been used for detection, determination and separation of water pollutants. Instrumental methods are highly sensitive to pollutants and rapid. Their sensitivity and versatility have made it possible to solve to complex

---

* For further study, see Coyne (1999), Patrick (1972).

problems. These methods are, however, costly and require sophisticated equipments as well as skilled operators to use them. On the other hand, sensitivity of non-instrumental methods is very low. These methods are inexpensive and simple. They can be used for analysis of a pollutant at fish culture areas. Therefore, there is a growing demand for the development of non-instrumental methods of pollutant analysis. The use of color reactions makes it possible to work out many sensitive and selective methods for detection and determination of pollutants.

Spot tests based on color reactions are extremely simple, inexpensive, and informative. They can be successfully used for field detection of pollutants. For quantitative analysis, absorption spectroscopy in the visible region is preferred for its simplicity. If the test material does not absorb in the visible region, it is converted into a colored compound which possesses chromophoric group or groups. Although methods based on colorimetry are among the earliest instrumental techniques, yet even today they are widely used owing to their overall utility. Paper chromatography technique is commonly used for the separation of pollutants from samples. It can be performed at the spot with minimum apparatus and reagents.

Like different inorganics and heavy metals, organic pollutants (such as bavistin, calixin, quinoline, simazine, 2,6-lutidine, acridine, azobenzene, amitrole, trimethylamine, etc.) are generated from the use of herbicides, fungicides, sewage, sludge, acaricides, and from pharmaceutical industries. They are the worst offenders of aquacultural peace and ecosystem balance. They cause disturbances in respiration and gastro-intestinal system, damage to kidney and liver and diminish growth of farmed fish. A number of methods has been employed for detection, determination, and separation of the pollutants. In contrast to developed countries, developing ones need simple and inexpensive methods for pollution monitoring and control.

As a generalization, two principal types of water monitoring such as chemical and biological are recognized. While the chemical monitoring essentially involves the analysis of toxicants in ecosystem (See Panel 20.1), the biological monitoring is characterized by the consideration of a vast variety of bio-indicator species.

Definition and Explanation of Biomonitoring

Simply defined, biomonitoring is the measurement of changes in the biological factors of a habitat on the basis of evaluation of the number and/or distribution of organisms or species. It focusses on biological effects at different levels of organization.

Some characteristic features of species are preferred in biological monitoring. These include (i) the species should be abundant so as to allow large samples to be repeatedly taken away without significant changes in the population features, (ii) a single species can then be used to monitor large areas, (iii) the preferred species should be long-lived so as to integrate the population over more than one year, (iv) the species should be easy to identify and to collect at all ages all the year round, and (v) the organism should be sufficiently large in size so as to provide sufficient tissue for analysis.

PANEL - 20.1
ANALYSIS OF TOXICANTS IN ECOSYSTEM*

Different heavy metals and pesticides are detected by various techniques (see below). While some of the techniques are relatively cheap and dependable, most of the methods are quite expensive but very rapid and reliable.

	Type / Name of Pesticide	Detected by
1.	Parathion and Malathion with silver nitrate-bromophenol blue reagent	Thin-Layer Chromatography
2.	Carbaryl (Sevin), Phorate, Malathion, Carbamates	Thin-Layer Chromatography
3.	Volatile pesticides	Gas Chromatography
4.	Residue analysis of chlorinated pesticides	Gas-Liquid Chromatography
5.	Organochlorine pesticides	Electron Capture Detection
6.	Metabolites of pesticides	Nuclear Magnetic Resonance and Mass Spectrometry
7.	Polychlorinated biphenyl	Gas Chromatography and Mass Spectrometry
8.	Identification and Characterization of pesticide compounds	Nuclear Magnetic Resonance
9.	Metabolites of 2,4-D	Mass Spectrometry
10.	Organic Pollutants	Paper Chromatography
11.	Heavy Metals	Flame Emission Spectroscopy, Inductively Coupled Palma-Atomic Emission Spectroscopy and Ring-Oven Technique

For analysis of heavy metals, 40 ml samples are digested with 5 ml of HNO_3 over low heat until 1.0 ml remained. A spectrophotometer is used to analyze the digested sample. For organic contaminants (such as phenolic compounds and polynuclear aromatic hydrocarbons), samples are first extracted with 200 ml of methylene chloride (CH_2Cl_2) followed by a 20 ml CH_2Cl_2 rinse. The method of extraction has been described by the United States Environmental Protection Agency (1985). Extracts are then concentrated to 1 ml in iso-octane and analyzed by gas chromatography and Electron capture detection. Water samples containing polycyclic aromatic hydrocarbons are extracted twice with 100 ml of CH_2Cl_2 and the extracts are then analyzed by gas chromatography and mass spectrophotometry by selective ion monitoring.

Organic pollutants containing a tertiary amino group can be detected by mild heating with citric acid in presence of acetic anhydride, that gives a red or purple color. Since water and various cations as well as anions interfere with the actual color produced, the method of detection has been rectified and applied this reaction for the analysis of polluted waters. Whatman No.1 filter papers are impregnated with 2 per

---

* For further study, see Burell (1974), APHA (1985), Marr and Cresser (1983), Rana (2005).

cent aqueous solution of citric acid and are dried at room temperature (26°C). A piece of this paper (1.5 cm diameter) is placed at the bottom of a 10 cc beaker and is placed on a heating block maintained at 165°C ± 2°C in order to eliminate the moisture. A drop of the sample is applied with the help of a micro-pipette of the centre of the filter paper and are allowed to dry for removing the solvent completely. Acetic anhydride (2-3 drops) is applied on the spot of the sample and finally heated again for 2 minutes. A violet or purple color is developed that indicates the presence of these pollutants. Strong influence of cations and anions are rectified by solvent extraction using chloroform. For this purpose, 20 ml of sample, 2 ml of aqueous 1M NaOH and 3 ml of chloroform are taken in a separating funnel. After shaking, the pollutant is detected in the chloroform layer.

For the separation of organic pollutants, ascending techniques of paper chromatography is applied. Whatman No.1 filter papers (14 cm × 3 cm strips) are spotted for the pollutants, hang in the glass jars (20 cm × 5 cm) and developed in various types of water and organic solvents. The pollutants are detected either by the method described above or by dipping the strip in the aqueous solution of phosphomolybdic acid as yellow spots or by dipping the strip in the solutions of copper iodide in acetonitrile (saturated) and saturated aqueous sodium carbonate respectively followed by irradiation with UV light as fluorescent spot.

Aliphatic organic compounds containing pyridine and quinoline are determined spectrophotometrically by simultaneous application of chloranil and cis-aconitic anhydric methods. In the former method, the sample is dissolved in toluene and then treated with 1 ml of 1 per cent solution of chloranil in toluene for 15 minutes in a boiling water bath. After cooling for 5 minutes, the absorbance at 610 nm is measured. In the latter method, the sample in 2 ml of toluene is treated with 1 ml of 0.25 per cent aconitic anhydride in toluene for 5 minutes in a boiling water bath.

The reaction is allowed to stand for 15 minutes and absorbance is measured at 500 nm. Though these methods are not general, certainly they could be used as auxillary tools in the ecosystem monitoring and control purposes.

Measurement of water quality by chemical analysis is not sensitive to toxic conditions and less frequent determinations of contaminant concentrations in waters require appreciable time delay from collection to analysis and explanation. In contrast to chemical tests, biological ones cater an instantaneous assessment of chemical toxicity that provides the chance for timely remedial operation.

The pollution indicator of an aquatic ecosystem is one which may exhibit vivid symptoms of effects on indicator species (either morphological or physiological) that dictates the possible presence of pollutants. These symptoms may be fairly distinct and contribute substantially to the qualitative estimation of pollutants; or mixture of pollutants, but generally, they do not provide an indication, and the availability of a particular toxicant must be ascertained mostly by physico-chemical methods.

* A regular biomonitoring programs in different aquatic ecosystems for evaluation of pollution load is cardinal not only to understand the environmental impression on aquatic life but also its effects on public health. This instruction about the current status of freshwater ecosystem health where intensive/semi-intensive fish culture systems are undertaken, will help to adopt suitable measures in due time to curb pollution, assigning to the conservation of fishery resources and management of fish culture industries.

* The biomonitoring program is maintained by many countries of the world to document temporal and geographical tendencies in levels of persistent aquatic contaminants that may denounce fish and other aquatic organisms. This program also provides some informations on the success of the regulatory action contempleted to decline level of toxicants in ecosystems. Such program is conducted by environmental specialists and toxicologists and must be periodically determined the level of potentially toxic substances in samples of fish, sediments, plankton, and bottom fauna collected from different types of aquatic ecosystems.

* Toxic evaluation of water pollutants involves analysis of chemical, physical, and biological parameters. While chemical and physical parameter provide an information related to the time of sampling that indicates the levels and sources of different pollutants, biological parameters (such as living organisms) provide appropriate full-time councellors of pollution because one can determine the effects over time and explore the presence of pollutants by analyzing the population structure of organisms. Changes in species diversity are also considered as a reliable and useful means of evaluating the biological effects of pollution. The species diversity indices are, however calculated from the abundance data of organisms and serve as an excellent indicator of pollution.

* In addition to conventional systems of water quality monitoring, various biological systems (species) are involved in this strategy as indicators which are very sensitive to chemical substances and useful for evaluation of fish culture ecosystems. The biotic resources are diverse and include a variety of phytoplankton, zooplankton, macrophytes, and other plants as well as animal communities. Some common representatives of biotic resource are shown in Fig. 20.6. Of them, however, many forms are sensitive to water pollutants or rather sensitive to changes in water quality. Some other forms appear to be tolerant to such water quality changes and thus act as bioindicator. Thus, to assess the effect of toxic substances originating in human activity, animals, plants, microbes, individuals, biotic communities, and aquatic ecosystems can be used as bioindicator. There are three distinct categories of tolerant aquatic species which are regarded as excellent indicators of pollution load in water (Table 20.2.)

To sum up, bioindicator species represent a wide spectrum of organisms serving as indicators of the environment. The use of these biological organisms as indicators of aquatic pollution is to use them, also serve as an index of their environment. Bioindicators indicate the general toxicity of the ecosystem. They are not able to assess the exact physical and chemical factors responsible for such toxicity.

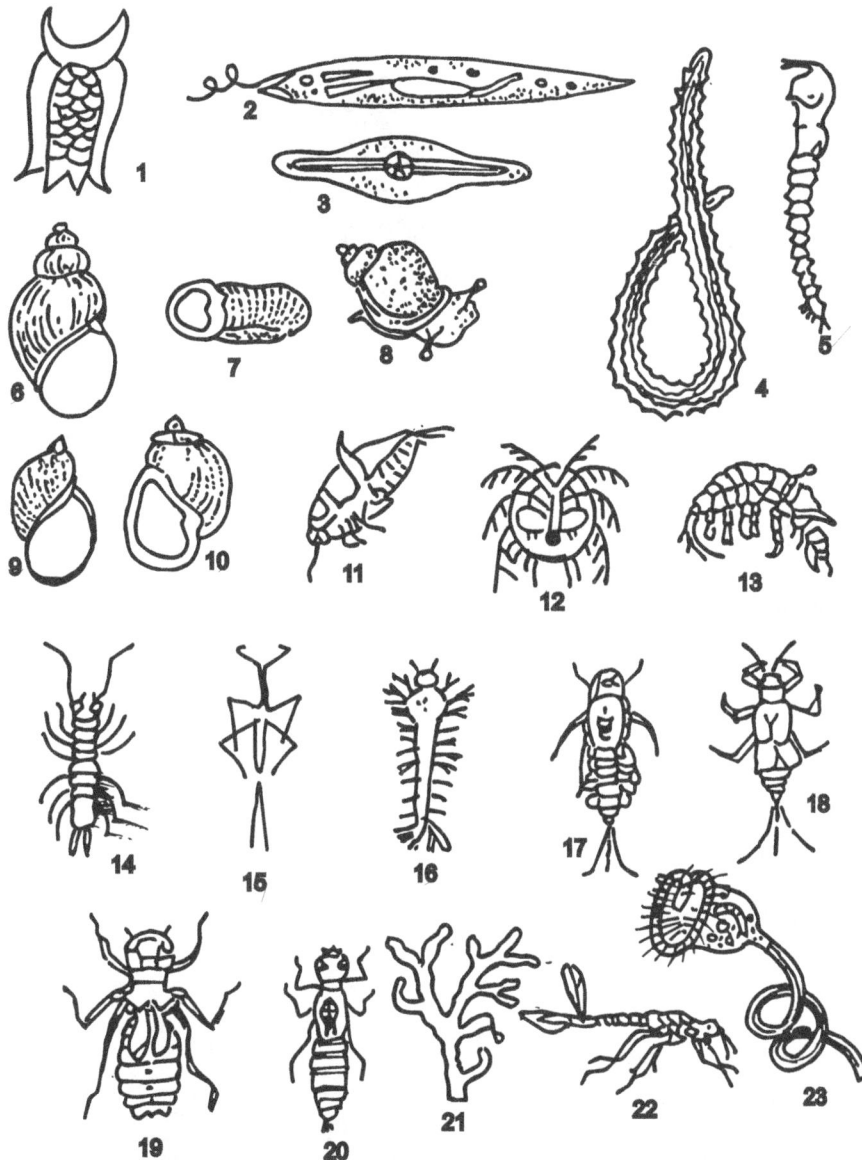

Fig. 20.6 : The Central Pollution Control Board has established the fact that among all the biological components of the aquatic ecosystem, several species of plankton and benthic macro-invertebrates are suitable for evaluation of water quality. Most of these animals especially aquatic insect larvae, share their biological life in freshwaters, while their adults fly over extensive areas in search of ideal freshwater ecosystems for reproduction, breeding, and laying eggs in suitable habitats and establish biological communities. Some common bio-indicator species are noted below : (1) *Brachinous;* (2) *Euglena,* (3) *Navicula;* (4) *Tubifex;* (5) *Chironomous;* (6) *Helisoma;* (7) *Physa;* (8) *Viviparus;* (9) *Lymnaea;* (10) *Columella;* (11) *Dytiscus;* (12) *Hydracarina;* (13) *Gammarus;* (14) *Asellus;* (15) *Ranatra;* (16) *Culex* larva; (17) Periphytic may fly nymph (*Cloeon*); (18) Benthic may fly nymph (*Caenis*); (19) *Helocordulla* (Dragon-fly nymph); (20) *Aeschna;* (21) *Zooglea;* (22) Damsel fly; (23) *Vorticella;*

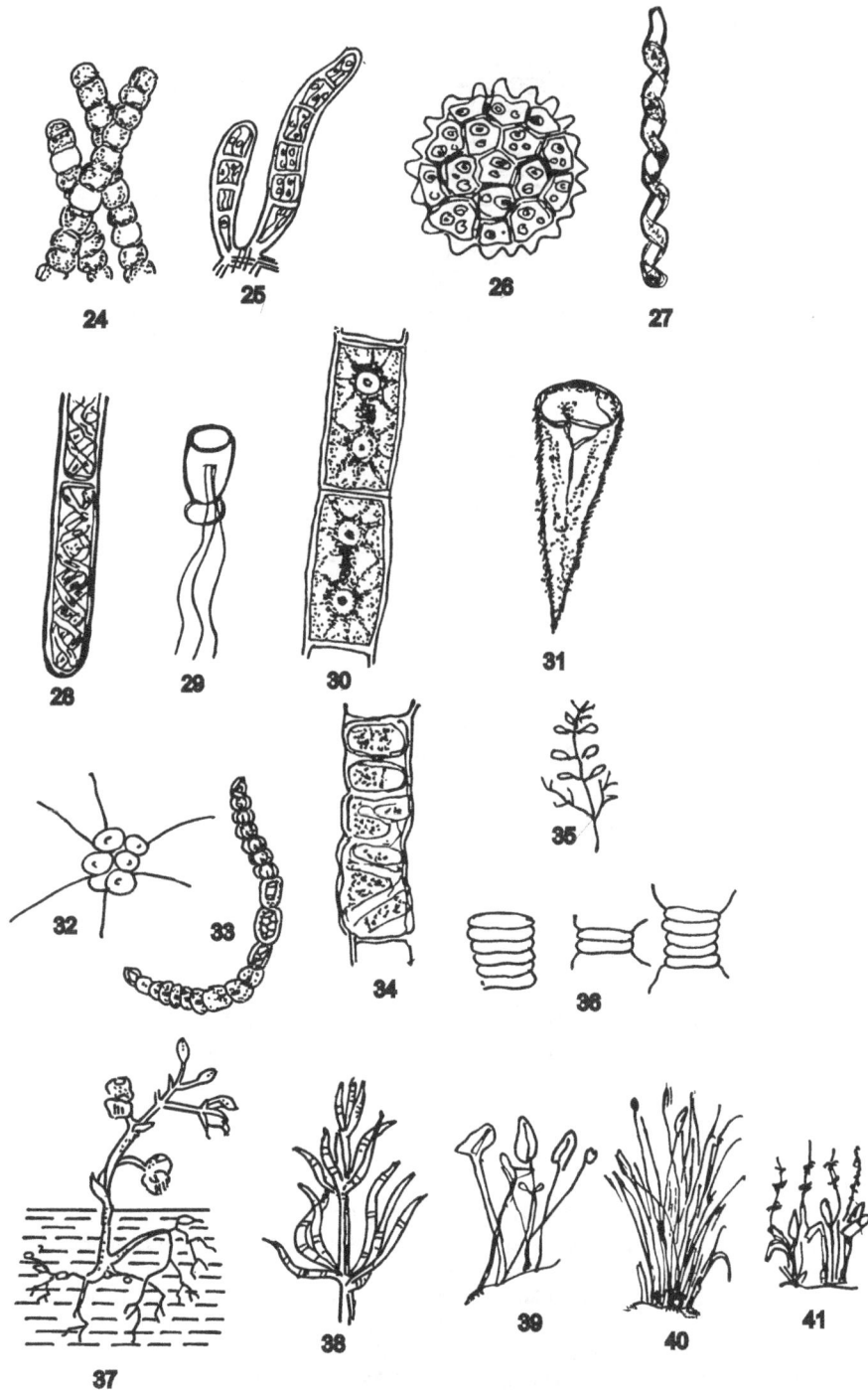

Fig. 26.6 : contd...

(24) *Oscillatoria;* (25) *Ulothrix;* (26) *Pediastrum;* (27) *Spirulina;* (28) *Spirogyra;* (29) *Filinia;* (30) *Cladophora;* (31) *Stentor;* (32) *Richterilla;* (33) *Anabaena;* (34) *Stigeoclonium;* (35) *Potomogeton;* (36) *Scenedesmus;* (37) *Utricularia;* (38) *Chara;* (39) *Nuphar;* (40) *Typha;* (41) *Sagittaria.*

Table 20.2: Some Indicator Species of Different Saprobic Level.

Oligosaprobic organisms (Indicator of scarcely polluted waters)		Mesosaprobic organisms (Indicator of moderate to highly polluted waters)		Polysaprobic organisms (Indicator of extremely polluted waters)	
Animals	Plants	Animals	Plants	Animals	Plants
Planaria sp.	Synedra sp	Stentor sp.	Oscillatoria sp	Tubifex sp.	Spirulina sp.
Larva of Peria sp.	Tabellaria sp.	Vorticella sp.	Nitzchia sp.	Rotaria sp.	Euglena sp.
Nothoica sp.	Ulothrix sp.	Brachionus sp.	Spyrogyra sp.	Chironomus sp.	Oscillatoria sp.
Vorticella sp.	Cladophora sp.	Stylaria sp	Scenedesmus sp.		
	Euastrum sp.	Cloeon sp.	Uronema sp.		
	Micrasterias sp.	Spirostomum sp.	Closterium sp.		
	Surinella sp.	Straliomys sp.	Melosira sp.		
	Bulbochaete sp.	Sphaenum sp.	Pediastrum sp.		
			Tebelloria sp.		

For quantitative method of water quality assessment, species diversity indices of different water bodies have been calculated which provide valuable clues for water quality monitoring. The diversity indices are estimated from the abundance data of the organisms and serve as a good indicator of pollution. Some diversity indices are noted below:

1. **Shannon-Wiener Index:** Species richness can be compared between ecosystems by counting the number of species either in suitable samples or in the whole community. Suppose two ecosystems each made up of two species. In the first ecosystem, there are 99 individuals of species A and one individual of species B; but in the second ecosystem, there are 50 individuals of such species. This indicates that the second ecosystem is more diverse than the first one. The formula for calculating the index H' is

$$H' = -\sum_{i=1}^{S} Pi \log e\, Pi$$

where   S   =   Number of species

   Pi   =   Proportion belonging to the ith species

The logarithms used can be to any base. As a generalization, most of the people use base e because it is easy and so far from 2. However, calculation of H' may be illustrated using some numbers from a pond ecosystem.

Species	Number of individuals	Pi	log e Pi	Pi × log e Pi
Daphnia sp.	50	0.507	- 0.5482	- 0.278
Cyclops .sp.	30	0.410	- 1.2171	- 0.499
Mesocyclops sp.	10	0.083	- 2.1452	- 0.178
Total	90	1.00	–	- 0.955

In the formula, the negative sum of Pi (log e Pi) is considered for practical reason of preferring to deal with positive numbers, H' for this set of data is then 0.955 (if the index is required binary form, the value of log e Pi may be converted to log 2 Pi by dividing it by 0.6931. For example, log 2 0.507 equals to −0.5482. However, for the bottom fauna, an H' greater than 3 indicates no or very slight pollution. Similarly, an index of 2 to 3 denotes light pollution and an index below 1 indicates heavy pollution.

2. **Simpson Diversity Index :** This diversity index can be calculated by the following formula :

$$D = \frac{N(N-1)}{\sum n(n-1)}$$

where   D = Diversity index

N = Total number of individuals

n = Number of individuals per species

E = Sum

Species	Number
A	20
B	12
C	75
D	56
E	77
Total	240

$$D = \frac{240(240-1)}{20(20-1)+12(12-1)+75(75-1).....}$$

$$D = \frac{57,360}{14,994} = 3.83$$

The value of D indicates the level of water pollution.

3. **Odums Species Index:** It is an excellent index which helps determine the level of pollution in both lentic and lotic water bodies. It is calculated as :

$$\text{Index value} = \frac{\text{Total number of sample present in the sample}}{\text{Total number of individuals of all the species}} \times 1,000$$

Some Examples of Ecosystem Monitoring

Since biomonitoring program and ecosystem impact assessment are the two elaborate subjects that are beyond the purview of this volume, we would rather restrict this sub-section to basic concepts of biomonitoring. In the recent past, industrial, agricultural, and technological developments have induced a variety of changes in fish culture ecosystems. Such changes in the pollution front and perturbation that cause severe damage to aquatic resources is a growing threat to fish production strategies.

In order to assess the changes caused by human interventions, adoption of cost-effective and most reliable monitoring systems are inevitable to predict and recognize the hazardous effects. As noted earlier, living organisms are the best indicators of aquatic ecosystems. Biomonitoring systems are, however, most sensitive and at the same time are also useful to fish culturists. Thus microbes, macrophytes, phytoplankton, zooplankton, and molluscs exhibit different degrees of sensitivity to pollutants and can also be successfully employed as bioindicators to assess and predict the change in ecosystem structure. Some common examples of ecosystem monitoring are noted below.

1. Various species of algae are excellent monitors of ecosystem pollution.

2. Some aquatic plants such as *Chara, Utricularia, Wolffia,* etc. prefer to grow in polluted waters.

3. Presence of bacteria such as *Escherichia coli* indicates water pollution.

4. Presence of diatoms indicate water pollution by sewage.

5. Migration of fish such as *Catla, Notopterus, Labeo gonius , L. bata, L. rohita,* etc. from contaminatd waters indicates industrial pollution.

6. Some species of fish (such as silver carp), Cladocerans (*Daphnia*) and algae (such as *Cladophora* and *Stigeolonium*) are used to monitor heavy metal and pesticide pollution of water.

7. Toxicants may alter the function of muscular, excretory, nervous and endocrine systems of fish. Such alterations may be investigated at physiological, biochemical, and morphological levels and can indicate the presence of toxic substances in different organs of fish.

Significance of Biomonitoring

Uninterrupted and regular evaluation of the effects of water pollution on sensitive organisms is authentic to biomonitoring programs. This is important for notifying about the possible damage to ecosystems, for evaluating the presence and distribution of pollutants and their effects over extensive water areas effects in time and space and for analyzing the trend/extent of water pollution.

The biomonitoring of aquatic ecosystem serves a better measure than with equipments. These methods assess the sink potential, produce warning signals for protection efforts, identification and distribution of pollutants, and rapid monitoring of pollutants. In all the cases, however, biomonitoring programs should be developed and enforced. Since the concentration of pollutants in ambient waters exhibits seasonal fluctuations, biomonitoring network should be planned accordingly because seasonal fluctuations will expedite the biomonitoring programs.

20.10 Safe Disposal of Toxicants

In the middle of the 20th century, the method for determination of median tolerance limit (TLM) was developed. But it was soon realized that the objective should be no mortality rather than 50 per cent mortality. Attempts were then made to endorse suitable factors to TLM data for deducing long-term safe concentration. Thereafter, three suggestions were made First, the safety factor should be derived from 11-month continuous flow bioassay tests. Second, it should be derived from methods that involved behavioral and physiological changes, and third, it should be estimated from simple bioassay tests.

It was proposed that long-term safe levels of toxicants could be derived from short-term and static bioassay data. However, for estimation of safety factors, the LC0 and LC100 values (concentrations at which respectively zero and 100 per cent of experimental animals die in 96h) were previously considered. But these values are not able to derive graphically or statistically, a safety factor using LC5 and LC95 values (concentrations at which respectively 5 and 95 per cent of animals die in 96h) has been derived from a concentration-mortality curve fitted by regression analysis. The safety factor is derived by dividing LC5 by LC95 and the safe disposal rate (SDR) is estimated by multiplying safety factor by LC50 (Table 20.3).

Table 20.3 : Estimation of Safe Disposal Rate of Ammonium Sulfate for Hatchlings of
***Labeo rohita* Using Static Bioassay Data at 33°C**

		Concentration (mg/L)
Static bioassay data	LC5	19.3
	LC50 (CL)	67.0 (50.0-82.5)
	LC95	115.0
Safe Application		
Factor Equation (SAFE)	LC5 / LC95	0.168
Safe Disposal Rate (SDR)	SAFE × LC50 (CL)	0.168 (0.082-0.254)

CL = Confidence limit

Safe Concentration

Many toxicologists prefer the term safe concentration of toxicants that can be derived from acute toxicity tests. Safe concentration - also called *presumably harmless concentration*, can be obtained by the following way.

$$C = \frac{48h\ LC50 \times 0.3}{S^2}$$

$$\text{where } S = \frac{24h\ LC50}{48h\ LC50}$$

C = Presumably harmless concentration

The European Fisheries Advisory Commission has recommended that one-fourth of 96h LC50 can be considered as safe concentration. But according to the view expressed by the International Joint Commission, 5 per cent of 96h LC50 to the most sensitive fish species should be considered as water quality criteria for degradable toxicants. Though different terms used for different toxicity tests differ considerably, the two terms such as Application Factor (AF) and Maximum Acceptable Toxicant Concentration (MATC) are widely considered for the purpose.

20.11 Fish Toxicity Test

Toxicity test is an indispensable component of water pollution assessment because studies on the fluctuation in physico-chemical parameters do not encourage the evaluation of toxic effects of pollutants on aquatic life. However, though the history of toxicity test dates back to the time of Aristotle (384-322 BC) when he observed the effects of sea water on freshwater organisms, the science of toxicology was developed in the early 1970s as a methodical exercise and since then man has been conscious about disastrous effects of chemicals on aquatic life. Toxicologists have now advocated the usefulness of toxicity test of chemicals and industrial wastes to fish for forecasting the potential damage to aquatic biota.

Necessary for Toxicity Test

Toxicity test is a very important criteria that provides some guidelines to a variety of questions that follow.

1. Short-term effects of chemicals aquatic biota.

2. Relative toxicity of toxicants to a test species.

3. Allowable discharge rates of toxicants.

4. Sub-lethal effects of toxicants on a partial or complete life cycle of fish.

5. Relative sensitivity of different species to a toxicant.

6. Agreeability of environmental factors for aquatic biota.

7. Optimum levels of environmental factors for aquatic biota.

Types of Toxicity Tests

Different types of toxicity tests have been developed from time to time. For reasons of the production of healthy fish, toxicologists have been forced to develop different methods for toxicity tests of chemicals which are beneficial to the survival of fish community. On the basis of several criteria, toxicity tests have been classified accordingly. These criteria include (1) Duration (long-term, short-term and intermediate), (2) Method of addition of test solution (static, renewal, recirculation or flow-through), and (3) Propose (effluent quality monitoring, relative toxicity test, relative sensitivity, growth and odor). Of them, short-term tests are very useful and widely used for exploratory tests and by this test method, the amount of effluent discharged is estimated. These tests rapidly estimate the toxicity, assessment of a toxicant to different species, and the evaluation of the relative toxicity of different toxicants

Short-term tests are advantageous for estimation of toxicant concentration to be used in long-term and intermediate tests. Long-term tests are conducted in two ways. First, the complete life cycle test and second, the sensitive stage of life cycle. These tests are used to evaluate the AF, MATC, and different sub-lethal effects such as anatomical, behavioral, histological, biochemical, and haematological.

The intermediate test is carried out when a toxicity test is discussed with a long life cycle organism or a longer life cycle stage which need additional time for determination of LC50 values.

Duration of Exposure

Toxic reactions may differ qualitatively depending on the duration of exposure. A simple exposure — or multiple exposures occurring over 1 or 2 days — represents aute exposure. Multiple exposures continuing over a longer period of time represent a chronic exposure. In fish farms where chemicals, feeds, fertilizers, and organic manures are used, both acute (e.g. accidental discharge) and chronic (e.g. repititve handling of a chemical) exposures may occur, whereas with chemicals found in the ecosystem (e.g. pollutants in water) chronic exposure is more likely.

20.12 Acute Toxicity Tests

These tests are very simple, cost-effective, ecologically significant, and have the greatest utility. These tests are carried out in one of the four methods that follow.

Static Bioassay Test

Simply defined, bioassay is the test that helps evaluate the relative potency of a chemical by comparing its effect on a living organism with that of a standard preparation. Although bioassays are not only used in pharmaceutical industries to evaluate the potency of drugs but also applied in toxicology to assess the state of the environment/ecosystem through the exposure of biological systems. Bioassay has been useful in biomonitoring of ecosystems.

A vast variety of organisms are used to study the toxic effects of chemicals in aquatic ecosystems. Use of algal and Tubifex bioassay tests for toxicity of water pollutants, *Daphnia* and *Cyclops* bioassay tests for ecotoxicological studies in aquatic food chains, and snails as well as *Mytilus* bioassay tests for toxicity of heavy metal pollution are apt examples.

Static bioassay methods are generally applied for determining the toxicity of any toxicant to different aquatic organisms. This test is used for short-term exposure and serves as first check to evaluate the acute toxicity of a chemical to a number of organisms. This test is useful if the chemical enters the aquatic ecosystem only once or periodically.

   1. **Equipments:**

   (i) Plastic tanks or collapsible pools are used for acclimatization of test fish for at least 10 days.

   (ii) Glass aquria of 10 litre capacity are used for acute toxicity tests.

   (iii) For water quality estimation, analytical equipments are used.

   (iv) For studying the toxicity of chemicals to zooplankton, glass jars (500 ml-capacity) are used.

   2. **Water Quality:** For toxicity test, clear and unpolluted water should be used. Routine analysis of physico-chemical variables such as dissolved oxygen, pH, alkalinity, temperature, and conductivity are carried out and all values are recorded. The American Society for Testing and Materials have recommended that two types of measures (such as biological and chemical) must be consulted. While the former measure includes survival of the test species through acclimatization and test period with no signs of abnormalities, the latter one embraces undetectable and low levels of priority chemicals.

   3. **Test Fish:** Generally uniform-sized juveniles and fingerlings, having an average weight of 5 g are considered for this type of test. Feeding is discontinued for 48 hours prior to the bioassay test. The number of fish and their weight must be limited to reduce oxygen depletion since over-crowding and gradual accumulation of waste products in the diluent water may induce stress in fishes.

   4. **Criteria For Selection of Test Species :**

   (i) The test animals should be available in adequate numbers for the test.

   (ii) The test animals should be capable of being held in healthy condition for at least 25 days.

   (iii) The test animals should portray a considerable trophic link in the ecosystem.

   (iv) The test animals should exist or be closely related to a species that appears in the receiving water for the toxicant being tested.

On the basis of above-mentioned criteria, following countries have been considered some species of fish as standard test animals for evaluation of the toxicity (Table 20.4).

Table 20.4: Some Important Species of Fish Used as Standard Test Animals for Toxicity Evaluation

Country	Fish Species
India	*Esomus dandricus, Gambusia affinis, Puntius ticto, Oreochromis mossambicus*
Japan	*Cyprinus carpio*
USA	*Pimephales promelas, Lepomis macrochirus, Rasbora heteromorpha., Salmo gairdneri Jordanella floridae*
Canada and Europe	*Salmo gairdneri, Jordenella floridae*
UK	*Rasbora heteromorpha, Cichlasoma migrafasciatum*

Though some species of fish have been recommended for toxicity evaluation, a large number of fish are, however, considered by toxicologists as test fish. It appears from several toxicological studies that toxicants produce certain identical toxic reaction patterns to almost all freshwater fishes, which differ from the most part only in their degree of reactivity. The behavioral reaction pattern that has demonstrated for standard test fish have also been demonstrated for the other species. Several of the biochemical and histological effects of test chemicals on different standard test species have also been observed in other species. Most of the differences in the toxicological data reported by a growing body of research appears to be more directly related to technical differences such as differences in the techniques followed, conditions of the test species, and the variability in the quality of the diluent water used.

5. **Dilution of Toxicant:** Toxicant dilutions are prepared by diluting the stock solution with diluent water. The stock solution of a chemical is prepared by dissolving the required quantity of chemical in either water or alcohol or any other suitable solvents as the case may be depending upon the nature of solubility of chemicals. For determination of the toxicity of less hazardous substances such as organic manures and chemical fertilizers, they are directly applied to the diluent water.

To prepare 1,000 mg/l stock solution of methyl parathion (metacid - 50%. W/V), for example, 2 g of the chemical is mixed with one litre of water. Different test media with required concentrations of the chemical is prepared from the stock solution.

6. **Test Procedures:**

(i) *Range Determination Test*: This test is carried out to determine the range of toxicity of an unknown chemical. Four fish per test container are exposed to several graded (3 to 4) logarithmic series of toxicant concentrations accompanying with controls. In these tests, fishes are allowed exposure to the toxicant for 24 hours.

(ii) *Definitive Test*: Ninety-six hour tests are conducted in 5 or 10 litre glass aquaria/ jars each containing known quantities of unpolluted water. For testing each concentration of the chemical, 2 to 4 test fish are exposed in each container and 8 to 16 fish per concentration. The mortality of fish in each container is recorded upto 96 hours at interval of 24 hours. Initial weight and length of each fish is recorded in a well- designed record sheet. From the results, Median Lethal Concentrations (LC50 values) for 24, 48, 72, and 96 hrs are determined using anyone of the following methods:

  * *Graphical Method*: The response curves are drawn on a graph paper by plotting per cent mortality (y) against toxicant concentration (x) for each exposure period. From the curves obtained, the concentration of a toxicant at which 50 per cent mortality occurs, is obtained that represents the LC50 for that exposure period.

  * *Probit Method*: In this method, the per cent mortality (y) is converted into per cent kill and plotted against toxicant concentration (x) in a double logarithmic grid. A straight line is fitted following the principle of least squares using the regression

equation $y = a + bx$, where a represents the y-intercept and b represents the slope of the line. The two constants (a and b) are calculated by the following formulae:

$$na + b\Sigma X = \Sigma y \qquad (1)$$
$$Xa + b\Sigma X^2 = \Sigma Xy \qquad (2)$$

where,

n = Number of observations

ΣX = Sum of X values

Σx^2 = Sum of x^2 values

Σy = Sum of y values

ΣXy = Sum of product of individual X and y values

The straight line thus fitted is intersected at probit 5 and the corresponding X value is find out which is the LC50 for the specific exposure period.

An example of a straight line fitting for the data obtained as per probit method on 96h exposure of *Labeo rohita* fry to different concentrations of urea is noted below:

Incorporation of the values in the equation

$$10\,a + b\,400 = 98 \qquad \ldots\ldots\ldots\ldots (1)$$
$$280\,a + b\,12810 = 2940 \qquad \ldots\ldots\ldots\ldots (2)$$

Multiply (1) with 'X' and (2) with 'n' values

$$[10\,a + b\,400 = 98]\,280$$
$$[280\,a + b\,12810 = 2940]\,10$$

$$\overline{}$$

$$2800\,a + 112000\,b = 27440$$
$$2800\,a + 128100\,b = 29400$$

$$\overline{}$$

Cancel out the equation

$$-16100\,b = -1960, \qquad b = 0.1217$$

Solve for 'a'

$$10\,a + 0.1217 \times 400 = 98$$
$$10\,a + 48.69 = 98$$
$$10\,a = 98 - 48.69$$
$$10\,a = 49.31$$
$$a = 4.931$$

Therefore, the regression equation may be written as $y = 4.931 + 0.1217\,X$

From the regression equation, the 'y' is calculated from known 'x' values for fitting the straight line and then the straight line is fitted by plotting calculated 'y' for the known 'x' values.

* *Dragstedt Behren's Method*: In this method, the cumulative mortality is determined at different concentrations and per cent mortalities are calculated from the cumulative mortalities. The cumulative number of dead animals is divided by the total cumulative number of dead and live animals and the value is multiplied by 100. The LC50 value is calculated using the following formula.

$$\text{Log LC50} = \text{Log A} + \frac{50 - a}{b - a} \times \text{Log 2}$$

Where,

A = Concentration of the toxicant having the percentage of mortality below 50 per cent

a = Percentage of mortality immediately below 50 per cent

b = Percentage of mortality immediately above 50 per cent

An example of observations made on *Labeo rohita* hatchlings exposed to urea is given in Table 20.5.

Table 20.5: Cumulative Mortality of *Labeo rohita* Hatchlings on 96h Exposure to Urea.

Concentration (mg/l)	Absolute live	Mortality dead	Cumulative live	Mortality dead	Per cent
8.0	15	0	61	0	0.00
10.0	12	3	47	5	9.62
16.0	10	5	40	13	24.53
20.0	8	7	31	21	40.38
25.0	8	7	24	29	54.72
30.0	7	8	19	40	67.79
35.0	7	8	12	52	81.25
40.0	6	9	5	65	92.86
45.0	4	11	3	76	96.20
50.0	2	13	1	89	98.89
55.0	1	14	1	98	98.99
60.0	0	15	0	112	100.00

$$\text{Log LC50} = 10 + \frac{50 - 40.38}{54.72 - 40.38} \times \text{Log 2}$$

$$= 1 + \frac{9.62}{14.34} \times 0.3010$$

$$= 1 + 0.2019$$
$$= 1.2019$$
$$\text{Antilog of } 1.2019 = 15.92 \text{ mg/l.}$$

RESIDUAL OXYGEN BIOASSAY TEST*

This type of bioassay test was developed in the early sixties and it is now considered as a rapid, quite sensitive, less expensive test, and feasible for on-site acute toxicity test in comparison to static bioassay method and can be used for assessment of acute toxicity of toxicants to aquatic animals.

This test is carried out in wide-mouthed glass jars of one litre capacity. Toxicant dilutions are prepared by compressed air to maintain dissolved oxygen concentration more than 8 mg/l. The loading of test fish in experimental jars depends on their size. In general, large-sized fish are kept in less numbers and smaller fish in large numbers. For best results, however, two sets of bioassays are conducted with fish and four fish loading density in jars of 1.5 litre capacity.

The concentration of residual dissolved oxygen is measured by either Winkler's titration method or oxygen meter, immediately after the death of fish in the test container. The time of death of fish in each jar is recorded. The principle of this test is that the survival time of fish is inversely correlated with the toxicant concentration. In other words, the higher the toxicant concentration the less will be the survival time remaining higher amount of residual dissolved oxygen. In the beginning of residual oxygen bioassay test, a range determining test is carried out with 4 to 6 concentrations of toxicants followed by definitive test.

The data of residual dissolved oxygen and survival time for experimental and control jar are plotted on semi-log graph paper. Dissolved oxygen and concentration of the toxicant are plotted on arithmetic and log scales, respectively. The concentration of toxicant in which the survival of test fish is first affected and the amount of residual dissolved oxygen in water increased is reported as *threshold limit* of the toxicant.

20.13 Chronic Toxicity Test*

The chronic toxicity test exhibits the effects of long-term exposure (exposure varies between 1 and 4 months) of fish to toxicant concentrations much lower than the lethal concentrations.There are two types of chronic toxicity test such as (1) complete life cycle test and (2) partial life cycle test. The former type of test is carried out by using hatchlings of test fish and the response parameters are tested to all subsequent life stages such as fry, fingerlings, and adult. By this method, the application factor (AF) and maximum allowable toxicant concentration (MATC) of toxicants and effluents are determined. The

---

* For further detail, see Ballard and Oliff (1969), McLaey (1976).

MATC is the maximum concentration of any toxicant that will not affect fish population in culture ponds. On the other hand, AF is the factor that designates toxicant concentration which is not dangerous for chronic or life time exposure of test fish. The AF is calculated by the following way:

$$AF = \frac{MATC}{96\,h\,LC\,50}$$

These tests are, however, also used to evaluate the effects of toxicants on various parameters of fish such as histological, morphological and behavioral changes, bioaccumulation of toxicants, haematological and biochemical alteration, feeding, survival, mortality, respiration, growth and reproduction.

COMPLETE LIFE CYCLE TEST

Test chambers generally considered for the growth of hatchlings or fry are of 18 X 15 X 18 cm in diamention where the depth of water is maintained to about 12 cm. In cases where tests are performed for reproduction, large-sized tanks (80 X 30 X 40 cm diameter) with 30 cm water depth are used. Tests are replicated 2-3 times along with controls. Prior to stocking the tanks with test fish, selected amounts of toxicants are mixed with the tank water. The toxicant concentrations and physico-chemical qualities of water (such as pH, hardness, oxygen, conductivity, and alkalinity) in all tanks are measured at interval of 7 days. Dissolved oxygen concentration and temperature of water in each tank are daily recorded.

Different data on test fish as noted above are recorded for each test chamber at the beginning, middle, and at the end of the test.

PARTIAL LIFE CYCLE TEST

Partial life cycle test is conducted at any life stage of fish and the selection of life stage depends on the type of parameters to be tested. For example, if growth rate is measured at regular intervals, fish are kept in different concentrations of toxicants with regular feeding and maintaining optimum water quality parameters. For each exposure period, a control should be accompanied to assess any significant difference between control and treatment. Fish growth rate is considered as a single parameter to evaluate the effect of toxicants on fish production potential in ponds. Other parameters of the test fish such as feeding and avoidance, bio-accumulation, reproduction, and histopathology are also assessed.

On the basis of the results of acute toxicity test, the range of toxicant concentration is considered for chronic toxicity test. Five to 10 sublethal concentrations of toxicants are considered for this test. Twenty to 30 fish are kept in different sublethal concentrations of toxicants for specific duration of time. Feeding of fish is continued for the entire period of study. At the end of the test, fish are sacrificed, their tissues are processed, and the amount of toxicants (especially heavy metals and pesticides) accumulated in target organs of the fish are estimated by various techniques such as Liquid Gas Chromatography and Atomic Absorption Spectrophotometry.

Studies on the effects of sublethal exposure of toxicants on the tissues of different organs of fish such as gills, kidney, liver, gonads, intestine, skin, and brain have shown encouraging results. At the end of the test, exposed fish are sacrificed, different organs are removed, sections are made, stained with specific dyes, and then examined with light compound microscope and Transmission Electron Microscope to observe histological changes and pathological lesions, if any, at tissue and sub-cellular levels.

The behavioral responses of exposed fish are considered for determination of the potency of toxicants. Though no specific method has been designed so far for testing behavioral changes, toxicologists generally use their own test designed for the purpose.

20.14 Ecotoxicology*

Ecotoxicology is the science of the study and analysis of toxic substances of different ecosystems and their mode of action on fish and aquatic organisms. The subject of ecotoxicology is, however, an outgrowth of the connection between chemistry, toxicology, ecology. In toxicology, dosage is the criterion which determines the toxic effects of a substance on fish and aquatic organisms. In ecotoxicology, different aquatic animals including fish are considered which not only exhibit varying degrees of sensitivity to a substance but are also exposed to it in different degrees. But no single measurement of the concentration of a substance in the ecosystem is enough in itself to assess the stress on the ecosystem. The way in which a substance is released into the ecosystem and the rate at which it is broken down, determine whether it reaches to an organism or affects it in any way generally termed as *ecotoxicoloagical effects*, are a function of bioavailability. Both potential effects and exposure must be considered in order to assess the ecotoxicological relevance of a chemical substance.

Public concern over large number of synthetic chemicals in the environment was exaggerated above all by the side effects of pesticides and the long-term effects of accumulated industrial pollutants. At the same time, it was less surprising to find out that many substances indeed had side effects than the forms in which they were manifested. The course of action of pollutants, their persistence in the ecosystem and the sensitivity to non-target species which, in many cases, were actually beneficial to organisms had not been anticipated.

CRITERIA FOR ECOTOXICOLOGICAL CONSIDERATIONS

Ecotoxicology is concerned with the toxic effects of chemical and physical agents on fish populations and other biotic communities within defined ecosystems. It includes the

---

* For further study, see Rombke and Moltman (1996).

transfer pathways of those agents in different trophic levels in a food chain and their interactions with the environmental factors. In order to understand the science of ecotoxicology in relation to fish production, the following processes should be considered.

1. Transformation of the pollutant(s) by physico-chemical processes at the time of transport from the point of discharge to the target organisms. Toxic substances are chemically transported into different biota and these toxicants have different behavioral norms and toxic properties. For most of the pollutants, the nature of these processes is not known.

2. Assessment of metabolism of different toxicants in aquatic organisms is badly needed. Toxicants are transported to target organisms via the food chain and the amount of accumulation of toxicants to different trophic levels should be calculated. Moreover, the doses of toxicants to receptors should also be assessed. The exposure of one or more target organisms and the type of exposure must be evaluated.

3. Assessment of the response of community, population or individual organism to the changed pollutant is cardinal.

4. The forms, amounts and sites of release of the pollutant(s) should be known.

The problems of aquatic ecosystem where fish culture is undertaken concern mainly haphazard management with limited water exchange and contamination of water by pollutants. Environmental contamination is an age-old phenomenon but until recently, this has dramatically changed the entire fish culture scenario. A problem associated with finfish culture is the turnover and removal of large amounts of nitrogenous compounds resulting from massive inputs of nitrogen-rich feeds. In addition to this, decomposition of uneaten feed, use of chemical fertilizers and organic manures, coupled with surface flow-off from agricultural fields have led to the accumulation of toxic intermediates in fish culture ecosystems to a large extent.

Pollution of rivers, ponds, lakes, and reservoirs in many Asian countries is a cause for concern, not only for farmers but also for mankind as a whole. This situation has caused considerable damage to fisheries and aquaculture around the world. Consequently, many species of fish have become obliterated from these water bodies. Environmental issues on aquaculture development are receiving considerable attention in the countries concerned, and numerous efforts are under way to ensure more sustainable development of aquatic ecosystems. Such efforts possibly originated from the widespread concern caused by the population explosion and massive use of pesticides in recent years. The feeling elicited by several organizations of high standards was to establish expert/scientific committee and to conduct conference/seminar/symposia on the ecosystem/environment at national/ international levels to evaluate the level of environmental contaminants and their risks to fish populations. The committee has a holistic view on fish culture industries, judicious use of aquatic resources, conservation of fish stocks, monitoring of fishing activities on the production side, research and development programs and fish marketing as well as export.

20.15 Applicability to Ecotoxicology

Different aquacultural activities are reflected through quality publications that provide a lot of informations on aquaculture issues and aquatic toxicology. Research publications principally focus on several informations such as biogeochemical cycles, dynamics of ecosystems, ecotoxicology, environmental monitoring, and environmental impact assessment. From aquaculture point of view, fish and aquatic plants/animals of human interest are most important and thus the effects of pollutants on fish health are principal to ecotoxicology.

When the effects of a chemical substance are investigated, quantitative evidence of its occurrence must be provided. This is the task of analytical chemistry. The detection of a substance is not an indication of its potential effects. Advances in technology have made it possible to detect inconspicuous quantities of substances in the residue analysis. In many cases, the quantities detected are lower than the known biological effect threshold. Consequently, the ecotoxicological relevance of residue data is still a subject of controversy.

Traditional toxicology is concerned with the impact on fish populations or on aquatic ecosystems. It is possible that an environmental event, while exerting severe effects on individual organisms, may have no important impact on fish populations or on an ecosystem. Thus the term 'toxicology' and 'ecotoxicology' are not interchangeable.

20.16 Input of Toxicants into Aquatic Ecosystems

On the basis of the intended use, a distinct sign of the environmantal concentration of a toxicant can be given, since the quantity used as well as the size of the area treated are known. A number of examples are known in which toxicants are found to occur even at long distances from the application sites (such as DDT and Atrazine). For environmental chemicals, a wide range of concentration is feasible in all media or compartments in a way which is arduous to predict. In addition to this, the condition of the ecosystem is intricated by the fact that secondary deposits, metabolic processes or built-ups may occur.

The use of a toxicant forms the decisive link between the compound and the target organism. The formulation, technique, and the ecosystem condition are the parameters which largely determine the actual effects and the possible side effects. Over-doses can inflict an inadequate amount of stress on the ecosystem.

In the tropics, toxicants used in the aquatic ecosystem are more common. However, in temperate latitudes, toxicants are only rarely used in the systems concerned. Apart from inadvertent discharges due to improper use, toxicants still enter the surface water or flow-off. Drainage water or groundwater are other paths of entry into the aquatic ecosystem.

20.17 Life Cycle of a Toxicant in an Aquatic Environment

Life cycle describes the phases of a toxicant's life, and the manner in which it produces the next product. However, toxicants have a life cycle involving a first step — the production (active ingredients and formulation aids) preceded by transport, application (exposure and uptake), effects and side-effects, fate (translocation and sinks), and disposal. The steps within each cycle are important for hazardous effects. Possible hazards to the ecosystem, however, include insufficient effect on the target organisms, development of resistance to chemicals, unexpected synergisms of several toxicants, damage to beneficial organisms, accumulation and remobilization, secondary effects by metabolites and residues.

EXPOSURE TO ECOSYSTEM

1. *Environmental Fate*: Exposure is comprehended to denote the concentration of a toxicant to which ecosystems are exposed at a certain time. For the plan of exploration, a difference is made between compartments and media of ecosystem (Figure 20.7).

The exposure of a chemical depends on its fate in the ecosystem. A difference can be made between the three partial processes which, however, are difficult to separate in actuality.: (i) transport/translocation (mobility), (ii) metabolism (transformation), and (iii) mineralization (degradation), In the first case, toxicants spread from one place to another within a compartment or move to a new compartment where they may continue to have an effect. Metabolites are developed either through photolysis or through micro-organisms. The fate and effect of metabolites are often very different from those of the initial substances. In other words, metabolites are more toxic than the original chemicals; but their hazardous effects may also be neutralized such as methylation of heavy metals. Heavy metals and some pesticides can be bound almost irreversibly in sediments thus eliminating their bioavailability.

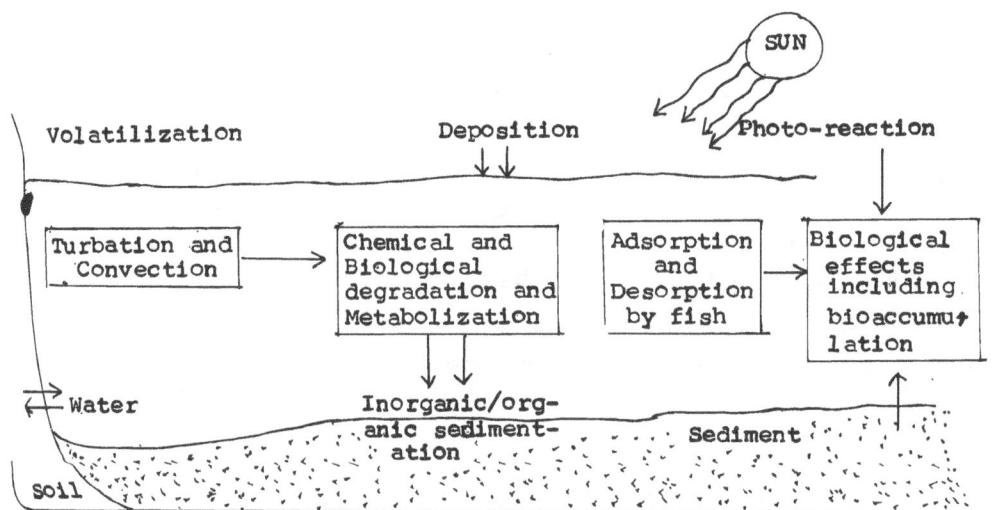

Figure 20.7: The fate and possible dispersion of a chemical in an aquatic medium.

2. *Sink*: Sink is comprehended to indicate a compartment or part of a compartment in which a toxicant temporarily or permanently loses the capacity for re-entry into different compartments. This can occur through adsorption in the pond sediment or deposition in fish body.

20.18 Three Approaches: Consumption, Retention, and Effect

The foregoing discussions remind us that three concepts of ecotoxicology have evolved through several decades of toxicity study. The first treats fish culture ecosystem toxicity as consumption of toxic substances by fish through food chain. 'The second treats fish and their food organisms as retention of toxicants within their body and the third considers as effects of toxicants on aquatic life and ecosystem in general. These conceptions illustrate the three approaches such as consumption, retention, and effect that can be used in studying ecotoxicology. However, since studies of adverse and chronic effects of toxicants on fish and aquatic ecosystems contribute equally to drastic changes in the physico-chemical factors followed by habitat destruction, these approaches cannot be neglected where intensive fish farming is practised.

20.19 Intake Versus Uptake of Toxicants

The term intake defines as the ingress of any chemical substance into the alimentary canal of fish through feed and the fate of the substance is controlled by the process called *absorption*. In contrast to intake, the uptake is understood to mean the absorption of the substance from surrounding water into the extracellular fluid for circulation and the substance undergoes metabolic processes.

TYPES OF EFFECTS

A toxicant is taken up by fish either directly or indirectly (through food). Toxicants must be absorbed in order to take full effect. Fish and fish food organisms can accumulate toxic substances into the body upto the level at which there is an observable effect. Accumulation takes place, for example, from consumption of a large quantity of contaminated fish food organisms.

Although there have been several cases in which persistent toxicants are accumulated in the food webs, no steady increase in the level of toxicants has been noted from one trophic level to another. This is due to the fact that many organisms occupy several trophic levels and that specific metabolism and elimination rates of toxicants have a significant effect on the growth of fish in ponds. Accumulation factor (the ratio between the concentration of toxicants in the ecosystem and the fish body) up to ten of thousands is possible.

Fish and fish food organisms have been shown to be absorbed through gills and body surfaces. Metabolites of nitrogen fertilizers are absorbed through gills, enter into the blood streams and cause severe health hazards. Pesticides and heavy metals are also entered into the body of fish through absorption and exhibit several behavioral, morphological, and physiological abnormalities. Fish in contaminated waters rapidly absorb a lot of toxicants from the alimentary canal via natural food organisms and are transported to liver where it is metabolized and excreted into the intestine; of course, the rate at which the metabolism of toxicants is not so rapid, only small amount of it is metabolized, excreted and substantial amount is permanently accumulated into the fish flesh.

ELIMINATION VERSUS RETENTION

The fact that there is consistent and heightened toxicant levels in fish fed on natural fish food organisms containing elevated levels of toxicants. This indicates that toxicant-contaminated diet is an important source of toxicant levels in fish. If toxicant-contaminated fish is, however, kept in a toxicant-free water, toxic substances will be eliminated in a short time but a small residual remains in the body, becoming measurable as a significant increase only after prolonged exposure to the toxicant contaminated food. A large number of pesticides have been found to be accumulated in fish flesh, liver, kidney, and brain in great quantity but with very low excretion rate. Methyl mercury is absorbed by the gills of fishes.

Toxicants which are accumulated in aquatic ecosystems are absorbed or adsorbed by large number of bacteria in the sediment and hence removed from the water and prevents them from dispersing. Fish food organisms consume these bacteria and so take the bound toxicants. These toxicant-enriched organisms are then consumed by fish and thus the toxicants are passed along the food chain where the diet is largely of these toxicant-enriched food organisms, the fish will, after a prolonged period of feeding, have heightened the toxicant levels in their tissues. The relative importance of the means of toxicant uptake is likely to vary according to the levels and forms of toxicants in the food, the water, and the abiotic-biotic factors of the ecosystem. The efficiency of mercury accumulation in fish at all trophic levels, for example, has been noted to be greater in low alkalinity than in high alkalinity lake/pond ecosystems.

The retention of toxicants in fish body is essential in determining the toxic effects. The amount of toxicants deposited can be measured by several methods (see Panel 20.1). The higher and lower values of toxicants deposited in the fish flesh and other target organs have been reported for heavy metals and pesticides.

ECOTOXICOLOGY IN TROPICAL AQUATIC ECOSYSTEMS AND MANAGEMENT*

Some basic conditions have been formulated for evaluation of the ecotoxicological hazard potential in the tropical aquatic ecosystems. These conditions, which correspond to the tiered procedure (A three stage test: laboratory, semi-field and field which are designed to minimize the amount of time and exposure that has to be invested in tests), are for all practical purposes still valid today.

---

* For further study, see O'Brien (1967), Piper et. al. (1986).

1. Data on the environmental fate and effects on organisms are to be collected under tropical conditions.

2. Effect tests are to be supplemented by accompanying with chemical residue analysis data and these studies should be carried out in laboratories in the tropics.

3. Data are insufficient for progresses of environmental hazard potential which can be validated with the aid of field studies.

4. Every ecosystem must be studied separately.

Criticism on large-scale use of some pesticides (such as methyl parathion, malathion, fenitrothion, and others) is becoming increasingly important in the tropics for the following reasons:

1. The ratio between the benefits of pest control and the anticipated side-effects is becoming exceedingly unfavorable.

2. Ecotoxicological side-effects are not taken into sufficient consideration.

Government organizations of many tropical countries have recognized the improper use of chemicals as an infrastructural problem. Various measures enable many farmers in the tropics to benefit from the advantages of chemical-based crop protection at high level of production. However,considerable improvements must still be made in order to cope with the use-related problems in terms of higher safety standards and awareness of possible side-effects. There is no economically viable mode of application which is also safe from the standpoint of ecotoxicology in small-holder subsistence agriculture.

As a generalization, the probability of exposure to toxicants is obviously higher in the tropics than it is in the temperate zone counterpart. At the same time, the fate of pesticides in the tropics is different from temperate latitudes. The fate of DDT, for example, under abiotic conditions in tropical countries is entirely different from the fate predicted on the basis of studies in the developed countries. The persistence of pesticides such as lindane and DDT, for example, is much lower due to more rapid metabolism rate of micro-organisms at higher temperatures. Additionally, photolysis plays an important role in the tropics due to stronger light effect than it does in higher latitudes. Although the lower persistence of pesticides in tropical ecosystems is considered positive at first glance, negative effects may occur even at sites which are far removed from the application site.

APPLICATION TECHNIQUES AND FORMULATIONS INFLUENCE THE EXPOSURE OF ECOSYSTEM

Large-scale applications of pesticides by helicopter and aeroplane to combat disease carriers are problematic in the tropics, since an entry in natural ecosystems leads to contamination of freshwater ecosystems with severe effects on fish and their food organisms. Under these circumstances in practice, the ecotoxicological effects may be far greater than those which were predicted in the laboratory or under controlled field conditions in temperate climates.

TROPICAL RISK ASSESSMENT

Despite all of the differences between the situation in temperate and tropical climates, there are only very few ecotoxicological methods as yet which are designed for tropical risk assessment. As a result of various research projects in southern Africa, India and their neighbouring countries, several methods have been evolved for terrestrial field studies. The data available for regulatory purpose may be applied to aquatic ecosystems. However, stringent action should be taken regarding the use of organo-chlorine compounds. With the exception of fish tests, monitoring programs are evidently carried out from time to time.

The lack or inadequacy of experience in utilization, serious ecosystem damage from the use of chemicals frequently occurs in tropical fish culture ecosystems. In addition to organo-phosphates, other substances such as synthetic pyrethroids which have been claimed to be harmless, have asserted with a view to subsequent proof that there are hundreds of cases of damaging the ecosystem integrity.

Due to lack of safety regulations of agricultural chemicals as well as the many small production sites which do not have much know-how in dealing with substances which are hazardous to aquaculture and fisheries, it must be assumed that a considerable hazard potential exists. Although the state of knowledge about the ecotoxicological risks of agricultural chemicals is minimal, one exception is the use of water hyacinth as an indicator of stress from heavy metals. Most recently, cases are becoming known in which developed countries have attempted to get rid the aquatic ecosystem of hazardous substances by exporting it to developing countries, which can also be assumed to do severe damage to the ecosystem.

Risk evaluation posed by hazardous substances in the tropics must take specific exposure conditions into consideration. But it should be kept in mind that further study on the effects of chemicals on fish and aquatic ecosystems as a whole should not be neglected.

RESEARCH ON ECOTOXICOLOGY

Research experts of United Nations Education, Scientific and Cultural Organizations (UNESCO) have provided a list of ecotoxicological research priorities and most of which are still valid today. However, recommendations for ecotoxicological research in the tropics are given below.

1. Assessment of degradation and metabolism of toxicants.
2. Tropical species should be selected on the criteria that are valid to test organisms.
3. Development of laboratory tests.
4. Semi-field methods should be adopted for fate and effect studies.
5. Effects of pesticides on non-target organisms should be evaluated in the field with consideration for the role of tropical beneficial organisms.

6. Long-term effect of toxicants on the biological productivity should be assessed if at all possible at the same sites.

7. Influence of aquatic systems due to input from terrestrial systems.

8. Special studies on ecosystem sinks should be conducted.

9. Ecotoxicological assessment and evaluation criteria should be prepared that are specially adapted to tropical conditions.

10. Mathematical models should be developed to assess the fate of chemicals in organisms or between ecosystem compartments.

To sum up, it can be said that there is a deficit of basic knowledge, appropriate ecotoxicological methods and corresponding assessment criteria in the developing countries. As in the developed countries, test ranging from pre-registration laboratory tests up to field monitoring usage should form the foundation ,for the definition of ecotoxicological measures. A decision should be made principally on the basis of knowledge gained by experience which have yet to be formulated as to which the tests required in the producer countries should be repeated for the relative assessment under regional conditions.

In contrast to the assessment criteria, which differ very much depending on the test systems and parameters, there are no specific evaluation criteria for tropical and sub-tropical aquatic ecosystems: the claim to foster the ecosystem integrity against damaging effects of chemicals holds true for all systems in the same way.

20.21 Acid Rain and Ecotoxicology

The problem of acid deposition and its possible effects on aquatic ecosystems has been recognized for centuries. In the late 1970s, animated by concern over possible deterioration of aquatic habitats, the United States Congress initiated an investigation on the potential problems as a result of acid deposition. However, the subject of acid deposition, though it is beyond the perview of this volume, is extremely complex and far reaching. It essentially involves detailed study of chemical, physiological, and ecological interactions across the state and national boundaries and encroaches on economic issues.

EFFECTS OF ACID DEPOSITION ON AQUATIC ECOSYSTEM

1. *Genreral Effects*: On the basis of a few lake or stream acidifications and unmanipulated waters, several generalizations may be drawn that follow but the effects of reduced pH on organisms and community dynamics are significant:

(i) High concentrations of H^+ can have a direct toxic effects on aquatic life. The common problem associated with low pH is impaired ability to balance sodium, chlorine and hydrogen. Ion regulation problems have been documented for fish and benthic invertebrates.

(ii) Aluminium changes its species with reduced pH. It is leached from acidified watersheds and can accumulate to potentially toxic levels in water. Its toxicity in water is complex and depends on concentration, pH, and calcium levels. The inorganic forms of aluminium (such as AlF_3-fluoride, $Al(OH)^+$ - hydroxide and Al^{+3} monomerite) are more toxic than organically-bound Al, especially to fish. Aluminium is thought to bind with phosphorus, rendering insoluble complex that becomes biologically unavailable. Because phosphorus is the limiting nutrient in oligotrophic waters, its binding may reduce primary productivity of water.

(iii)   Other metals may become more toxic with reduced pH. While aqueous levels of lead may increase because of concomitant deposition of lead with acidified minerals, cadmium may increase in waters because of leaching in watershed soils. Elevated concentrations of mercury in fish may be due directly to increased deposition or mobilization of mercury or indirectly through enhanced methylation of the metal. Methyl-mercury is the form most readily assimilated by organisms, and reduced pH may influence the rate of methylation by microbial processes.

(iv)   Concentrations of base anions ameliorate the effects of pH toxicity. Calcium is low in acid-sensitive waters. Lethal pHs for fish and benthic invertebrates are often higher in soft water (less than 100-150 ueg/1 Ca) than in waters with greater calcium. Low levels of calcium increase cell membrane permeability to ion and water exchange and hence the toxic effects of H^+. Molluscs and crustaceans are sensitive to low levels of calcium and often the first species to disappear with acidification.

(v)   Lake clarity often increases with reduced pH. Increase in transparency of water may be due to changes in the chemical composition of dissolved organics to flocculation of organo-aluminium complexes rather than to decreased phytoplankton abundance.

(vi)   Humic substances (suspended and dissolved organic compounds) have complex relations with pH. At pH 5.0, these compounds are weak buffers that can help maintain pH. Surface waters that are high in dissolved organic compounds are less sensitive to different metals and alter their availability to organisms. Ability can be decreased or increased depending on type, substance and concentration of humic substances. Alumunium toxicity, for example, is often lower in water with moderate levels of dissolved organic compounds (5-10 mg/l) .

(vii)   Species richness tends to correlate with pH. Decreased richness has been noted in phytoplankton, zooplankton, periphyton, fish, and macro-invertebrates.

(viii)   Species composition of water may change without any apparent reduction in biomass or productivity. Sensitivity to reduced pH varies among species, and acid-sensitive species are more likely to disappear early in the acidification of surface waters. In general, primary productivity, nutrient turnover or rate of decomposition are measurably influenced by acidification.

2.   *Specific Effects*: Direct effects of reduced pH on fish differ with species, life stages, and interaction with other chemicals. Cleavage eggs have been found to be the most sensitive life stage of several fish in temperate regions. Eyed eggs are more tolerant and are able to survive in low pH (5.0). Mortality again increases just

prior to, and during hatching. At 100 mg/l aluminium has been found to decrease in the mortality of cleavage eggs at lethal pH.

The effect of reduced pH on fish interferes with ion regulation. Fish exposed to high levels of H^+ concentrations, which may ultimately be fatal. At 2 mg/l calcium, no added aluminium, for example, and pH 4.5 net exflux and influx of sodium has been found to balance in brown trout. Fish tend to be more sensitive to Al^{3+} and $Al(OH)_2^+$ than other forms of inorganic, monomeric and insensitive to organically-bound aluminium.

For most species, toxic pH levels are around 5.0, but some cyprinids are more sensitive and may die at pH 6.0. Salmonoids are moderately sensitive, but exhibit a wide range in tolerance from around pH 6.0 for Atlantic salmon to 4.5 for brook trout. Most sunfishes are also moderately sensitive and are able to live in waters with pH less than 5.0.

Sub-lethal effects of acidification in fish include behavioral changes, reduced growth, ventilation problems, deformations, and bone abnormalities. Juvenile brook trout when exposed to pH 5.5, for example, exhibited depressed range of movements, prey capture success, and aggression levels.

Fathead minnows experience many problems at pH less than 7.0. Adult minnows have been found to cease spawning, hatching success, and survival of fry. Fish in acid treatments demonstrated delayed yolk-sac absorption and a host of histological anomalies in gills, kidneys, and livers.

Population and community effects have been documented through whole lake experimental acidification in Ontario, USA. The lake pH was reduced from 6.8 to 5.0 with sulfuric acid over 8 years. Recruitment failure of fish occurred at the following pH levels: Fathead minnow (*Pimephales promelas*) - - 5.0, Lake trout (*Salvelinus namaycush*) - - 5.6, White sucker (*Catostomus commersoni*) 5.0, and Pearl dace (*Semotilus margarita*) - - 5.1. Recruitment failure of rock bass (*Amblopites rupestris*) at pH 5.6 seemed to occur in little critical pH below which species richness declines or absence of particular species becomes apparent is possibly between pH 5.0 and 6.0.

ACIDITY OF AQUATIC ECOSYSTEM CAUSES POOR FISH PRODUCTION

Although sulfur-containing fertilizers have long been recognized as essential for plankton and fish, their excessive use causes aquaculture pollution to a large extent. There are four major sources from which fish culture ecosystems can be supplied with sulfur: (1) organic matter, (2) soil minerals, (3) atmospheric gases, and (4) sulfur-containing fertilizers/ substances.

1. Sulfur Oxidation and Acidity

As the oxidation reactions for sulfur shows, oxidizing sulfur is an acidifying process. For every sulfur atom oxidized, two hydrogen ions are formed. By adding sulfur, this acidifying reaction is utilized to reduce the alkalinity of water and to reduce the pH of soils. The acid-forming reactions must also be taken into consideration when choosing a sulfur-containing fertilizer, since some materials (such as ammonium sulfate) contain elemental sulfur and might lower soil pH unfavorable.

The addition of atmospheric sulfur to aquatic ecosystems through precipitation over extensive areas increases soil acidity. Some of this acidity also results from nitrogen oxides as they react in the atmosphere to form nitric and nitrous acids. Sulfur and nitrogen are jointly responsible for this problem. The pH of this (also called *acid rain*) may be 4.0 or even lower compared to about 5.6 for natural precipitation. These low pH values harm plants and aquatic organisms including fish.

2. Extreme Soil Acidity in Fish Ponds

The acidifying effect of the oxidation of sulfur-containing compounds can bring about extremely acid soil conditions, for example, when brackish water is drained in fish ponds and put under cultivation. Sulfates in the water are reduced to sulfides, generally as iron and manganese sulfides, in which forms they are stabilized.

When high-sulfide lands are drained in aquatic ecosystems, the sulfides and/or elemental sulfur are quickly oxidized, forming sulfuric acid. The soil pH may drops as low as 4.0 (or even 3.0), a level unknown in normal pond soil. Obviously, natural food organisms cannot occur under these conditions. In addition to this, fish are succumbed to diseases. Furthermore, the quantity of limestone needed to neutralize the acidity is so high as to make this remedy completely uneconomical.

20.22 Evaluation of Ecotoxicology

To control discharges of a variety of toxicants into different freshwater bodies, pollution discharge elimination system should be designed. The main focus of this system is to evaluate the effects on the basis of a set of technological requirements and agreement to water quality criteria for toxicants. These tests are based on toxicity of individual pollutants, and may not be plentiful to completely protect aquatic life that receive toxicants in mixtures of unexplored toxicity. In assessing the hazard of composite discharges, it is more pertinent to orderly evaluate the toxicity of water (where several contaminants are discharged simultaneously) than to evaluate the degree of pollution based on the criteria for individual pollutants.

At present, tests using sensitive aquatic animals have been recommended by the United States Environmental Protection Agency for biological evaluation of toxicity of contaminated aquatic systems with toxicants. However, the importance of biological testing and identification of toxic conditions in aquatic ecosystems should be emphasized than in chemical analysis of contaminated waters. The biological testing is important to all attention to the potentially widespread contamination of the region with pollutants and associated by-products.

ECOTOXICOLOGICAL EFFECTS ON FISH AND ECOSYSTEM

The presence of pollutants even in small amounts in water may directly or indirectly affect the fish populations. Non-toxic chemicals may become responsible for eliminating some species indirectly by depleting the dissolved oxygen concentration of the water body. Some sensitive species such as Bass and Trout which require high levels of oxygen, are first eliminated. Further reduction of dissolved oxygen level in water causes mass mortality of carp and catfish. A variety of fishes have now become limited only to a few species. The most economically important species of fish such as *Rita rita, Colisa fasciatus, Chanda nanga, Pangasius pangasius, Ophiocephalus punctatus* and *Notopterus* spp. was exuberant in the Gangetic plains of India even before 20 years ago. At present, however, these species (and also other species) are almost in the verge of extinction. The cause of this depletion in fish diversity may be ascribed, at least partly, to the increasing concentration of toxic substances in water and sediment.

ECOSYSTEM CONSIDERATIONS

Certain chemical and physical features are known to be important for estimating the potential hazard involved for ecosystem toxicants. In addition to information regarding effects on different organisms, knowledge about the following properties is central to predict ecosystem impact: the *degradability* of the substance; its *mobility* through soil and water; whether or not *bioaccumulation* occurs; and its transport and *biomagnification* through food chains. Toxicants that are poorly degraded (by abiotic and biotic pathways) exhibits ecosystem persistence and thus can accumulate, resulting in tissue residues. When the pollutant is incorporated into the food chain, biomagnification occurs as one species feeds upon others and concentrates the toxicant. The pollutants that have the widest ecosystem impact are poorly degradable; are relatively mobile in water and soil; exhibit bioaccumulation; and exhibit biomagnification.

BIOACCUMULATION AND BIOMAGNIFICATION

Aquatic life has the capacity to consume a lot of toxic and other undesirable substances but they fail to eliminate certain non-biodegradable chemical substances and some other forms tend to accumulate in aquatic plants and animals for a long period — a phenomenon known as *bioaccumulation*. In this case, the consumption of long-lasting contaminants by an organism exceeds the latter's ability to metabolize or excrete the substance and the chemical accmulates within the tissues of the organism. Some organo-chlorine compounds (such as DDT and polychlorinated biphenyls), for example, are partly insoluble in water but highly soluble in fat and therefore, accumulate in fish. This is not the end point of incident, rather, it is the starting point of the extent of concentration of toxicants. Following bioaccumulation, these compounds shift into other trophic levels via food chain (Table 20.6)

Gradual increase in the concentration of toxic substances in the food chain at successively higher trophic levels is known as *biomagnification* or *biological magnification*. The concentration of a contaminant is not detectable in water. Hence it may be magnified hundreds or millions of time as the contaminant passes up the food chain. Table 20.6 indicates that the magnification is compounded in fish and birds by extensive deposition of body fat in which DDT residue accumulates. The end result of the use of DDT is that the total population of fish and detritus feeders are entirely eliminated.

Table 20.6 Concentration of DDT in Aquatic Food Chain.

Item	DDT Residue (mg/l)
Water	0.0000185
Phytoplankton	0.006
Zooplankton	0.05
Steelhead minnow	0.94
Pickerel (Predatory fish)	1.30
Needle fish (Predatory fish)	2.07
Herring gull (Scavenger)	6.00
Merganser (Fish-eating duck)	22.80
Cormorant (Feeds on large fish)	26.00

The biomagnification of DDT (Table 20.6) and polychlorinated biphenyls (PCBs) (Table 20.6A) respectively in the marshes of Long Island and Great Lakes is illustrated by the residue values obtained from different published literatures. Thus, the biomagnification for PCB and DDT in the food chain, beginning with the plankton and ending with the herring gull/cormorant, is nearly 50,000 and 52,80,000 fold, respectively. Domesticated animals and humans may eat contaminated fish, resulting in their residues in these species as well.

Table 20.6A: Concentration of PCB in Great Lakes

Source	PCB concentration (mg/l)	Concentration relative to plankton
Phytoplankton	0.0025	1.0
Zooplankton	0.123	49.2
Rainbow smelt	1.04	416.0
Lake trout	4.83	1,932.0
Herring gull (scavenger)	124.0	49,600.0

Source: Environment Canada (1991)

1. *Mechanism of Biomagnification*: Biomagnification takes place when the pollutant is both lipophilic and persistent. Due to their lipophilic nature, they are partioned from the surrounding water into the lipids of prokaryotes and eukaryotes, and their levels in microbial cells may be one to three orders of magnitude higher than those in the surrounding water. In the food chain, chemical is neither excreted nor degraded to considerable levels. Thus the levels of such chemicals increase as they are transferred to higher trophic levels. Therefore, the top carnivore and

large fish becomes loaded with pollutants that exceeds environmental concentration by a factor of about 10^4 or even more. A biomagnified pollutant may cause death or serious problems to aquatic organisms.

DEPLETION OF DISSOLVED OXYGEN

Under normal conditions, the dissolved oxygen (DO) content of water at 26°C varies between 6 and 10 mg/l. The water with a DO content below 4 mg/l is considered as polluted that results in fish mortality. In general, the DO in natural waters is influenced by (1) the photosynthetic acitivity of the producers in water, (2) the ambient temperature, (3) the degree of aeration, and (4) the consumption of DO by fish and decomposers.

Discharge of industrial effluents and sewage in huge quantities definitely deplete the content of DO because aerobic decomposition of the organic matter tends to exhaust appreciable amount of DO from water. Under extreme conditions, the DO drops below 3 mg/l or even to zero. such condition causes massive destruction of aquatic ecosystems.

EUTROPHICATION*

Eutrophication refers to an aging process in ponds and lakes during overly rich in dissolved nutrients. Over a long time period, freshwater ecosystems classically exhibit a natural progression from oligotrophic (few nutrients) to a eutrophic (rich in nutrients) condition (Table 20.7).

Table 20.7: General Features of Oligotrophic and Eutrophic Aquatic Ecosystem

Feature	Oligotrophic	Eutrophic
Depth	Deep	Shallow
Summer oxygen in hypolimnion	Present	Absent
Algae	High species diversity with low density and productivity dominated by green algae	Low species diversity with high density and productivity dominated by blue-green algae
Blooms	Rare	Frequent
Plant nutrient influx	Low	High
Fish yield	Low	High
Fish species present	Trout and Whitefish	Perch, roach, carp, tilapia, catfish, snakehead

The main factors that result in eutrophication, involve the use of nitrogen and phosphorus at high rates on crop lands. Eutrophication generates acute economic and

---

* For further study, see Mason (1981) and Harper (1995).

ecosystem problems, however. The main effects of eutrophication on receiving ecosystems are noted below.

* Decrease in species diversity and change in dominant biota.
* Increase in algae, fish, and plankton biomass.
* Increase in turbidity and sedimentation ratio.
* The life span of the pond/lake is shortened.
* Development of anoxic condition.

Nitrates and phosphates limit the productivity of aquatic ecosystems. Additional nitrate and phosphate, therefore, favors an increase in rapidly-growing planktonic species. Although a substantial amount of this planktonic species is consumed by the consumer organisms, the excess plankton enters the decomposition pathway. Decomposition is an oxygen-demanding process. Dissolved oxygen levels may be reduced below those necessary for the survival of fish. In extreme cases, the death of fish (and also other aquatic species) and their subsequent decomposition can inflict a further oxygen demand, making the condition of ecosystem increasingly worse. Bacterial decomposition of organic wastes also requires oxygen and with heavy loads, the content of its oxygen may reduce below the level where many species of fish can no longer survive. As the condition of water bodies becomes anaerobic due to increased oxygen depletion, the breakdown products become reduced, and many of which produce offensive odors.

Excessive nutrient levels in fish culture systems can also cause growth of water-loving plants such as *Eichornia, Pistia, Azolla, Lemna,* etc. These notorious plants can impair fishing, fish spawning, and production.

1. *Monitoring of Eutrophication* : Eutrophication can be monitored biologically and chemically. This provides the opportunity for remedial action before sudden and catastrophic fish kills in ponds/lakes occur. Oscillations in phytoplankton species help indicate eutrophication. Eutrophic waters exhibit high abundance and low species diversity of Phytoplankton.

The most widely measure of eutrophication is a biochemical oxygen demand (BOD), which measure the rate of oxygen depletion by aquatic organisms. This is a test (See chapter 26) of the amount of oxygen neccessary for bacteria to break down organic matter in the pond water over a period of 5 days. Though the BOD values for different waste-waters and eutrophicated ponds are extremely variable, water for fish culture should have a BOD less than 15 mg/l (range may vary from 10 to 20 mg/l). In fish culture, however, this test is normally not important, but in some cases (when BOD is used in conjunction with other water quality indicators) it may account for up to 50 per cent of the total BOD. Therefore, BOD is an approximate rather than precise guide to understand the quality of water.

The sum and substance of this chapter is that pollution is a serious threat which is more neglected by all countries of the world. The response from environmentalists to technologists for production has mostly bemused. The dangers of pollution of fresh-water ecosystems either accidentally or deliberately, are multiplying across the globe. To save freshwaters for better utilization, we have to force polluters through mass democratic movements like reclaimed campaign for pollution control. Unfortunately, it has not been possible for us to save freshwaters from pollution. We have totally failed to do so. Possibly, the human society is such an irrational, poorly evolved species that we cannot overcome our tribalism and act in our own self-defence. But when the alternatives are a barren freshwater environment that is high in level of pollutants, it can be said that we ought to find out the cause of this situation. It may be recalled that if pollution control is not totally stopped, there lies before many species of freshwater edible fish the risk of extinction.

20.23 Conclusion

Five conclusions that follow can be drawn from this chapter. *First*, two principal factors of a fish culture ecosystem — water and soil — are the most useful to fish farmers. Since these factors are the most important things of advanced stage in fish husbandry, it is very difficult for a farmer to cultivate fish in thier ponds unless water and soil are in good quality. But it is unfortunate that most of the water bodies are severely contaminated with a vast variety of chemicals that have tremendous toxic potential to fish and aquatic ecosystems. It is, therefore, neccessary to make the best use of water for fish culture. *Second*, though the use of agricultural chemicals in developing countries is cardinal for crop production and disease control management strategies, much efforts should therefore be given to their rational use rather than restriction or prohibition on the use of chemicals. At the same time, readily degradable chemicals such as organophosphates, pyrethroids, and carbamates should be recommended and enforced. Even if new compounds are available in the market, these should be properly assessed that are relatively environment-friendly and suitable for integrated pest management programs. *Third*, while scientific and technological advances have resulted in an enormous increase in the production and harvest of fish, pollution of fish culture ecosystems has shown negative impacts on fish culture development strategy. This situation has attracted the attention to toxicologists and aquaculturists resulting in pollution and toxicology-related investigations on freshwater resources and aquatic life of commercial importance. Suitable devices should, however, be adopted for safe disposal of toxicants, their elimination as well as detoxification which will help prevent fish and ecosystems from destruction. *Fourth*, worried over large-scale contamination of toxic substances in freshwaters, different government organizations have decided to keep a close tab on the toxic compound mitigation projects. Also, trainning and motivation programs will have to be organized to create awareness about the need for the removal of contaminants. Most of the aquafarmers are not aware of the need for using toxicant-free water for fish culture. Spreading of awareness is the only remedial measure, however. *Fifth*, though it is beyond possible to prevent fish cultute ecosystems from pollution totally, the pollution and toxicity problems would be minimized to a greater extent by adopting certain precautionary measures such as development of ecosystem control and management. At the same time, regular monitoring programs must be built-up. To keep the ecosystem free from any kind of pollutants and to produce robust fish from such water bodies, adequate knowledge on the subject concerned must be developed that is used in managing waters for fish culture. The knowledge is definitely based on the assiduous involvement of toxicologists and aquaculturists across the world.

References

APHA. (American Public Health Association). 1985. Standard Methods for the Examination of Water and Wastewater. Washington, DC.

Abbasi, S. A. 1989. Experimental Aquatic Ecology and Water Pollution. Pondichery University, Pondichery, India.

Alexander, M. 1994. Biodegradation and Bioremediation. Academic Press, San Diego.

Anna, M. F. and F. Harding-Barlow. 1987. Advances in Modern Environmental Toxicology. *Ed.*: M. A. Mchlanan. Princeton Scientific Publishing Co. Inc. p. 87-128.

Ballard, J. A. and W.. D. Oliff. 1969. A rapid method for measuring the acute toxicity of a dissolved materials to fishes. *Water Res.* 3 313-333.

Bhart, R. and M. S. Dkhar. 1998. New Trends in Microbial Ecology. International Society for Conservation of Natural Resources, Varanasi, India.

Burrel, D. C. 1974. Atomic spectrometric analysis of heavy metal pollutants in water. Ann Arbor Science Publishers, Ann Arbor.

Coyne, M. S. 1999. Soil Microbiology, Delmar Publishers, Albany, New York, 462 p

Handa, S. K. 2000. Principles of Pesticide Chemistry. Agrobios (India), Jodhpur.

Kondaiah, K. and A. S. Murty. 1994. Avoidance behaviour test as an alternative to acute toxicity test. *Bull. Environ. Contam. Toxicol.* 53: 836-843.

Marr, I. L.and M. S. Cresser. 1983. Environmental Chemical Analysis. International Text Book Co., New York.

McLaey. D. J. 1976. A rapid method for measuring the acute toxicity of effluents and other toxicants to salmonoid fish at ambient room temperature. *J. Fish. Res. Bd. Canada.* 33 : 1303-1311.

Harper, B.1995. Eutrophication of Freshwaters: Principles, Problems and Restoration. Chapman and Itali, London.

Mason, C. F. 1981. Biology of Freshwater Pollution, Longman.

Motsumura, F. 1985. Toxicology of Insecticides. Plenum Press, New York.

Murty, A. S. 1986. Toxicity of Pesticides to Fish. CRC Press Inc. Boca Raton, Florida, USA.

O'Brien. 1967. Insecticide Action and Metabolism. Academic Press, New York.

Patrick, R. 1972. Aquatic communities as indices of pollution. Load Indicators of Environmental Quality (Ed.: W. A. Thomas). Plenum Press New York.

Piper, R. G. *et.al* 1986. Fish hatchery management III pp. 3-59. American Fishery Society, Washington, DC.

Rana, S. V. S. 2005. Biotechniques : Theory and Practice. Rastogi Publications, Meerut, India.

Revera O. 1979. Biological Aspects of Freshwater Pollution. Pergamon Press, New York.

Rombke, J. and J. P. Moltman. 1996. Applied Ecotoxicology. Lews Publishers, London.

Sprague, J. B. and D. E. Durury. 1969. Avoidance reactions of salmonoid fish to representative pollutants. pp. 169-179. *In:* Advances in Water Pollution Research.(*Ed:* S. H. Jenkis). Pergamon Press, New York.

Tewari, R. 2000. Role of Microbes in the Management of Environmental Pollution. APH. Publishing Corporation, New Delhi, India.

Trivedy, R. K. 1998. Ecology and Pollution of Indian Rivers. Ashis Publishing House, New Delhi, India.

Trivedy, R. K. 2001. Aquatic Pollution and Toxicology. ABD Publishers, Jaipur, India.

Wittman, G. T. W. 1983. Toxic Metals. *In:* Metal Pollution in the Aquatic Environment. (*Eds:* Forstner, U and G. T. W. Wittman), Springer Verlag, Berlin.

Yasue, S. 1986. Impact of Water Pollution on Agriculture and Fishery in Japan. *In:* Proceedings of the International Symposium on Environmental Pollution and Toxicology. Hisar, India, PP. 95-110.

Questions

1. Discuss the ecological aspect of toxic substances.

2. Discuss how fish culture ecosystems are contaminated with toxicants.

3. What are the biological effects of toxic substances?

4. Mention different types of effects of toxic substances generally observed in fish when they are drained into aquatic ecosystems.

5. What are heavy metals? Discuss how they affect fish culture ecosystems.

6. What are the steps that should be considered for protection of fish culture ecosystems from toxicity?

7. Name some common pollutants that are harmful to fish. Discuss how pollutants are removed from aquatic ecosystems.

8. What is biomonitoring? Why it is so essential in fish cultivation? Discuss how biomonitoring programs are carried out. What is the significance of biomonitoring ?

9. What do you mean by the safe disposal of toxicants? How safe levels of toxicants are estimated?

10. What are the different types of fish toxicity test? Why they are so important in fish culture industry? Which test is most suitable for the purpose and why?

11. How acute toxicity tests are performed? Distinguish between acute and chronic toxicity tests.

12. The LC50 of a chemical (say copper sulfate) is 1.45. What does it mean? Discuss how LC50 value of any chemical is determined.

13. Distinguish between partial and complete life cycle tests.

14. What is ecotoxicology ? What are the criteria for consideration of ecotoxicology ?

15. Discuss the fate and dispersion of a chemical in an aquatic ecosystem.

16. What are the conditions that are essential for the assessment of ecotoxicological hazard potential in tropical ecosystems? Give some examples of ecosystem monitoring to predict and recognize the hazardous effects.

17. Discuss why hazard potential of tropical aquatic ecosystems exists, compared to temperate ecosystems.

18. What are the recommendations for ecotoxicological research in tropical regions? Why such research is inevitable for precise evaluation of the toxicity of chemical substances on fish and aquatic ecosystems?

19. What is acid rain? How it is related to fish and ecosystem toxicity?

20. What are the effects of acid deposition on fish and aquatic ecosystem?

21. Discuss how ecotoxicology is assessed.

22. Distinguish between oligotrophic and eutrophic conditions of aquatic ecosystem. What are the effects of eutrophication on aquatic ecosystem?

23. Discuss how ecosystem toxicity greatly affect fish survival, growth, and production.

24. *Distinguish between*:

(a) toxicant and pollutant, (b) toxicology and ecotoxicology, (c) degradable and non-degradable toxicants, (d) antagonistic and synergistic effects of toxicants, (e) organophosphorus and organo-chlorine pesticides, (f) bioaccumulation and biomagnification.

25. Writes notes on the following:

(a) concentration factor, (b) residual toxicity, (c) adsorption, (d) tolerance limit of toxicants, (e) Shannon-Wiener Index, (f) bio-indicator species, (g) safe concentration of toxicants, (h) static bioassay test, (i) Dragstedt - Behren's method, (j) ecotoxicological effects, (k) elimination and retention of toxicants, (1) eutrophication, (m) degradation of chemicals in ecosystems.

21

Fish Culture in Relation to Public Health

Health may be defined as the state in which the mental and physical activities of the body are adjusted satisfactorily to the physical, mental and social environment. According to World Health Organization (WHO), health is defined as a state of complete physical, mental, and social well-being and not merely the absence of disease or infirmity. Health issues are, however, inter-linked with the process of rapid development of aquaculture and agriculture. Contamination of freshwater ecosystems with sewage, solid wastes, fertilizers and pesticides causes severe human health hazards. Since management of fish culture ecosystems is an important technical tool to reduce toxicant contamination and pathogen transmission risks, different management protocols are rated with respect to their pathogen and toxicant elimination potential. Management of fish culture in general will definitely break the routes of disease transmission considerably.

One of the most serious hygienic risks in fish culture industry is the spread of pathogenic micro-organisms to man. Viruses, fungi, protozoa, helminth eggs, and intestinal parasites are examples of sources of infection which may be spread through aquaculture. Besides these, other aquatic animals especially snails and copepods, surface flow-off, and agricultural activities around the aquatic ecosystems are the principal of infection and contamination. It is essential to note that almost all forms of aquaculture starting from culture to processing strategies increase the risk of health status not only of farmers but the society as well.

21.1 Water and Human Health are Closely Related to Each Other

Waters which are used in fish culture must be free from flukes and tapeworms and contain faecal colliform below the acceptable levels and the international consultative groups have formulated guidelines for the microbial quality of fish product.

The potential health risks and current epidemiological evidences for actual risks from pathogen transmission through fish culture must be seriously evaluated. Water quality standards required for fish culture should be compared with the microbial quality of natural waters from which fish are harvested. Adequate steps that follow must be taken for reasons of public health protection and for acceptance of contaminated polluted aquaculture.

1. Minimizing the level of excreted pathogens in the water where fish are grown.
2. Maintaining an adequate personnal and institutional hygiene during harvesting, handling, processing and transport, and marketing.

3. Treatment for conservation of raw products prior to sale and consumption, or cooking prior to consumption.

Contamination of aquatic ecosystems with pollutants and pathogenic micro-organisms both from humans and animals is omnipresent and any fish, whether grown in or harvested from fish culture ponds, will carry indicator organisms or be potentially contaminated by pathogens. Most regulatory agencies,therefore, do not stipulate toxicological standards for freshly caught fish, but specify and enforce standards for processed products.

QUALITY OF WATER

For sustainable fish culture, water must be free from pathogens and agricultural chemicals that are detrimental to human health. Human populations depend very much upon the water of streams, rivers, ponds, and lakes. These waters are frequently putrefied with sewage or human wastes, industrial wastes, and agricultural chemicals.

SEWAGE (HUMAN WASTE)

The sewage is generally used in aggregate perception for exhausted supplies of industries, homes, polluted waters, or societies. The solid wastes of digestive system are called faeces and considered as the most important common pollutant of aquatic ecosystems. These pollutants are teemed with intestinal pathogens. Micro-organisms produce bad taste, odor, color, and algal blooms.

INDUSTRIAL WASTES

Industries always use enormous quantities of water and hence their waste waters are heavily loaded with a vast variety of toxic substances most of which are not impressionable to biodegradation. The chemical composition and quantity of industrial and domestic wastes generally varies from hour to hour and even day to day. Toxic substances are known to be highly toxic to fish and fish food organisms.

CONTAMINATION OF WATER WITH DISEASE-PRODUCING ORGANISMS

In rural areas, fish pond water is often contaminated with faeces and untreated waste-waters that results in some important diseases of man (Table 21.1). Amoebic dysentry, cholera, and other enteric diseases are caused by the consumption of contaminated fish and water. In general, water is contaminated with shigellas and salmonellas. Shigellas are restricted to human as hosts. These bacteria cause bacillar dysentry which is termed as *shigellosis* and is characterized by inflamation of the lung, intestine, mucus, and pus.

Salmonellas is the group of organisms that are closely related to another and should be considered as a single species. However, over 2000 antigenic types of salmonellae occur. Salmonellas may cause three kinds infections: Septicemia, Gastro-enteritis, and Enteric fever (typhoid and paratyphoid fevers). Each of these has certain characteristic features. Infection by *Salmonella* spp. that affect the gastro-intestinal tract is termed as *salmonellosis*.

Table 21.1 : Relative Health Risks from Untreated Wastewater and Use of Excreta in Fish Culture

Pathogen	Relative amount of execess frequency of infection or disease through reuse*	
	In developing countries	In industrialized countries
1. Intestinal nematodes (*Ascaris, Trichuris*)	High	Least
2. Bacteria (Salmonella, Shigella, bacterial diarrhoea)	Medium	Medium
3. Viruses (Polio, Hepatitis A, Viral diarrhoea)	Least	High
4. Trematodes (Flukes) and Cestodes (Tapeworm)		
(a) Schistosomiasis	From high to zero	Least
(b) Taeniasis		
(c) Chlonorchiasis		

* Excess frequency is the attributable to the particular route of transmission occurring over and above the frequency already existing in the specific community.

Source : Strauss (1996)

21.2 Health Risks From Human Pathogens in Fish Culture*

About more than 30 excreted pathogens cause deleterious effects on public health. The aquacultural use of wastewater and excreta may lead to an actual public health risk only if (1) an infective dose of an excreted pathogen reaches the pond, or if the pathogen multiplies in an intermediate host living in the pond, (2) this infective dose reaches a human host through contact or comsumption of the aqua-products, (3) this host becomes infected, and (4) this infection causes disease or further transmission. While the first three constitute the so-called *potential risk* the last one is termed as *actual risk.* If the actual risk does not occur, the risk of public health remains potential only. The potential risk is assessable by microbial investigations, where the actual risk is evaluated through health impact studies.

The actual public health risk occurring through wastewater use in fish culture is divided into three types such as (1) consumer risk, (2) operator's risk, and (3) worker's risk. Depending on the type of health protection measures used and enforced, none, all or only selected risks may occur. Risks can be reduced or eliminated if proper health protection measures such as waste treatment, hygienic practices in fish processing industries, and adequate personal hygiene are adopted.

BACTERIAL CONTAMINATION IN FISH

Reuse of wastewater in fish culture is receiving scientific attention and its potential to recycle nutrients in productive and socially-acceptable system has been realized. It is

---

* For further study, see Strauss and Blumenthal (1990), Mara and Cairneross (1989).

prevalent in some Asian countries where human excreta and conventional sewage are used for fertilizing fish ponds. Public health is the most important concern in wastewater-fed fish and there is a lack of knowledge concerning the possibility of getting water-borne diseases from the consumption of wastewater-fed fish. A number of bacteria has been recorded from various organs and muscles of fish exposed to bacteria-infected fish culture ponds. At the same time, a number of chemical substances can be accumulated in fish. As the degradation rate of these chemicals is often slow, their persistence in fish tissues is of great concern to human health.

The risk of bacterial contamination in fish flesh increases when pathogens in the pond water reach high numbers. The contaminated fish transmit pathogens to its handlers, consumers, and fish farmers. The World Health Organization has a tentative bacterial guideline for fish ponds which is below $10^3/100$ ml . The potential fish muscle infection by bacteria increases when fish are exposed to pond water containing faecal coliforms greater than 10^4 - 10^5 /100 ml.

The environment of wastewater-fed fish culture ponds can be stressful for fish and stress can reduce the threshold value of penetration of bacteria into the muscle of fish and increase their susceptibility to diseases. Obviously, it is cardinal that wastewater and solid waste loaded with bacteria should be treated prior to discharge into fish ponds in order to produce uncontaminated and robust fish.

EPIDEMIOLOGY*

When the accumulation of pathogens in a fish pond becomes greater, severe infection occurs. The degree of infection is influenced by the amount and quality of waste substances to be used for fish ponds. The epidemiological evidence for disease transmission associated with the use of excreta and wastewater for fish ponds, however, clearly indicates the transmission of certain trematodes (flukes), bacillary dysentry, bacterial diarrhoea, and hepatitis A virus diseases in man. The risk of diarrhoeal disease associated with the use of excreta in fish culture ponds in some Asian countries is significant. There is vivid evidence from several hepatitis A outbreaks reported from the United States that infection is due to the consumption of finfish and shellfish grown in polluted waters. Passive transport of pathogens by finfish and shellfish grown in excreta-fed and wastewater-fed ponds must, therefore, be considered as an important transmission route.

21.3 Pathogens and their Transmission

GENERAL CONSIDERATION ON THE TRANSMISSION OF DISEASES

Small bodies of water where fish culture is undertaken by small and marginal farmers, are of interest and, at the same time, headache to the public health departments because they provide a medium for the transmission of certain human diseases. In most remote

---

* For further detail, see Prein (1990).

areas, latrines are emptied into ponds which are stocked with fish. Many rural people void stool and urine near fish culture ecosystems which are drained into the pond water with adverse effects on fish. Water-borne pathogens usually cause a number of intestinal infections. Such infections are generally acquired by the consumption of polluted waters containing human faecal matter from healthy carriers or patients. When human faeces pollute the water bodies, the outbreaks of intestinal diseases tend to be the epidemic type, and the source of water is the common factor.

Disease-infected fish flesh may be intestinal diseases but can be other types as well. With regard to their epidemiology, they can be divided into two major categories: (1) In food-borne intoxications such as Salmonellae and Faecal Coliforms fish poisoning, the causative micro-organisms produce toxin in fish flesh; when a person consumes the food, the toxin is ingested and gives rise to the disease and (2) In food-borne infections, the causative organisms are ingested, subsequently grow within the body and cause damage the health status.

Different types of infectious microbial diseases (such as virus, bacteria, fungus, and protozoa) and other parasitic diseases (such as helminths and copepods) are capable of destroying host tissues and multiplying within the fish body. The resultant pathology and the cause of disease may depend on such factors as infective dose, resistance and virulance of individual host as well as host nutrition and other environmentally influenced variables.

Pathogenic micro-organisms cause several diseases in farmed fish. They may be carried passively on the scales and in the inner organs of fish, and shellfish or crustacea. Many helminthic pathogens play a minor role in developed countries but they are endemic in most tropical countries and have, via waste-fed fish culture systems, various potential transmission patterns.

TAPEWORMS

The members of the class Trematoda of the phylum Platyhelminthes, the flukes are highly specialized group of parasites, possessing the most complex life cycle of any metazoan (Figure 21.1). Various types of trematodes such as *Fasciola buski*, *Paragonimus westermani*, *Fasciola hepatica*, *Distonum rathousii*, *Opisthorchis sinensis*, various species of the genus *Dicrocoelium*, *Schistosoma haematobium*, and *S. japonicum* infect about 300 million people world-wide. Their eggs with contained larvae are voided with the urine and/or faeces, and if they reach water, the ciliated miracidium larvae penetrate into the interior of a freshwater snail and gives rise to sporocysts from which cercariae are developed. The cercariae become free, and if the opportunity present itself, may enter the human body. Among different types of flukes, intestinal, liver, lung, and blood flukes are, however, very common and cause diseases of man. The following examples will illustrate this point.

1. *Paragonimus westermanii* (Lung fluke) : Human infection caused by the oriental lung fluke, *P. westermanii* (Braun), is very common in Japan, Korea, India, China, Africa, and the USA. The life cycle passes through two hosts: Man (definitive host) and snail of the genus *Melonia* sp. (first intermediate host), as well as crab or crayfish

Eggs enter water with feces

Oral sucker

Becomes adult in
human liver
Intestine

Swallowed by
aquatic snail

Miracidium
Egg contain-
ing ciliated
larva

Ovary

Testes

Sporocyst
Young
redia

Human

Snail

IN

Fish

Swallowed with
raw fish
Metacercaria

Water

Encysts
and
loses tail

Young cercaria

Swallowed with
raw fish

Redia

Leaves
snail

Attached to fish

Cercaria

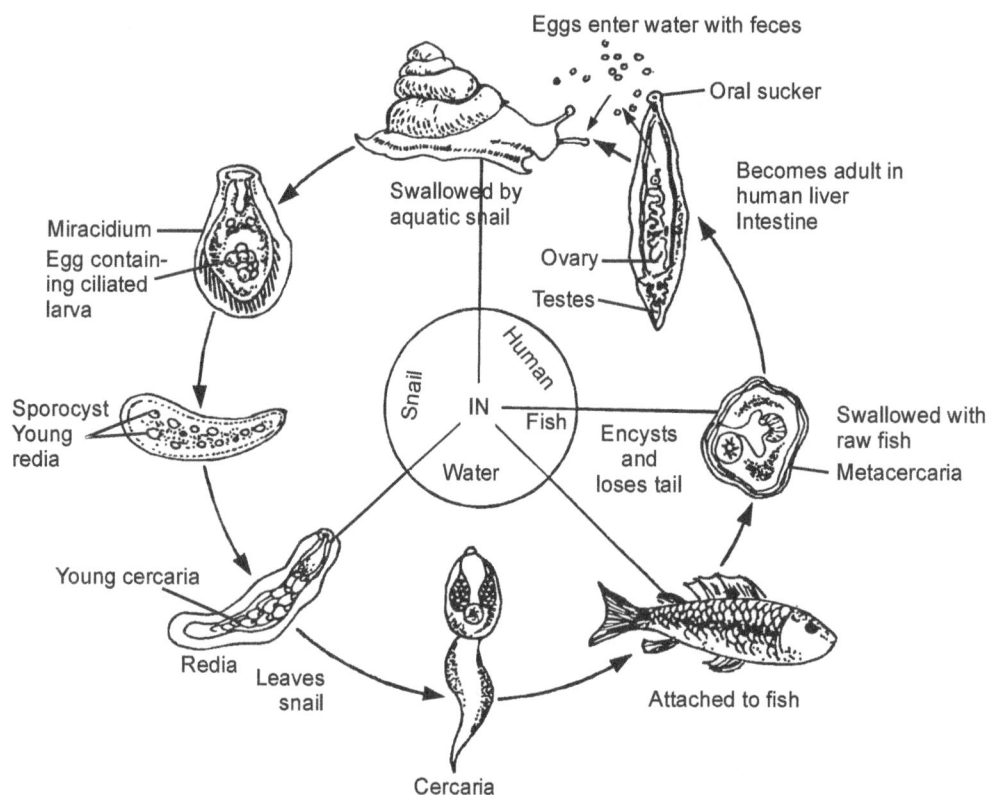

Fig. 21.1 : Life cycle of a Chinese liver fluke of man. The first ciliated larva (miracidium) enters into the body of first intermediate host, which in most species is a freshwater snail (such as _Planorbis_, _Lymnaea_, _Bulinus_, etc.). The principal features pertain to the reproductive system. Note the cercaria larva is attached to fish body. When raw or poorly-cooked fish are consumed by man, the metacercaira becomes adult in human liver. Fish acts as second intermediate host whereas man serves as a primary host. Though most of the flukes are minute, those parasitic in fish average close to half an inch in length.

(second intermediate host). The fluke causes mild or severe illness called _paragonimiasis_ which is characterized by intermittent cough, blood-stained sputum slight fever, and mild anaemia. Infection of man is, however, caused by the eating of raw or improperly cooked flesh of an infected crab or crayfish. The infection may also occur by pond water containing detached cysts from the infected crab/ crayfish.

2 _Fasciolopsis buski_ (Intestinal fluke) : The large-sized elongated intestinal fluke (7.5 cm in length and 20 mm in breadth), _F. buski_ (Odhner) is widely distributed in southeast Asia and in the Oriental region. The life cycle passes through two different hosts: Man (definitive host) and small coiled snail of the genus _Segmentina hemisphaerula_ and _Hippetis contori_ (intermediate hosts). Metacercarium larvae are liberated from the second intermediate host and encysted on aquatic plants (particularly water chestnut) and when eaten raw, result in the transmission of

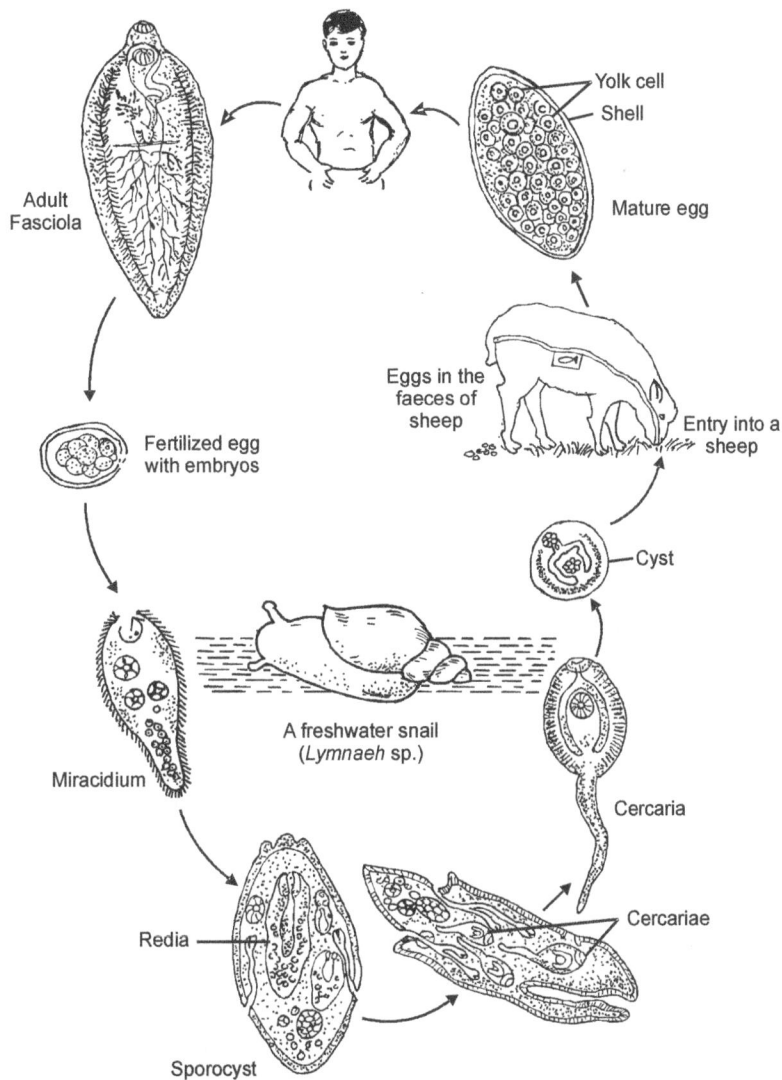

Fig. 21.2 :   Life cycle of a digenetic trematode, *Fasciola hepatica*. The parasite rests in the liver and feeds on blood of the liver causing a serious damage to liver and causing a disease known as *liver rot*. Sometimes immature liver fluke obstructs the bile ducts leading to obstructive jaundice. The parasite completes its life cycle in two different hosts such as (i) Primary host (sheep, goat, man, etc.) and (ii) Intermediate host (a freshwater snail of the genus *Lymnaea truncatula*).

disease. The disease is characterized by mild anemia, vomiting, chronic diarrhoea alternate with constipation, abdominal pain, and swelling of face and limbs.

3. *Fasciola hepatica*: (Liver fluke) : The leaf-like liver fluke, *F. hepatica* (15-30 mm in length), is an important endoparasite of man, goat, horse, etc. and is distributed in India and some other Asian countries. Hepatic infection has also been noted in South America, the Caribbean area, and Europe. It rests on the liver and feeds

on blood of the liver causing a serious damage to liver and causing a disease known as *liver rot*. Immature *Fasciola* sometimes get stranded within the bile duct and hamper bile secretion. When bile ducts are obstructed, it leads to obstructive jaundice. It completes its life cycle in two hosts: Man and sheep (primary hosts) and a snail of the genus *Lymnaea* or *Bulinus* or *Planorbis* spp. (intermediate hosts) (Figure 21.2).

4. *Clonorchis sinensis* (Chinese Liver fluke) : The oriental liver fluke,*C. sinensis* (Cobbold), is endemic in certain regions of Southeast Asia and China. This tapeworm is transmitted via fish and compulsory intermediate hosts. It is an important cause of biliary disease. The fluke infects the bile ducts of humans and the disease caused by this fluke is termed as *clonorchiasis*. Infection is caused by eating uncooked freshwater fish containing encysted larvae. Early symptoms include loss of appetite and diarrhoea; later there may be signs of cirrhosis of the liver.

5. *Intestinal and Blood Worms* : In Central and Eastern Europe, USA, Japan, and Central Africa, the human tapeworm, *Diphyllobothrium latum* (Linn.) and the blood worm *Opisthorchis sinensis* (Linn.) are transmitted by freshwater fish eaten raw in fish salads or insufficiently cooked. The adult lives in the intestine of fish-eating mammals, including humans. The largest tapeworm infesting humans, it may reach a length of 15.2 to 18.3 metres (average 9.0 m in length and 2.5 m in width). The eggs develop into ciliated larvae that are eaten by copepods. The larvae pass through several stages in the copepods, and develop further after the copepods are eaten by fish, and finally encysting in the fish muscle. People acquire infection by eating raw or poorly cooked fish that contains cysts. The cysticercus (or bladder worm) stage of *D. latum* which is elongated and solid, occurs in the Pike and certain other freshwater fishes. Patients often repor abdominal pain, loss of weight, digestive disorders, progressive weakness, and symptoms of pernicious anemia because the worm absorbs ingested Vit. B_{12} from the gastro-intestinal tract. It produces anemia and gastro-intestinal disturbances in man. The disease is endemic worldwide especially in temperate zone where partly-cooked or eating raw fish flesh is popular.

The *Opisthorchis* is a parasite of the cat, dog, and man. The larvae are found in the freshwater fish *Leuciscus* sp. and *Idus* sp. Larval trematodes, *Metagonimus yokogawai* and *Heterophyes heterophyes*, infective to man, especially in the Middle East. Intermediate hosts are snails and fish such as mullets, *Mugil cephalus* and *Mugil japonicus*, found in brackishwaters. In heavy infections, chronic intermittent diarrhoea, nausia, and minor abdominal discomfort sometimes occur. The genus *Heterophyes* sp. is found in the Far East, Egypt, and Palestine. *Metagonimus* sp. occurs particularly in Japan, Korea, China, and Eastern Siberia. The blood flukes (several species of *Schistosoma haematobium*, *S. mansoni*, *S. japonicum* and *S. Intercalatum*) are widespread in the oriental regions, Middle East, and much of Africa. Adults live in blood vessels of visceral organs. Eggs make their way into the bladder or intestine of the host and are discharged in the urine of faeces. Eggs hatch into miracidia which enter snails and transform into sporocysts. These develop daughter sporocysts, which give rise to fork-tailed cercariae. These leave the snail and enter the final host directly through the skin or mucus membrane. The genus *Schistosoma* is a typical water-related disease of stagnant waters. They cause a serious illness called *schistosomiasis*. The species *S. mansoni*, *S. Japonicum* and *S. intercalatum* are responsible for intestinal disorders whereas *S. haematobium* causes urinary tract infections. The flukes alternate between human host and a type of freshwater snail (Figure 21.3), and so people who bathe and wash in pond water contaminated with immature forms of the fluke called *Cercariae*, are likely to be infected.

ROUNDWORMS

In contrast to the flatworms, all roundworms (Phylum: Nemathelminthes, Class: Nematoda) are elongated and slender in form. Many forms are highly developed parasites, occurring specially endo-parasitically in both invertebrate and vertebrate animals. A few representatives of nematodes are known to cause human diseases such as *dracunculosis* and *ascariasis*. The guinea-worm or 'fiery serpent' (*Dracunculosis medinensis*) infection afflicts an estimated 30 million people across a band of North-Eastern and Sub-Saharan Africa, the Arabian Peninsula, India, Russian Federation, Pakistan, Brazil, Afganistan, and Iran. Guinea-worm infection is transmitted solely through freshwater ponds teemed with the most common water flea of the genus *Cyclops* spp. Although *Cyclops* is a very important fish food organism present in freshwater fish culture ponds, it acts as a vector of guinea-worms and when their larvae are ingested by *Cyclops* they afflict man. Dracunculosis is a painful disease which can incapacitate as many as 50 per cent of population during epidemics. Another roundworm of the genus *Ascaris* which is the most common helminthic pathogen in developing countries, may be carried out passively by fish. It is, however, likely that other larval roundworms and tapeworms from freshwater and brackish water fish may occasionally infect man. These are considered as very dangerous parasites of man. The transmission cycle of passively carried pathogens such as bacteria, viruses, protozoa and helminth eggs which use fish as intermediate hosts are shown in Figure 21.4.

21.4 Control of Infection

The factors involved in controling infection give clues to methods of reducing infections. These involve effective sanitary measures and adequate disposal of human wastes, treatment of wastewaters, and use of sanitary methods for production and, handling of fish. Carriers of endemic organisms must not be permitted to participate in the handling and preparation of fish. Detection of carriers, inspection of fish and fish processing plants for the occurrence of pathogens and identification of untreated/ improperly treated water sources are important functions of state and local public health authorities.

In acidic ponds, fishes are very much susceptible to various infectious diseases. There are a number of chemicals and sources of carbonate of lime which are effective in controlling fish diseases. The effective application rate of chemicals generally vary depending upon the degree of intensity of infections and the type of fish species to be reared.

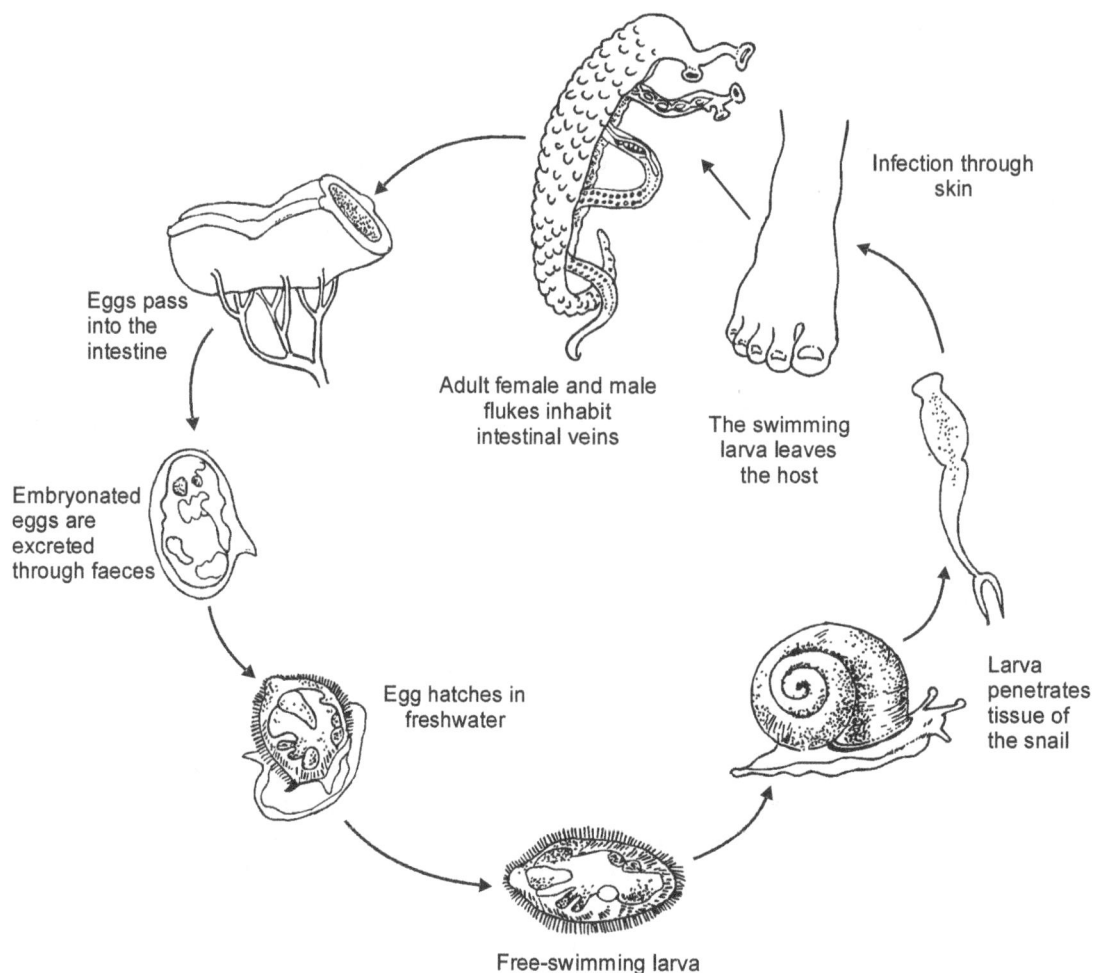

Fig. 21.3 : **Life cycle of a blood fluke, *Schistosoma*. Adult forms are long and thin. This allows them to travel through their host's blood vessels. They have hooks and suckers to attach themselves to the host's tissues. It alternates between human host and a type of freshwater snail, and so people who bathe and wash in infected ponds or rivers are likely to be infected. Flukes damage the liver, lungs and muscles. The consequences can be serious. The first signs of an infection are an ichy rash and a tingling where it has penetrated the skin. Later, flu-like symptoms develop, including fever, chills and aches. Blood flukes infect millions of people worldwide, causing a serious illness called *Schistosomiasis*.**

21.5 Effects of Water Pollution on Human Health

Chemical contanination of fish culture ecosystems leads to serious health hazards. Toxic metals, pesticides, fertilizers, and their different metabolic products are taken up by the human through consumption of contaminated fish. The body burden of these toxicants increases day by day and as a result, a variety of complex physiological as well as pathological diseases manifest in course of time.

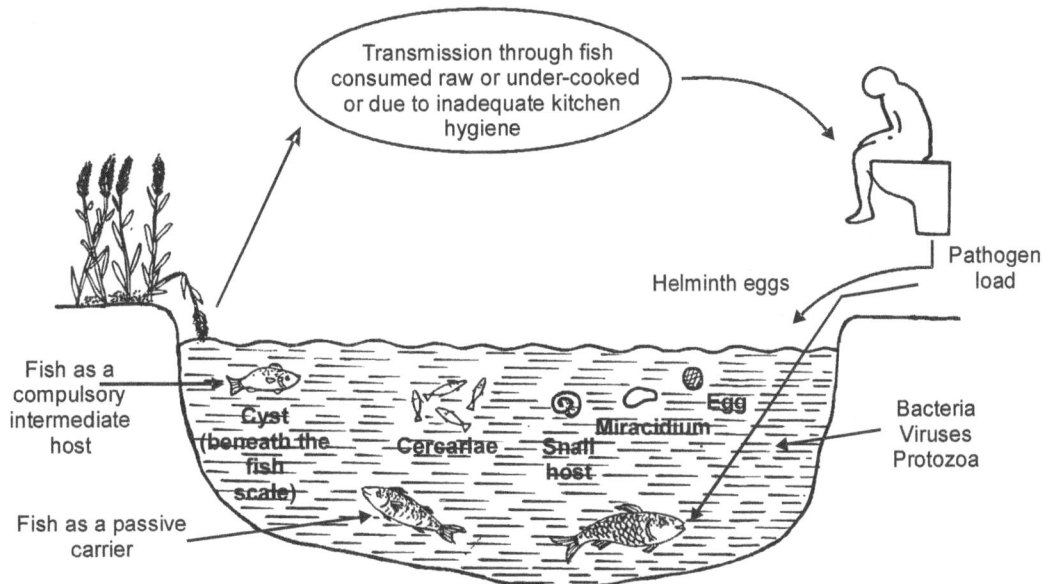

Fig. 21.4 : Transmission of excreted infections through fish.

FERTILIZER POLLUTION

Application of different types of nitrogen fertilizers in fish culture ponds in quantities far in excess of those taken up by fish and fish food organisms can result in contamination of both surface and drainage waters. Nitrates and phosphates are the chemicals most often involved. Nitrate contanination can occur in both surface and flow-off and drainage waters, and excessive levels of phosphates generally occur only in surface flow-off.

The effect of nitrogen and phosphorus on water quality is more serious. If nitrate accidentally contaminates the drinking water and the level remains above about 10 mg/l, it causes human health hazards. Many fish culture ponds in rural areas have been found to contain nitrates significantly above this safe limit.

ODOR POLLUTION AND HUMAN HEALTH

In almost all types of fish culture ecosystems, there do occur a number of biodegradable substances which are odorous or volatile and incessantly lose molecules to the water. They are referred to as *osmogenes* and their presence is not desirable in view of the fact that they produce odor and smell. Upon inhalation in gaseous form, these osmogenes are toxic to eyes, skin and may cause irritation in the respiratory epithelium, which ultimately may result in the loss of appetite, giddiness, nausea, anorexia, and several other problems.

A number of micro-organisms such as *Micromonospora* (heterotrophic bacterium), *Anabaena* (blue-green alga), *Trichoderma* (fungi), etc. produce quite a good number of compounds that can generate odor. Microbial decay and anaerobic decomposition of organic substances in fish ponds constitute the most important source of odor pollution that result in the production of various acids, alcohols, methane, hydrogen sulfide, ammonia, and a number of organic acids.

PESTICIDE POLLUTION AND HUMAN HEALTH

Indiscriminate use of pesticides in agriculture as plant protection agents for boosting production, apart from being an occupational hazard in the developing world, has been posing a serious threat to human life. A vast variety of pesticides are drained into aquatic ecosystems through surface run-off, accumulate in aquatic organisms (bioaccumulations) and move into other trophic levels through food chains. This gradual but steady increase in the concentration of toxic substances in the food chain at successively higher trophic levels is termed as *biomagnification*. Organic decay in the pond mud decreases the dissolved oxygen content of the water body which expedites bioaccumulation and hence biomagnification.

A variety of fish and fish products (if not properly processed and preserved) have been found to contain pesticide residues in large amounts. The level of DDT residue in finfish and shellfish, for example, varies from 200 to 250 mg/l and 500 to 850 mg/l, respectively. In most of the cases, residues may exceed the tolerance limits as prescribed by the World Health Organization. It is reported that the level of accumulation of DDT in the body tissue of an average Indian is highest in the world, varying between 12.8 and 31.0 mg/l. Substantial quantities of benzene hexachloride, endosulfan, dieldrin, lindane, and malathion residues and many other pesticide residues have been noted in fish body.

About 160 pesticides are used in India and their consumption is around one lakh tonnes per year (data for the year 2009). Some of them leave behind residues in finfish and shellfish and thereby produce ill-effects when the concentration exceeds safe tolerance level. Many pesticides are suspected as carcinogenic and teratogenic.

The example entitled 'Endemic Familiar Arthritis in India' among the poor people of Malnad, Andhra Pradesh, has brought deleterious effects of pesticides into light. The victims suffered from pain in the hip and knee joints and even lost the ability to stand erect. The cause for this kind of symptoms was attributed to the fish, prawn and crabs harvested from paddy fields sprayed with pesticides for prolonged periods.

Another fatal water-borne disease caused by arsenic is termed as *arsenicosis*. Arsenic may be present in soil and water as a common environmental pollutant. This constitutes one of the dreaded consequences of disease in recent years, drastically affecting more than 90 million people all over the world. The metal gets accumulated in fish body if fish ponds are contaminated by the metal. Since the metal commonly occurs in some chemical fertilizers, insecticides, fungicides, and herbicides, there is every possibility to contaminate fish culture ecosystems through surface flow-off. Arsenic is concentrated in the liver, kidney, spleen, nails, hair, etc. It seriously affects the blood vascular system, resulting in anaemia, leucopaenia, oedema, skin cancer, etc. and ultimately leading to painful death.

HEAVY METAL POISONING

Concern over the possible build-up of heavy metals in aquatic ecosystems from large applications of sewage sludge, phosphatic fertilizers and pesticides has promted research on the toxic effects of these chemicals on fish. Serious attention has been given to zinc, copper, lead, nickel and cadmium which are present in significant levels in sludge, pesticides, and fertilizers. These elements are generally bound by soil constituents and consequently they do not leach from the soil. Heavy metals are concentrated in muscles, fatty tissues and liver as they move up the food chain until they accumulate in fish to levels that may be toxic to humans.

21.6 Use of Antibiotics in Fish Culture and Potential Human Health Risk

The use of antibiotics at recommended rates in fish culture along with supplementary feeds is widely known. There are some antibiotics which are known to destroy or inhibit the growth of pathogenic micro-organisms. At present, however, more than one dozens of banned antibiotics (such as Chloramphanicol, Nitrofurans, Neomycin, Nalidixic acid, Sulphamethoxazole, Dapsone, Dimetridazole, Colchicine, Chloroform, Chlorpromazine, Metronidazole, Clembuterol, Diethylstilbestrol, Sulfonamide, Glycopeptides, Ipronidazole, etc.) are being used in fish culture. But the presence of their residues in fishery products may create an alarming situation in international markets. The use of antibiotics in aquafarms is a matter of serious concern because their residual effects can lead to human health hazards. The hazardous effects are so serious that the European Union has issued a red alert on imported aqua-products from major exporting countries such as China, Thailand, and Vietnam and imposed a ban on their products. This situation has pushed other Asian countries (especially India) into a serious retrospective mood to locate and prevent all malpractices related to the use of antibiotics, visualizing that the nation may not be far behind to face the burden of a similar ban on the products.

Considering the problems faced by farmers concerned, the use of antibiotics in aquafarms should be completely prohibited to ascertain human health conditions and for assuring smooth exports. Concerned farmers have to face the challenge by assuring yield of antibiotic-free fish and fisk products. Government and non-government organizations as well as the industry have an important role to face the hassles in this regard.

21.7 Growth of Toxic Algae in Ponds Versus Human Health

Blue-green algae (Cyanobacteria) exhibit an enormous diversity of arrangements and shapes, from unicellular rods or cocci (such as *Gleocapsa* spp. and *Merismopedia* sp.) to long trichomes (such as *Oscillatoris* sp, *Anabaena* spp. etc.). They are widely distributed in freshwaters, marine waters and also in soils.

Another division of algae which is termed as Pyrrophycophyta, includes the dinoflagellates — a diverse group of biflagellated unicellular organisms which constitute an important component of fresh, marine, and brackish-water impoundments. Due to fast swimming, they are rapidly distributed all over the aquatic environment.

These algal forms, however, grow well in both shallow and deep waters where nutrient loads are very much redundant. Some algae such as *Anabaena, Microcystis, Nodularia*, etc. produce toxic compounds (such as neurotoxins and hepatotoxins). These substances may cause human illness and damage nerve cells and liver. Herbivorous fish (especially grass carp, common carp, tilapia, gourami, silver carp, *Puntius*, tench, snappers, sea bass, etc.) consume these toxic algae as food. They are important food-fish for man, although in some situations their flesh may be poisonous, due to the build-up of toxins ingested from algal-eating fishes which the herbivorous fishes in turn have eaten. Some toxins (such as ciguatera) are known to be highly soluble in fat, ultimately deposited in fish flesh and thus it passes through the food chain to humans (See Panel 21.1).

21.8 Use of Animal Wastes as Fish Feed and Human Health Risk

In intensive and semi-intensive culture systems, fish are regularly fed with balanced supplementary feeds. To reduce the cost of fish production, animal wastes such as poultry viscera, blood, hatchery waste, prawn waste, and silkworm pupae are used as fish feeds. These wastes are well mixed with some locally- available ingredients such as rice bran, wheat bran, mustard oil cake, groundnut cake, and soyabean meal. If animal wastes are not properly processed before mixing with these ingredients, waste materials might exhibit a number of health risks. The health risk involves spreading of pathogenic micro-organisms, unsanitary condition within residential areas, and generation of foul odor. Therefore, the effects of animal wastes on human health must be evaluated before their use over extensive areas is permitted.

21.9 Management of Fish Culture Ponds is Inevitable to Reduce the Risk of Human Health

Most of the developing and under-developed countries have enormous freshwater potential. This generous amounts of water could be effectively utilized for fish culture. But unfortunately, due to lack of adequate pond management, substantial quantities of water have become useless for fish farming.

It has been recognized that several factors that follow contribute to the public health issues while fish culture is undertaken on commercial scale and adverse aquatic ecosystem is one of the most serious constraints in this regard.

1. Pathogen loads should be reduced to an acceptable level following the adoption of suitable measures. Reduction of pathogens even at zero per cent level may not be feasible.
2. Control of eutrophication.
3. Control of aquatic weeds helps remove molluscan and insect populations.
4. Fallow and unutilized ponds harbor a number of disease-producing organisms. Hence, such ponds should be converted into productive ones for fish culture.

PANEL - 21.1
FISH INVOLVED IN ALGAL POISONING

Ichthyotoxic fishes have been divided into three groups: (1) *Ichthyotoxic fishes*: those fishes which produce a toxin that is restricted to the gonads. Most members are freshwater species. Roe produced from these fish is poisonous. (2) *Ichthyohemotoxic fishes*: those fishes which have a toxin in their blood. Some freshwater eels (and several marine fishes) make up this group. (3) *Ichthyosarcotoxic fishes*: those fishes which contain a toxin within their musculature, skin or viscera, which when ingested produce deleterious effects. This type of poisoning is identified with many marine fishes only.

Many freshwater and marine fishes have been implicated in algal poisoning. The toxin is a metabolic product of algae. Since many of these fishes are edible and many are valuable food fishes in many parts of the world, algal poisoning is not only the most common but also the most treacherous form of ichthyotoxism.

Most tropical fishes involved in algal poisoning are surface forms; a few are bottom-dwelling forms. Most toxic species are either carnivorous or benthonic algae feeders. There appears to be a positive correlation between the degree of toxicity and the amount of fish in the diet. There is also a tendency for the smaller fish of a species to be less toxic than the larger fish of the same species. The liver and musculature are usually the most poisonous parts of the fish.

It appears certain that fish poisoning is associated with the food-chain relationship of the fish. It is suspected that the poison originates in a benthonic organism, form which it is transferred directly to herbivorous fishes and indirectly to carnivorous ones. While both carnivorous and herbivorous fishes can cause algal (and also ciguatera) poisoning, the former fishes are more toxic, and in some areas are the only fishes adequately toxic to cause poisoning in man. The herbivorous species feed on a toxic alga (or protistan), but the evidence favors the blue-green algae as the most probable common source for the ciguatera toxin. Toxicologists have discussed the possible role of these algae in the poisoning.

When poisonous algae are consumed by the fish, toxins are deposited mainly in flesh, liver, and brain. A number of observations have been noted in the vagotomized dog by toxicologists following injection of the toxin. These observations indicate that the toxin has a direct effect on the heart, an effect which is in part responsible for the cardiovascular crisis. In man, however, toxins interfere with the nerve impulse transmission by altering the cell membrane and sodium channel polarization. Clinically, there is tingling of the lips, tongue, and throat with nausea, vomiting, diarrhoea, paralysis, and numbness. Other symptoms arterial blood pressure, phrenic nerve potentials and suppresses the indirectly elicited contractions of the diaphragm and often reduces the directly stimulated contractions.

5. The hoof prints around the water bodies are the most fertile breeding grounds for mosquitoes. According to an estimate, one hectare shallow water-spread may produce in a single night about 6,25,000 female mosquitoes able to transmit malaria. Such water-spreads are stocked with mosquito larvae-eating fish. Thus the risk of malaria spread can be reduced.

6. Other management practices include the reduction in stressful conditions of ecosystems, development of new effective chemotherapeutic which leave minimal residues, development of quality control procedures and techniques for screening fish products for chemical residues and pathogens, and studies on the host-pathogen- environment relationship.

USE OF LARVIVOROUS FISH IN FISH PONDS EFFECTIVELY DESTROY MOSQUITOES AND THEIR LARVAE

Small- or medium-sized fish ponds may assist in the spread of malaria through breeding of different species of *Anopheles* mosquitoes. Anopheles mosquitoes are notorious vectors of malaria caused by four species of protozoa belonging to general *Plasmodium*. In many countries of the world, a large number of ponds, tanks, and ditches are stocked with different species of mosquito larvae-eating fish. A number of indigenous and introduced species of fish are known, at one stage or the other, to consume mosquito larvae. The indigenous species belongs to the following genera:

Oxygaster spp., *Barillius* spp., *Laubuca* spp., *Rasbora, Daino, Esomus dendricus, Punctius* spp., *Wallago attu, Alpocheilus* spp., *Notopterus* spp., *Ophiocephalus* spp., *Colisa* sp. *Etroplus* sp., *Anabas* sp., *Mugil* spp., *Mystus* spp. *Therapon* sp, *Ambassis* spp. and *Glossogobius guiris*. On the other hand, a few introduced fish such as *Ctenopharyngodon idella, Carassius auratus, Labistes* sp. and *Gambusia affinis* have been noted to be effective to control mosquito larvae. Some larvivorous fish species are shown in Figure 21.5.

1. *Control of Mosquito Larvae*: For control of mosquito larvae, dense surface vegetation must be removed followed by the removal of predatory and weed fishes from ponds. Cleaning of ponds is very important which permits the larvicidal fish to propagate normally. The stocking rate of larvivorous fish usually depends upon the fish species to be reared and their ratio. The most effective stocking rate of *Aplocheilus panchax* for example, has been found to be three per 900 ^{2}cm of the breeding area of mosquitoes. In some situations, more than one larvivorous fish in different ratios may be recommended for stocking.

Large, clean,and unshaded fish culture ponds with an exposed stretch of water are not likely to increase the breeding of the dangerous species of *Anopheles*. Small-sized unshaded ponds are safe if fish are stocked permanently.

SANITARY ASPECTS DURING FISH CULTURE

Most of the farmers do not take adequate care during the entire culture period and consequently, the rural residents are victimized due to severe contamination of water

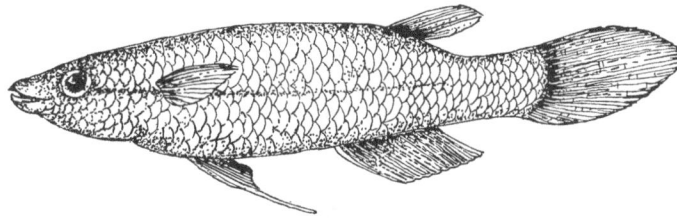

Aplocheilus lineatus (Valenciennes)

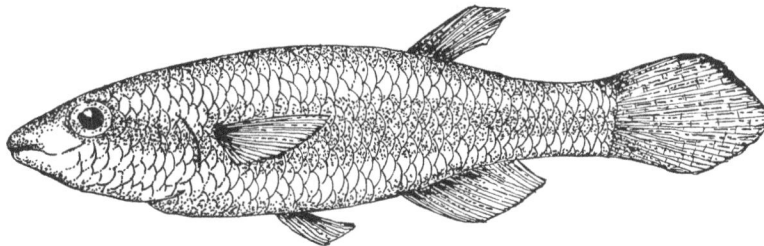

A. panchax (Hamilton)

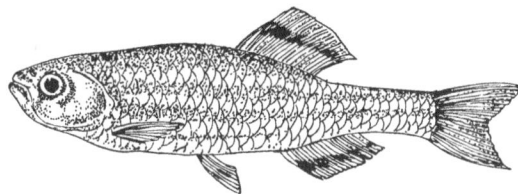

Danio aequipinnatus (McClelland)

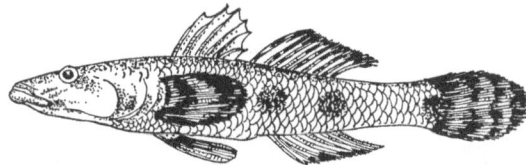

Glossogobius guiris (Hamilton)

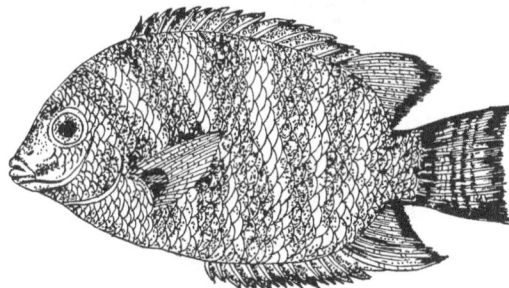

Etroplus suralensis (Baird-Girard)

Fig. 21.5 : **The ability of larvivorous fish to move in shallow and weed-infested water bodies is extremely important. Their capacity to breed freely in confined water areas allows to adjust their activities rapidly in response to changes in their environment. In their ability to flourish both in shallow and deep waters and to escape their natural enemies, larvivorous fish are the most pertinent to any group of fish. An ideal larvivorous fish must be surface-feeder, hardy, absolutely worthless and not so important for the table. Some representatives of larvivorous fish are shown in this figure.**

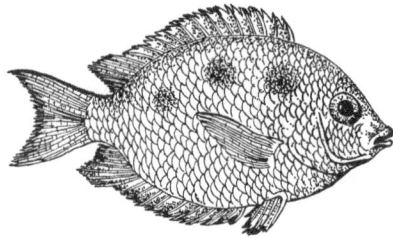

Etroplus maculatus (Bloch)

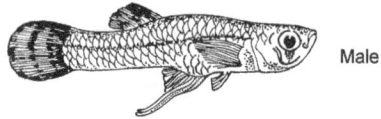

Male

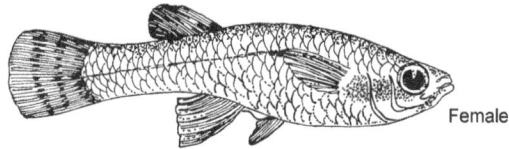

Female

Gambusia affinis (Baird-Girard)

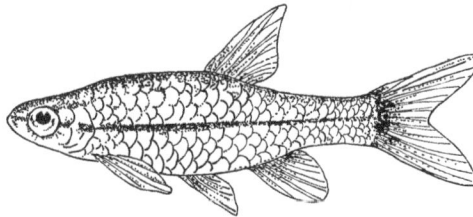

Rasbora rasbora (Hamilton-Buchanon)

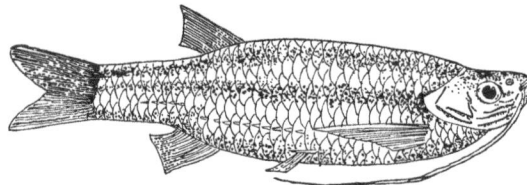

Esomus dandricus (Hamilton-Buchanon)

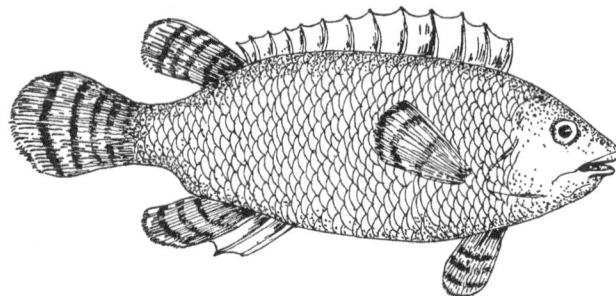

Nandus nandus (Hamilton)

Fig. 21.5 *contd...*

bodies. More than 70 per cent of the world diseases, specially in the developing countries are linked with water. However, improvement of sanitary conditions may reduce the risk of health status considerably. The improvement of sanitary systems include: adoption of wastewater treatment and recirculatory systems for reuse in fish culture, protection of water from contamination, growing awareness among farmers' community and control of diseases and parasites.

It should be pointed out that although the use of human waste and waste-waters for fish culture is prohibited in many countries of the world, systems still exist and in some cases the use is on the increase for high fish production. Public health concerns are of highest importance. Of course, very little is known about the true health risks associated with waste-water reuse and treatment of waste-waters is necessary considering the present rate of expansion in fish culture.

A growing body of research has shown that the ingestion of contaminated fish from polluted or wastewater-fed fish culture ponds in summer and winter seasons results in outbreaks of human digestive disturbances. Some causative agents have been identified as *Vibrio* spp., *Escherichia coli*, *Salmonella* spp, *Shigella* sp., bacillus dysentry, bacterial diarrhoea, *Ascaris* sp. *Trichiuris* sp., *Schistosoma* spp., *Chlonorchis* sp., and *Taenia* sp. These organisms have, however, been reported from the body surface, skin, and digestive tracts of fish. Most of the cases disclose severe or mild illness in fish infected with pathogens.

HAZARDOUS EFFECTS OF SNAILS IN PONDS AND THEIR CONTROL

As members of the class Gastropoda under the phylum Mollusca, the snails constitute the main food item of carnivorous fish. The destructive effects of gastropod populations are often very serious. Not only they cause an imbalance in the availability of other fish food organisms but also spreads several diseases in man. It has been estimated that about 40 per cent of the total calcium in one hectare pond is removed by gastropod populations.

In some situations, molluscan populations exceed food fish populations. The mismanagement of fish ponds has greatly exacerbated molluscan populations. In monsoon and winter seasons and in highly acidic ponds, results on the effects of imolluscan populations on fish growth and production are highly discouraging.

Excessive multiplication of gastropods in ponds is a concern to most tropical and sub tropical countries of the world. Most aquatic ecosystems have been subject to severe damage by the growth of snails, decline in pond productivity and fish production as well as outbreaks of several diseases.

Different parasitic trematodes cause several diseases such as Fascioliasis, Schistosomiasis, Amphistomiasis and Dicrocoeliasis in man and cattle which are the main hosts. Parasitic trematodes spend part of their life cycles on one or other species of gastropods which serves as intermediate hosts and encourage the spread of above-mentioned diseases. In Central Africa, East and South Africa, India, Egypt, all far Eastern countries, South America, and China, millions of people are affected by the parasitic trematodes. Therefore, premanent reduction of snail populations from pond, reservoirs, and swamps is necessary to check the spread of such diseases. It has been found that the introduction of several species of catfishes such as *Wallago attu*, *Chrysichthys mabusi* which are closely related form of *Mystus* spp. in India, *Pangasius pangasius*, *Clarias batrachus*, etc. serve to combat gastropod populations.

Rearing of high yielding duck breeds (such as Indian Runner, . Khaki campbell, Nageswari, and Sylhat mate) in fish culture ponds is the most important feature in the control of snail and insect populations. This system of fish farming is widely practised in Poland, East Germany, China, Hungary, Asia and to some extent, in India. However, duck keep the pond water clean by controlling aquatic insects and snails. These organisms are used as food by ducks and hence serve as an effective biological control of a number of human diseases.

21.10 Conclusion

Fish culture ecosystems and contaminated fish are a much more important source of disease of human beings that can be avoided through adequate management systems. Contamination of fish with chemicals and waste materials presents two results. *First*, they produce harmful effects by accumlating the chemicals in fish body and *second*, pathogens which invade fish culture ecosystems through wastewater-fed system, infect the fish flesh. These two results are increasingly pertinent because bioaccumulation of toxic substances in substantial amounts and their detrimental effects on human population through consumption of contaminated fish should not be disregraded.

It should be remembered that small, neglected, and shallow water bodies provide shelter for a variety of insects, snails, and aquatic plants. This situation is important in view of the spreading of various types of diseases to human beings. Insects and snails must be controlled by the introduction of carnivorous fish into the ponds and at the same time no vegetation should be allowed to grow in ponds.

Though pathogen loads can be reduced to acceptable levels if adequate measures are adopted, zero risk may not be attained. However, it would neither be inevitable nor feasible due to omnipresence of pathogens and for economic reasons, sewage or wastewater must be treated before their widespread use in fish culture is permitted. At the same time, adequate institutional hygiene during handling and processing of aqua-products, adequate personal hygiene, use of food preservatives, and cooking of fish products at suitable temperatures are some of the important health protection measures to be taken into account. Thus, public health can be reliably protected. The quality of the products developed from different fish culture systems would have to agree with the hygienic regulations for fish as food under natural conditions.

The contamination of fish and water with pathogens and toxic substances can be effectively controlled by sanitary measures such as proper disposal of human wastes, prevention of putrefaction of water, and use of sanitary methods for production and handling of fish. Carriers of enteric organisms must not be permitted to participate in the handling and preparation of fish. For removal of pathogens and/or contaminants from live fish body, depuration studies should be conducted in ponds for several months prior to consumption. Regular inspection of fish culture ecosystems and fish processing plants for occurrence of enteric pathogens and identification of improperly treated waters are important functions of state and public health departments.

References

Blumenthal, U. J., B. Abisudjan and S. Bennett. 1991. The risk of diarrhoel disease associated with the use of excreta in aquaculture and evaluation of microbiological guidelines for use of human waste in aquaculture.*In* Proceedings of the Third Conference of the International Society for Environmental Epidemiology, Jerusalem, August 11-15, 1991.

Chanler, A. C. and C. P. Read. 1981. Introduction to Parasitology. John Wiley and Sons, Inc. New York, USA.

Cheng, T. C. 1986. General.Parasitology. Academic Press, London.

Edwards, P. and R. S. V. Pullin. 1990. Wastewater-fed Aquaculture. *In* Proceedings of the International Seminar on Wastewater Recycling and Reuse for Aquaculture. Calcutta, 6-9 December, 1989.

Feachem, R. G. *et. al* , 1983. Sanitation and Disease — Health Aspects of Excreta and Wastewater Management. John Wiley and Sons, Inc. New York, USA.

Liang, Y., R. Y. H. Cheung and M. H. Wong. 1995. Reclamation of Wastewater for Aquaculture: Trace Metal and Bacterial Contamination in Fish. *In*: Recycling the Resource (*Eds*: J. Staudenmann et. al.),Transtec Publications, Brandrain 6 - Switzerland.

Mara, D. M. and S. Cairmeross. 1989. Guidelines for the safe use of waste-water and excreta in agriculture and aquaculture, World Health Organization, Geneva.

Niu, S. and B. Ling. 1991. Health Assessment of Night Soil and Wastewater in Agriculture and Aquaculture (WHO Technical Services Agreement WP/ICP 7 RUD 7001/RB/88-310). Chinese Academy of Preventive Medicine. Institute of Environmental Health and Engineering.

Prein, M. 1990. Wastewater-fed Fish Culture in Germany. *In*: Proceedings of the International Seminar on Wastewater Recycling and Reuse for Aquaculture. Calcutta. 6-9 November, 1989.

Smyth, J. D. 1994. Animal Parasitology. Cambridge University Press.

Strauss, M. and U. J. Blumenthal. 1990. Human Waste Use in Agriculture and Aquaculture Utilization Practices and Health Perspectives. IRCWD Report 08/90, EAWAG/SANDEC. Duebendorf, Switzerland.

WHO. 1989. Technical Report Series, 778. World Health Organization, Geneva, Switzerland.

Questions

1. What do you mean by the term health? How fish culture is related to public health?

2. How does fish culture pose a formidable hazard to human health?

3. Discuss the impact of fish culture on the degradation of aquatic ecosystems and human health.

4. Discuss how fish culture ecosystems are closely related to human health.

5. Discuss how pathogens are related to human health risks.

6. Name some important organisms that are responsible for human disease. How these organisms spread the disease and cause human health hazards?

7. What are the effects of water pollution on human health? State how chemical and biological pollutants are controlled.

8. What is larvivorous fish? Why these fishes are important in fish culture?

9. What are the sanitary aspects that should be considered at the time of fish culture to reduce the risk of health condition considerably?

10. Discuss how the use of animal wastes and banned antibiotics is related to human health status. Why the use of antibiotics should be regulated in fish culture industries?

11. What are the steps that are involved in the management of fish culture ponds? Why the pond management is so important?

12. Discuss how human health is related to (a) toxic algae, (b) snail and (c) antibiotics.

22

Harvesting of Fish

With the discussion on different aspects of aquatic ecosystem in relation to fish culture and the description of the management strategy in Volume 1 and the preceding chapters of this Volume, we can now proceed with the study of the post-culture technology task. And it would be logical to embark on the study of post-culture technology. To put such technology in perspective, it would be essential to have a succinct idea on this aspect undertaken by an entrepreneour. Our discussions on aquaculture technology are spread over a series of four chapters (Chapters 22 through 25). In the present chapter, we shall discuss the overall harvesting process of post-culture technology how fish culture strategies undertake the harvesting job at the commercial level so as to take care of its long-term interests in aquaculture businesses. In chapter 23 on processing and preservation that follows, we shall move down to the next post-culture technology and discusses how a given aquaculture business unit of the farm takes on its marketing job, how taking its broad cues from the farm overall business plan and it sets as well as develops its marketing strategy. In Chapter 24 on transport, we shall go into the brief discussions on transport strategy formulation of fish and fish seed. The success of fish culture strategic planning and other related aspects depends on the ability of the farm to adopt certain core competencies and competitive advantages. This task is elaborated in the 25th Chapter in post-culture technology series. And a very brief discussion on case study regarding trout and salmon marketing will complete the series.

22.1 General Considerations of Fish Harvesting

The term harvesting encompasses all aspects of man's occupation of the aquatic animal in the seas and in inland waters all over the world. A very wide variety of methods is employed. Men were huntered fish from different water bodies before they became agricultural farmers, and harvesting of fish from any water resource is, therefore, one of the oldest occupations of mankind. Increasing human populations have made the efficient harvesting of food from different water resources more and more important. Fish is a vital source of food in most countries of the world where the land is desolated, mountainous or inexorable and agriculture cannot be easily developed; and fish are intensively or extensively cultured and harvested on commercial basis and processed for the market. Moreover, modern aquaculture and fisheries are not only limited to harvest fish but also include many other harvests from inland and marine resources such as crabs, lobsters, clams, oysters, mussels, shrimps, prawns, cephalopods (cuttle fish, devil fish, squids), crayfish, and sea weeds. It must be presumed that different aquatic resources are exhaustible repository of fish and other aquatic animals. Indeed, there are vast areas of the inland waters which have very few fish. Many factors affect the productivity of freshwater resources. These include depth, siltation from surface flow-off, temperature variations, pollution, over-exploitation, and disposal of wastes. Consequently, expansion of many commercial fishing grounds are suffering from toxic and carcinogenic components present in a variety of agricultural chemicals. And yet, pollution and over-fishing are becoming more frequent due to loopholes in norms and regulations enforced by the government organizations.

It should not be over-estimated that harvesting of fish is comparabl to that of mining and it is a robber industry. If men harvest fish at a faster rate then nature can replace them, there will consequently be very few left. With rapid, highly-mechanized crafts and extremely efficient and deadly weapons for harvesting fish, many regions of the world are already faced by a great decline in the annual catch. Harvesting of fish far in excess than requirements and wasteful killing of immature fish must be checked, not just by individual countries, but as fish realize no national boundaries, on an international basis.

There is a boundary to how much the inland waters are abused can receive destructive fishing practices and pollution. Undoubtedly both have to be checked if the inland water biodiversity of commercial fish is not destroyrd beyond salvation. Vested interests responsible for this destruction have to be punished if they persist in their actions. However, fish protection and conservation strategies in world fisheries must be promptly acted upon by all nations if the supply of fish is to be ascertained for current and future generations.

22.2 Location of Harvesting Grounds

Although fish may be harvested in the seas and in freshwater in any parts of the globe, the world's major commercial harvesting grounds are located in the cool waters of the northern hemisphere. Commercial harvesting is little developed in the tropics or in the southern hemisphere. The importance of the northern hemisphere in harvesting can be evaluated by the fact that it is a land hemisphere with a large population and with vast freshwater resources teemed with plankton and fish. Moreover, fish are abundant in certain areas than in others because of the availability of plankton. The reasons for the concentration of harvesting grounds in the northern hemisphere are outlined below.

PLANKTON

Plankton is a collective term for the millions of microscopically small organisms. Phytoplankton are tiny plant organisms drifting about on the water or near the surface. They form the food of the microscopic animals (zooplankton) which are, in turn, devoured by fish. The supply of plankton to fish as food depends on various factors that follows.

1. *Shallow Water*: Since phytoplankton forms the base of the food pyramid and depend on sunlight for their existence, they can well develop in shallow waters. The best

harvesting grounds are thus located where plankton of different kinds are most abundant.

2. *Cool Water:* Plankton grow and survive well in cold or cool waters Therefore, subtropical and temperate waters are relatively rich in plankton.

3. *Minerals:* Several minerals and nutrients are extracted from water by plankton for nourishment. These nutrients are brought from the land carried by rivers. Therefore, plankton concentrations are very high in waters and soils where nutrients are most abundant.

COOL CLIMATE

It is needless to mention that cold water life is best developed in waters with a temperature lower than 22°C. Tropical waters are hot for most part of the year and the fish species are generally of less commercial value. In most cold water habitats, valuable fish such as trout, salmon, haddock (white fish), mackerel, scale carp, etc. thrive well. Moreover, fish cannot be kept for long period in moist, hot, and tropical conditions and are vulnerable to deterioration. In contrast to this, temperate, and coldwater habitats as well as cold winters provide natural ice for preserving the fish.

22.3 Harvesting of Fish

The harvesting process can be done in two ways. *First,* by dewatering the ponds and *second,* by not dewatering the ponds. Generally, small and medium-sized ponds are dewatered completely and then fishes are captured by simply hand-picking. On the other hand, large-sized deeper ponds and lakes which cannot be drained completely, special devices are adopted for the purpose.

During harvesting the fish stocks from ponds, following precautions should be taken: (1) feeding schedules of fish should be withdrawn 2 or 3 days before dewatering and (2) fish should be harvested at the time of cool weather, particularly in the morning when dissolved oxygen content of water is very low.

FISH HARVESTING FROM DRAINABLE PONDS

Fish harvesting from drainable ponds is very easy if there is a harvesting sump. In a nursery or fry-rearing pond, it is necessary to have a harvesting sump to avoid injury to fry or fingerlings at the time of harvesting. Harvesting sump is a place into which water is drained at the bottom of the pond.

The action of dewatering ponds is accomplished at a rate appropriate for the purpose which is dependent upon the size of the outlet and the drainage channel and the fish aggregated in the harvesting sump. From the sump, fishes are collected by loading equipments and if necessary, with the help of a net. If the harvesting sump is too small for the quantity of fish, draining and seining are performed simultaneously to make the fish harvest in good conditions and more effective. Some of the fish may be seined and the remaining stocks are caught in the sump. When the fish stock in the sump is marketed, freshwater is sprayed in the sump to avoid the mortality of fish.

FISH HARVESTING FROM UNDRAINABLE PONDS

Ponds which are not able to drain the entire amount of water, it is necessary to drain water with the help of diesel-operated pump set. Ponds are irregularly drained to reduce the cost of production. When multiple harvesting and stocking of fish are undertaken, commercial fishing gear is employed. Seine net is most common and is used for fishing in farmed ponds. This type of fishing is suited for harvesting many species of fish particularly tilapia and carps.

Generally for efficient harvesting of ponds by seining, the shape and size of ponds would have to be kept in mind. The length of a seine net is about one and a half times the width of the pond and the depth is about 2-3 times the pond depth. The mesh size of the net depends on the size of the fish to be harvested. Very fine meshed seines generally make hauling it more difficult due to high resistance in water. Therefore, a longer mesh size is used for fingerling fishes. Fine meshed seines made of mosquito nets are used in nursery ponds.

Generally wooden poles are attached to both ends of the seine so that net can be comfortably dragged and stretched vertically. Along the upper margin of the net, a stout rope passes the head rope. The head rope of the seine has floats at regular intervals. Floats of the conventional Indian types are made of soft wood, cork, plastic or other suitable floating materials. When big floats are required, sealed wooden drums, kerosine tins, hollow balls of glass, plastic or aluminium are used in mechanized nets. The lower margin of the seine has another rope termed as *bottom rope*. The bottom rope is provided with sinkers made of lead, iron, stones or pieces of brick. The bottom rope has a propensity to sink into muddy bottoms which permit the net to roll up and resulting in loss of fish caught in the net. For this reason, the use of mud lines is considered. Mud lines consist of a number of thin ropes linked together made of a jute material that absorb water. Mud lines glide on the bottom without excavating or elevating the mud. This process helps reduce escape under the seine.

Before netting, the level of water is reduced considerably by partial drainage to aggregate the fish and make the fish caught easier. Seining begins at the deeper region of the pond and completes at the shallow part. After reaching the shallow area of the pond, the net is sealed and lifted with the haul.

In contrast to mechanized seines, the unmechanized ones require considerable labor if the catch is massive and hence various types of elevating devices have been developed. Highly mechanized dip nets are used in large farms. Soft rubber conveyor belts are used effectively, particularly when the fish are lifted to high elevation. To avoid any damage of the fish, specially-designed fish screws for lifting fish with water have been devised.

22.4 Time of Dewatering The Ponds

Dewatering the ponds, whether partial or total as the case may be depending on the type of net used, depend in part on the nature of the farming systems. In several Asian countries, air-breathing fish culture ponds are drained out in the autumn or in the summer when the pond water remains at minimum level. During this time, most of the ponds remain more or less dry. Since harvesting of air-breathing fish from ponds/tanks are quite difficult, complete or partial dewatering is necessary to ascertain the catch of total stocks. If it is necessary to drain out the fattening ponds (particularly carp culture ponds) at the end of the culture period, dewatering is performed in the autumn or before the onset of monsoon when the ponds have to be prepared for the next crop. Trout fattening ponds are also allowed to dewater all the year rounds; of course, according to the demand of the market.

It should be kept in mind that ponds are not dewatered during summer season when the temperature of water is high enough. If ponds are, however, drained in summer season, it should be carried out at dawn when the weather is cool.

22.5 Methods of Dewatering

Dewatering the ponds should be slow and regular in order that the fish can follow the lowering of water. Rapid dewatering is not desirable because fish populations will be confined to the floating or sub-merged weeds and to the floating water level. In many cases most of the fish will get concealed in the pond mud and will be lost.

Draining the medium- or large-sized ponds are executed by one or two diesel-operated 3 HP engines. A suction pipe is fitted with the engine and the mouth of the pipe having a wire mesh is immersed into the water. A delivery pipe is also attached to the engine proper. Draining continues for several days or even several weeks depending on the size and depth of the pond.

The cultivation of fish in paddy fields require good outlets for rapid draining of flood waters or during harvesting the fish. The diameter of outlets depends on the volume of water flow. During harvesting, the mouth of the outlet pipes is provided with a bag net within which fish are caught and then lifted.

22.6 Fish Harvesting Without Draining

In cases where it is practically impossible to drain water completely such as in nursery ponds, large stocking ponds and lakes, fish under this condition are harvested for the following purposes: (1) intermediate fishing, (2) fishing out carp fry from nursery ponds, (3) regular capture of fingerlings and brood fishes for the table, and (4) fishing from periodical checking of fish health and growth.

Fish harvesting without draining can be done with a variety of methods such as long lines, lining, trap nets, seine nets, cast nets, drop nets, bag nets, drag and trawl nets, barrier nets, dip nets, and trawl nets. Different types of nets (so called *gears*) have their own local names in different linguistic areas of the country. For example, net is called *Jal* in West

Bengal and Maharashtra, *Jalo* in Orissa, *Vala* in Kerala and Andhra Pradesh, *Valai* in Tamil Nadu, *Bale*, in Mysore and *Jharia* in Gujarat. However, carp and tilapia fry are caught with the aid of scoop net or cast net made up of nylon mesh. Before netting operation, fry or fingerlings are concentrated in front of the area from which the net is supposed to be stretched.

In cases where carp and trout fishing are carried out regularly in small quantities for domestic consumption, a scoop net/drop net/cast net is used for the purpose. A substantial quantity of fish biomass is aggregated at one corner of the pond by broadcasting artificial feeds in those areas where they will consume.

22.7 Freshwater Fishing

Fish have well adapted to virtually all types of aquatic environment. Since they live in such diverse habitats as oceans, seas, polar waters, rivers, streams, springs, ponds, lakes, swamps, reservoirs, derelict waters, and thermal springs that heat up to 44°C, suitable crafts and gears are essential for commercial fishing. Various types of locally-made crafts and gears are in use for capturing different kinds of fishes including crustaceans. Some gears are particularly used for catching crabs, lobsters and prawns; while some are used for catching fishes. At present, highly-mechanized crafts and gears are being extensively used in commercial fishing.

Fish are caught in many different ways: they can be captured by hands, with a spear or with a arrow (Figure 22.1a,b) or with a barbed dart for striking and killing large fish. These methods of fishing, though most primitive and uneconomic, still exist in many places of the world. But the more advanced methods of fishing include trapping, netting, and lining. Commercial fishing is based on traditional techniques and the absolute difference lies on the scale of actions. Commercial catches must be massive to reduce costs. Several techniques of fishing assure that this is possible and large crafts are used by carrying with large gears.

Fishing in large lakes, rivers, reservoirs, seas, and oceans is carried by highly mechanized fishing vessels. The vessel has a cold-storage facility, echo-sounder, position finder, radar, purse seiner, electrical steering, radio-telephone, and some other expensive modern equipments used to track fish stocks both day and night and in some cases, vessels are so well equipped that the bulk catch is quickly processed and fishing vessels do not have a return to harbor so repeatedly. One or two auxillary engines help to drive two generators for electricity. Each vessel carries farmers, experts, pilots, and other technical staffs and their numbers depend on the size and type of vessels used in fishing.

Most of the fishing vessels used in many Asian countries for inland fishing are not so well equipped as mentioned above. However, large or small plank-built boats are used for fishing. Small types of boats called *canoes* are used for fishing in rivers, estuaries, lakes and backwaters. Large-sized boats are constructed by planks with frames and ribs. Canoe is a narrow keel-less open wooden boat with pointed ends and propelled with a paddle.

(a)

(b)

Fig.22.1 : Primitive methods of fishing (a) with a spear and (b) with an arrow

Plank-built boats are durable, light, and economic. Small and large boats are propelled by paddling with oars. The size of the caones varies between 4.0 and 15.0 metres in length.

Although the caone is simple and primitive in structure, it has some advantages. It can be dragged ashore after use with or without difficulty for repairing. In rough weather, it will not sink and the oarman can easily manage to reach the shore.

At the beginning of the 1980s, motorization of boats was the first attempt to adopt the improved inland fishing. Such mechanization permits to reach the fishing spots and return to the shore in a short time, but the gear used for the purpose is still traditional. This has resulted in low yield of fish. Unless gears are well-designed, the additional costs required for motorization may negate the benefit derived from this kind of fishing strategy. However, this type of mechanized device is not at all used in pond fishing where cast nets or drag nets are widely used and simple type of boat may or may not be used. In order to catch different types of fish, fishermen must employ suitable gears and must know the habits of the fish concerned.

Production of fish in the third world is expected to increase in the future. And the technology advancement will further enhance harvesting capabilities. Some fishing crafts being launched or have started in motion in several regions of Asia aimed at improving commercial fishing ventures.

Freshwater fish are caught in rivers, streams, ponds, lakes, reservoirs, and paddy fields. They are caught to meet the demand of local people and are seldom exported. Different types of freshwater fish such as carps, tilapias, catfishes, murrels, trouts, freshwater prawns, and crabs are caught. Salmon is the only exception. While most salmon are caught near the coast, some may reach the higher reaches of rivers and be caught by rod and line or other methods. Many commercial fish species in many lakes and rivers are caught by gill-nets, seines, dip nets, traps, and lines. Fresh-water fishing is only commercially significant when there are major rivers or large inland lakes and reservoirs. Tonle Cap in Kampuchea, Caspean sea (a freshwater lake), Lakes Titicaca and Victoria, Issyk-Kul of Tien Shan, Alakol and Balkhash of Kazakhstan, Sevan in Armenia, Qinghai in China; rivers Erquishi of Western China, Yarlung of Tibet, Chu in Kyrgyzstan/Kazakhstan and reservoirs Pong and Gobindasagar of Himachal Pradesh (India) are some of the most important examples of economically-viable fishing grounds.

In Europe and North America, freshwater fishing for trout, roach, tench, carp, perch, pike or salmon in rivers and lakes is very popular and is a part-time or hobby occupation. Inland fishing is more important in the Russian Federation and in East Asia. The sturgeon, highly valued for its flesh, is fished chiefly in the Caspean sea. In China, Japan and some other Asian countries, inland fishing is extremely important. In India, Bangladesh, and Pakistan, inland fishing is also extremely important. In these regions, carps, murrels, catfishes, and tilapias are commercially harvested.

22.8 Fishing Device

In general, two types of methods are extensively used for harvesting fish such as conventional methods and non-conventional methods. While the former methods cover the use of different types of gears (nets), the latter ones include the use of explosives and poisons.

CONVENTIONAL METHODS OF FISHING

1. *Cast Net*: It is a simplest type of net operated by one farmer by throwing it over the water surface. The net is small and used in shallow areas of the pond, lake, and river. The net is more effective in muddy and sandy bottom having no rugged surface.

The net is conical in shape (Figure 22.2.), forming a circle when spread out. The foot rope along the circumference of the net has lead or iron sinkers attached all round about 15 cm intervals. A string is passed from the centre which helps operate the net.

Fig. 22.2 : Cast net

A portion of the net is puckered up in each hand and thrown in the air with a strong swing to spread it. The net is then dropped on the water surface and sinks down encompassing the fishes within its circumference. The central string is held in hand and the net is slowly pulled up.

Large-sized cast net having 15 to 23 m in diameter with sinkers at the edges of the net is gathered into wrinkles on the boat. The net is operated from a long narrow boat that has the capacity to carry at least five fishermen. The net is kept at full length on one side of the boat and dropped as it drifts enclosing a circular space. When the central string tied to the boat is stretched, the net is hauled up. The net catches large-sized fish.

2 *Scoop Net*: The scoop net has two bamboo poles, each 2.5 m long and are tied at one end to form a V-shaped structure (Figure 22.3). Near the tied angular end, 40 -cm from it, a small cross piece of 50 cm long is tied. On the 'V', a piece of net is fixed having a mesh size of 0.9 to 1.6 cm. At the wider end of the poles the net is fixed by a string. At the narrower end, the net is fixed to the small cross-bar. At the posterior end, the net is continued as a narrow and long baggy cod end.

The net is operated in the rainy season by a farmer from a boat or by walking through water like a duck. The apical portion of the net is held in one hand and the tail portion in the other. By holding the net in such a fashion, the net is gradually scooped over the bottom till the fish stocks are collected in the tail end. A large scpop net is called as *hoop net* which has a very solid frame. Juveniles of catfish, trout, mahseer and major carps are collected from different river systems. scoop net is used in shallow and running waters.

Fig. 22.3 : Scoop net.

In some other countries, however, the scoop net has a wooden handle about 1.5 m long. The handle is fixed to a circular wooden rim made up of two pieces. The first forked piece is connected with the handle and the other is curved or semi-circular. A conical net is fixed on the wooden rim. The net can be used in fast-flowing water which has even in rainy season. It is operated by one person by holding the handle, dipping the net in water and quickly lifted it out.

3. *Trap Net*: Trap net forms one of the earliest devices to collect fishes, prawns and crabs. Several types of traps are used in different areas. Some traps are operated by hand to enclose the fishes from above, others are used to trap fishes while passing through the currents. In some other traps, baits are kept to attract fishes and to enter into the traps. Trap nets are rigid, mobile, or fixed and vary considerably both in shapes and diamentions.

The net is a conical cage-like, made of fine reeds with a wider mouth at one end (24 cm in diameter) and gradually tapering at the opposite end to a length varying between 1 and 3 m (Figure 22.4). The narrow end is tied with rope and can be opened and closed as and when required. At the mouth end, the reeds are fixed all around and their free ends are covered inwards to form a one-way entrance. The opening of the net has a diameter close to the size of fish trapped.

The operation involves immersion of several trap nets in water one behind the other with some spaces in between. On either side of the mouth ends, dry weeds are fixed by means of bamboo sticks. The open ends of the net are directed towards the water current. The traps are set in the evening and opened at dawn and catches are taken out by opening

Fig. 22.4 : Trap net.

the conical ends. During the rainy season, substantial amounts of prawn and fish are caught. Traps are generally used for river and stream fishing.

4. *Fyke Net*: These nets are more or less similar to trap nets though they are flexible and formed by a net attached to hoops. Fyke nets are cone-shaped structure and are held in position among three pegs. Two pegs are settled on each side of the mouth and are injected into the soil. The third peg is injected right behind the net. The inner side of the net is furnished with two or more flexible necks and are easily carried when the hoops are pulled together. Hoops are rigid circular bands used for fastening casks. Casks are barrel-like containers generally usd for the storage of fish. Fike nets are used to catch pike hatchlings from spawning beds.

The efficiency of fike nets and trap nets is assisted by placing elastic or rigid wings made from flexible twigs. The fish strike against the wings, move along the side and enter the trapping net.

5. *Drop Net*: The net is mounted squarely and suspended at its corners by two pliable hoops fixed in a cross at the top and attached to a pole. It is dragged up and down at regular intervals from a boat or from the banks of the water.

6. *Gill Net*: Gill nets are widely used all-over the world to catch every kind of fish because the structure of the gear is very simple and one of the most cheapest fishing gears. The net consists of a rectangular net with sinkers and floats on its bottom and top, respectively in order to spread the net in the water column. The net is fixed at the bottom of water adjacent to the mud in a vertical position in places

where fishes abound. When fishes are passed through the net, they are entangled in the net by their opercula.

There are three basic types of gill net depending on the type of fish to be caught and the conditions of the fishing ground. These include fixed gill net, drift gill net, and encircling gill net. (Figure 22.5). However, surface gill nets catch fish stocks which swim near the surface of the water. Either one end or both ends of the net is fixed to the bottom by anchors to prevent the net from drifting away when the net is operated. This gill net is operated in shallow waters. Bottom gill net is fixed on the bottom with anchor to catch bottom-dwelling fishes. The total sinking power of the net, rope and sinkers is larger than the buoyancy of float.

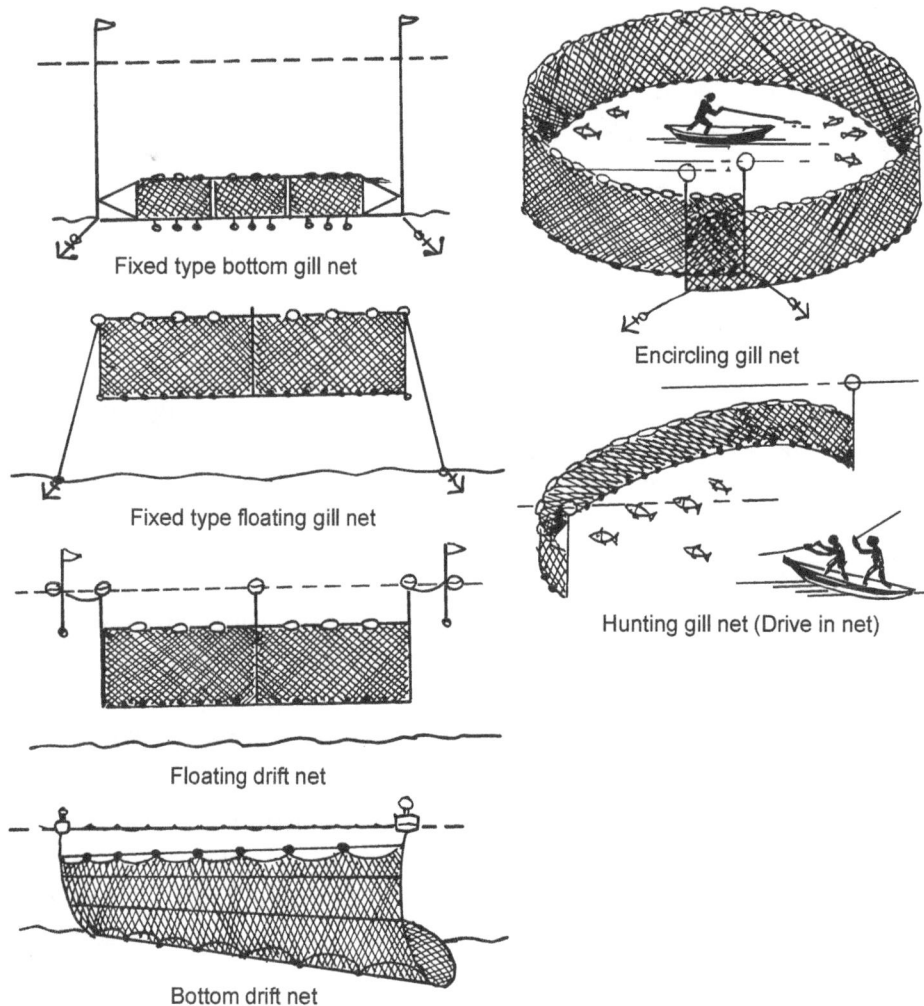

Fig. 22.5 : Various types of gill net.

(i) *Features of Gill Net*: The gear should be such that fish which are entangled or enmeshed in the net should not be able to escape. For this reason, the material of netting cord, mesh size, hang-in-ratio and tension are some of the important factors which must be taken into consideration. In early times, some natural fibres such as hemp, cotton and silk are used. Flexibility of the cord witn high breaking strength is also necessary for gill net fishing operation. Nylon fibres composed of multi-filament or mono-filament is more effective in all respects such as breaking strength, flexibility, elongation, durability, color, and high efficiency of catch. It does not decay in water.

Adequate elongation of netting cord is necessary for gill net to catch fish. Normally, the elongation ratio (it is the elongation at the breaking point divided by original length) is about 25 t0 30 per cent.

To hang the gill net with a proper mesh pattern that will receive the fish, the net is sewed to a rope that is shorter than the net length in a suitable ratio. The ratio at which the net is shortened is termed as *hang-in-ratio*. In general, this ratio is directly related to the body of the fish to be caught.

When the gill net is suspended in the water, a floating force and a sinking force comes into play on the net. In order to held the net in suspended condition at a desired level, a balance between this two forces must be maintained. The weight and pulling forces of the fish caught in the net must also be taken into account. A gill net holds the fish which have become enmeshed by means of the tension exerted by the legs of the meshes. If the tension is not sufficient, it becomes easier for the fish to work themselves free of the net. Gill nets used to catch bottom-dwelling fishes must be stretched with greater tension. In contrast to this, drift gill nets should be set with as little tension as possible to catch fish in large numbers, allowing a large *hang-in-ratio* and plenty of slack in the net overall.

The catch of fish has been increased by using a thinner and less visible twine in net construction. The elements involved in invisibility are color and transparency. The twine used in the net does not offer a contrast to the environment in which it is used. For example, dark-green twine is used for nets catching salmon in the dark middle-depth waters.

(ii) *Design Specifications for Effective Gill Netting*: The aim of gill netting is to catch the fish to enmesh itself in the net. When the fish enters a mesh in the net head, the thread of the net must have adequate flexibility to allow the fish to continue to move through the mesh to a point where the tension at the mesh threads equals to the muscular strength and flexibility of the fish, after which the fish is held in place and caught. In the case of large-sized fish or fish with body shapes which make it difficult for the fish to enter a mesh eye, the catching efficiency of the net is to make the fish entangle themselves in the net rather than enmeshing in one mesh.

Materials that are required for a gill net besides the body net material, include ropes, sewing thread, floats, sinkers, buoy, buoy line, anchors and anchor ropes. All netting and ropes are made of synthetic fibres. Sinkers are made of lead and floats usually of plastics. Synthetic fibres are superior to the natural fibres with respect to durability and light weight.

7. *Drift Net*: Some gill nets are made use of drifting over bed surface to catch fishes and prawns. Drift gill nets do not require an anchor. These nets, however, hang vertically in the reservoirs, rivers, seas and lakes like a tennis net. Each drift net is made up of 16 to 25 pieces, each with 3 to 5 m long and 2 to 3 m high with mesh size of 5 to 6 cm. Head and bottom ropes of the net are fitted with floats and sinkers respectively at intervals to keep the net in position (Figure 22.6) and are usually placed just a few metres below the water surface where the fish will swim into the net. The fish are entangled by their gills and are not able to move either backwards or forwards because their heads are caught and their body is too large to get through. Drift nets are efficiently used to catch pelagic fish (such as salmon, sardine, trout, catla, silver carp, skip-jack tuna, herring, mackerel, and anchovies) on a large scale. For harvesting fish in huge quantities, drift net measures 55 by 14 metres and as many as 90 nets and a number of vessels may be used together to form a continuous curtain stretching for about 3 kilometres.

8. *Drag or Trawl Net* : Bag nets can be towed by moving crafts and hence they are called as *drag* or *trawl nets* (Figure 22.7). However, it is a bag-shaped net whose mouth is kept open by otter boards or head-beams and is by far the most efficient method of catching demersal fish (bottom-dwelling fish such as common carp, mrigal, tilapias, haddock, cod, plaice, catfishes, sole, etc.). Its mouth has floats at the top and sinkers at the bottom, and the net is made stronger at the 'cod end' in which fish are caught. The net is dragged along the bottom by a boat or a trawler for about 2 hours, at a speed of about 8 kilometres per hour. When the master of the vessel will be fully pleased that an adequate amount of fish has been

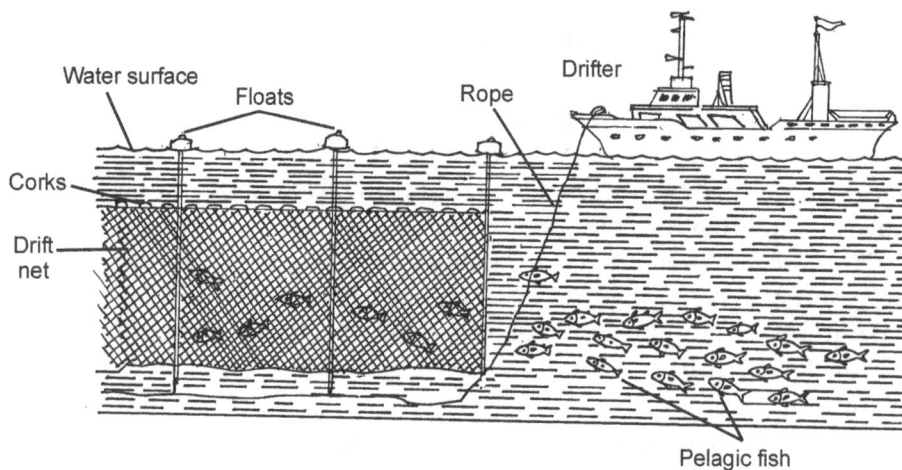

Fig. 22.6 : Drift net

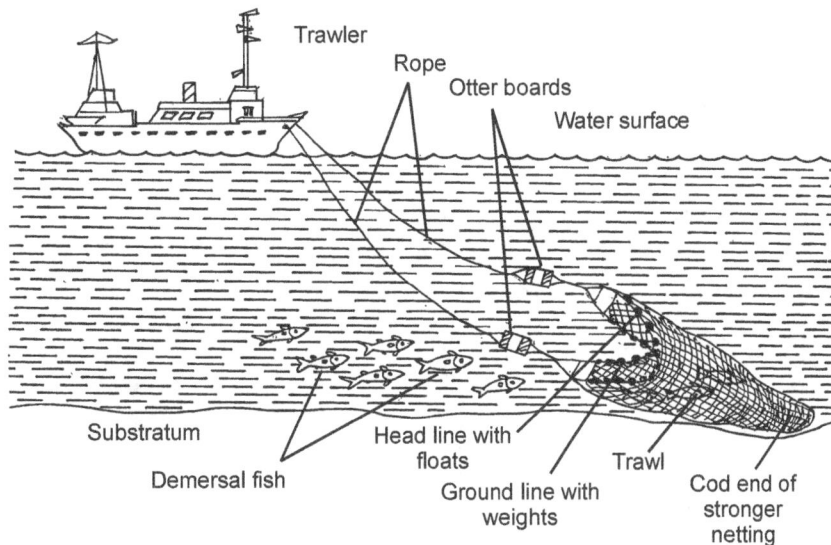

Fig. 22.7 : Trawl net.

trapped within the net, the boat stops and the net is hauled in. The normal length of a trawl net is about 45 metres. Rough weather interrupts fishing operations and endanger the lives of fisherman.

Small-sized drag net measures 6.0 to 15.9 m long with several spreader sticks (10 to 20 in number) of 0.6 to 0.7 m height. Mesh size varies from 0.6 to 1.3 cm. These nets are generally used in shallow ponds and tanks.

Four men drag the net in water by wading in shallow regions of the pond. When the net is pulled in, the force of water makes the net a circle and gradually forms a bag-like structure. Such harvesting operation forms a substantial catch of fishes and prawns.

9. *Seine Net*: Seine nets such as purse seine and haul seine have intermediate features between drift and trawl nets. The purse seine has a narrow conical end and 'wing' of netting rather than otter board (Figure 22.8). It is small and cheaper than a trawl and takes a small catch but is more suitable to use by small-sized boats. The fish are directed into the net by the wings which are fastened to long ropes. The ropes are then drawn to a stationary boat, forcing the fish into the centre of the net. By closing the bottom with a drag string, fish are prevented from escaping beneath the net and the seine is finally hauled on board.

The haul seine is similar to that of a drift net, which is kept vertically in the water like a wall by floats and sinkers on the top and below respectively. After encompassing a shoal of fish, the net is dragged to the bank at both ends. Fish are trapped in the net and then collected immediately.

10. *Dip Net*: A dip net is composed of cod end, main net, side net, selvedge net. It is designed to scoop the shoal of fish which is attracted around the boat by the fish-

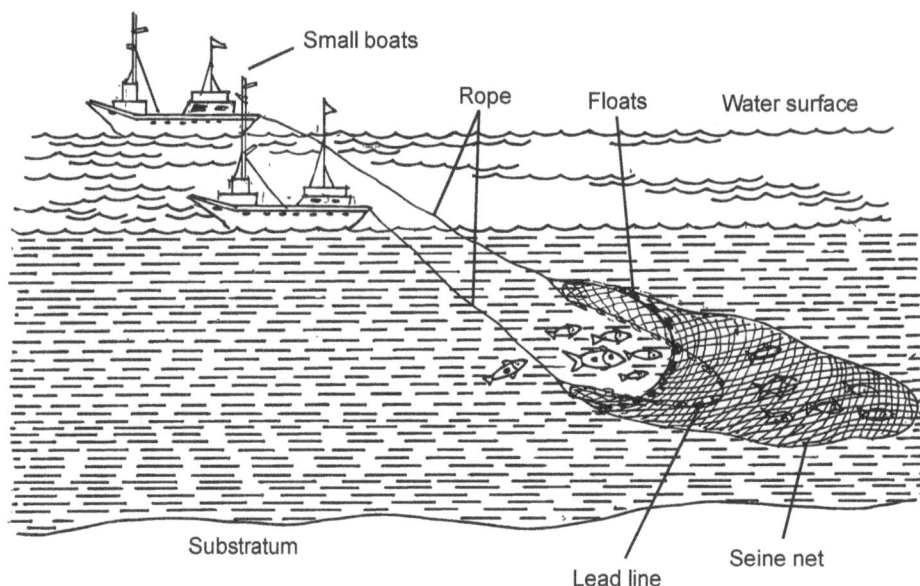

Fig. 22.8 : Seine net, though similar to trawl net, is smaller in size and may be used by farmers in small boats.

lamp illumination or by baits. The net is provided with a number of small sinkers on the sinker line of the net and heavier sinkers to which several ropes (at least 6) are attached to the front end of the net for hauling purposes.

At the opposite end, the net is connected to a bundle of bamboo poles to keep the upper part of the net afloat on the surface of water. On each side of the upper part of the net end, some floats are attached which help to prevent the escape of fish during hauling operation.

The bamboo poles are used for stretching out the net during operation. These are taken off from the net when the net pulled towards the boat.

(1) *Method of Operation*: Motorized fishing boats are used for effective operation of dip nets to catch huge quantities of fish (quantity may vary from 100 to 500 kg). In case of small catch, the net is operated by one man without a boat. However, the net is set in fishing premises where fish abound. The operation is performed at either side of the boat to make the operation easier and more convenient.

The net is operated with fish attracting lanps or baits to attract fish during night and day respectively. These luring devices help to concentrate on either side of the net where the net is set. Hauling lines are dragged by using the line hauler until the front part of the net is hauled up to the surface of water to prevent fish from escaping. The bamboo poles are taken off from the net or pulled with the net towards the boat so that it will be easier to scoop the fish being caught.

Though there is no specific design of the boat used in dip netting, the boat should have some features such as shallow draft, shallow freeboard for easy operation, resistance against wind to maintain the relative position between the net and the boat during fishing-and subsequent holding the fish stock.

11. *Lift Net*: Lift net is an extremely diversified fishing method that is used all over the world in rivers, lakes, marshes and even in open seas with or without motorized boats. The net is set horizontally under water and then lifted suddenly to catch fish that have either wandered naturally into the water above the net, or have been attached there by light or by scattering baits in the water.

The lift nets can be divided into two groups: (i) Floating lift nets — set in the surface of mid-water and (ii) Bottom lift nets - set on the bottom of water. However, different types of lift nets which are shown in Figure 22.9 are extensively used throughout Asia.

Fishes caught by lift nets include *Osteobrama cotio, Esomus dendricus, Puntius*, catfishes, snakeheads, *Glossoglobius* sp., *Notopterus, Ambliopharyngodon mola, Anabas* sp., etc. The catch ranges in weight from 2 to 5 kg/net/day. Net size varies from 1.5 X 2.5m to 2 X 3 m.

12. *Line Fishing*: Though line fishing was more important before the invention of the nets, it is still commercially employed in certain types of demersal fishing, particularly where the floor of the fishing area is rugged and are likely to damage trawl nets. Lines are of two types such as (i) Hand line and (ii) Long line. The lines are made up of synthetic mono-filament with diameter varying between 0.45 and 0.60 mm.

(i) *Hand Lines*: Each hand line is provided with a single baited hook, cast from the boat or deck and drawn up by individual farmer when the float indicates that the fish is hooked. Its commercial importance is limited by the fact that it is slow and uneconomical for deep water fishing. It is, however, most popular for sport-fishing.

(ii) *Long Lines*: Long lines are provided with as many as 100- 4,000 hooks. Each line is provided with at least 6 hooks fixed to different snoods of 10 cm. All hooks are baited and thrown with violently to a distance of about 15m from the bank of the water body. Such hooks and lines are set in series at a gap of about 0.5m in between. Each line is fastened around an empty can of 10 cm in diameter and 14 cm in height for recovering the hooks, and the cans are held down with weights (such as stones) to prevent the sweeping away. Supporting ropes are tied on two bamboo stacks parallel to the bank at the height of about 4 to 45 cm from the water surface. This supporting system helps prevent the recovering can from being carried away by the fish.

The baits generally used for line fishing include worms, cuttle fish, mussels, small insects, shellfish, small pieces of fish, and bread and flour. But this method of fishing is not recognized in developed countries where different kinds of insect-like baits have been developed from the feathers of cock or hen. Well-developed baits are produced from the necks of cock or hen that has great demand in fishing industries in many Asian countries. High quality artificial bait consists of fermented rice (800g), rice bran (200g), wheat flour

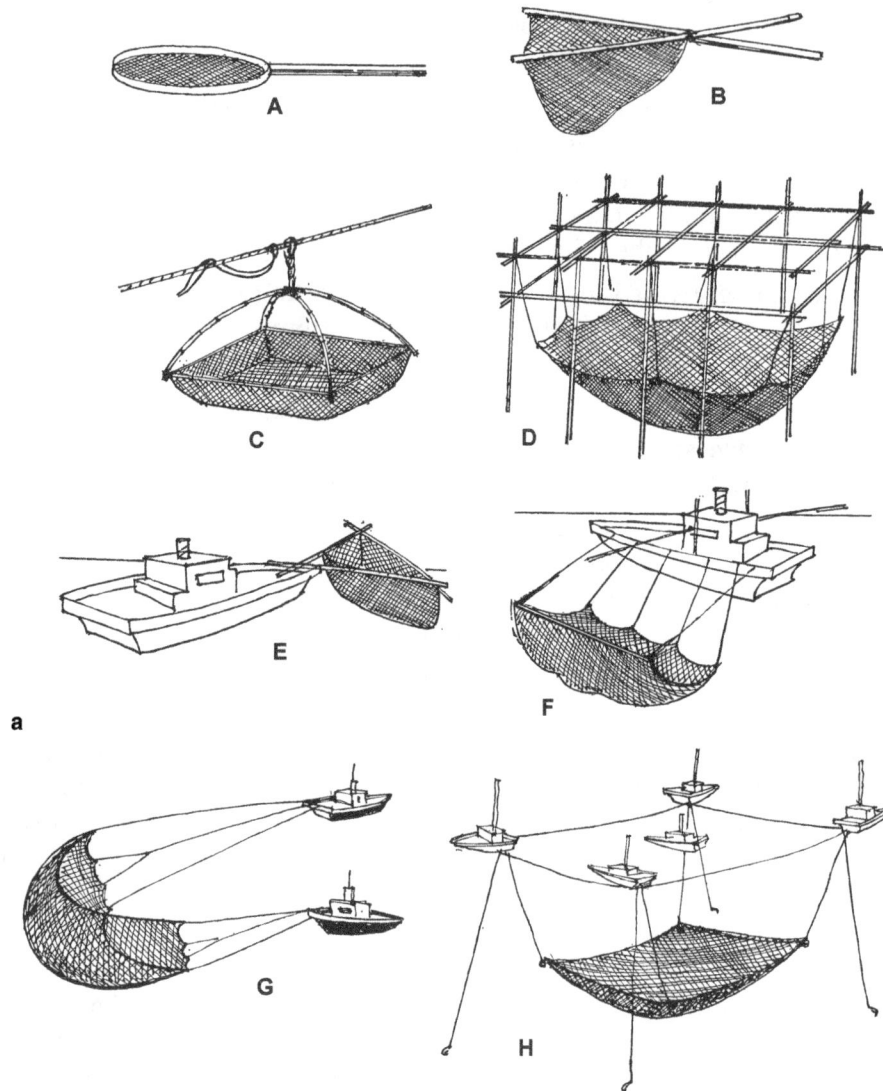

Fig. 22.9 - Different types of lift net. A-D – Hand-operated net not using boat;
E-F – Lift net using one boat; H – Lift net using four boat

(200g), honey (5 tea spoonful), ginger (a small piece), common salt (half tea-cup) and powdered sesame (50g). All these ingredients are mixed, cooked and then cooled. The mixture is moulted spherically with a diameter of about 20-25 mm.

The hooked lines loaded with fish are pulled slowly and reaches the shore and scooped out of the water using a scoop net. Fishes are kept either by tying them beneath their opercula by means of a rope and released in the water or by keeping them inside a cage made of bamboo splits (diamentions vary between 0.5 X 1.0 X 0.5 m^3 and 1.0 X 1.8 X 1.0 m^3)

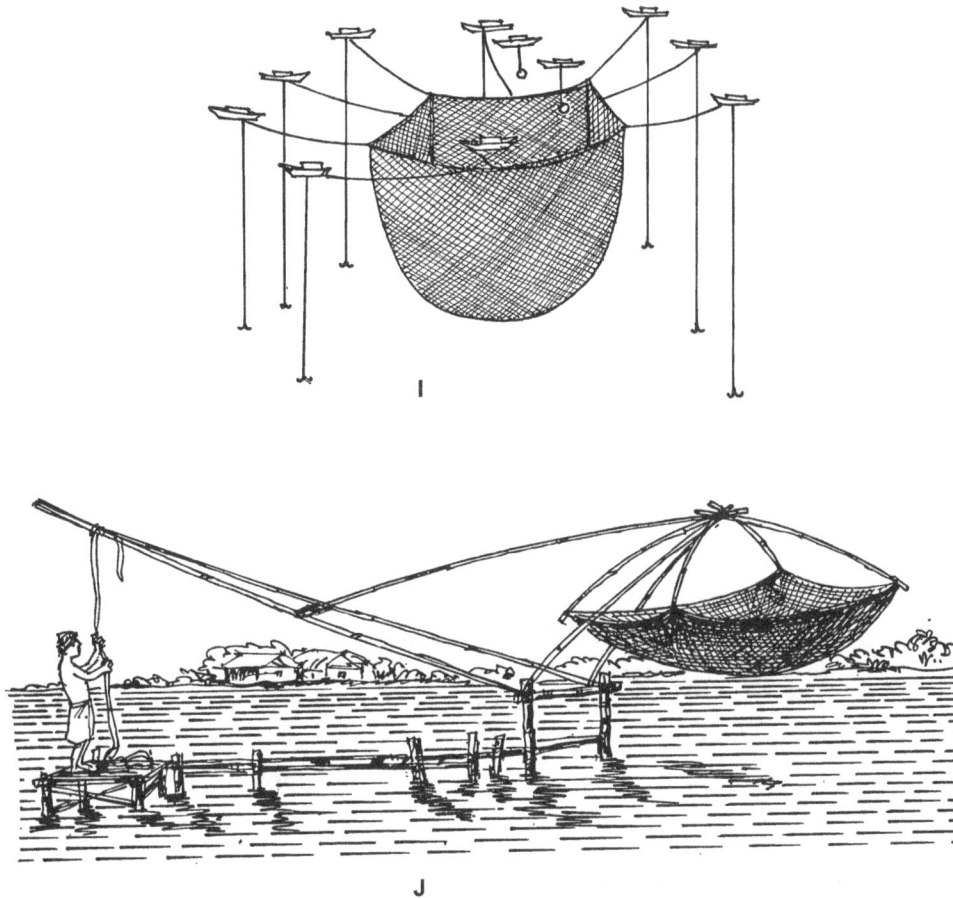

Fig. 22.9 *contd...* – **I. Lift net using eight boat; J. Farmer-operated lift net.**

which is partly immersed in water. Several species of freshwater fish such as silver carp, grass carp, common carp, mrigal, rohu, calbasu, *Labeo gonius, Catla catla*, etc. are caught by this method. Depending on the seasonal variations, the total catch per day varies from 50 to 400 kg.

13. *Fish Spearing*: Fish are speared at night with the aid of a torch light or a lantern. These lighting equipments are used to attract the fish. The fish which come to light are hit on their heads with a spear. Two to 4 persons are generally involved in spearing.

NON-CONVENTIONAL METHODS OF FISHING

1. *Use of Toxicants*: Crushed leaves and fruits of *Adhatoda vasica* and *Randia dumentorum* are commonly used in ditches for catching fish. Leaves of *Sapium insigne, Agave americana* and barks as well as roots of *Dalbergia* sp. are also widely used for poisoning fish. Other plants such as *Edgeworthia gardneria, Polygonum flacidum,*

P. hydropiper, Ficus punila, Acacia pennata and *Bassica latifolia* are used as fish toxicants. These plant products are crushed and thrown uniformly over the water. In some Asian countries (such as India and Bangladesh), mahua oil cake, tea seed cake and derris root powder are widely used during pond preparation and as a result all fish are harvested. The harvested fish can be consumed without any health hazards. In cases where the water spread is large enough, several pesticides such as benzene hexachloride, aldrin, thiodan, and malathion are also used at recommended rates for complete harvesting.

2. *Use of Explosives*: In this method, hill-stream fishes such as *Tor* spp., *Garra* spp., *Glyptothorax* sp., *Noemacheilus* and some other large-sized fishes are harvested. Before applying explosives or any substance which can easily explode, fishermen broadcast rice bran or oil cakes into the water to attract and congregate fish biomass in a particular area. Fishermen wrap the explosive in a cloth, ignite it and push it violently in the area where fishes are concentrated. The stunned or killed fish are then collected by a scoop net.

22.9 Electro-Fishing

For decades, the use of electro-fishing equipments for efficient harvesting the fish stocks from large-sized freshwater ponds, lakes, reservoirs, and rivers has been synonymous. There is great variation in techniques and performance of harvesting which is due to the wide variability of water conditions, fish size and species, and sampling. In large farms, electric seine has been successfully used for removing predatory fishes from ponds and to use the ponds for fish culture.

PRINCIPLES OF ELECTRO-FISHING

The basic principle of electro-fishing is to generate an adequate electrical stimulus in fish near the electrodes that pemits easy capture by gear. The electrical stimulus immobilizes the fish stock and collect them from considerable depths or distances.

Fish are susceptible to alternating current than to direct current and consequently, in the radius of action of an alternating current, electro-fishing divice operating at a specific voltage is greater than a direct current at the same voltage. Pulsed direct current is more effective than a direct current and combines the desirable forced swimming response (*termed as elctrotaxis*) of direct current with the larger radius of action associated with a current.

FACTORS INVOLVED IN ELECTRO-FISHING

For electro-fishing system, some minimum currents per electrode exist which cause fishing from some distance from the electrode. This minimum current principally depends very much on the nature of current (alternating/direct/pulsed direct currents), shape of the electrode, fish species, conductivity and temperature of water and many other factors. Of them, only the first two types are under the control of the operator. Other factors are compensated for by variations in the operation of electro-fishing methods.

1. *Fish Species*: Some species of fish such as northern pike are difficult to harvest by electro-fishing because of the strong swimming capability and natural tendency to escape danger by fast-swimming coupled with a high sensitivity to electric fields. Since carps are very much susceptible to alternating currents, they can be easily harvested by this fishing method. Mortalities of carp as a result of AC electro-fishing are higher than those of direct current or pulsed direct current. Mortality is associated with regions of high current density close to the electrodes. Bottom-dwelling fish are not sensitive to alternating current shocks than those species live in mid- and surface layers of the water.

2. *Fish Size* : Individual variation is observed among fish even though they are of the same species and have similar length. Large-sized fish species are more sensitive than a given electrical shock. Fish, however, absorb power as a function of body length. Also, the greater resistance of small-sized fish, the reduction in response to shocks. Thus a greater proportion of large-sized fish are harvested.

3. *Conductivity*: As a generalization, an inverse relationship between voltage and conductivity of water exists (see below) and it is highly significant in the success of electro-fishing. Water having low conductivity makes it arduous to attain adequate currents to generate electro-fishing responses. For extremely high water conductivity, pulsed direct current or alternating current is more effective than that of the direct current.

Power (Kilo watt)	Voltage (V)	Conductivity (micro-mhos/cm)
2.0	460	10 - 40
3.5	460	20 - 40
4.5	460 - 390	40 - 60
4.5	390 - 320	60 - 100
4.5	320 - 230	100 - 160
4.5	230 - 160	160 - 350
4.5	160 - 100	350 - 700

The presence of bottom materials with high conductivity tend to short circuit the current out of the water and consequently the effectiveness of electro-fishing greatly reduced and sometimes the power source is over-loaded. Hence the direct current between the bottom and electrodes should be avoided.

High turbidity of water and excessive direct current or direct current (DC) aquatic macrophytes reduce the immobilizing capabilities of alternating current electro-fishing. In this case, direct current and pulsed direct current have been found to be more effective.

DESIGN AND OPERATION OF ELECTRO-FISHING

A lot of problems with instrument limitations and lack of operating guidelines exist for such fishing operations. Therefore, proper guidelines for such fishing device must be developed by adjusting the variations in extexnal conditions. A number of concepts that follow are useful in guiding the development of electro-fishing equipments.

1. *Design*: The main components of a fixed electro-fishing systems involve four major sub-systems that follows:

(i) *Boat and Mechanical Components*: Selection of mechanical components such as motor and boat selection (metal, wooden or fibre-glass boats) should be made depending on their availability, convenience, and ease of maintenance. Metal boats have the advantage of grounding of different electrical equipnents through their attachment to the boat. Special care should be taken to ground the wooden boats to assure electrical safety. However, boats should be constructed in such a way that they can effectively and conveniently carry the complete system.

(ii) *Electrical Device*: This device involves a number of components such as the main generator, an auxillary generator and metering, disconnect switches and overload protection, voltage control and power conversion equipment, and a safe and reliable system of inter-connecting the various components.

An apparatus of quickly disconnecting the power should also be provided as an essential component of the safety device.

* *Main generator*: In case of alternating current (AC) electro-fishing, a three-phase AC generator is used. This generator can be utilized by suitable electrode designs. If pulsed direct current or direct current (DC) is used, a transformer is necessary which permits a voltage controlled D.C. source. The auxillary generator and battery system essentially consists of generator, voltage regulator, meters, and battery can be integrated with the main generator to form a compact unit.

* *Metering*: For AC electro-fishing, one voltmeter and a set of three ammeters are necessary. To monitor the generator performance, a second voltmeter is required to indicate the actual output voltage.

  For DC operation, a voltmeter and an ammeter are necessary. In case of pulsed DC, two voltmeters — one for average voltage and the other is to read peak voltage are useful.

* *Disconnect switch and overload protection*: A disconnect switch and overload protection must be arranged, since the electrical load vary considerably because of fluctuations in the conductivity of water. Hence, circuit breakers are placed over fuses for protection of the main power circuits.

* *Voltage Control*: The output voltage is directly controlled by the use of transformers which provides a fit between electrode resistance and the generator feature is cardinal for operating over a wide range of water conductivity.

(iii) *Safety Systems*: A number of safety systems that follow must be kept in mind.

* Different equipments should be selected and designed to avoid potentially hazardous situations.

* Net handles which could form an electrical connection between the water and an operator and power connectors with metallic parts which could become energized through insulation failures must be avoided.

* A low-voltage system should be used for all electrical functions.

* A quick disconnect system is inevitable.

* All metallic parts in the craft are joined electrically.

(iv) *Electrode Systems*: For an effective electro-fishing electrode system, the following requirements should be kept in mind.

* An effective electric current distribution in the water should be established.

* Large current densities should be avoided which are potentially harmful to fish.

* Changes in the conductivity of water should be adjusted.

* Unnecessary physical disturbance to water should be avoided.

* Electrode systems should be easily disassembled and assembled.

Each electrode is surrounded by a region of steadily decreasing current density which includes a perception zone at large distances, an effective zone at moderate distances, and a danger zone close to the electrode. In general, two types of electrodes such as spherical (rings, square plates) and cylindrical (long thin cylinders and long thin plates) are commonly used in electro-fishing. While increase in the radius of a spherical electrode causes a large, decrease and increase in the danger and effective zones respectively' increased electrode radius for a cylindrical electrode causes a moderate decrease in danger zone and a moderate increase in effective zone. Spherical electrodes generally have superior electrical properties over the cylindrical electrodes but have several mechanical disadvantages.

GUIDELINES FOR ELECTRO-FISHING OPERATION

The following guidelines should be noted at the time of electro-fishing operation.

1. The water conductivity should be measured which helps adjust the voltage source and change the size of electrode to confirm the conductivity to ascertain maximum current densities.

2. Metering devices should be followed to ascertain that adequate currents are being provided.

3. While high current produces maximum effectiveness, 80 per cent of this value generally permits for normal conductivity variations.

4. Large electrodes and low voltage operation are superior. High voltages are used at low water conductivity.

5. The safety switching system can be used and all electrical systems should be in good working conditions.

6. To improve the success of electro-fishing, adquate knowledge on the topography, water areas, habitat, and habit of fish species is inevitable.

* *Alternating Current (AC) Operation*: Though fish stocks are not attracted by A.C., such current has the largest range of action particularly in shallow and clean water bodies having a gravel or sand substrate. Sometime better performance can be obtained at reduced output if fish are being stunned too far from the electrodes. It is operated at high boat speeds.

* *Direct Current (DC) Operation*: Fish stocks are attracted to the anodes without stunning. It is operated at very low boat speeds that permits the fish to swim up to the anode and concentrate there until netted.

* *Pulsed D.C. Operation*: (i) This operation is advantageous when water is turbid with dense vegetation, (ii) Low pulse rates (5-15 pps) or higher pulse rates (40-80-120 pps) are effective because fish are stunned as they move towards the anodes.

Carps, trout, bull heads, and small- as well as large-mouthed bass respond well to higher pulse rates and will approach close to the anodes before being stunned. On the other hand, yellow perch, bluegills, walleye, and white as well as yellow bass are easily stunned and lower pulse rates are enough to bring them close to the anodes. Optimal pulse rate is required for each species, however.

22.10 Preservation of Gears

Gears are made of either natural or synthetic fibre twines. Natural fibres are cotton, hemp, sisal and coir derived from vegetable substances. Cotton fibre is obtained from the seeds of the plant *Gossypium herbaceum*. Hemp is the fibre extracted from the stem of the plant *Crotalaria juncea*. Sisal is the fibre extracted from the leaves of the plant *Agava rigides sisalane*. Coir is prepared from the fibre of the fruits of cocoanut plant, *Cocos nucifera*. All these plants are, however, widely cultivated in many Asian countries.

Synthetic fibre twines are also widely used in fishing industries. These twines are made up of polyamide (nylon), polyester (terene), polypropylene and polyethylene. Now-a-days, nets are generally made in nylon and without knots.

Natural fibre twines when exposed to water for prolonged period are invaded by fungi and bacteria resulting in the loss of strength. Therefore, it is essential to ensure that nets are kept in good condition. For this purpose, natural fibre twines must be treated with preservatives (Panel 22.1) to resist microbial attacks. Nylon nets have several advantages over natural fibres; they are very light in weight, easily handled and easily dried up in the air. At the same time, insects do not cause any damage to the net and have high resistance to abrasions.

After harvesting, the nets must be thoroughly washed with clean water which helps remove mud and other dirty materials. They are then suspended in a shaded place for drying in the air (Figure 22.10). Nets should be protected from strong sunlight, rain and rodents. Rodents are very dangerous which completely or partially destroy the nets.

PANEL - 22.1

USE OF PRESERVATIVES FOR GEAR PRESERVATION

The preservation of gears is performed by using a number of preservatives which are extracted from vegetable materials. Vegetable extracts are called tannins and are obtained from the bark of *Odina wodier* (Kalasam), *Acacia arabica* (Babul), nuts of *Terminalis chebula* (Myrobalan), and *Diospyros embryopteris* (Panachikka fruits).

The preparation of tannins involves crushing of plant barks seeds, and fruits. After crushing, the material is boiled with water in a metal vessel till a concentrated solution is obtained. The residue is also squeezed, liquid portion is then collected and filtered. The net is dipped in the solution for a day or two and dried in the sun. If necessary, a second dipping is performed. In many cases, tannins are fixed by immersing the net in 1 per cent solution of ammonia and copper sulfate. The net is further strengthened by dipping it in coal-tar or in combination with creosote and coal-tar.

Chemical preservatives such as coal-tar, copper sulfate, ammonia, kerosine oil, brown cuprinol, B. C. green cuprinol, cunimine, etc. are also used for preservation of fishing gears. Treatment essentially involves dipping of gears in chemicals for 30 minutes and drying them in the sun. The use of only chemicals is highly effective in protecting the strength of nets for a long period. Coal-tar is by far the best preservative and hence widely used by Indian farmers. In some situations, however, both chemicals and tannins are also used in various combinations.

Fig. 22.10 : A net is considered as an inseparable part of farmer's life. After harvesting the fish, the net is thoroughly washed with clean water. It is then suspended on bamboo poles for drying in the sun or in a shaded place. But drying on the ground is not recommended because a number of minute apertures that develop in nets when vehicles are moved or walking over the nets. Appearance of apperture s drastically reduce the longevity of nets. When nets are not used, they should be stored in clean and dry conditions.

22.11 Harvesting of Some Species of Fish

In most tropical and sub-tropical countries where many freshwater fish of commercial importance abound, inland water resources have enormous harvesting potential. The techniques of fish harvesting followed by farmers vary not only from country to country but also from region to region. Harvesting techniques are also vary depending on the type of fish species reared. In this section, harvesting of some common freshwater food fishes has been noted in a general way.

The harvesting of fish generally performed at the termination of the draining; of course, if the depth of the pond is low. Utmost care is taken in capturing delicate fish (such as fry and broodfish). At the time of draining, fish may be kept temporarily in ponds or tanks until restocked or being sold. These tanks are supplied with freshwater. If such fish stocks are transferred to the market or from these tanks to other tanks/ponds, it is necessary to keep them in hanging nets. In hanging nets, fish are allowed themselves to clean in the tanks and remove the mud covering from their body and gills. Cleaned fish are more attractive to consumers.

HARVESTING OF CARP

Harvesting of carps such as Indian major carps, common carp, grass carp, silver carp, and big-head carp is performed from culture ponds round the year. For small catches, cast nets are generally employed whereas in commercial catches, large drag nets are used. Harvesting is done in the early morning when the dissolved oxygen level drops below 3 mg/l. Nets of different mesh sizes are used. Mesh size depends on the type of culture. In nursery ponds, fine meshed drag net is used for transferring fry to rearing ponds. Fry should not be harvested on a bright,cloudy or sunny day as heat or oxygen depletion may cause heavy mortality. Large mesh-sized drag nets are used for harvesting fingerling ponds. Fingerlings are either harvested to transfer to the fattening ponds or to the market.

HARVESTING OF TILAPIA

Harvesting of tilapia depends very much on the seed stock used and the climatic conditions in the region. If mono-sex male fish are cultured in tropical ponds, the duration of rearing can be adjusted according to the selected size for the market. If the hybrids or sorted stocks along with some females or if unsorted stocks are cultured, harvesting is not a very arduous task. Partial harvesting is done by seining, but differences in catches between hybrids and species are significant. The specie *Tilapia nilotica* and *T. aurea* avoid seines by placing on their side on the pond bottom and hence repeated seining are necessary if substantial quantities of the stock are to be harvested. In contrast to this, the species *T. hornorum* and. *T mossambicus* can be easily harvested.

HARVESTING OF CATFISH AND MURRELS

Seining and draining methods are employed for harvesting catfishes such as *Anabas testudineus, Clarias batrachus, Heteropneustes fossilis, Mystus* spp. *Channa* spp and *Ictalurus* sp. (Channel catfish) from ponds.

Since these fishes are cultured in small ponds under Indian conditions, draining is an effective method adopted by farmers since it allows better management of pond soils; but in ponds where refilling requires pumping, it involves an additional cost. There also appear some practical differences in synchronizing the termination of dewatering with the delivery time of fish. Seining has an advantage in this respect and it also permits partial harvesting.

Harvesting of these species is very arduous and therefore farmers prefer to rear them either in cages or in cement cisterns. This type of culture facilitates easy harvesting of the entire stock.

HARVESTING OF EEL

Repeated harvesting and stocking of eel at regular intervals are practiced in eel culture. Generally, the capture of eel is based on the trapping method using a knowledge of the eel ecology and biology. However, for capture of table-sized eel, a scoop net is used in the feeding areas where fish are assemblaged at the usual feeding time. Before harvesting, feeding schedule is withdrawn so that the eels assemblage at the feeding spot. At the time of harvesting, freshwater is introduced to avoid a reduction in oxygen levels. A seine net is hauled from deeper part of the pond towards the inlet, and finally the harvest is removed by small dip nets. Seining is repeated several times which helps aerate the pond. Harvesting is done in summer months.

In winter months, complete harvesting is done by draining the ponds. Generally draining is performed in the moming when the eels will not burrow into the bottom soil and will swim out. Buried eels in the mud can be provoked by using a T-shaped wooden scraper.

It should be added as a note that it is very profitable to sale the harvested fish in live and fresh conditions in the market and therefore, harvesting techniques have to be adopted for capturing them without injury as far as possible.

(i) *Harvesting of Eel by Pots*: A cylinder-shaped pot is used to catch eel in several waters. The entrails of fish are put in the cylinder as a bait before it is lowered to the bottom. After 1 or 2 days, the cylinder is hauled again and the catch is removed. The fishing operation involves raising the cylinders from the bottom by means of a line hauler, removing the lid, holding the funnel in place, taking out catch, keeping new bait, replacing the funnel and lid, and stacking the cylinders on deck. When a full long line is hauled, the line is taken to a new ground for re-laying.

The pot is made up of an anti-deteriorant-treated polypropylene netting stretched over a stainless steel rod frame (318 cm diameter). The front of the pot is fitted with a funnel-function entrance while the back is equipped with a bag opening for removing the catch. On the sides and top of the pot, moss is stuck into the netting to darken the interior of the pot. The contrast of light and shadow created by moss is effective in attracting eel to the pot. In a standard-sized pot, as much as 11 kg of eel have been caught. Eel and freshwater bream fish are caught by this method.

HARVESTING OF SHRIMP/PRAWN AND CRAB

Shallow waters are important habitats for finfish and shellfish. In many countries of the world, these species are harvested by pot method. The diameter of the front and back ends of a shrimp pot measures about 80 and 60 centimetres, respectively. The length of the pot measures about 50 cm. Similarly, the front and back ends of a crab pot measures about 140 and 85 cm, respectively, and the length of the pot vary between 70 and 120 cm. A brown netting stained with coal-tar is used for these pots and the standard netting used is made up of either nylon or polyethylene mono-filament. The size of the mesh is 12 cm.

The baiting of the pots in done before the fishing boats leave the port and the pots are stacked on boat at collapsed state. On arriving the fishing ground, the pots are erected and thrown into the water one by one. The pots are hauled again several hours after setting. About 100 pots are attached to one long line and about four lines are set in one operation.

In contrast to gill-net fishing method, pot fishing has two advantages. *First*, fish pot does not damage the catch and the catch is kept alive even if the pot is left in the water for several days and *second*, it requires less hand-labor than freeing each species from the gill-net so that the entire haul can be sold as live condition and bring a higher market price.

(i) *Harvesting by Traps*: Small fish traps are used around the world to catch crab, prawn, shrimp, lobster, eel, and all kinds of reef fish. A small boat using at least 30 traps and landing 30 kg of fish or shellfish per day may make a profitable living. While some small boats set only 5 or 10 traps (or pots), many large modern vessels set hundred traps or more at a time.

(ii) *Harvesting by Divers*: Dive fishing to collect shellfish is a primitive method which is now rarely practised. It is practised only a few places in the world. They use simple fishing gear (shellfish rake) and collect shellfish especially crabs, clams and edible oysters. Both make and female divers are actively involved in this type of fishing.

22.12 Conclusion

Harvesting technology is a multifaceted activity that involves fishing grounds, crafts, gears, and labor and have a very strong influence and limited by the harvesting capabilities of the farmers. In most Asian countries, fishing crafts and gears are so traditional that it is not always possible to catch fish in large quantities. In order to get the advantage of greater catch per unit time of harvesting and to find out and capture of fish stocks with certainty, fishery scientists are actively involved in search of designing gears, crafts and other harvesting-related equipments.

Most of the harvesting methods followed by many countries are traditional and such fishing is performed in water bodies where the abundance of fish is very low. But in cases where water bodies are large and are teemed with a variety of fish species, commercial fishing should be executed through mechanization of fishing boats for long hours of continuous operation. The development of new systems of mechanization would definitely have greater fishing power. If it is feasible to construct units which could successfully look for control and capture of fish below the surface water and to operate them for fishing, it would be possible to harvest large quantities of fish if sufficient quantities are present.

Mechanized fishing crafts are deployed mainly for (i) the capacity to move over an extensive water area, (ii) the capability to detect fishing grounds, (iii) the ability to execute fishing operation, and (iv) the capacity to transport the catch. Depending on the depth of harvest, however, harvesting is done by crafts alloted permanently-anchored factory ships or by automated underwater vehicles operated from ship or shore regions. These equipments not only make it possible to catch more fish at lower cost and with less effort, but also open the track to the exploitation of new fishing grounds.

Reference

Biswas, K. P. 1996. Harvesting Aquatic Resources. Daya Publishing House, New Delhi, India.

Questions

1.   What will be the harmful effects if excessive harvesting of fish is performed?

2.   State how the harvesting grounds are located.

3.   How fish are harvested? What steps should be taken at the time of harvesting? How ponds are dewatered at the time of fish harvesting?

4.   Discuss briefly how fish are harvested from perennial water bodies.

5.   Discuss different methods by which fish are harvested from fresh-water resources.

6.   Why different types of fishing gears and crafts are used in different water bodies? Discuss.

7.   What steps should be taken to improve the fishing from large water bodies?

8.   Discuss how fishing gears are preserved. Why it is so important?

9.   Briefly discuss how carps, tilapia and catfish are harvested from ponds.

10. Write notes on the following:

     (a) Non-conventional methods of fishing, (b) Long line, (c) dip net,

     (d) Seine net, (e) Drag net, (f) cast net, (g) Gill net.

23

Processing and Preservation of Fish

Immediately after harvesting, sorting, and handling, utmost care should be taken to preserve the fish flesh and their products. Since the marketing of harvested fish and fish products in as fresh a state as feasible is the main target in any aquaculture industry, only very simple processing and preservation techniques are employed for aqua products except for those that are considered for export.

23.1 Overview

Harvested fish stocks are more assailable to spoilage than meat and eggs. As a general rule, fish flesh is quickly invaded, digested, and spoiled by the growth and development of micro-organisms which are luxuriant on the skin and in the intestine. Certain enzymes equally contribute to the dissolution and oxidation by atmospheric oxygen is a supplemental process of deterioration.

Consequently, methods of processing and preservation to counteract these processes must have been indispensable to the utilization of fish as food. The principal techniques of preservation of fish employed to resist bacterial and other forms of spoilage are few in number. Preservation by drying, smoking, and salting with common salt in different combinations is bactericidal or bacteriostatic in varying degree. Use of vinegar, sugar, and certain spices and herbs have a similar effect. Fermentation itself can also be a means of preservation if the quality is subjected to strict regulation and if the process of decomposition is adequately controlled. Cooling by means of ice reduces the multiplication of micro-organisms and hard-freezing interrupts it altogether. All these processes are widely practised in many countries of the world but have their certain characteristic problems and limitations.

By the Middle Age (500-1,500 AD), it is possible to distinguish that certain methods of treatment possessed specific advantages in a given geographical, social, and technological environment. Several processing devices have become fixed over long periods of time. Methods of salting and drying whether alone or in combination are apt examples.

Different methods practised are extremely limited by natural, biological, and physical factors, for example, and have been largely beyond processor's control. Several minor local variations also exist by which a process is adapted to geographical and climatic environment. This seems to account for differences between the processes of salted and dried fish produced in different regions of the world.

When a certain threshold is attained, the old techniques of preservation are often replaced over a short period by new ones which comply with the new plights. Heavily-salted fish which are prevalent among the general public for consumption in certain regions of the world (such as Britain), for example, curtailed sharply in the middle of the 18th century, principally because transport networks permitted large-scale and widespread inland distribution of fish in more robust, lucrative, and fresh condition. By preserving fish in a mouth-watering form, fresh and robust condition altered the popular posture towards it and thus created vast demand which took place during the succeeding century.

Canning which has emerged as a vital method of fish preservation, is now being widely practised around the world; but some objections have been raised whether heating during canning drastically affects the palatatibility of the Product. It is claimed that heating essentially removes the odor, alters the taste and hardens the fibres. On the contrary, canning retains much of its nutrient potential (such as vitamins) and this process requires steam, iron boilers, tin-plate, complicated machinery and the factory system — all of which are part and parcel of the industrial era.

There is no definite single reason that can be cited as instance to account for the longevity value of certain techniques following the rejection of others. Various patterns of mild-smoke-curing that have developed during the industrial era is apt example. The British adopted 'cold-smoked' products to the total omission of 'hot-smoked' ones, cooked in the kiln which are represented at present. On the other hand, over most of the Northern Europe, there is a good preference for hot-smoked products.

The growth of applied research into the technology of fish preservation in the later half of the 20th century is striking. Before the advent of modern science, there were no idea to counteract spoilage of fish by adopting different methods of preservation. Most of the people assumed that the methods adopted were inaccurate and often couched in elegant-semi-spiritually symbolic language of a type which is still to be met with in the older methods. But when the nature and function of micro-organisms and spoilage of food substances were fully understood, preservation experts had been compelled to develop certain methods by which fish could be preserved without any deterioration. The belief was that the oxygen in air was hostile to preservation.

The most important factor that should be kept in mind is that the destruction of all micro-organisms by thorough heating is a sine-qua-non of canning industry. Keeping the processed fish products away from micro-organisms and air by means of hermetically-sealed containers, it is possible to prevent sterilized products from subsequent re-infection.

Since the end of the 19th century, various organizations across the world had been devoted considerable attention to the problems of fish processing and preservation. But unfortunately, little effort was directed to throw light on the scientific facts underlying the practical viewpoints of the preservation of fish as food. Due to the technical backwardness of the fish industry and the simple as well as primitive nature of its basic processes of fish preservation, the consciousness of the importance of the application of science and technology and its subsequent development in the area concerned was late. But poor financial plights of the aqua-industries have stimulated interest in the feasibility of planned technological advancement investigated through research. At the same time, scientists have been realized that the improvement of aqua-industry lies by and large in the study and application of best possible state-of-the-art methods of fish preservation at a minimum cost.

Rapid change in the aquaculture industry has been one of the aspects of development of the sector concerned. Radical changes have occured in the middle of the 20th century when several factors were combined together and influenced practically all the elements of the industry concurrently. Many aqua-industries of the world are right now in the anguishes of such change. The reforms have changed the industry structure in businesses of aqua-products, radically and rapidly. As the various elements that contribute processing industry structure, such as, the investment pattern, processing capacity of the plant, technology profile, etc. were touched by the developments, have resulted in its rehashing in the concerned industries.

Development of quality and zero-defect aqua-products reinforced the change in industry structure. Individual Quick Frozen (IQF) and the radiation processing are the two latest editions in fish processing industries of the world. Successful freezing of fish through IQF has benefited to a large extent by an amalgamation of three widely divergent disciplines of bio physics, bio chemistry and refrigeration engineering. With the high investment capacity and technology clout, however, several processing industries are now producing quality fish products for domestic consumption and export.

23.2 Principles of Fish Preservation

Preservation of fish is the ultimate strategy in aquaculture industry not only for long-term storage but for marketing and export as well. Therefore, sound preservation is necessary that makes the industry more profitable. Adoption of a number of techniques with the principal aim of ascertaining quality products following the principles of preservation, meeting the requirements of the rural and urban populations are obviously the key step towards preservation.

Preservation may be defined as the management and technology for the utilization and consumption of different varieties of fish and fish products by farmed animals including humankind, so that the preservation may develop sustainable benefit for the present and future generations. Simply defined, preservation refers to the rational use of either fresh or preserved fish and fish products to provide nutrition to humankind. The main principle is definitely to preserve essential nutrients that can be derived from the preservation technology at minimum cost. In certain cases such as Individual Quick Frozen (IQF), preservation is a cost factor if fish farms have to enter the international markets. The ultimate goal of preservation is, therefore, to protect the nutrient contents of fish that helps stop spoilage of fish flesh from bacterial contamination.

Fish preservation necessitates a holistic approach involving the adoption of a number of common strategies and a few recent techniques and the results obtained so far are highly encouraging. The ever-increasing demand for fishery products and the consequent damage to huge quantities of landed fish owing to the lack of adequate preservation facilities has been forced the farmers to follow certain principles relating to preservation. Let us now get acquainted with some of the principles that are very related to preservation.

1. *Cleaning*: The landed fish contains a lot of undesirable substances such as clay, sand and microbes. These substances cause rapid spoilage of fish particularly in the tropics. Moreover, netting causes severe injury to many fishes that results in mild/ severe bleeding of the fish. All these circumstances make the condition of fish more unhygienic. Therefore, cleaning of fish with fresh and unpolluted cold water greatly prevents fish spoilage.

2. *Gutting*: Immediately after cleaning, fishes are cut along with their mid-ventral side and their alimentary canals are removed. Gutting helps remove micro-organisms present in the gastro-intestinal tract and the enzymes of visceral organs. Removal of enzymes and micro-organisms prevent autolysis and decomposition, respectively.

3. *Increasing the Temperature*: Moderate to high temperature decomposes enzymes that initiate autolysis and as a result enzymatic activity is ceased. Moreover, temperature under pressure also destroy micro-organisms that are known to cause spoilage. Therefore, under special conditions, increase in temperature ensures a long-term preservation

4. *Decreasing the Temperature*: Similar to increasing the temperature, decreasing the same up to -40°C destroys several strains of bacteria and at the same time makes them inactive. This controls fish rancidity to a great extent.

5. *Dehydration*: Since moisture content of fish flesh is highly favorable to bacterial growth and multiplication, complete removal of moisture greatly enhances the quality of fish fillet. The duration of fish preservation depends upon the degree of dehydration.

6. *Use of Salt*: Common salt (sodium chloride) is one of the most important chemical substances that brings about the removal of moisture from fish flesh by osmosis followed by the entry of salt into tissues and the salt concentration increased to a saturation point. Salting of fish has, however, lost some of its appeal to the western countries and is also becoming less important as a method of preservation, consequently, salting has greatly reduced the demand for these products in daily diets in the western world.

Salted fish for export has been produced by Canada, Ice-land, Denmark, Norway and other neighbouring countries while France and Portugal are supplementing their growing domestic production. Southern Europe, Latin America, and parts of Africa are the main consuming centres for salted fish.

Reduction in the extent of salting of fish is evidently the growth of freezing operations. A number of quick-freezing plants have been opened for operation in recent years in many countries of the world.

7. *Use of Preservatives*: A number of drugs and chemicals have been found to be effective that prevents fish from spoilage. Some important preservatives include ascorbic acid (prevents rancidity of fatty fishes), chemical ice (ice that contains small amount of sodium nitrite), aureomycin (an antibiotic), phenolic components (such as cresol, pyrogallol, guaicol and catechol) and acetic acid which are bactericidal.

8. *Antibiotics as Preservatives*: For the purpose of fish preservation, the use of some antibiotics at suitable rates may be desirable which helps in curbing spoilage of fish flesh as a result of bacterial and fungal infections in fish consignments. The most commonly used antibiotics for fish preservation include aureomycin, terramycin, chloromycetin, tetracycline, hydroxylamine, etc. Though antibiotics are being widely used in fish preservation, some are known to be highly toxic to human health. several countries have presented guidelines for their use as fish preservatives which are conducive to human health.

9. *Chemical Preservatives* : A number of chemicals are also being widely used as fish preservatives (See Chart 23.A). In early times, boric acid and salicylic acid were used for preservation of Pacific salmon. However, some common and effective chemicals that are now being considered as good preservatives include sodium benzoate, carbon di-oxide, sodium hydrochloride, mixture of sodium acid phosphate (3 per cent) plus sodium benzoate (0.25 per cent), sodium chloride and sodium nitrite, etc. Almost all these chemicals have industrial use in many countries of the world.

23.2 Fish Preservation by Cold

One of the best methods of preservation of dead fish in a fresh condition is the lowering of temperature and thus fish are kept quite a long time. Refrigeration does not permit fish to be preserved for a period of 2-3 weeks. On the other hand, deep freezing helps preserve fish for several months.

In the process of refrigeration, fish are preserved at around 0°C which prevents decomposition for a limited time. The most cheapest, and widely adopted method of fish preservation is the use of ice carried out in suitable containers. The ice and the fish (in 1 : 1 ratio) are mixed in alternate layers in containers (Figure 23.1) which permits the cold water from the melting ice to flow easily through layers of fish. Ice has a very large and rapid cooling capacity for a given volume or weight. In addition, ice will keep the fish cool, moist, glossy, and hence controls deterioration. About 30 per cent of ice by weight of the fish is necessary to bring the temperature down to 0°C.

Fish can be preserved in ice in good condition for 2-10 days depending on the species and size of fish and distance to be travelled at the time of marketing. Fatty fish, such as, sardine has a limited shelf life of 2-4 days in ice. Freshwater fish can be preserved in ice

---

CHART - 23.A
CHEMICAL CONTROL OF MICROBIOLOGICAL DETERIORATION

Since the beginning of the 20th century, a number of preservatives such as salicylic and boric acids, sodium hypochlorite, hydrogen peroxide, benzoic acid and its esters, sulfur dioxide, etc. were used in preserving fish for many years and the keeping quality of whole, eviscerated fish and of fillets could be improved if they were kept in these chemical preservatives. These preservatives were effective in controlling bacterial spoilage of fish. The use of inorganic and organic acids to curb bacterial spoilage of fish has also been recommended. Although organic acids (especially acetic acid) were very useful in preservation of acid-cured and salted fish products, it has been pointed out that they can have little use in successful preservation of fresh fish, since acidification of fish muscles tends to bring the proteins to their insoluble isoelectric points* with a resulting in serious impairment of their water-holding properties.

In the mid 20th century, it was thought that the application of any preservative to fish should be carried out as soon as possible after capture and killing. There are three general methods of application of preservatives to fish which have been routinely employed in actual practice: (i) incorporation in ice used to ice fish, (ii) immersing whole or eviscerated fish in, or spraying them with concentrated solutions of the germicide and then storing them in ordinary ice, and (iii) addition to refrigerated sea water in which fish are stored. Fillets cut from fish may be dipped.

Incorporation of Germicides in Ice Blocks: Germicides can normally be more readily incorporated in rapidly-formed flake-type ice than in block ice. Since most ice additives tend to migrate to form a core at the centre bottom part of ice blocks, the problem of proper distribution of ice additives has often been raised. Although this has not been a problem with benzoic acid ice and certain sulfa compounds, it is certainly true for the tetracycline antibiotics. Although sodium nitrite does tend to move to a core, it is occluded to an extent enough to ensure a fairly even distribution once the block ice is crushed and mixed with fish. Effectiveness of some chemical preservatives in inhibiting bacterial spoilage of fish is shown below:

Compound and Method of Application	Approximate Effectiveness
0.002% sulfanilamide incorporated in ice	Moderate
0.002% sulfathiazole in ice	Moderate
0.1% dehydroacetic acid in fish cakes	Slight
0.7% chloroform solution, dip treatment	Good
1% sodium benziote, dip treatment	Good
0.1% benzoic acid or 0.09% ethyl para-hydroxybenzoite, dip treatment	Slight
0.1% benzoic acid, incorporated in ice	Moderate
0.1% boric acid	Moderate

* Having no net electric charge or potential difference.

Application of Preservatives to Acid-cured, Salted and Smoked Fish: A variety of acid-cured products is prepared from fish. The preservation of these depends on (i) the starting product of good bacteriological quality, (ii) the maintenance of proper chill temperatures during preparation and subsequent storage, (iii) maintenance of a low pH, (iv) pasteurization in the case of certain products, and (v) proper use of certain chemical has prepared an index of harmlessness for several chemical preservatives commonly used in food (such as formic acid, benzoic acid, orthochlorobenzoic acid, parachlorobenzoic acid, parahydroxybenzoic acid, and boric acid; ethyl and propyl parahydroxybenzoites, sulfur di-oxide, hexamethylene-teramine, potassium bromate, etc.). Among different preservatives, however, boric acid, formic acid, sulfur di-oxide, and sorbic acid are on the approved list of preservatives for use in fish products.

Combinations of preservatives are also used in foods and has emphasized the desirability for proper pharmacological tests of each combination of preservatives before any such combination is permitted as a food additive. It is assumed that there is often some synergistic action between certain preservatives, a smaller concentration of a mixture being more effective than a larger concentration of either single preservative alone.

Microbiological spoilage of salted fish can be controlled by the use of preservative mixture (3% sodium acid phosphate plus 0.25% sodium benzoite). Sorbic acid has also proved useful in control of the development of yeasts and molds on lightly-smoked fish.

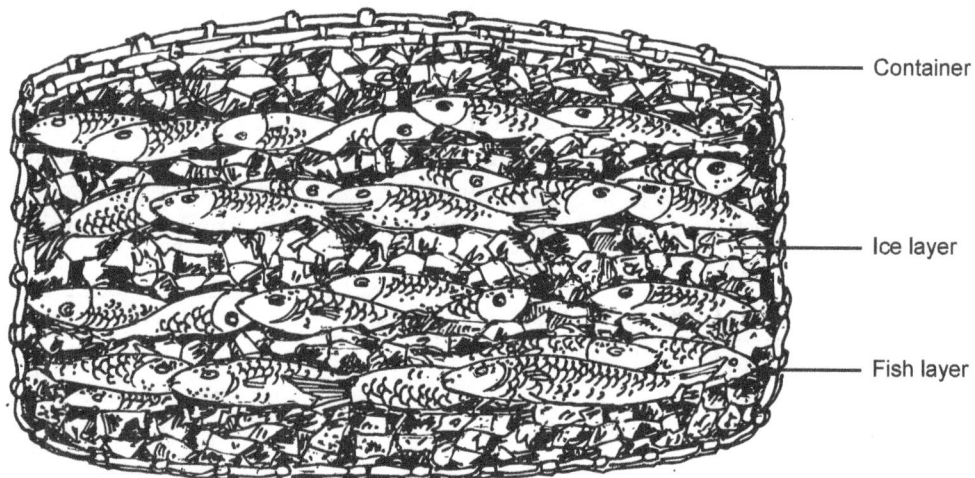

Fig. 23.1 : **Internal view of fish container made of bamboo slits or wood showing alternate layers of fish and crushed ice. Different or same species of fish are placed in layers. Fish layers are kept in cooled condition to prevent the fish stocks from spoilage. It is a very simple and common method of fish preservation for long-distance transport of table-sized fish because it ensures the preservation of fish flesh which is deteriorated if they are not properly packed with ice.**

for longer period compared to marine counterparts. Similarly, tropical fish are kept for prolonged period in ice than the cold water species.

Deep freezing involves the preservation of fish at temperature of at least - 18 °C. This will permit preservation for a long time. Before freezing, the small fish are thoroughly washed with water and the heads, viscera, fins, and gills of larger fish are removed and again washed and are deep frozen. Fresh fish are preserved in good condition.

TYPE AND QUALITY OF ICE

Various types of ice such as block ice, rapid-block ice, flake ice, tube ice, and soft ice are widely used in fish preservation. Each type of ice has its own advantages. Block ice is crushed before icing fish. It provides a large surface area per unit mass and mask more fish for rapid cooling.

Flake ice can also mask large quantities of fish for a given weight. In contrast to crushed ice, flake ice does not cause any rupture of fish during packaging. The benefits of flake ice include: (i) fast cooling rate, (ii) have no sharp edges, (iii) if stored at -8°C, flake ice could not freeze, and (iv) it is easy to process and supply. Soft ice is used in the form of slush which makes the ice excellent contact with the fish. Soft ice does not form air pocket as can occur in other types of ice.

It should be kept in mind that ice has a potential source of contact with fish during preservation. Good condition of fish can be maintained if the ice used is of superior quality. Therefore, ice should be made in ice plant from good quality water. According to prescribed norms, the bacterial range varies between 10^2 and 10^3/g.

23.3 Drying and Salting of Fish

DRYING

Drying is one of the oldest and most prevalent methods of preservation of fish. Non-availability of proper infrastructure for fish preservation at most of the landing centres in rural areas of tropical and sub-tropical countries has forced the farmers to adopt this method. Many countries produce stock fish in bulk quantities which is made from commercially important fish species and is dried only in the sun and wind without proper salting. While small-sized fish are allowed to dry in the sun, larger ones are splitted, salted and then dried in open air. Drying is of two types such as (1) Natural drying and (2) Artificial drying.

1. *Natural Drying* : Drying of fresh fish in the sun is a very simple, cheap, and primitive method of fish preservation which is extensively practised in India and adjacent countries. Sun drying of small- and medium-sized fish involves the elimination of moisture contents of the fish that inhibit the enzymatic autolysis and bacterial decompostion.

Before sun-drying, fishes are thoroughly washed with clean water and then they are suspended in the open air (Figure 23.2). The process has several disadvantages. First, it is

Fig. 23.2 : Drying in the sun is still one of the chief methods of preserving fish throughout the world, especially in Asia and Africa, where much of fish is landed *in toto* from suitable ponds, lakes and rivers and mostly preserved by primitive methods, involving combinations of sun drying, hot smoking, and salting. Whole or splitted fish are suspenced on ropes to accelerate the drying in the sun. In many instances, the process involves solely in taking out the intestines, removing the head and tip of the tail and exposing the fish to the sun. Several hundred tonnes of fish are dried in the sun each year.

a very slow process and results in loss through putrefaction and second, large and fatty fish are not allowed to dry in the sun because such fish are prone to bacterial decomposition which is continued in deeper zone of the body of fish. Sun-drying is, however, highly effective only in dry and well-aerated climatic conditions when the intensity of sunlight is moderate to high.

2. *Artificial Drying*: Artificial drying of fish has been introduced in a simplified form in many Asian and African countries which makes the product superior to the sun-dried fish. Artificial drying machine contains a drying tunnel equipped with fans together with an oven where fish are dried most effectively within 25-36 hours contrasting to at least 15 days needed to dry in the sun during the rainy season. However, this technology needs further improvement to all areas of the world.

Countries like Japan, USA, Portugal, Norway, and Iceland produce huge quantities of dried fish and almost the entire product is exported, the principal markets being Italy and West African territories. Due to the opening up of the extensive markets of the under-nourished in Latin America and Tropical Africa, the demand for dried fish is mounting considerably.

Dried fish products are essential foods for the poor in many under-developed countries, despite their minor commercial importance. These food items are common in the daily diet of the poor in many countries such as Portugal, Italy, Greece, Angola, South and West Africa, the Rhodesias, Mauritius, Madagascar, India, and Bangladesh.

Techniques of improving the primitive fish drying processes are usually   practised in Africa and other tropical regions. Sun-drying of fish is the traditional African methods, but this process largely depends on the vagaries of weather and is not satisfactory during the monsoon season.

* *Use of Heat in the Drying Process*: Fish are thoroughly washed with clean water and then spread on wire flakes contained in compartments traversed by pipes in which steam or hot water is permitted to circulate. After the fish is warmed, currents of cool and dry air is forced over and under the fish on these flakes for at least 3 hours. By alternate heating and cooling, 2 per cent of moisture per hour is removed from the fish. The evaporation of products is carried out from the compartments through suitable ventilators in the roof.

Large farms generally use artificial dryers in which the above-mentioned process is employed or a current of heated air is passed over fish or flakes. Either hot water or hot air or steam boilers are used, together with electrically-operated fans. In this case, however, optimum temperature, relative humidity, and air velocity must receive serious attention while such drying is carried out.

* *Method of Improvement of Sun-dried Fish*: To improve the quality of sun-dried fish during storage, radiation processing is generally adopted which could be applied to semi-dried fishes to control microbial contamination and disinfection.

Sun-dried products are packaged in polyethylene bags. All packets are exposed to gamma radiation dose of 5.0 KGy at 26°C using a $^{60}C_O$ Food Package Irradiator (dose rate 0.05 KGy/ min.). Irradiation at this dose will effectively destroy the bacterial and fungal growth of storage products resulting in appreciable extension in storage life up to 6 months at room temperature.

The technical limitation of drying methods in Tropical Africa and in many Asian countries is the lack of protective packaging. The dried products, if not properly packed, cause insect infestation. Sometimes larvae may account for a substantial portion of the dry matter. Such infestation is not desirable while human health protection is considered and hence deserves special attention.

SALTING

Salting is one of the oldest methods for fish preservation and still it is followed by many oriental countries and the product has good demand in domestic markets. Salted fish which are also produced in some developed countries are exported to developing ones. The main consuming centres for salted fish are located in Latin America and parts of Africa. In many Asian countries, salted fish have experienced almost a boon due to the greatly expanded freshwater fishery activities. The future of this product hinges on a combination of drying and salting, making it one of the best items in the world food market.

Contribution to a reduction in the extent of salting of fish is undoubtedly the growth of freezing operations. A number of quick freezing plants have been developed for operation in countries where export potential is great. More emphasis on freezing and less on salting is being given to a year-round operation.

1. *Types of Salting*: In many oriental countries, three types of salting that follows are adopted.

(i) *Dry Salting Method*: This method is principally based on the forced reduction in water activity of fish to the extent of preventing the activity of enzymes and micro-organisms. This method essentially involves two simultaneous operations such as salting and drying. While salting permits partial removal of moisture that results in the increasing activity of enzymes and micro-organisms, drying helps remove moisture to the maximum extent. Thus, this combined process ensures the product a longer shelf life. Lean fishes are generally used for dry salting.

The processing steps involve (i) the removal of fish fins, gills, scales, and viscera, (ii) washing them with clean chlorinated water (at the rate of 3.0 mg/l), and (iii) salting. Large-sized fish are cut into pieces before salting. For salting of small fish, they are stacked between layers of the salt.

For salting, a fish: salt ratio of 4 : 1 is recommended. Fish are unifomrly rubbed with salt and all cavities are filled with salt. In some cases following the removal of viscera from fish, the abdominal cavity is filled with salt. Fishes are, however, kept in alternate layers with the common salt in a container having solid base and perforations on all sides. All the gaps in between layers of fish are filled with salt. The stacked fish is covered with a wooden plank on which a heavy weight is placed for at least 24 hours. The brine liquid is discarded. The individual fish is then dipped in clean water to remove excess salt from the body. Fishes are then dried in the sun for several days or weeks. The dried products are then packed in polythylene bags.

Dry salted fish products do not lose much of its nutrient potential and the process is less expensive. While moderate salting requires the temperature around 16°C, normal atmospheric conditions for strong salting are highly effective to make the product more lucrative to consumers.

(ii) *Wet Salting Method* : This method (also called *Ratnagiri method*) is very popular in India where fishes are splitted, gutted and washed with clean water. About 33 kilogram of salt is required for each 100 kilogram of fish to be cured. Fifty per cent of the total quantity of salt (16.5 kg) is rubbed on the fish. On the first day, fishes are stacked to a height of 1 m. On the next day, 25 per cent of. the remaining salt is again rubbed, fishes are stacked in such a way that the top fish become the bottom one. On the third day, the remaining salt is applied and the fishes are again stacked. The entire stack is kept for at least 12 days in this condition and then removed for marketing. The product is very moist since it is not dried either in the sun or in the dry and cool shady places.

(iii) *Pickling With Salt and Tamarind* : This method is very popular in peninsular India which involves the gutting, washing, and salting. Common salt with a small piece of Malabar tamarind is thrust into the abdomen of fish. The fish are arranged in layers in a barrel with sprinkling of salt and tamarind between layers. Heavy weights are kept on the top of the pile and the barrel is temporarily closed. After 3-4 days, the brine is allowed to flow off through a small aperture which is situated at the bottom of the barrel. When the layers shrink considerably, they are pressed down and the barrel is again packed with more fish from other barrels to make the shrinkage of fish more effective. Thus the barrel is completely filled with fish. About 45 kg of common salt is necessary to preserve 100 kg of fish. The product is exported to neighbouring countries.

23.4 Smoking and Canning of Fish

Fish for commercial smoking constitute only 10 per cent of the world catch of food fish. Smoking in the Western countries such as UK, Russian Federation, West Germany, United States, The Netherlands, and Poland has lost their importance in the menu chart. Nonetheless, tradition and the desire for diet variation give a clue to the production of smoked fish in the Western Europe. Smoking still plays a significant role in this region. A variety of fish species are smoked.

Many primitive societies consider smoked fish as a delicious food item where considereble quantities of fresh fish are available. Smoked products constitute a staple item in their local diet. Smoking permits fish to be transported and delivered within a much wider range than otherwise would be the case.

Many Asian and south African countries have developed an export market for their smoked fish. They are sold to their fortunate countries. All excess fish not immediately consumed goes into this production.

1. *Methods of Smoking*: Fish are generally hot-smoked or cold-smoked. Hot smoking is carried out in fresh fish or on fish which are rapidly salted in brine. Fish are first gutted, cleaned and brined for about 10 minutes. They are then hot-smoked in smoking kilns. Well-dressed fish are first brined by immersing in a brine solution for 2-5 hours depending on the size of the fish followed by smoking at different temperatures, starting with 35-40°C for an hour, followed by 50°C for another hour and finally for an hour at 75-80°C. Gradual increase in temperature helps uniform drying. The smoked fish are then cooled down to 4-10°C before packaging to prevent the formation of moulds.

Cold smoking of fish involves drying followed by light salting and exposed to a forced draught kiln where temperature of 38°C is maintained throughout the period after which the real smoking starts at a tamperature not exceeding 28°C.

In most cases, medium- or large-sized fish are gutted, heads are removed, cleaned, and the sides are made neat by cutting away unwanted parts from the neck to the tail or the vent. The sides are salted by placing them on a layer of salt. After 10-12 hours, surface salts are washed out. Salting the fillets is also carried out by immersing them in 10 per cent brine solution. Brined or salted fish are dried in cool air and smoked at ambient temperature not exceeding 28°C.

Smoking may be industrial or artisanal. Industrial smoking includes galleries with a special smoking device and a system which forces the smoke to circulate. In case of artisanal smoking, a chimney with a hood, a hearth, and twigs are used for the purpose. A small construction made of stone is also used for smoking of fish.

CANNING

Although both freshwater and marine fishes of commercial importance are equally considered as raw materials for canning industry, more emphasis is given towards canning for marine fish. A large number of freshwater fish in Europe, England, Asia (such as pike, eels, pike-perch, white fish, trout, catfish, etc.) are widely considered because these fishes are very suitable for canning.

Canning is a very complicated process (See Chart 23.B) which essentially involves costly equipments and technical know-how and consequently, the products are very costly and at the same time have lucrative markets in Western countries. However, fresh fish are brought by the cannery industries from the harvesting places, thoroughly washed with fresh cold water to remove sand, clay, and other unwanted substances. Fishes are then gutted, their heads, scales and fins are removed. Large-sized fish are cut into desirable sizes, further washed with water which are followed by dip treatment with brine to remove blood and to provide the proper degree of firmness. Pieces are then exposed to steam to remove adhering substances which are very difficult to remove with water.

The cleaned fishes or pieces of known quantity are kept in hot cans. These open cans are passed through an exhaust chamber that receives steam under pressure. Exhaustion allows the product to expand, removes air or gas bubbles from the containers, and creates a steam atmosphere in the remaining vacant part of the container. Each container is then vacuum sealed. The sealed cans are then subjected to steam treatment under pressure at 7 kg/6. 5 sq. cm and at temperature ranging from 110 to 120°C to prevent the canned products from infection with pathogenic micro-organisrns. Finally, the containers are tested whether the product is zero defect.

Though heating, in some cases, severely/moderately changes the composition and characteristic features of the product, a long-time exposure (8-10 hours) to high temperature (70-80°C) following the cooled air or cold water treatment has been found to be beneficial in averting from these changes.

CHART - 23.B
A flow chart for canning fish from raw material to storage

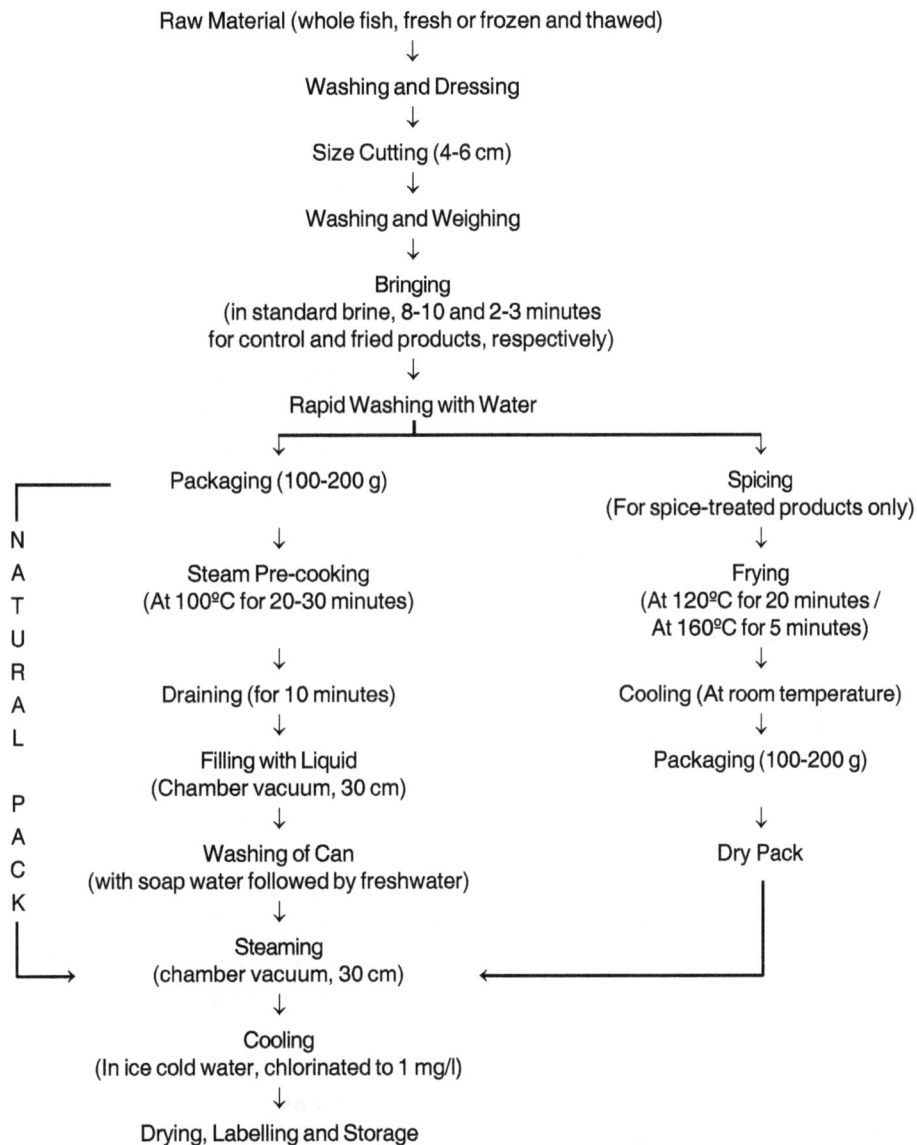

Raw Material (whole fish, fresh or frozen and thawed)
↓
Washing and Dressing
↓
Size Cutting (4-6 cm)
↓
Washing and Weighing
↓
Bringing
(in standard brine, 8-10 and 2-3 minutes
for control and fried products, respectively)
↓
Rapid Washing with Water

Left branch (NATURAL PACK):

Packaging (100-200 g)
↓
Steam Pre-cooking
(At 100ºC for 20-30 minutes)
↓
Draining (for 10 minutes)
↓
Filling with Liquid
(Chamber vacuum, 30 cm)
↓
Washing of Can
(with soap water followed by freshwater)
↓
Steaming
(chamber vacuum, 30 cm)
↓
Cooling
(In ice cold water, chlorinated to 1 mg/l)
↓
Drying, Labelling and Storage

Right branch:

Spicing
(For spice-treated products only)
↓
Frying
(At 120ºC for 20 minutes /
At 160ºC for 5 minutes)
↓
Cooling (At room temperature)
↓
Packaging (100-200 g)
↓
Dry Pack

Although canning procedures have well been standardised for several species of marine fish, freshwater fish, on the other hand, are not generally considered for the purpose. It is hoped that through modification of canning process, it would be feasible to can several species of carp, catfish, tilapia, etc. in the near future involving partial drying followed by smoking prior to canning.

The United States is by far the leading fish-canning nation of the world. About two-thirds of this canned fish is, however, devoted to pet food. The American packs of tuna and salmon constitute about three-sixths of the total world production of canned fish. Second in a rank comes Japan followed by Russian Federation, Germany and others. The continuous demand for canned tuna and salmon has dramatically encouraged the local fishing and canning not only indigenously but also in other countries, producing to cater either frozen or fresh fish as raw materials to the canneries. A medium-sized cannary has a daily output of more than 200 tons of such product.

Next to tuna and salmon, a variety of other species like anchovies, sardines, herring and similar species are also canned particularly in Japan, France, Portugal, the United States, and Canada. These countries have established sardine and herring canning industries. Several small clupeid species are being canned in different Latin American countries (such as Argentina, Peru, and Venezuela), Australia, and Asia.

Canned fish are frequently packed with vegetables and sometimes seasoned with soya sauce and other flavoring matter. Ginger, red pepper and certain condiments are commonly used spices in canned fish, and such products are served as a side dish to a rice meal.

23.5 Drying and Scalding of Fish

In rural areas, a wide variety of freshwater fish such as *Anabus* sp., *Ophiocephalus* spp., *Colisa* sp., *Lepidocephalus* sp., *Mastacembelus* spp., *Puntius* spp., *Ambliopharyngodon mola*, *Macrognathus* sp., *Esomus dandricus*, carp, and tilapia are generally exploited for preparation of various dried and scalded fish products.

PROCESSING METHODS

The processing centres receive the fresh dead fish stocks that vary between 50 and 500 kg per day. In general, large-sized fish are roasted. The processing activity is seasonal since it lasts for 8 months (september to April). The entire consignment is segregated species-wise. Fishes are then spread on the iron net (4 X 4 square feet in size), fixed to a wooden frame with four handles. Fire is placed below the net with the help of dry jute sticks or wood. After 10-15 minutes when one side is dried, another iron net is placed over it and is immediately over-turned for drying the other sides of fish. It takes about half an hour to make the drying process effective. The dried fishes are then spread over a long iron net for cooling, after which fishes are placed in a smoking chamber to remove moisture. The products are then allowed to dry in the sun to remove the excess moisture. Finished products are then packed in plastic or jute bags for marketing. In most Asian countries, the products are sold at Rs. 120.00-170.00 per kilogram. In order to produce 1 kg of dried fish, about 3 kg of fresh fish is required. Moreover, one processing unit can produce daily about 100 kg of dried fish products. The product has good taste and hence market value.

Since these products have good market potential, production of export- related products requires sophisticated methodology to improve the products hygienic condition. In addition to this, quality control measures have to be adopted and at the same time packaging methods should also be improved to attract the buyers.

23.6 Loss of Nutrients During Processing

Losses which occur in fish fillets and whole fish are due to the formation of free liquid or drip. This loss occurs in unfrozen fish, but is certainly greatly accentuated by freezing and subsequent thawing. Light brining of fish fillets has been found to prevent the formation of drip, but this brining procedure accelerates the development of oxidative rancidity in frozen fish. Lightly-brined fillets should, therefore, be preserved under conditions which are designed to minimize the development of oxidative rancidity.

Frozen fish are susceptible to a number of undesirable alterations which are characterized by physical desiccation, denaturation of the muscle protein, and development of oxidative rancidity. Nutritive value of fish may either be unaltered or altered to a negligible extent by prolonged storage in the frozen state.

Development of oxidative rancidity has an adverse effect on fish, since oxidized fats are known to have a marked destructive effect on certain of the essential growth factors such as biotin, vitamins A and E. They may also exert a direct toxic action and destroy certain enzyme systems.

Desiccation of frozen fish in cold storage rooms is a serious problem with ice-glazed unwrapped fish. The important effects on nutritive value are entirely indirect, since loss of surface water is accompanied by undesirable external appearance such as loss of weight and 'freezer burn'. Proper design of cold storage rooms is the best method by which desiccation is prevented or minimized. In specially-designed cold storage rooms, a high relative humidity is maintained. Desiccation can also be prevented following good glazing procedures. A very large number of inorganic and organic materials are used as glazing substances which has non-cracking or reducing potential. Each fish or block of frozen fish is glazed to prevent oxidation.

Salting and smoking are generally accompanied by protein losses that varies from 6 to 30 per cent. Excessive heating of fish in an air-flow drying apparatus may decrease the potentiality of amino acids in fish flesh.

Cooling of fish products below freezing point causes water to leave the fish flesh and freezes out has higher salt concentration. The range of temperature between -6 and -20° C is the zone of formation of maximum ice crystal. If fish are preserved at very low temperatures, dehydration of muscle is greater resulting in maximum damage to the product.

23.7 Individual Quick Freezing (IQF) of Fish

In the preceding pages, we have brought out in a capsule the different methods of fish preservation. The importance has several implications particularly in rural areas where fishes are transported to short distances or preserved for a few hours or days. When fishes are exported, a series of important steps are taken into consideration to make the product attractive to buyers. These steps are carried out,by a well-knit technology by which fishes are individually frozen at -35° or even -40°C. Generally high-value species such as shrimp, prawn, lobsters, eel, salmon, trout, tilapia, etc. are preserved by the following methods and this is predominantly based on export-related industry.

PRINCIPLES OF IQF

Different items of aqua-products are individually wrapped in polyethylene sheets and frozen in blast/funnel or shelf freezers. The polyethylene wrapper acts as water vapor barrier which helps retain the moisture content of fairly intact at their original levels. These products are bulked packed and exported. Individual wrapping of small-sized fish is not economical and if they are frozen in the conventional freezers without wrapping, their moisture content is lost considerably, thus adversely affect the textural properties of the frozen products. For this reason, the IQF technique of aqua-products are now being widely used for the development of export market and hence the product has global acceptance in developed countries in view of the increased freezer cabinets available at homes and catering establishments.

TYPES OF IQF

1. *Air-blast Freezer*: In this type of freezer, fishes are kept in a single layer on a conveyor belt which moves continuously counter to the air-flow at -35°C maintained at the velocity of 10 metre/second. To reduce the length of the freezer, double or triple freezers are designed. The speed of the conveyor belt can be adjusted to suit the freezing time and type of the product. Freezing capacity of 100 kg/hour is available and freezing rates of 5 to 30 minutes per hour is acceptable in this freezer. These freezers ere suitable for non-sticky products since sticky products tend to stick to the surface of the belt when frozen superficially and transfer of products from one belt to another which leads to loss of the product weight. In continuous freezers, however, stainless steel mesh or chain-link belts or plastic belts with interlocking links are generally used. While selecting different conveyor systems of freezing, some important points would have to be kept in mind: (i) they should be flexible, (ii) they should be suitable for direct contact to fish, (iii) they should cause minimum damage and (iv) they should allow rapid transfer of heat from the product.

2. *Fluidized and Semi-Fluidized Freezing*: The term' fluidization' means inducing a semi-weightlessness in the product by suspending it in air at such a velocity and pressure as to assure its rapid freezing. In this system of freezing, however, fish are frozen in a tray. The bottom of the tray is perforated through which cold air is passed under pressure. Due to the upward lifting action of the air, the product behaves like a fluid and is sustained through the length of the freezer for freezing the product and the product is claimed to cause less weight loss during freezing.

In a semi-fluidized freezer (Figure 23.3), conveyor belt system is used for freezing the fish in the initial stages that helps prevent fish from sticking either with each other or to the conveyor belt. High speed cooled air is blown from below the belt. When the product

Raw fish in semi-fried state

First conveyer belt

High speed air
blast

Air blast

IQF Fish fillets   Second conveyer belt

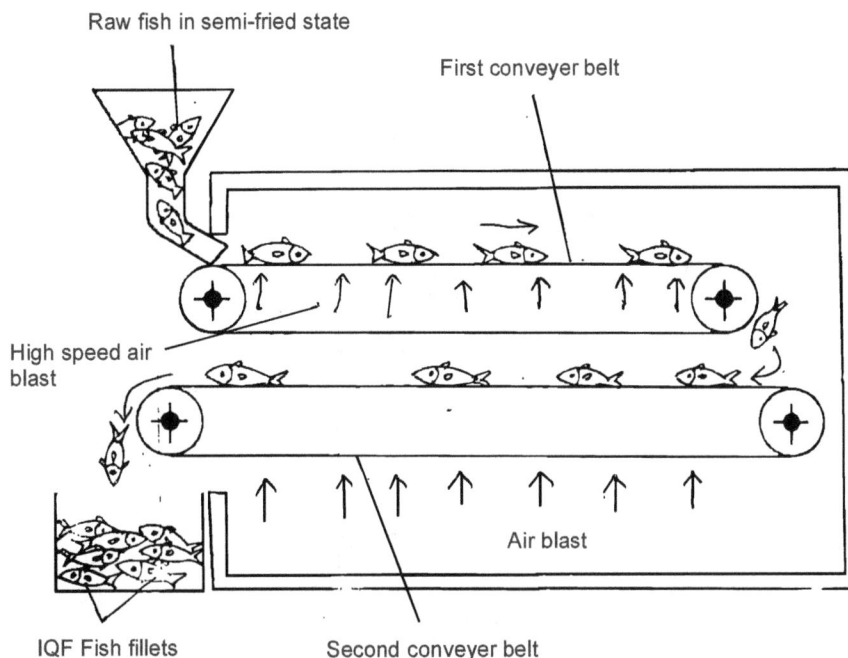

Fig. 23.3 : Diagramatic representation of semi fluidized air blast freezer.

is hardened, they are transferred to a second belt for further freezing. In this type of freezing, the IQF products are superior in quality, yields increased due to loss of body fluids and there is no shrinkage in the shape of products. Under this freezing method, shrimp products that vary in number between 25 and 500 per kilogram are frozen within 8 minutes and those of fish fillets within 15-20 minutes. The processed products are packaged in consumer-friendly packages for marketing.

3. *Surface Contact Freezer*: In this type of freezer, fish are frozen in contact with refrigerated surfaces such as plates, drums or belts. In the belt type freezer, the fish are frozen in the top surface of a stainless steel band, the underside of which is sprayed with refrigerated brine (Figure 23.4). The time of freezing can be reduced further by installing an air blast freezer system to cool the upper surface.

In the drum type freezer (Figure 23.5), a revolving drum with a refrigerant inside operates in an insulated housing. The wet fish is fed to the lower surface of the drum when it sticks to the cold surface of the drum. As the drum revolves at the suitable speed, (the process of freezing is completed within one revolution of the drum) the fish gets frozen and is removed from the upper by a doctor blade adjacent to the drum.

4. *Liquid Nitrogen Freezing*: In this type of freezing, the fish is passed on a continuous stainless steel wire mesh conveyor belt running through an insulated refrigerated tunnel of length of about 10 m (Figure 23.6). The tunnel consists of three zones as follows.

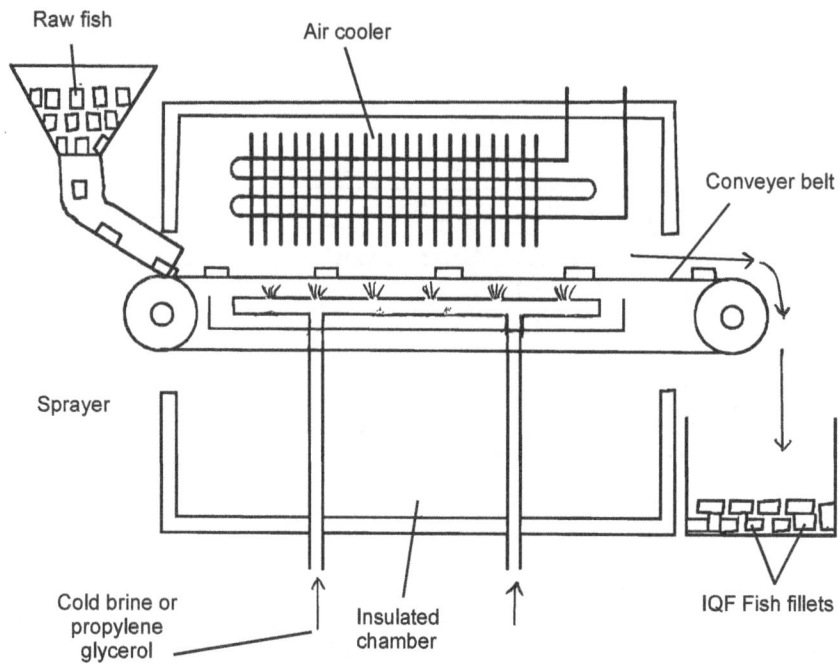

Fig. 23.4 : Diagramatic representation of surface contact freezer (Belt type)

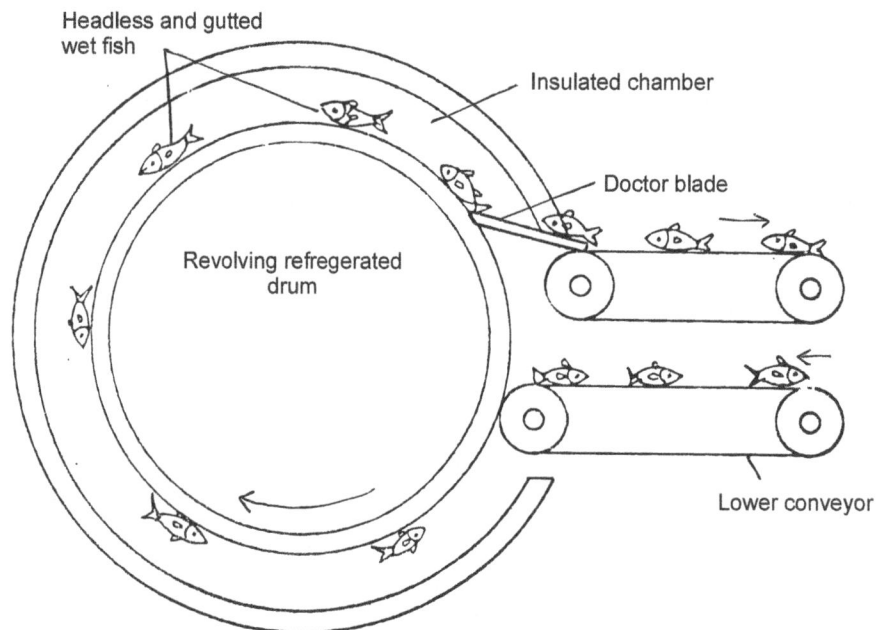

Fig. 23.5 : Diagramatic representation of surface contact freezer (Drum type)

Raw, headless and gutted fish

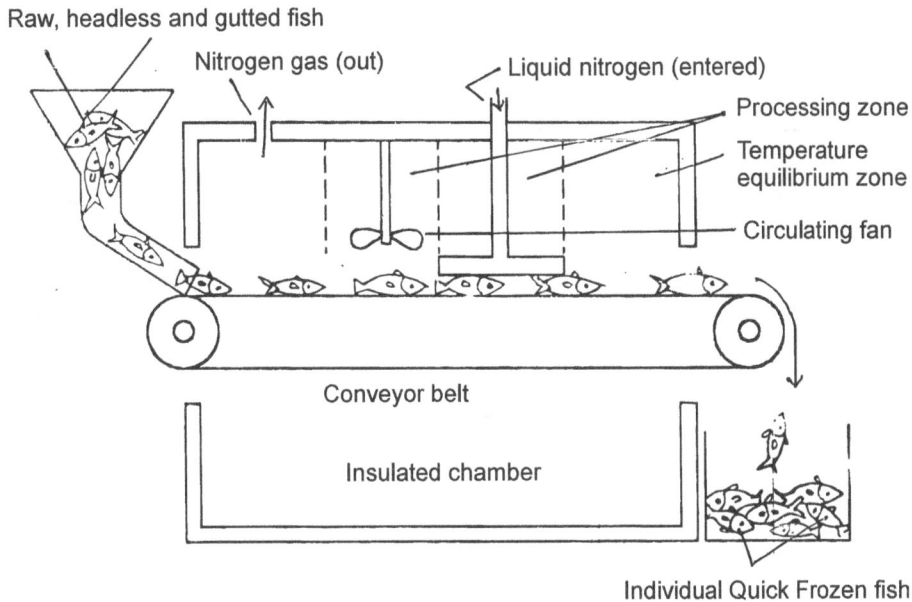

Fig. 23.6 : Diagramatic representation of liquid nitrogen freezer.

(i) *Pre-cooling Zone*: In this zone, gaseous nitrogen flows counter-current to the flow of fish by means of high speed circulating fans. The fans extract the maximum refrigeration energy from the fish. An exhaust fan at the cooling deck removes the spent gas.

(ii) *Freezing Zone*: In this zone, high velocity fine droplets of liquid nitrogen are sprayed through a spray headed on the fish which cause ultra-rapid freezing.

(iii) *Temperature Equilibrium Zone*: This zone is separated by multiple flexible nylon curtains and equipped with counter-rotating high speed circulating fans. This brings the temperature of fish to an uniform level of -40°C before discharging them from the freezer. On the basis of the thickness of the fish and the thermal properties, the speed of conveyor is adjusted from 30 cm to 3.5 m per minute.

Advantages of Liquid Nitrogen Freezing System

The most important advantage of liquid nitrogen freezing is the quick absorption of heat from the fish body due to its low boiling point (-196 °C) and its expansion into gaseous form. The liquid nitrogen expands 650 times of its original volume at the cryogenic temperature. Moreover, the liquid nitrogen is colorless, odorless, sterile and inert and immiscible with the product even while in contact. The liquid nitrogen is easily available at an economic price and at the same time the process is less capital outlay and the maintenance cost is very low.

5. *Immersion Freezing*: In this type of freezing, different refrigerants such as solution of sodium chloride, dichloro difluoro methane (Freon 12) and mixture of glucose

and common salt are used. Twenty three per cent sodium chloride solution is used which has a freezing point of -21°C. At the time of freezing, the cleaned fish/ prawn/shrimp are taken in a nylon net bag or perforated tray and then immersed in the circulating brine with frequent agitation which prevent them from adhering together and the products are frozen at -18°C. Limitations in this process involve the non-attainability of temperature below -18°C as occurred in other freezers and the uptake of salt by the fish which results in flavor and textural changes.

In some other cases the product is immersed in a liquid Freon-12 which is held in an insulated tank when the products get partially frozen. These partially-frozen products are then transferred to a horizontal conveyor belt and freezing is completed with a spray of liquid refrigerant in large amounts. Though this type of freezing has the advantage of rapid freezing as in liquid nitrogen freezers, this type of freezing is not approved in some countries. It is, however, an expensive process compared to air blast freezing in view of the high losses of the refrigerant.

23.8 Glazing

The term glaze refers to vitreous or any substance fused on to the surface of the product to form an impervious attractive coating. In the most sophisticated air blast freezers, built-in arrangements are provided for glazing the frozen products with spray of cold water near about the exit point of the conveyor belt. In other types of freezers, the frozen products are taken in wire mesh baskets, immersed in ice water for a few seconds, lifted out and shaken in the basket to prevent the products from sticking together. The entire process is repeated until the required amount of glaze gets deposited on the product. Some importing countries urge on up to 30 per cent glaze for the stability/safety of the product during storage and transportation.

23.9 Packaging of the IQF Products

Some farms pack the IQF finfish/shellfish in bulk quantities of 10-20 kg in polythene bags and further in corrugated fibre board master cartons. The importers repack the product into consumer pack, with their own brand names and then excepting for the higher prices offered, the identity of the country of origin of the product goes into difficulty. Therefore, it is necessary to resort to consumer packages of 100/200 g by first sealing then in polythene bags which have low oxygen and water vapor transmission rates and mechanical stability at low temperatures. These, in turn, are kept in attractively printed cartons of wax or poly-coated duplex board, stacked in wax or poly-coated fibre board master cartons up to gross weights of 30 kg, fastened with synthetic straps and the packs are stored at -30 °C until delivery.

Since IQF products have larger surface area facilitating quick thawing and adhering of individual pieces, in-line packaging system is desirable for automatic filling and sealing operations. In some developed countries, shutter type packaging system is adopted where individual pieces are separated by an inter-leaving wax paper or plastic flim to prevent adhesion. Some of them are also flexible, laminated, and printed bags and tray type cartons with heat-sealing lid. Polyester trays for pre-cooked products capable of withstanding temperatures upto 140°C are also employed. Automatic file and heat-seal machines are commercially available with capacities of 200 per minute.

23.10 Cautions Taken During Processing

Several cautions that follow should be undertaken while harvested fish are taken into consideration for processing.

GENERAL CAUTIONS

1. Immediately after harvesting, the catches must be washed with clean freshwater and mixed with ice in the ratio of 1 : 1 by weight and shelves or boxes and kept in fish holds having adequate drainage arrangements. Generally polythene sheets are used for packing and so arranged that ice-melt water containing blood, slime, dirt, etc. does not trickle over the others. For long-distance transport, vehicles are well-equipped with refregeration equipments and must be pre-cooled to an air temperature of 5°C below before loading begins.

2. The initial processing is carried out in a well-protected shed near the processing plant where clean freshwater is available in adequate quantities. Smooth, cemented, and sloping floors are necessary so that water can easily flow through the drainage system. Floor walls should be rounded to facilitate easy cleaning. The place where fish stocks is received must be separated from the area where grading and packing programs are executed to avoid cross-contamination. Adequate ventilation is necessary in the processing plant; and if necessary, exhaust fans may be installed. Joints for doors, windows and grills for drain pipes should be properly furnished to prevent rodents and insects from entering.

3. Potable and chlorinated (at the level of 10 mg/l) water must be available in adequate quantities for processing. Generally, 1 kg of fish requires about 10 litres of water.

4. Peeling, deveining, gutting, descaling, and other processes should be carried out on wooden tables which are covered with stainless steel or aluminium sheets ,and other non-reacting materials. When the work is finished for the day, all equipments must be washed with detergents followed by disinfection using sodium hypochlorite (at the 100 mg/l rate) for 15-20 minutes before final washing.

5. After processing, it is necessary to rid the processing plant of waste materials at regular intervals.

PERSONAL CAUTIONS

1. Factory staffs and workers should maintain a high standard of cleanliness. All dress materials should be cleaned while on duty.

2. Hands from elbow down must be cleaned with soap followed by disinfectants using chlorine (200 mg/l strength). Likewise, legs should also be washed and disinfected before entering the processing room. For this purpose, a disinfectant bath is made near the self-closing entrance door.

3. Excessive talking, spitting, sneezing, and using tobacco should be avoided.

4. Consumption of any kind of food items must be in a separate room away from processing areas.

5. Medical check-up of factory staffs should be made with degree of regularity and the personnel having communicable diseases and injuries should not allow to enter the processing places.

6. Hands of the workers in the processing areas are susceptible to certain lesions and skin eruptions due to constant contact to shellfish, finfish, water, and ice. Such eruptions cause a burning sensation immediately after contact to soap and chlorine. To avoid this situation, gloves may be used. Use of an effective and antiseptic cream has been found to provide overnight convalescence of the injured skin.

23.11 Other Potential Processing Techniques for Fish Preservation

PRODUCTION OF FROZEN WHOLE FISH

For the production of frozen whole fish, harvested fish stocks are thoroughly washed with fresh and unpolluted water or chlorinated water (2 mg/l) which is then followed by weighing, sorting, again washing, salting, addition of glaze water, freezing, packaging, strapping, sealing, labelling and storing in cold storage at -20°C. Fish are, however, arranged in a standard-sized tray for freezing or duplex cartons lined with polythylene sheets. If one kilogram of fish is weighed, a calculated additional weight (5 to 20 per cent) is added to compensate for the loss of weight of the product during the process of dehydration. When the arrangement of fish is over, glaze water is added in the tray. The trays and cartons are arranged in the contact plate freezer for freezing or the individual fish can be frozen using IQF machine.

Packing is done in master cartons made of corrugated fibre board externally lined with wax. Ready-to-shop master cartons are then strapped and sealed followed by labelling and storing.

PRODUCTION OF SEMI-DRIED AND DRESSED FISH

These product styles are followed for high-value fish species. The successive stages involved in the production of semi-dried and dressed fish include washing (for removal of about 80 per cent bacterial load), weighing, sorting, washing, removal of scales and viscera, and removal of gills. Thus, semi-dried fish are produced which are either frozen in IQF machine or head and fins are removed to produce dressed fish before freezing. The frozen products are packed and then stored.

COOK-CHILL PROCESS FOR LEAN FISH

This process is highly suitable for the production of value-added fishery items. The process that follows is apt for domestic marketing of fish.

Fresh fish are beheaded, eviscerated and steaked. The steaks are washed with clean water which is then followed by immersion in 10 per cent solution of sodium chloride for 2-3 hours at ambient temperature. The steaks are drained and a further dip treatment for 2-3 minutes in sodium tripolyphosphate is followed. The salted steaks are then kept in stainless steel trays and are allowed to cook in steam for 15 minutes. The steaks are then allowed to cool and then packed in polythylene or polypropylene pouches each containing 100g steak. The packed products are stored at less than 40°C. This process, however, offers convenience, freshness and quality to the product.

PRODUCTION OF FISH FILLETS

The term fillet refers to boneless pieces of fish flesh. Two fillets are obtained from a fish and before filleting, each fish is eviscerated. After evisceration, fish is placed on a board. A deep cut (up to the vertebral column) is made from behind the base of the pectoral fin up to the tail along the line of dorsal fin. The cut is given in such a way as to clear the fillet from ribs. The knife is held parallel to the rib bones. Similar cut is given through fin bones. A further cut is given over the edge of the ribs towards the tail. The fillet is then trimmed to remove, if any, fins and belly flap. The process is repeated on opposite side of the same fish.

After removal of the fillet, skinning process is carried out which starts from the tail to the head by using a knife. Skinning fillets are kept in a dip solution that contains polyphosphate (4-8 per cent), ascorbic acid (0.2 per cent), citric acid (0.1 per cent), and butylated hydroxy anisole (0.01 per cent). After dip treatment, fillets are packed in polythylene-lined trays in 0.5,1.0, 2.0 or 5.0 kg blocks. Blocks are frozen at -40°C and then stored at -20°C.

SOUS-VIDE PROCESSING

In this process, the food is prepared, seasoned, vacuum sealed and then cooked at controlled temperatures followed by chilling immediately at below 4°C. The treatment provides the products a shelf life of about 4 weeks. The products are, however, retained in super-market shelves for much shorter duration.

COOK-FREEZE PROCESSING

The technology involves the addition of sauce in frozen fresh raw fillets and cooking the product prior to consumption. In several cases (such as farmed carps and trouts), fish fillets are prepared by pickling in sodium chloride and acetic acid for 7 days which is transferred into a second pickle of the salt, molt vinegar and sugar for 14 days. These products have high demand in markets.

BREADING AND BATTERING

It is a very popular method of fish preservation and is a very tasty product which is produced from diverse finfish and shellfish. The cleaned raw material is pre-dusted with flour or dry butter which is followed by freezing before or after frying. Fried products can be heated in the microwave before consumption.

MODIFIED ATMOSPHERE PACKAGING

In this process, skinless and boneless fish fillets are preserved. The normal air that contains within the package is replaced by a mixture of nitrogen, carbon di-oxide and oxygen in the ratio of 25 : 35 : 40 respectively. This mixture retards the growth of micro-organisms. In cases where fatty fish are processed, oxygen in the pouches is completely replaced by nitrogen. The entire products are stored below 3°C.

PRODUCTION OF MINCED FISH

Minced fish are generally prepared from white-fleshed low fat content of small varieties of fish species. After beheading, gutting, split opening, the material is passed through a meat-bone machine in which bones are separated from the meat. The meat is then chopped in a mincer. The meat is weighed in 500g and made into blocks. The material is frozen at -40°C and then stored at -20°C.

Minced fish are considered as starting material for the production of different formulated products. Fish sausage, fish hamburgers, patties, breaded products, and various paste products (such as fish balls, kamaboko, chikuwa, satsuma-age (cake), hampen and sumae are apt examples.

PRODUCTION OF FISH PASTE

Harvested fish are properly dressed and washed with clean water. Meat is collected either by manually or by using a meat picking machine. In this machine, however, the dressed fish is pressed against a perforated surface to such an extent that only meat is passed through it leaving the bones and skin behind. This meat is termed as *picked meat*.

The picked meat is ruffled in a container containing clean freshwater and then allowed to settle. The supernatent is removed and the process is repeated several times to ascertain that undesirable substances (such as blood, pigments, fat, water-soluble proteins, etc.) have been completely eliminated. The meat is pressed using any suitable press. The meat is packed and stored at 0°C.

The minced meat is well mixed with various food ingredients according to the recipe for the product. This can be done using either a silent cutter or a stone grinder. While latter type is equipped with a set of cutting blades that revolve within the cavity of the bowl, the former one is provided with 2 or 3 pestles that revolve within the cavity of the morter. In both the cases, however, fish meat and food ingredients are ground into a fine and viscous mass, which is termed as *fish paste*. Different ingredients such as common salt, starch, sugar, phosphate, coloring agents, spices, soyabean soil, and egg white are added in suitable proportions to make the product attractive to consumers.

The paste is moulded into desirable shape using hand moulds or machines. The product is heated in different devices such as steaming, broiling, boiling, and frying depending on the type of products manufactured. The products are then cooled, packed by wrapping in a parchment paper and stored in frozen condition.

PRODUCTION OF SURIMI

Since the beginning of the 1980s, advances in the production of surimi and surimi-based products throughout the world have radically changed the feeding habit of a large section of populations. This high-tech material has excited the imaginations of different categories of consumers. As a corollary, nutrient consumption has significantly increased. Interest on this high-tech material has generated among industrial enterprises due principally to the rapid expansion in the popularity of surimi-based products.

Surimi is a Japanese term and is known in Asian dietary from quite a long time ago, but recorded from Japan in 1100 AD and the most popular use of surimi in Asia has been the fish ball or Kamaboko. Simply defined, surimi is a minced fish meat that is mechanically deboned, water washed and mixed with sugar and cryoprotectants to obtain a good shelf life during storage in frozen condition. '

1. *Characteristic Features of Surimi*: The principal features that follows are considered for the development of aqua-product as food.

   * Low-value, less utilized and under-utilized fish species are selected as raw materials.

   * Frozen surimi has a shelf-life of more than one year and contains functional protein ingredients.

   * By changing the appearance and the quality of raw surimi through adoption of processing technologies,. a variety of products can be produced.

   * State-of-the-art technology permits mass production of surimi.

2. *Turning Minced Meat From Fish into Surimi*: Previous discussions remind us that most of the fresh- and marine water fish species are processed into fresh or frozen fillets and whole dressed fish. Other products include steaks, nugget, and value-added products which accounted for about 30 per cent of the total products sold in 1990s. The dress-out yield of fish when processed as fillets is about 55 per cent, generating about 65 per cent by-products. At present, about one million metric tonne of surimi-based products are manufatured annually. These products have excellent market potential in Japan, USA, Korea, and in the European countries.

Most fish processing countries of the world have increasingly shown interest in converting by-products or processing wastes into edible value-added food products. Several processors have investigated the recovery of minced meat from filleted fish frames using a mechanical meat-bone separator. This minced meat can further be processed into surimi. Surimi has little or no flavor of the original fish and hence can be used as an intermediate raw material for the production of various surimi-based products particularly Kamaboko, Chikuwa, Nampen, and Satsuma-Age (See Panel 23.1).

3. *Processing of Surimi*: The surimi manufacturing process essentially involves washing the mince, draining to eliminate excessive water, straining or refining to eliminate remaining skin or bones, screw-pressing for dewatering, mixing the mince with cryoprotectants, packing, and freezing. The estimated production cost for producing surimi from mince recovered from different sources of fish using the standard commercial techniques is probably higher than that of commercial surimi made from Alaska pollock. The process by which surimi is manufactured requires large volumes of water to remove blood, pigments, lipids, and water-soluble proteins. Reducing the water required would not only lower the cost of production but also reduce the space required for waste-water treatment. Lower manufacturing costs would definitely encourage fish processors to further invest and engage actively to extend the manufacture line to include surimi.

(i) *Raw Materials Used For Surimi Production*: Surimi is prepared from low-value white flesh fishes having low fat content. Such flesh has excellent taste. Although many marine water fish species (especially *Trichiurus* spp. (ribbon fish), *Epinephelus* spp., *Nemipterus* spp. (threadfin bream), *Priacanthus* spp. (snapper), *Jobnius* spp., *Otolithus* spp.,*Pennabiar* spp. (croakers), *Sphyraena* spp. (barracudas), sardine, mackerel, and Alaska pollock) are widely used as raw material, fresh- water fish species with good gel-forming ability and white meat with low fat content can be used for surimi production. It should be kept in mind that only a few species of fish particularly Alaska pollock can able to satisfy the demand of world markets.

(ii) *Treatment of Raw Materials*: Treatment process essentially involves beheading, evisceration, cutting out of the vertebral column, and production of boneless fillets. After landing, however, fish are kept at low temperatures. Generally, ice water slurry is apt for lowering the temperature. The time required to lower the temperature depends on the size of the fish used. It takes about 20-30 minutes, for example, to decrease the temperature of fish having 10-20 cm in length from 20 to less than 8°C.

* *Mincing/Meat-Bone Extraction*: Commercial production of surimi requires extensive use of extraction machines by which the meat content of fish is separated from the bones. The machines work on the principles of compressing the fish fillets against a perforated steel plate or screen into the interior of a drum, with mesh size of 3.5-4.0 mm, while leaving bones, cartilage and skin on the exterior. This mesh size helps prevent the skin from passing through the drum. A cooling arrangement in the machine with ice is maintained to reduce the heat generated during operation and also to maintain the minced meat at a low temperature. In addition to this, a sheer force is built to make the extraction of meat more effective.

PANEL - 23.1

Surimi-based Products

As we have discussed in the text, minced meat is converted into raw surimi which is followed by the production of frozen surimi. Through dressing of raw material (fish) followed by water bleaching and dehydration, mincing, grinding with salt, sugar, and monosodium glutamate (MSG), and heat processing, a number of products which are termed as *surimi-based products* are manufactured. While about 90 per cent of surimi-based products are in the form of various types of fish cakes, imitation crab, and other shellfish analogues which are termed as *Kamaboko*, 10 per cent products are represented by fish ham, fish burgers, fish sausage, and fish balls.

A number of surimi-based products are intensively prepared in Japan and in other countries. These are traditional fish paste products. Although paste products were originally prepared from fish, the majority of manufacturer at present prepare them from surimi.

Different special products have in different countries of the world play a far greater role in the effective utilization of available fish resources. Leading in this respect is Japan, with an assortment of fish paste, sausages, and related products such as Katsuobushi (dried product but inferior in taste and flavor), Tsukudani, Tsukemono and Sikokara (seasoned products), Huge quantities of these products are produced industrially. Large-sized manufacturing plants have a daily output of more than 360 tons of such products.

Some of these products have a long historical tradition. Kamaboko is especially mentioned in Japanese documents of 400 years ago. But for a century, did an assortment of Kamaboko products appear on the scene. Any fish available and suitable can be used for making of Kamaboko.

1. *Kamaboko*: Kamaboko is a steamed fish paste. The paste is formed either manually or by a machine into a unique shape on a wooden board. It is made by grinding fish meat and adding salt and sugar as seasoning. The product is finally steamed at 20-40°C for 20-30 minutes depending on the type of raw materials used. After steaming, the product is moulded in varying shapes. Sometimes it is baked and is coated with a colourless substance for good appearance. The steamed product is termed as *mushi kamaboko*.

2. *Chikuwa*: Ground fish meat moulded around the brass rod is rotated when it is conveyed through a long furnace of 18 metres. By passing over a very long furnace for 4-5 minutes, the meat gets broiled. The furnace consists of 3 sections: (i) Initial section (6 metres in length with medium heat), (ii) Central section (6 metres in length with very strong heat), and (iii) Final section (6 metres in length with weak heat).

After broiling, the brass rods are removed and the finished product is wrapped in parchment paper and packed in wooden cases or cartons. The product is termed as

Gelation kamaboko. Generally the length of Chikuwa is 19 cm with outside diameter of 3 cm.

3. *Fish Ball*: Finely-ground fish meat with additives such as starch, sugar, salt, polyphosphate, etc. is shaped into small balls of about 20g. Balls are then boiled in water till the meat coagulates. The product is then cooled under air, fried in oil at 180°C till it develops golden yellow color and then frozen. The product is packed and stored in frozen condition.

4. *Hampen*: Hampen is a product of ground fish meat with crushed rice flour and yam. Seasoned ground fish paste is mixed with 1 per cent rice flour and 5 per cent yam and is kept in wooden circular and square-shaped moulds. The mould is then soaked in water to remove the shaped ground fish paste. It is then boiled in water at 90°C for about 8 minutes till it floats to the surface. The product is then removed from water and then cooled, packed and stored. A rectangular-shaped hampen is about 6cm^2 in length and 300 g in weight.

5. *Satsuma-Age (Fried Fish Cake)*: It is a product of fried kamabokco with different shapes and features. Other products such as vegetables, shrimp, and minced fish are mixed with the surimi paste to make the product delicious. The paste thus prepared is moulded into 50 g pieces of various shapes such as flat, square, rectangular, circular, and triangular. The shaped fish paste is then fried in oil at 180°C till it floats on the surface of oil. They are then cooled, packed, and stored.

The basic compositions of Kamaboko, Chikuwa, and Fish ball in terms of per cent weight are given below.

Product						Per cent weight of fish paste							
	Fish meat	Salt	Starch	Cane sugar	Sweet sake	MSG*	Egg white	Iced water	Pep-per	Cori-ander	Chilli powder	Ginger	Laurel**
Kama-boko	80.0	2.8	4.0 (Potato)	8.0	3.2	1.0	1.0	-	-	-	-	-	-
Chi-kuwa	86.0	3.0	6.0 (Wheat)	-	4.0	1.0	-	-	-	-	-	-	-
Fish ball	80.0	1.8	8.0	1.25	-	0.15	-	8.0	0.40	0.15	0.08	0.09	0.08

* Monosodium glutamate

** Any of various kinds of shrub with dark-green glossy leaves

6. *Fish Burger*: It is a round patty of seasoned minced fish, fried, or grilled and typically served in a bread roll.

7. *Fish Sausage*: It is a short cylindrical tube of minced fish encased in a skin, typically sold as raw and grilled or fried before eating. Generally, low-value and under-utilized white fleshed low or medium fatty fish can be used for the production fish sausage. After separation of the meat from bones and skin either by manually (in the case of

large fish) or by meat picking machine (in the case of small fish). The picked meat is then chopped by mincing machine. Finally, the meat is blended in the silent cutter to a fine paste. At this stage, various additives (See below) are added in suitable amounts.

Composition of fish sausage (30 kilograms raw material)

Composition	Weight	Spices(powder)	Weight
Fish meat	21.0 kg	Pipper	30 g
Fat	1.2 kg	Chilly	120 g
Starch (binder)	2.7 kg	Coriander	90 g
Sugar	450 g	Garlic	30 g
Salt	750 g	Ginger	30 g
Spices	300 g		
Onions (blanched)	300 g		
Sodium ascorbate	6 g		
Sodium succinate	6 g		
Polyphosphate	51 g		
Colour solution			
(such as poncean 4R, carmosine, fast red, etc.)	39 cc		
Furyl furamide	30 g		
Crushed ice	2.787.kg		
Total:	30 kilograms		300 grams

The prepared fish paste is stuffed in synthetic casting, boiled at 90 °C for 1 hour and is cooled at a temperature of 15°C for 20 minutes. The product is reboiled at 100°C for 1 minute to make the surface of the product smooth. Finally, the product is dried and then packed in a cellophane paper.

In order to improve the shelf life of the product at room temperature, Furyl furamide, sorbic acid, ascorbic acid, etc. may be used as preservatives. In cases where preservatives are not used, it can be stored at 2°C for 3 months.

Crushed ice is used to maintain low temperature throughout the process.

Deboned and deskinned fillets provide fresh minced meat since membrane, blood, and other undesirable substances have been completely eliminated.

* *Washing and Dewatering*: Thorough washing of beboned fillets with clear, refrigerated and soft water is one of the most important steps that should be considered in surimi processing industries since it will ascertain to obtain a odorless and colorless product. Some of the associated problems such as color, taste, and

odor are developed in surimi. While two-thirds of minced meat consists of myofibrilar proteins which are essential components in the formation of gel structure, the remaining one-third is composed of blood, fat, and myoglobin which obstruct the final quality of the surimi gel. All these problems can, however, be minimized or eliminated if meat is washed with utmost care. The volume of water required for washing and the number of washing cycles depends very much on the type of washing and washing unit, freshness of fish, and the quality of the surimi to be obtained.

Dewatering process involves the use of calcium chloride and common salt (0.4 per cent) in the minced meat. The salt facilitates the removal of water from the meat. Depending on the scale of production, the process requires very simple equipments (such as pails and nylon mesh), intermediate equipments (such as manual process), and large-sized stainless steel tanks and screw-press. The screw-press is highly suitable when large amounts of meat are considered for dewatering. Washing with sodium bicarbonate has been found to be highly effective in removing the fat from the mince. For batch processing, a hydraulic press is used. The screw-press which generally has 0.5-1.5 mm platform, helps eliminate moisture to the tune of at least 88 per cent and makes the product exactly similar to that of a fish fillet and an excellent raw material for the production of surimi-based products. Since substantial amounts of fine particles are removed when screw-or hydraulic-press is used, dewatering process can be easily carried out.

A decanter centrifuge machine is now widely used which significantly prevent the loss of functional proteins from the minced meat.

* Straining: Straining helps remove the remaining scales, connective tissues, and small bones to obtain a superior quality of surimi. After washing, however, the minced meat is forced through a strainer (mesh size varies from 1.0 to 3.2 mm). The strainer performs uninterrupted filtration of the meat and they are provided with ice jackets to absorb the heat generated during operation.

* *Stabilizing the Meat With Cryoprotectants*: After straining, the minced meat is mixed with the cryoprotectants. A cryoprotectant amalgamate contains sugar, sorbitol, and phosphate (in the form of sodium tripolyphosphate or pyrophosphate). Mixing is performed with the help of a grinder or mixture. These substances help to extend the shelf-life, textur strength, and moisture retention capacity of the product. At present, silent cutters are used because they uniformly distribute cryoprotectants at a rapid rate and less temperature is generated during chopping operation.

* *Improving the color of Surimi*: Appearance plays a critical role in product acceptance. The surimi is thoroughly mixed with titanium dioxide (0.6 per cent). Such treatment has been found to enhance the whiteness of surimi that has good acceptance by consumers. Soyabean oil is commonly used in surimi-based products as a texture modifier, color enhancer or processing aid. Besides these, a few colors such as

carmine[1], paprika[2], annatto[3], beet juice extract[4], caramel[5] etc. are widely used in surimi industries.

* *Flavoring*: Since the taste of surimi-based products is one of the most important factors for consumer acceptance, use of suitable flavors is critical to the success of marketing of the product. Several flavors such as amino acids, peptides and other organic acids are used which provide the purity, good stability, and clear appearance of the product.

* *Use of Salt*: Use of common salt (at the rate of 3 per cent) in particular to the ground fish paste helps increase the gel strength. This is due to the fact that the salt solubilizes myofibrilar proteins which is cardinal for the formation of meat-gel texture.

(iii) *Packaging*:. The minced meat (in the form of 10 kilogram) along with cryoprotectants is wrapped in plastic and then quick frozen in a air-blast or contact plate freezer for about 3 hours or until the core temperature reaches at -25°C.

Surimi-based products are tested with a metal detector and then packed following improved packaging methods such as vacuum and inert gas packaging and are adopted for consumer acceptance. A variety of becteriostatic techniques of the packaged products are also followed to control the growth of micro-organisms. The product can be kept without any deterioration for more than one year.

(iv) *Treatment of Wash Water*: Huge amounts of water is required for the production of surimi. To minimize the cost of water, wash water should be treated for further use by employing counter-current flow technology. In this procedure, clean water is only used to clean the final product. Then this water is recycled and used to clean the next dirtiest product and so on. This procedure typically results in saving of 50 per cent of the water used since water consumption will be reduced by 50 per cent.

The amount of water which is required for fish washing, contains hazardous substances and should not be recycled. Water-soluble proteins should, however, be recovered by ultra-filtration or flocculation which is able to reduce the pollution hazard.

---

1 It is a vidid purple red powder obtained from the dried bodies of the female insect of the genus *Coccus cacci*. The powder is highly stable to heat, light, and oxidation.

2 The color is extracted form the pod of *Capsicum annum* plant that contains three essential naturally-occurring pigments such as beta-carotene, capsanthin and capsorubin. These pigments together produce a bright- to red-orange color.

3 It is extracted from the seeds of the plant *Bixa orellano*. The pigments produce yellow to orange color range which is due to the presence of caretonoids norbixin (water-soluble form) and bixin (oil-soluble form). It is pH sensitive and trasnforms from the yellow-orange to the red color.

4 Beet juice extract contributes to a bluish-red color which is produced by a compound known as *betamin* and is stable at a higher pH range.

5 It is prepared by heating high grade carbohydrates. The pigment is soluble in water and develops a color that range from golden-brown to nearly black.

(v) *Quality Assessment*: Assessment of quality of surimi is one of the most important factors for overseas markets. Quality includes whiteness*, gel strength**, color, and flavor; of course, these are greatly influenced by the type of fish species used, processing methods followed, moisture content and freshness of fish. Organoleptic tests, folding test, and teeth cutting test are some of the additional tests carried out to assess the quality of surimi.

FISH PROTEIN CONCENTRATE (FPC)

The FPC is .a stable product very suitable for human consumption and is prepared from whole raw and non-fatty fish or cooked fish meat. The preparation of FPC involves weighing the raw materials followed by pressing (removal of fins, scales, head, and entails), washing with clean freshwater, meat picking (carried out in a meat picking machine), weighing (picked meat), cooking (for 10-15 minutes), washing with cold water, pressing, mincing, hydrolyzing (using 1N HCl in 1:1 ratio and keeping overnight), press drying at 20°C, treatment with acetic acid (1 per cent), drying at 70°C, and packaging in tin or glass containers.

In case of fatty fish, however, the meat is cooked with an equal quantity of water containing acetic acid (9.5 per cent by volume). The mixture is kept for an hour and the fish oil is skimmed off from the top layer. The slurry is filtered through canvas bag and then pressed to eliminate water. The fat is first removed from pressed cakes by using ethyl alcohol which is followed by the second time removal with a mixture of hexane and alcohol. The defatted mass is pressed in a screw-press and then dried under vacuum, powdered, and packed. The finished product is tasteless, colorless, and odorless and can be kept up to 2-3 years in room temperature with little changes of flavor. The standard chemical composition of FPC (in terms of per cent) is noted below.

1. Moisture 10
2. Protein 67.5
3. Pepsin digestibility 92
4. Lysine 6.5
5. Total lipids 0.75

REFRIGERATED SEA-WATER SYSTEM OF PROCESSING

Refrigerated sea-water (RSW) system is used on-board fishing vessels where fish are maintained at 0° or -1°C. Sea water cooled by mechanical means is circulated through tanks installed on-board fishing vessels. Fishes are kept in the tanks and since they are less denser than the sea water, will float on it. Alternately, fish can be suspended in perforated vessels

---

* It is measured by using a whiteness meter and is expressed as per cent Lovobond and it is compared with the standard whiteness of 93 per cent Lovobond pure whiteness.

** It is expressed in g/cm as the product of the force required to break the sample.

immersed in RSW. After sufficient cooling, fishes are stored. Commotion is provided in the tank by pump, mechanical stirrer or compressed air. Chilling in RSW is more rapid than in melting ice because of the intimate contact between the cooling medium and the fish. In this process, huge quantities of fish are packed without any physical damage to the fish. Uptake of salt by the fish is one of the most significant disadvantages; of course, salt uptake principally depends on the size and species of fish, whether the fish to be preserved is whole or gutted. Since the medium of RSW is more favorable for the growth of micro-organisms to a greater or less degree, some micro-biological problems have been noted in several cases. Of course, fish preserved in RSW have greater firmness and better appearance.

To increase the shelf life of fish held in RSW, carbon di-oxide gas is passed through water which is generally termed as *modified RSW*. Bubbling of carbon di-oxide through water decreases the pH, inhibit bacterial growth and hence increased storage life of fish.

CHILLED SEA-WATER SYSTEM OF PROCESSING

This method involves the mixing of ice with the sea water in the ratio of 1 : 2. The mixture is often termed as *slush ice* which helps fish to preserve for prolonged period of time. This chilled sea water has the capacity to establish an intimate contact between the cooling medium and the fish.

During melting of ice, the salinity of water drops below the normal level which is necessary to adjust the required level by adding sodium chloride salt. This system of preservation is relatively inexpensive and can be operated even by unskilled persons. In contrast to RSW where the water is circulated by pumping to ensure well-mixing and cooling, the chilled sea water system of preservation does not require any pumping device. The movement of the vessel is adequate to ensure enough mixing.

23.12 Use of Enzymes in Fish Processing Industry

Enzymes are organic biocatalysts which have the capability to promote a specific reaction under the chemical condition that prevails on all living organisms. This unique character of enzymes has been exploited to elevate specific chemical conversions in different food processing industries. Traditional processing and preservation of fish in many countries of the world have been carried out by using endogenous enzymes extracted from huge quantities of discarded fish which are not economic to preserve. Gradual mechanization of fishing around the world and subsequent development of fish processing industries have accelerated research on the use of certain specific enzymes for the manufacture of some fish products such as fish feed, fish silage, fish sauce and other potential application of enzymes such as skin removal, 'roe' production, manbrane removal, descaling of fishes, deskinning of tuna and herring, loosening of shells from shrimp body before industrial peeling, pre-treatment of fish skin for collagen production, and enzymatic cleaning of scallops.

The commercial exploitation of protein concentrates derived from cheap pelagic fish and fish offals is a part of biotechnology. The use of biotechnology in fish processing and product development relates to judicial use of hydrolytic enzymes generally termed as *fish proteases*. These are highly soluble in water and consequently can be used to supplement the nutritional value of liquid food-stuffs.

A variety of endogenous fish enzymes such as pepsin, trypsin, chymotrypsin and peptidases are now being widely used for the preparation of raw materials which are directly used for fish processing. Some potential applications of enzyme biotechnology in fish processing industries are summarized below.

FISH SILAGE

It is a liquid product prepared from fish or parts of fish. Generally organic acids such as formic acid, lactic acid, etc. are widely used for the preparation of silage. The preservation effect of organic acids becomes active at higher pH level.

Use of acid can be avoided by adding bacteria, *Lactobacillus* along with sugar to fish or fish wastes. Bacteria convert sugar into lactic acid which preserves the fish and creates favorable conditions for the silage. In addition to the formation of acid, some strains of *Lactobacillus* produve antibiotics which acts as preservatives. The bacilli helps prevent fats from oxidation.

FISH SAUCE

For preparation of sauce, the fish consignment is passed through rigor mortic in the initial stages of fish spoilage. Prior to salting, the initial number of bacteria has been reported to vary between 105 and 112 cfu/g. It has been noted that the bacterial proteinase activity is highly favorable when spoiled flora approximate to 105 cfu/g, however.

In under-developed countries, the preparation of fish sauce by autolysis is very common. It is generally produced through bacterial fermentation which involves a combination of salting and enzyme hydrolysis. The ratio of salt and fish has been calculated as 1:3 and 1:2 for freshwater and marine species, respectively. Fresh and cleaned fish are dumped on a concrete floor where salt is thoroughly mixed with the fish. The salt-fish mixture is then transferred into the fermentation tank or vat and is filled up to 5 cm below the brine. The tank/ vat is then exposed to the sun for 1 to 3 months for ripening. During this process, huge amounts of self-brine is produced and special device is taken to keep the fish in submersed condition. For this Purpose, wooden plank and heavy weight are used. During salting, fish tissues are hydrolyzed by fish gut enzymes. These enzymes provide nutrients for bacterial fermentation. The time required for fermentation varies between 2 and 10 months depending on the type of fish species used. After fermentation, the liquid which is termed as *Nampla*, is drained out and a saturated brine (10 per cent common salt) is added to the residue to extract the left-over soluble matters. The product is filtered and blended with sugar and monosodium glutamate to bring the product a mouth-watering form.

The use of salt in fish sauce production is inevitable because salt helps prevent the growth of putrefactive micro-organisms from fish decaying. Generally, 20 per cent salt is used for the purpose because at this rate of salting, the growth of such micro-organisms is inhibited. It should be added as a note that salt along with strong enzymatic activities reduces the time of fermentation to a great extent.

Current technology of fish sauce production essentially involves the use of strong fish gut enzymes which helps reduce the fermentation time drastically. The more the protein and liquid hydrolyses with compounds having low molecular weights the better the quality of the fish sauce. Several species of bacteria such as *Staphylococcus saprophyticus*, *Micrococcus variance*, *Paracoccus halopenitrificans*, and *Pediococcus halophilus* are known to produce flavor and acid in fish sauce. The principal overseas markets for fish sauce and *Nampla* are Canada, USA, European Union, Australia, Chili, and Morocco.

REMOVAL OF SKIN

Since it is very arduous to remove the skin of fish by mechanical means, several methods have been developed in recent years to remove skin using enzymes. The method of deskinning for example, involves steam treatment at 60°C together with a mixture of proteases and carbohydrases for tunas and skates, use of fish pepsins at low pH for herrings, and use of weak solution containing papin and pepsin for squids.

REMOVAL OF SCALE

During processing and preservation, fish scales are gently removed by spraying enzyme water after incubating in a solution at 0°C.

MEMBRANE REMOVAL

The black membrane of cod's swim bladder, collagen membrane of cod's liver and intestine from scallop muscles are removed by fish proteases.

23.13 Use of Radiation in Fish Preservation*

Irradiation is a process by which fish and fish products are. exposed to gamma rays and X-rays (5 MeV)[1] having shorter wave-length (less than 300 mm) and higher energy that cause ionization by removing electrons from the outer shell of atoms and molecules. Generally ionizing radiation that is emitted by radio-isotopes Cobalt-60 and Caesium-137 are used in fish preservation.

PRINCIPLES OF IONIZING RADIATION

When ionizing radiation is used in food, the water molecules are broken down and free radicals and hydrogen peroxide are generated. Free radicals are highly reactive

---

* For further study, see Venugopal (1999).

1 The energy of all types of radiation is measured in electron volts, eV or normally MeV.

chemical species from which the effects of irradiation are developed. The genetic material of fish pathogenic and spoiled bacteria — the DNA (deoxyribonucleic acid) is completely destroyed by irradiation thus improve food safety, minimize losses and extend the shelf life of products. A comprehensive survey conducted on fishery products have shown that irradiated products are safe for human consumption.

CONTROL OF BIOLOGICAL HAZARDS THROUGH IRRADIATION

The biological hazards in fish include insects, viruses, parasites, and pathogenic micro-organisms. In many Asian countries, millions of peoples are habituated with the consumption of partially-processed fish, partially-cooked or raw fish items. But if aqua-products are to be exported to developed countries, it is necessary to implement the Hazard Analysis and Critical Control Point (HACCP) concept (See Panel 23.2) along with hygienic handling and processing in order to protect the customers against health hazards. The safety issues associated with aquacultural products have recently been highlighted by several international bodies such as WHO (World Health Organization), FAO (Food and Agricultural Organization), Codex Alimentarius Commission, US Food and Drug Administration, International Consultative Group on Food Irradiation, NACA (Network of Aquaculture Centres in Asia-Facific), etc. Fishery products which are allowed exposure to ionizing radiation have good market potential in developed countries. Ionizing radiation is an emerging technology to improve the quality of a variety of food items, however.

It should be noted that the internal market of fish and fishery products depends mostly on low-quality and low-value fish, but high value fish species after processing following standard norms are generally exported to developed countries. In most cases, fishery products are rejected by the buyers of several developed countries such as Japan, Europe, and USA which is often attributed to the presence of pathogenic bacteria, toxic substances, and parasites.

23.14 Salient Features of Radiation Processing of Fishery Products

Radiation processing of fishery products has the following characteristic features.

1. Radiation treatment is a low-energy process.
2. It is a cold process since no rise in temperature of the product occurs.
3. It is a cost-effective process. Different radiation sources used for the purpose are easily available.
4. It causes very few chemical and physical changes than cooking, canning or freezing.
5. Since it is a physical process, there is no residue in the treated products.
6. High penetrating power of the radiation can easily destroy parasites and pathogenic bacteria from the fish body.
7. Ready-to-export frozen block can be treated before shipment.
8. Radiation of shellfish and chilled fish helps extend the shelf life by destroying micro-organisms which are responsible for spoilage.

PANEL- 23.2

HACCP Concept in Fish Products

The HACCP is the abbreviation for the Hazard Analysis and Critical Control Point approach to safety of fish products which was developed by Pill bury Corporation, USA, to provide a means of assuring the safety of food items produced for the space exploration program in the early 1970s. This approach is, however, designed as a system of analyzing in food manufacturing process with a goal of reducing the likelihood of unsafe food being produced. The program generally involves the following steps, such as (1) hazard analysis, (2) identification of critical control point from the hazard analysis step, (3) establishment of critical control point criteria, (4) monitoring the critical control points (control points can be physical, chemical, micro-biological and/or sensory attributes), (5) establishment of protocols for critical control point deviation (whether the product quality deviates from the standard desired), and (6) periodical examination of the product to determine the effectiveness of the program to improve product quality. Critical control points are those steps of production where loss of control of the process could result in an unacceptable loss of product quality. In addition to the development of aqua-product quality, however, the HACCP principles are now being widely used in the manufacture of quality food for domestic animals, finfish, and shellfish. Most of the finfish and shellfish processing industries which produce and export a variety of quality products for human consumption, have been recommended by the government as ISO : 9001-2000 (International Quality Standard) and HACCP (Food Safety) certified organizations.

23.15 Use of Low Dose Gamma Irradiation for Control of Biohazards

The potential application of low dose gamma irradiation for complete destruction of bacteria and parasites is considerable. It has been estimated that the irradiation of any food item upto an overall average dose of 10 KGy* does not cause any health hazards. The dose of radiation generally varies with the type of biohazards. A dose of 1 KGy, for example, eliminates parasites from the product whereas a dose varying between 4 and 6 KGy is necessary to destroy all pathogenic micro-organisms in frozen fish and shellfish. For most of the parasites, the Minimal Effective Dose (MED) for their elimination is in the range of 0.1 - 1.0 KGy.

23.16 Use of Ion Wind in Fish Preservation

Electrically-charged air particles which are blown as ion wind help to destroy the microbes. Charged air particles are generated by a high voltage generator which are discharged from a conductor maintained at higher electrical potential. These negatively-

---

* The SI unit for the measurement of radiation dose is the Gray (Gy). A dose of 1 Gy is the absorption of one Jule of energy per Kg of food.

charged air particles are attracted towards the object which is maintained at positive potential. In this method, high quality and flavored fish are produced and the product is safe for human use.

23.17 Advantages and Disadvantages of Radiation Processing of Aqua Products

ADVANTAGES

1. Though radiation processing involves passing fish products through a radiation field but the food itself never comes in contact with the radioactive substances. Electron beams, X-rays and gamma rays used do not induce any radioactivity in processed foods.

2. Processed food is quite safe for human consumption and does not exhibit any toxicological hazards.

3. Though radiation processing products exhibit very little chemical changes, these changes have not been found to be detrimental to human health.

4. Radiation treatment has very little or no effect on proteins, fats, carbohydrates, and minerals. Vitamins exhibit varied sensitivity to fish products. For example, niacin, vitamin D, and riboflavin are more stable to radiation; while vitamins A, E, C, K, and B_1 are relatively sensitive. According the international expert committee, the radiation processing does not include any nutritional problem in aqua-products.

DISADVANTAGES

Irradiation treatment essentially involves high capital investment and requires a critical minimum capacity for economic operation. Moreover, skilled manpower and enforcement of safety standards are the other pre-requisites.

To sum up, irradiation of farmed and exported fish and fish products can immensely enhance its hygienic quality and helps protect the consumers when these fishes/fish items are consumed as raw, semi-cooked or picked. Several countries still have strong reservations about irradiation since they feel that the irradiated fish products are carcinogenic. But it can undoubtedly be stated that the entire product has been found to be as zero defect. Irradiation must be done under the most stringent control by technically-skilled personnel. The radiation treatment plant may be of mobile type or centrally located near the ponds. The treatment, therefore, can eventually results in an increased consumer confidence in the products and hence enhanced international trade.

23.18 Appraisal of Fish Processing and Conservation

Adequately processed fish and fishery products stand a chance to enter the market throughout the world. At present, the demand for these products is mostly confined to the limited regions. Hence it is important that suitable and advanced technologies must be adopted that will play a vital role in the economic development of any developing country. Recent trends in the development of value-added fish products in several Asian countries including India have given an impetus for the development of export-related marketing. These trends are mainly due to high rate of fish production following adoption of modern technology. However, there is wide scope for producing value-added aqua products in future. Introduction of fish revolution and production of quality fish products to make them appealing globally to diverse consumer segments have reflected in the form of two important variables such as increased demand for various types of products and increased final output.

23.19 Conclusion

Fish are excellent sources of nutrients, vitamins and minerals that can be properly preserved through different methods for human use. Generally high-value finfish and shellfish species are preserved for overseas markets. At the same time, low-value fish species are also preserved for domestic consumption and market.

The use of the state-of-the-art technology for fish preservation can serve two purposes. First, the technology helps retain the fish as fresh as possible with minimum losses in flavor, color, odor, taste, form, nutrititive value, and digestibility of flesh or increase in the quality of fish as a whole and resist them from spoiling. Second, the processed and preserved fish have enormous market potential in many countries of the world. Although the second of these purposes is increasingly pertinent to meet the demand of consumers because huge quantities of fish are captured from natural and culture water bodies, the foreign currency earnings from the sale of quality fish products cannot be disregarded.

The application of simple and low-cost methods for processing and preservation of fish for domestic consumption and export are sound practices that have characterized by successful fish farming strategies through decades. It is not less important to day in view of the high demands of preserved fish needed to feed the ever-increasing human population. Judicious methods of fish preservation should be seriously followed to the extent that retention of the quality of fish will permit.

References

Balachandran, K. K. 2001. Post-Harvest Technology of Fish and Fish Products. Daya Publishing House, Delhi, India.

Chakrabarty, R. *et. al.* 1998. Advances and Priorities of Fisheries Technology. (Eds. Balachandran.*et. al*). Society of Fisheries Technologists (India), Cochin, 274 pp.

Cutting, C. L. 2002. Fish Processing and Preservation. Agrobios (India), Jodhpur.

FAO. 1997. Codex Code for Fish and Fishery Products. FAO, Rome, Italy.

Govindan, T. K. 1984. Handling and Transportation of Fresh Fish in India. Harvest and Post-Harvest of Fish Society of Fisheries Technologists (India), Cochin, 353 pp.

Lanier, T. C. and C. M. Lee. 1992. Surimi Technology. Marcel Dekker Inc., New York.

Lyer, T. S. G. 2000. GMP In fish landing and quality assurance in seafood processing. Society of Fisheries Technologists (India), Cochin.

Martin, A. M. 1994. Fisheries Processing- - Biotechnological Applications. Chapman and Hall, London.

MPEDA (Marine Products Export Development Authority). 1995. A Handbook on Fish Processing, Quality Control and Marketing. MPEDA Kochi, India.

Nambudiri, D. D. 2006. Technology of Fishery Products. Fishing Chimes, Visakhapatnam, India, 140 pp.

Pigott, G. 2001. Aquatic Food Product Technology. Hoaworth Press, London.

Shamasundar, K. 2002. Fish Processing. In: Sustainable Fisheries for Nutritional Security (Ed. T. J. Pandian). National Academy of Agricultural Sciences, New Delhi, India, p. 250-271.

Trim, D. S. and C. A. Curran. 1983. A study of solar and sun-drying of fish in Equador, I. 60. Technical Products Institute, London, 44 pp.

Venugopal, V., S. N. Doke and T. Paul. 1999. Radiation processing to improve the quality of fishery p.roducts. *Critical Rev. Food Sci. Nutri.*, 39 : 351-440.

Vogt, M. M. and J. R. Bot1a. 1990. Advances in Fisheries Technology and Biotechnology for Increased Profitability. Technomic Lancaster, PA.

Questions

1. Distinguish between processing and preservation. What are the principles of preservation?

2. How fish are preserved by cold?

3. Discuss the process by which fish are preserved by drying and salting.

4. Discuss how fish are smoked and canned for preservation.

5. State why fish nutrients are lost during processing.

6. What do you mean by Individual Quick Frozen (IQF) of fish? Mention different types of IQF generally employed in fish processing. ,State why IQF techniques are employed for export of aqua-products.

7. Explain how fish are processed through IQF.

8. Discuss the role of enzymes in fish processing technology.

9. Discuss the importance of irriadiation technology in fish preservation. What are the characteristic features of radiation processing ?

10. Discuss different techniques of fish preservation other than IQF and conventional methods. Why irradiation technology is globally accepted? '

11. Irradiation and IQF techniques are equally important if fish processing industries are improved to produce quality aqua-products for export. Discuss this statement in relation to world fish marketing and trade.

12. Discuss why processing and preservation are so important in fish culture industry.

13. What steps should be taken into consideration during fish processing?

14. Write notes on the following :

(a) IQF, (b) Use of salt and ice in fish preservation, (c) Use of preservatives in fish preservation, (d) Drying, (e) Dry and wet salting, (f) Smoking, (g) Canning, (h) Air-blast freezer, (i) Liquid nitrogen freezing of fish, (j) Immersion freezing, (k) Glazing, (1) Surimi, (m) Sea-water system of processing, (n) Fish silage and sauce, (o) HACCP.

24

Transport of Fish and Fish Seed

Transportation of live and freshly-killed fish as well as fish seed from one place to another is a very important aspect of sustainable fish culture industry. Fish farmers carry hatchlings, fry, fingerlings, and brood fishes for breeding, cultivation, and marketing. Considerable quantities of fish are transported by farmers, small trucks, bi-cycles, three-wheeler van rickshaw fitted with inflated tyres, rail, and aeroplane. For quick and long distance transport of high-value commercial fish species, the last two types of transport are inevitable. Improved and quick transport facilities not only reduce the risk of fish mortality considerably but also encourage the farmers to cultivate fish in their ponds/tanks because the success of fish culture and their marketing largely depends on good transport facilities.

Distribution of different life stages of fish in aluminium/plastic vessles is an old practice in many countries of the tropics and sub-tropics. The system of transport is traditional in these regions and still exists. In developed countries, much improved methods of transport are adopted to reduce the risk of fish mortality. Lack of transport facility in remote areas is a serious threat to fish culture industries around the world.

24.1 Patterns of Fish Seed Trading for Transport

Well-organized trading and transportation systems are in part responsible for the marked expansion of fish culture industries in the growth of rural economics especially in the developing nations. Poor trading facilities, inaccessibility to many remote areas, and traditional transport systems in most of the Asian countries are the main factors responsible for dramatic losses of fish seed at the time when they are available in large numbers. During the past two decades or so, little advances in fish seed trading systems and their transport to different regions have changed this situation. Mortality of fish seed has been reduced to some extent. Several techniques have helped to check the mortality of fish stocks. In countries where fish culture is critical and where the trading as well as transportation systems are entirely not adquate to meet the demand of farmers, the fish culture extends to be among farmer's most dangerous problems.

In India, substantial quantities of hatchlings, fry and fingerlings of carps (such as *Labeo rohita, Catla catla, Cirrhinus mrigala, Labeo bata, L. calbasu, L. gonius,* grass carp, siver carp, common carp, *Pundtius javanicus,* and *Aristichthys mobilis*), catfishes (*Heteropneustes fossilis,*

Clarias batrachus, C. gariepinus and *Mystus* spp.) and murrels are available during both rainy and winter seasons. Though small quantities of these stocking materials (about 20 per cent of the total) are collected by farmers from natural freshwater resources, most of the seeds (80 per cent of the total) come from different local breeding centres. Farmers bring the fish seeds into the market either directly or through small traders. The traders sell the entire consignment to the customers either on advance order or by furnishing a price decided after an agreement concerning sale or purchase. Traders however, sale the consignement to farmers either on cash payment or to their known farmers with an arrangement to pay the cost on instalments. While the hatchlings are sold by volume, fry and fingerlings by weight. Volume is measured using a container called as *bati* which contains a volume of 250 ml and the number of hatchlings contained in the container range from 50,000 to 60,000. Valuable fish seeds (such as *Mystus* spp *Clarias* sp, *Heteropneustes* sp. and prawn juveniles) are sold by counting. It is important to note that the extent of fish seed trading strategy varies in different seasons of the year and the strategy is noticeable in rainy season. Unless transport systems are improved, the fish trading network will never provide ample opportunities to farmers and at the same time they will lose interest in their farming activities.

24.2 Methods of Fish Transport

Fish have a range of environmental needs which are changed depending on their life stages. For optimum survival of fish during transport, these environmental needs must be satisfied. When fish are confronted with an environmental challenge during transport, such as low oxygen or high temperature, they normally escape to more element conditions. This discretion is at best partially available to fish in transport. Catfish, for example, prefer to remain at the bottom of the container of low oxygen during transport. As a generalization, fish must be transported within their range of environmental tolerance. However, during transport system certain environmental factors such as temperature, pH, dissolved oxygen, and carbon di-oxide play an important role that determine the survival and activity of fishes in the container. Two methods of transport of live fish and fish seeds are generally followed such as open system and closed system.

OPEN SYSTEM

In this system of transport, hatchlings, fry, and fingerlings are carried out in earthen aluminium vessels known as *hundies*. One farmer carries two hundies on his shoulders (Figure 24.1). Two-thirds of each hundy are filled with fresh unpolluted pond water and then loaded with requisite quantities of fish seeds. Loaded hundies are then suspended on either side of a split bamboo sling which is carried on shoulders of the farmer. Some amounts of pulvurized soil are mixed with water of each container. During transport, the mouth of the containers is covered with a piece of cloth known as *gamchha* or a perforated lid to prevent fish seeds from spilling to the outside during splashing of water. While carried on the shoulder, water in containers gets disturbed during splashing and jerky movement of the bamboo slings. When farmers proceed long distances, halt for relax, and

**Fig. 24.1 : Low-cost short distance transport of hatchlings and fry. This system of transport
exists in remote areas.**

stir the water of the containers by hands. Local farmers generally prefer to carry the
containers on their head for short distance transport as it is economical.

In many cases, oxygen is supplied to the containers from a rubber tube packed with
oxygen. Two narrow and elongated delivary tubes which are originated from oxygen-filled
tube, are inserted into the containers (Fig. 24.2). Thus oxygen is supplied to the water of
the containers by controlling the knob situated between the Oxygen-filled tube and the
delivery tube. This system also permits long distance transport with reduced mortalities.

1 *Types of Containers* : Hundies or containers are generally of two types such as (i)
    the larger one of 25 cm diameter and 32 litre capacity used for transport by rail
    or truck and (ii) the smaller one of 20 cm diameter and 23 litre capacity carried
    by farmers. The number of fry contained in small and large vessels are 50,000
    and 75,000 respectively. The vessels are filled with pond water (prefereably as
    fresh as possible) taken from spawning grounds. The water is placed in such a
    way that a little air space is left above the surface of water. However, three-wheeler
    vans and bi-cycles are also used as a popular means of transporting fish seeds
    (Fig. 24.3). Each van can sustain four hundies at a time. Big farmers or a group of
    farmers carry fish seed containers by trucks and travel a long distance.

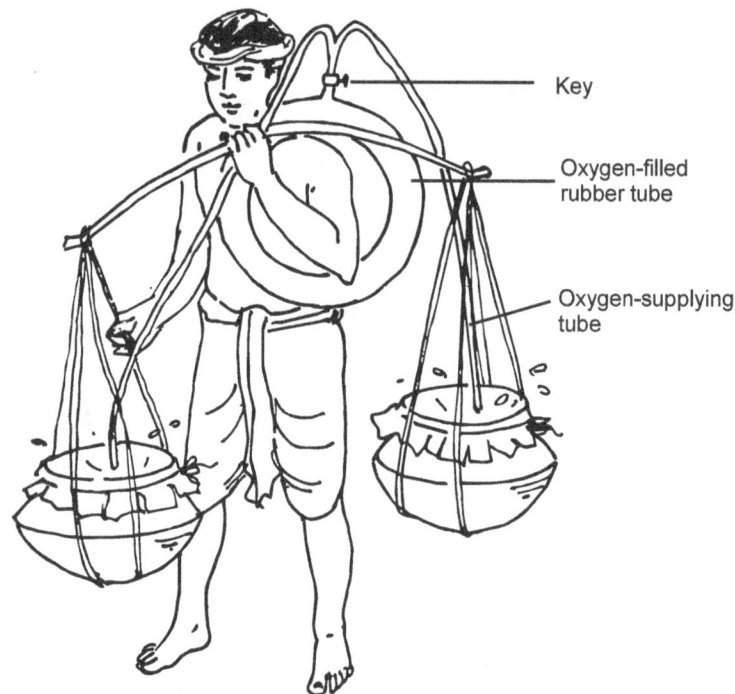

Fig. 24.2 : Low-cost short- or long-distance transport of hatchlings and fry. This system exists in many parts of India and adjacent countries.

Now-a-days improved open metal containers are widely used. The mouth of the container can be closed with a perforated lid. The capacity of each container is 53, 20, and 40 cm in diameter at the base, mouth, and height, respectively. During journey, the container is covered with woollen materials or kept moist.

The State Fisheries Department, Orissa, has devised an open transport semi-insulated motor van extemporized with a small semi-rotary pump. The distribution tube of the pump is provided with two rows of holes at an angle of 45° to each other which helps produce two oppositely-directed sprayers over the water surface. By this device, fry can be transported up to a distance of about 500 kilometres with 95 per cent survival.

2. *Role of Pulverized Soil During Transport*: At the time of tranport, about 60 g of pulverized red soil is mixed with water of the container. The containers are periodically flustered with sprayers. This permits atmospheric oxygen to dissolve in water and aggregation of fish seeds within the contaipers is thus avoided.

Addition of pulverized red soil (or any other soil) helps coagulate the suspended organic pollutants and thus reduces the pollution of water. Moreover, dead and infected fry are precipitated at the bottom of the vessel and are covered by the soil. Bottom sediments are periodically removed with a cloth rope. The water of the vessel is partially removed. This system permits transport of hatchlings and fry up to a duration of 30 hours.

Fig. 24.3 : The system of bi-cycle/three-wheeler van rickshaw for short or moderate distance transport by farmers for local haulage effectively serve to transport fish seed. Farmers' themselves have a special role in local distribution. Small farmers operating the fish farming industry have to be realized that bi-cycles play a major role in secondary transport. They are cheap, available in plenty and are ideal for rural roads.

Pulverized soil and activated charcoal tend to absorb carbon di-oxide and ammonia from water and consequently mortality rate significantly reduced.

3. *Transport of Fingerlings*: During transport of fingerlings, special care is undertaken to avoid mass mortality. Transport of fingerlings in small packing units are not economical as severe mortality occurs during transport. Fingerlings are, however, kept on stervation so as to evaccuate the stomach. This prevents water from contamination with faecal matter. Open canvas container (1 m × 1m × 1.25 m) and galvanized iron drums (48 cm × 30 cm) each of 180 litre capacity are used in many states of India.

In some far Eastern countries, open containers have been developed for transport of fingerlings and adult fish. Water-tight and tar-coated bamboo baskets of 15 litre capacity are used in Indonesia. Water is continuously agitated and the polluted water is periodically removed from the container. In many cases, specially designed carrier boats with apertures divided into several chambers and other contrivances to facilitate continuous circulation of freshwater are generally used. Each boat may carry about 10 million fry at a time with 70 per cent survival rate.

In cases where large-sized fingerlings and adult fish are transported, truck-mounted open tanks (1 m × 1 m × 1.25 m) with mechanized aeration or circulation of water are used. Since large-sized fish are transported through this system, special care should be taken to avoid shock hazards.

CLOSED SYSTEM

In this system, air or oxygen is supplied by oxygen cylinder into an enclosed space above the water of the container. The container is tightly closed so as to prevent the contact of environmental air with the contained water.

1. *Transport of Hatchlings and Fry*: For long duration transport, hatchlings and fry are transported using sealed metal container charged with oxygen. The container is made of galvanized iron and measures 45.7 cm × 35.4 cm with two air-tight openings at the top. While the one opening is used for oxygen supply, the other to let out the displaced water. For a journey of 12 hours duration in winter months, about 200 fry or 100 fingerlings of 7.5 to 10.0 cm in length may be kept in the container.

Polythene bag (thickness 0.062 cm) measuring 840mm × 610 mm X 650 mm has been successfully employed for transporting hatchlings and fry. The polythene bag is placed in a tin container of 18 litre capacity and one-third of the bag is filled up with water (6-7 litres) in which requisite number of hatchlings (about 35,000) or fry (about 2,000) are kept (Fig. 24.4). The air of the remaining portion of the bag is liberated by squeezing the same and then two-thirds portion of the bag is charged with oxygen by a tube fitted with the oxygen cylinder. The remaining 10-15 cm portion of the bag is fastened with a thread. It has been calculated that this tin can sustain about 2,000 fry (1-2 cm long) and can be

Rubber tube

Oxygen
cylinder

Polythene
bag

Can

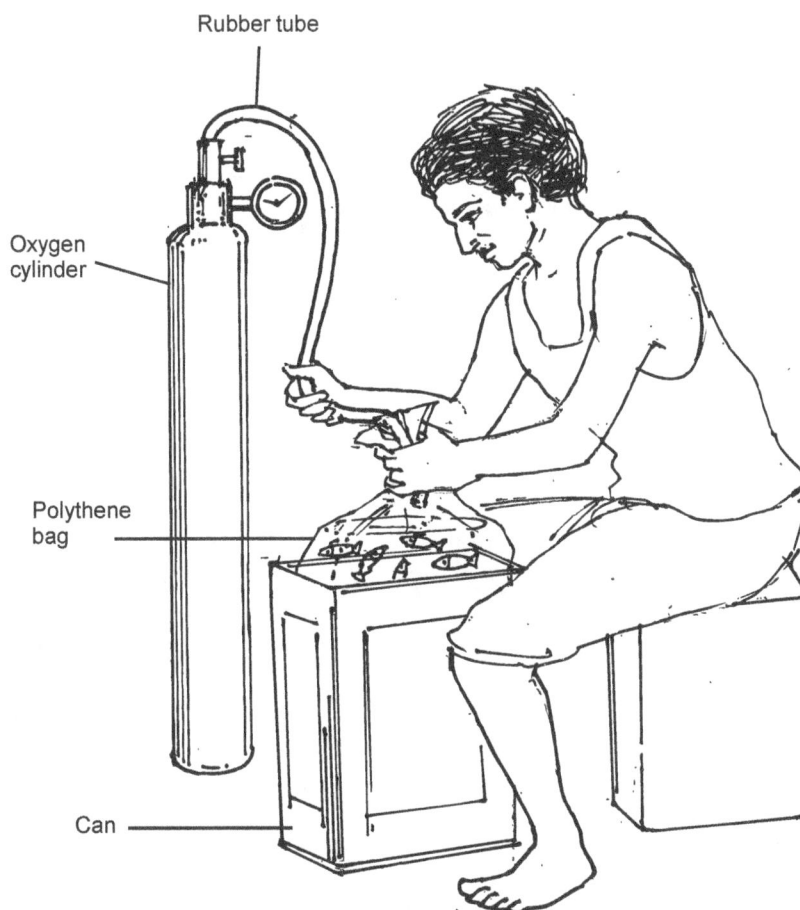

Fig. 24.4 : Modern method of transport of hatchlings and fry in polythene bags charged with oxygen.

transported without any difficulty for over 15 hours. The number of fry/hatchlings may vary depending on the time and distance to be travelled. Though this method of transport is most efficient with regard to survival, the cost involved is very high especially for small farmer and may negate the benefit derived.

It has been estimated that during 12 hours journey, greater quantity of oxygen (1,680 ml) is required for transport of 50 fingerlings of *Labeo rohita* (110-125 mm in length) at 30-32°C as compared to *Cirrhinus mrigala* where only 475 ml of oxygen is necessary to cover the same distance and time. On the other hand, oxygen requirement of grass carp fingerlings is much lower as compared to those of silver carp. Only 25 fingerlings of silver carp (100-150 mm in length) as against 90 fingerlings of grass carp (60-90 mm in length) could be transported at temperature varying between 24 and 27°C.

2. *Transport of Fingerlings and Broodfish*: In India, two models of closed system have been designed for transport of live fingerlings and broodfishes such as Mammen's model and Patro's model.

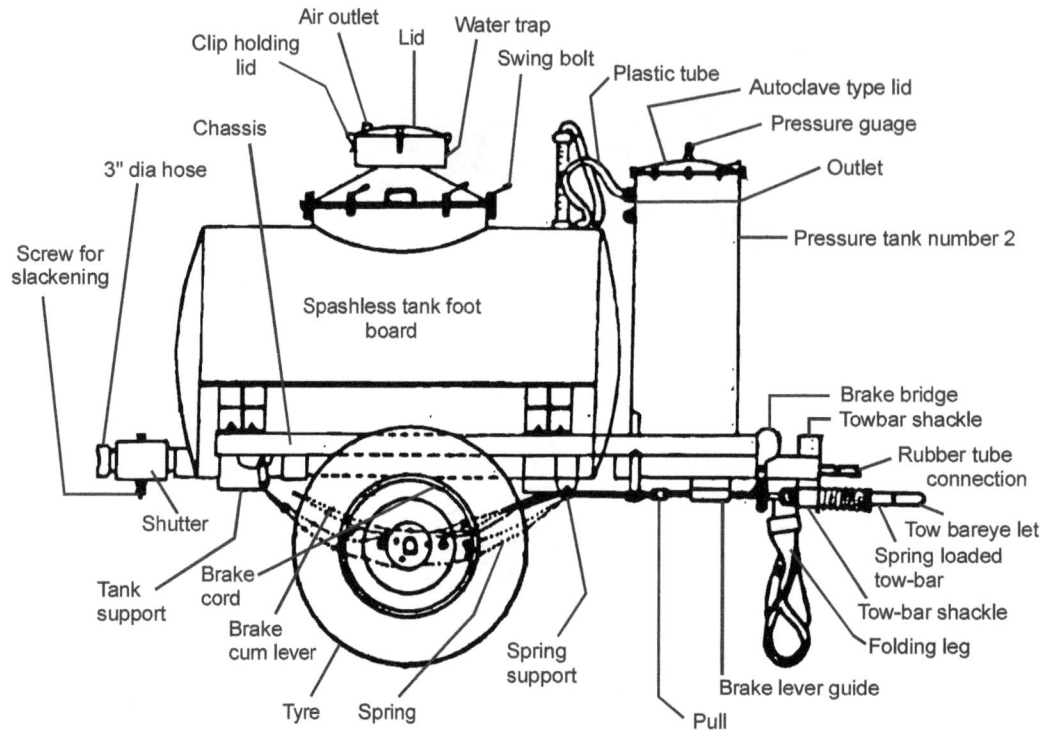

Fig. 24.5 : Structural details of a 'Splashless Tank' degi designed by Mammen (1967).

This is the most modern system of long-distance transport of fry and fingerlings. Though the model van carry huge quantities of fish, the cost of operating such models is moderately high. But the cost may appreciate the benefit derived from fish production.

(i) *Mammen's Model*: Mammen termed fish container as *splashless tank*. This is a petrol tanker design of 1,150 litre capacity with an autoclave type lid (Figure 24.5). An aeration system is provided for supplying air from a compressor machine which is operated by the engine of the transporting carriage. For emergency supply of oxygen, a cylinder is kept along with the tank. The inside of the tank is lined with U-foam. This lining helps prevent fish from injuring at the time of transport. To increase the dissolved oxygen concentration of tank water, a sprinkler is fitted. A total weight of about 250 kg of live fish can be transported in the tank. The total ratio of fish to water in this type of tank is about 1 kg of fish/4.5 litre of water.

(ii) *Patro's Model*: Patro's carrier as shown in the Figure 24.6, is more or less similar to that of a laboratory gas supply design type. The carrier consists of two barrels — the larger outer and the smaller inner barrels. The outer barrel of 120 cm diameter is open from top and the inner one is closed from the top. During transport, smaller barrel fits inside the larger one. An air vent with an oxygen valve is fitted

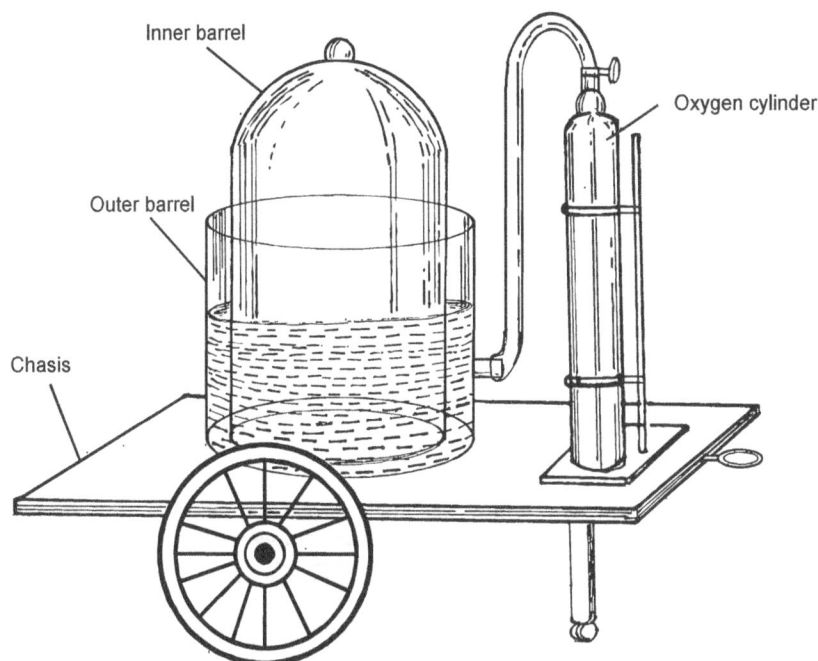

Fig. 24.6 : Structure of a Patro's Model. This is also the most modern system of long-distance transport of brood fishes.

at the top of the inner barrel. The outer chamber is previously filled with water along with fish and under high pressure, oxygen is entered from a cylinder into the outer tank. The ratio of fish to water is 1 kg: 20 litre. The supply of Oxygen to the tank lasts for 5 hours. If road facility is available, the vehicle can run for more than 200 kilometres.

EXPENSIVE MEANS OF TRANSPORTATION

Distress sales in rural areas are further facilitated by the fact that the means of transportation to farmers are not adequate and far from satisfactory. His transport vehicle, though fitted with inflated tyres, continues to be the chief source of traction. Although other mechanized means of transportation have also fairly developed over a period of time to carry landed fish over long distances, they are so costly that it is not always possible to reach the market in due time. Freight charges form a large share of the price of fish. High freight charges deter the farmer from taking his stock to a large market.

24.3 Determination of Packing Load During Transport

The success of fish transport depends upon the selection of exact number of fish. If adequate quantity of fish is kept in each packet/carrier, the mortality rate can be drastically reduced. The number of fish to be transported depends upon several factors such as type of species, size and weight, amount of water, duration of transport, and type of transport (Tables 24.1 and 24.2) . Therefore, it is very difficult to select the exact number of fish to be transported. If the required amount of oxygen is known for a particular species of fish, the exact number of fish to be transported can be easily calculated by the following formula:

$$N = \frac{(DO - l) \times V}{(O_2) \times h}$$

where   N   = Number of fish packed
           DO = Amount of dissolved oxygen in the water (mg/l)
                = Amount of oxygen left in the container at the end of trnasport
           V   = Volume of water (in litre) in the container
           O_2 = Rate of oxygen consumption (in mg) by individual fish
           h   = Time required for transport (per hour)

Table 24.1 : Number of Transported Fish (carp) in a Plastic Container Having 8 litre of Water Chargd With Oxygen Under Different Water Temperature

Life stages of fish	Temperature					
	25.0°C			30.0°C		
	Duration of transport (hour)					
	6	12	24	6	12	24
Spawn (8 mm)	12,000	6,000	3,000	10,000	5,000	2,500
Fry (8-40 mm)	600	350	175	500	300	150
Early fingerlings (40-80 mm)	175	100	50	150	80	10

Table 24.2 Number of Carp Fish Required to Transport for 12 hour Packed in the Plastic Container Charged With Oxygen. Average Number is Given in Parenthesis.

Fish size (mm)	Number of Fish	
	In 50 litre of water	In 20 litre of water
10	500-25,000 (15,000)	1000-10,000 (5,500)
20	500-20,000 (6,700)	500-5,000 (2,200)
30	200-10,000 (3,000)	200-1,000 (600)
40	100-700 (400)	100-500 (330)
50	50-500 (250)	70-300 (225)
60	50-400 (200)	50-200 (80)
70	50-300 (150)	20-100 (70)
80	50-200 (90)	25-50 (40)

Source: Fisheries Research Centre, Hyderabad

24.4 General Rules for Transportation

For successful transport of different life stage of fish, the following rules must be taken into consideration.

1. It is advisable to transport fish during cold weather; but in summer season, transport of fish should be done either at dawn or in night.

2. During transport, clean and well-oxygenated water must be used.

3. Since fish are very susceptible to various factors such as temperature, oxygen, shock, etc., special care should be taken so that fish can be transported within very short time. Transport vehicles should be kept as a stand-by at the place where fishes will be transported.

4. Before transport, fishes should be kept under starvation for 24 hours. This prevents pollution of water caused as a result of excretion of metabolic waste products by fishes.

5. During transport, care should be taken so that the temperature of water does not increase too much.

6. Only requisite number of fish should be packed and transported.

7. Same species of fish should be transported at a time.

8. At the time of transport, it is necessary to change the water of the container frequently with utmost care. To increase the content of its oxygen, occasional shaking of surface water with hands must be performed that makes the water more saturated with oxygen.

9. After reaching the destination, fishes should be cautiously released into the pond. Fishes should be acclimatized, if necessary, for several hours by keeping them in a hapa.

24.5 Causes of Fish Mortality During Transport

Various types of infectious diseases, bad handling of fish stocks, high stocking density in containers, accumulation of waste products such as ammonia, physical injuries are solely responsible for high mortality of fishes during transport. Different physiological factors such as high carbon-di-oxide tension and deficiency of oxygen in the transporting medium are also involved. These factors act simultaneously, however, and drastically affect fish survival in containers.

EFFECT OF AMMONIA ON OXYGEN CONSUMPTION

When fishes are transported through closed system, huge quantities of ammonia are excreted by fishes and concentration of ammonia further increased through bacterial decomposition of nitrogenous waste products and dead fishes. It has been noted that inspite of the presence of adequate amount of oxygen in the sealed atmosphere of the container, fish mortality occurs in presence of ammoniacal nitrogen at high levels (more than

20 mg/l). Hatchlings can withstand a concentration of 2.5 and 15 mg/l of dissolved free ammonia and dissolved ammonia in the form of inorganic salts, respectively. As a generalization, the oxygen content of blood decreases along with increase in the concentration of ammonia in water. Consequently, the concentration of carbon di-oxide in blood significantly increased since ammonia conflicts with the carbon di-oxide - oxygen exchange ability of blood with the water of the container. Decrease in dissolved oxygen following the increase in temperature of water, however, decreases the endurance of fish to ammonia.

OXYGEN CONSUMPTION IN RELATION TO OXYGEN CONTENT

Generally the rate of oxygen consumption by fish in a closed vessel decreases along with decrease in the dissolved oxygen concentration of water. The following example will illustrate this point. At temperature, pH, dissolved oxygen and carbon di-oxide ranges of 26.3-28.7°C, 7.0-8.0, 3.2-6.0 mg/l and 1.9-5.2 mg/l respectively, the average rate of oxygen consumption per 100 number of hatchlings of *Catla catla* and *Cirrhinus mrigala* (5mm in length and 0.8 mg in weight for each species) has been estimated as 0.12 mg/hour. Similarly, the rate of oxygen consumption by each species has been noted as proporationately to its body weight.

TEMPERATURE EFFECTS ON METABOLIC RATE DURING TRANSPORT

The role of temperature on the metabolic rate is very complex since temperature may alone be a lethal factor and toxic products that are developed in the packing units may change lethal thermal limits. Temperature generally acts as a controlling factor through its effects on metabolism which, in turn, sets the limits on greatest efficiency. Essentially, transported fishes are ectothermic. This means that the body temperature is almost exactly the same as that of the environment. This is because the rate of heat exchange with the surrounding is admirable or that even large-sized fish exhibit very little temperature lag when subjected to rapid temperature changes. Ectothermic also means that the rate of metabolism in fish sustains nearly two-fold increase with every 10°C increase in temperature. This is generally termed as Q_{10}. This generalization excludes the significant fact that the rate of temperature change may have as great an effect on fish as the absolute amount of change. Fluctuations in temperature result in an overshoot in metabolism. This overshoot may be ascribed to alteration in physical activity adequately exhibits a transient Q_{10} of far greater than two. Thus, acclimation to a change in temperature accompanies to harmonize the direct effect of temperature on fish.

The aquatic medium has some other chemical and/or physical factors that may influence the survival of transported fish. Dissolved oxygen concentration is one of these parameters and oxygen concentrations are in turn influenced by temperature. The oxygen level in water decreases with a rise in temperature so that, for instance, imbrued water holds nearly 13.2 mg/l oxygen at 4°C and 8 mg/l at 30°C. This significant decrease in dissolved oxygen can be permitted by fish, however, when combined with other factors such as alkalinity and pH, it may prove to be fatal. *Second*, polluted substances when accumulated in water, produce a biological oxygen demand by micro-organisms particularly plankton and bacteria will have this demand potentiated by increase in temperature. The effect of a change in temperature on fish mortality may be mediated through the availability of dissolved oxygen at a constant temperature and exhibit little effect on mortality unless these decreases are severe.

When live fishes are transported, oxygen in water is consumed by plankton, bacteria and organic matter for their respiration or oxidation. Obviously clean, well-aerated, and bacteria as well as plankton-free water should be used at the time of packing of fish in containers. Of course, it is not always feasible to completely remove bacteria from the medium because of the fact that the medium becomes teemed with bacteria immediately after decomposition of dead fishes in the container. The bacterial load in the transported medium along with the waste products developed by metabolic activities of fishes create polluted environment which makes it difficult to survive well for the fish.

Fish usually reply to even minute stimuli with hyperactivity. Consequently, lactic acid tends to accumulate in their tissues and severe 'oxygen debts' are created. It takes a long time for fish to overcome this condition even in their natural life in rivers, streams and other habitats. But during transport, it is often observed that after handling, transport and liberation even in oxygen- rich water, many of the surviving fish die during the next few hours of release for no definite reason. For this reason, in modern live fish transport techniques, the use of sedatives finds an important place.

On account of hyperactivity, the bigger fish suffers from injuries which may cause death due to external infections. It is also possible that some bigger fry under crowded conditions of the packed unit might accidently kill or cause injuries to others which may latter lead to severe infections and death. Selection of uniform-sized fish and species composition helps, to some extent, avoid the risk of mortality considerably on this account.

EFFECTS OF CARBON DI-OXIDE ON OXYGEN CONSUMPTION

The oxygen requirement of a fish is regulated by the fish itself and by many factors in its environment. These relationships have been the subject of several investigations. However, the effects of carbon di-oxide on the oxygen consumption of various life stages of fish clearly indicate that the carbon di-oxide influences the rate of oxygen consumption in two ways. *First*, it is the consequence of alterations in the utilization on the rate of oxygen consumption and *second*, it influences the utilization of the water respired. The effects of carbon di-oxide on oxygen consumption are not conspicuous until severe situations eventuate. Increase in partial pressure of carbon di-oxide results in increase in oxygen consumption due to increase in ventilation and transport. This is an endeavour requires to recompense for the declined utilization. Further increase in carbon di-oxide level reduces the active metabolic rate and may prove to be destructive to fish.

Reciprocal influences of oxygen and carbon di-oxide on various life stages of Indian major carps (hatchlings, fry, and fingerlings) have shown that the larvae and fry (8 cm in length) and fingerlings (15 cm in length) can able to survive for 24 hours or more in water charged with oxygen at the rate of 0.5 and 1.0 mg/l, respectively. Tests on the effect of different combinations of carbon di-oxide and oxygen levels on hatchlings, fry and fingerlings of carp have shown that hatchlings (4-5 mm) can withstand 0.5 mg/l oxygen with 2.5 mg/l carbon di-oxide but with 1 to 4 mg/l oxygen, they can withstand 7.5 mg/l. carbon di-oxide. Similarly, fry (6 to 8 cm) can withstand 0.5 mg/l oxygen having 15 mg/l carbon di-oxide. When oxygen level significantly increased from 1 to 4 mg/l, fry can tolerate 125 mg/l carbon di-oxide but 50 per cent of fry are destroyed in 150-175 mg/l carbon di-oxide. Further increase in carbon di-oxide level at the rate of 175-250 mg/l is dangerous to fish. The fingerlings (10-15 cm in length) can tolerate 1 mg/l oxygen with 2.5 and 5.0 mg/l, carbon di-oxide. However, in a test conducted for 56 and 72 hours, in waters having 1.2-3.0 and 5-6 mg/l oxygen and 75-110 and 100-1,500 mg/l carbon di-oxide, respectively, fingerlings were observed to consume oxygen at the rate of 14.6 mg/l and withstand 5 mg/l oxygen with 150 mg/l and 2 mg/l oxygen with 110 mg/) carbon di-oxide for 72 hours without any mortality.

In a closed-system transport method where water is charged with oxygen, carbon di-oxide forms a limiting factor and fish mortality is principally due to bacterial load in the medium. During transport, a few percentage of fish is died and consequently, bacterial populations instantaneously increased and utilize oxygen already present in water. According to an estimate, bacterial populations increased from 200 to 100 million/ cc during 24 hours journey. The gas carbon di-oxide as low as 5 mg/l has been found to be toxic if oxygen level is maintained between 0.1 and 0.5 mg/l.

Carbon di-oxide exhibits antagonistic effects on the potentialily of the ammonia toxicity to fish because only unionized ammonia is toxic and low pH increases the degree of ionization of ammonia. At high pH, there is a gradual reduction in the ionization of ammonia. Carbon di-oxide levels at rates varying between 20 and 60 mg/l, for example, suppress the toxicity of ammonia solution by 40-70 per cent.

24.6 Use of Drugs, Antiseptics, and Antibiotics in Fish Transport

The use of drugs, antiseptics and antibiotics in fish transport is undoubtedly of practical importance. As noted earlier, oxygen concentrations of water in transport vehicles maintained either through mechanically or manually are important for long distance transport. But in addition to this method, a number of other techniques exist which should be kept in mind while transporting fish stocks for long / short distances. The use of these substances immediately after packing discourages damage to fish seeds due to infection and other reasons that follow.

DRUGS

The use of drugs during transport of live fish is very essential. Though their use in India is very limited, they are widely used in many European countries and USA. Particularly different drugs and chemical substances are used when broodfishes are transported. It should be noted that chemical substances must be used in appropriate levels with utmost care otherwise there is every possibility to die off all fishes during transport.

1. *Reasons for Using Drugs and Chemicals* :

(i) Chemicals absorb/adsorb carbon di-oxide, ammonia, and other dissolved organic substances.

(ii) To increase the number of fish for transport by reducing their hyperactivity.

(iii) To reduce the death rate of fish due to excessive strain and hyperactivity during transport.

(iv) To prevent fungal and bacterial diseases during transport.

ANTISEPTICS AND ANTIBIOTICS

During fish seed transport, however, anaesthetic materials, antiseptics and antibiotics are also generally used. Their use to reduce the metabolic rate of fish has not been comprehensively beneficial. To be competent, these chemicals have to be carefully chosen and the dosage should be closely controlled. Most of the workers have done extensive research on this aspect. They have considered these chemical substances for inducing stillness during transport. These chemicals include sodium amytal (50-170 mg/l), tricaine methene sulfonate (1/10,000), and tertiary amyl alcohol (0. 45 mg/l of water). Some of these compounds may, however, be harmful to a large extent. Sodium amytal, for example, has the disadvantage of being affected by calcium content of the water and tricaine methene sulfonate tends to break down the motionlessness of fish. Sodium amytal should, therefore, be used only in soft water. Most of the chemicals at recommended dosages are considered as desirable and have instantaneous potency, are readily soluble in both salt water and freshwater ecosystems, encourage deep sedation within 5-10 minutes and are consistent with calcium. For deep sedation, tertiary amyl alcohol and chloral hydrate respectively at the 2 mg and 3 g/ gallon rates are generally recommended. Previous reports have clearly shown that some drugs such as uratane, taiouracil, quinaldine, and hydroxy quinaline at the rates of 100, 10, 5-10, and 1.0 mg/l respectively have been found to be equally effective for fish transport. To prepare requisite concentrations of drugs for use in transport of fish stocks, they are first dissolved in water and then mixed with the transported medium. Some drugs such as novocaine, barbital sodium, and amobarbital sodium are also used intra-muscularly for narcotizing fish during transport at the rates of 50, 50, and 8 mg/kg of fish body weight, respectively. It should be added as a note that the application rates of drugs generally depend upon the type of fish species to be transported, their size and weight, and distance to be travelled.

24.7 Quarantine Measure

A basic important aspect in attempting to prevent induction of infectious diseases, parasites, and aquatic insects during transport of fish consignments is that mild/severe infection of fishes can never be prevented. It is not even practical to completely prohibit the infection of fish. This situation has been forced to adopt some precautionary measures such as prophylactic and quarantine regulations involving the use of antiseptic, antibiotics and germicides. The aim of fish quarantine is to prevent the entire fish stock from mortality and the spread of pathogenic micro-organisms. During transport, short-duration chemical treatments of fish are recommended. An instantaneous bath before packing for transport may help in restraining the infectious diseases from further spread in fish populations. In most rural areas, some chemicals which are easily available such as acriflavin (10 mg/l), copper sulfate (0.5 mg/l), potassium permanganate (3 mg/l), methylene blue (2 mg/l), methyl parafynol (1 mg/l), chloral hydrate (2 mg/l), sodium chloride, formalin, and chloromycetin are widely used for fish bath.

24.8 Effects of Over-stretch in Transport

The effects of over-stretch on fish during transport cause death due to severe muscular activity such as swimming against swift currents and floundering in the box. Though there is no vivid explanation regarding the exact cause of death, an explanation involves the acute acid-base disturbance following the accumulation of lactic acid in large amounts which results in fish death during transport. The lactic acid level is enough to attenuate the acid-combining capacity of the blood and consequently its oxygen-carrying capacity must be greatly affected. Stern muscular activity quickly decreases the maximum swimming capacity of fish but it is important to note that the recovery of such activity is slow. Thus, fish kept in a exhausted condition may be at a disadvantage apart from any mortalities that may have resulted from over-stretch.

24.9 Principles of Aeration*

Since excretion of waste materials and their fermentation is a great problem in fish transport, the transport of oxygen from gas to water has received much consideration. When water is stirred so that the level of oxygen is essentially uniform throughout its volume, the rate of transfer of oxygen from gas to water can be expressed by Bohr's equation that follows.

$$\frac{dc}{dt} = K (C_a - C)$$

where C = concentration of oxygen in the water at time t, C_a = the concentration of oxygen in equilibrium with the gas phase and K = specific rate of aeration. These values refer to saturation with air and is multiplied by 5 when pure oxygen is used for aeration. The K value can be expressed by use of the expression:

$$K = f \frac{A}{V}$$

Where,
A = Area of the gas-water interface
V = Total value of water
f = The rate of the travel of the gas through the interface

---

* For further study, see Lagler (1956), Hutchinson (1957).

When the water of the container is stirred, the surface of the gas-water interface varies continuously. Also, the constant f value also fluctuates greatly which is due to the degree of turbulence of the water since the value of f is greatly influenced by contaminants, the constant K should be determined for each individual piece of transportation equipment using water in which the respiration of fish will take place. While the value of f is in the range of 1-10 in the plastic-bag system, it is in the range of 5-100 for systems in which aeration is achieved by oxygen or diffusion of air or recirculation of water through jets. From the relation K to f, it has been noted that the area of the gas-water interface is very important, the most efficient aeration can be achieved when the gas is presented to the water in the greatest number of bubbles. When the bubbles are issued by a diffuser having adquate capacity, they will coalesce, and the efficiency of the system will increase.

When air is allowed to diffuse through water, the values for the exchange of carbon di-oxide and oxygen are the same. The values of f are also similar when the water is aerated by stirring. If oxygen is used instead of air, the carbon di-oxide will be rapidly removed. When hatchlings are transported in the plastic bag system, carbon di-oxide will build up in the water that can be temporarily eliminated by using some organic buffers such as tris (hydroxymethyl) and aminomethane and has been reported a fifty-fold increase in buffering capacity at 20 g/ gallon with maximum hydrogen ion absorption between pH 7.5 and 8.8.

It is not possible to keep the ammonia below toxic levels by aeration. Possibly, the ammonia is neutralized by the generation of carbon di-oxide during respiration of fish. A dangerous increase of ammonia if occurs, can be removed by ion-exchange resins.

24.10 Transport of Freshly-Killed Fish

Fishes are one of the most important perishable food items. They are harvested from different aquatic resources such as freshwater, marine, and brackish waters and exhibit wide variations in their nutrient composition such as proteins, fats, carbohydrates, vitamins, and minerals.

Fish consumption in most developing countries has been restricted to the areas of production and in some cases, the entire harvested fish or in part are transported to nearby localities. The transportation of freshly harvested and killed fish is a common practice around the world, but certain difficulties arise because the harvested fish undergo rapid decomposition. Decomposition starts in fish flesh as a result of physical, chemical, and post-mortem alterations. However, it should be kept in mind that at the time of transport, it is necessary to ascertain that the transported fish will not only find a ready market for high returns but also should reach the consumer in an acceptable condition. The quality of freshly-killed fish should be properly maintained at the time of transport which depends upon how the fish was handled, sorted, packed, preserved, and transported. The primary responsibility of assuring the quality of harvested fish lies on the initial harvesting practices and manpower involved.

In order to arrest decomposition, harvested fish should be transported at low temperature (between 0 and 4°C). This will definitely inhibit the development of bacteria.

To maintain such temperature range, fish are kept in crushed ice. To reduce the wastage and spoilage of harvested fish and to distribute quality fish stocks to markets/areas located far away from the production farms, adoption of necessary methodologies for handling and transport of fish must receive primary attention.

CONTROL OF FISH DECAY

Decline in quality of fish flesh followed by the loss of nutrients through bacterial decomposition in tropical countries is more vigorous. In regions where temperature is high for months, the potential for nutrient loss through decay is also high. Such condition exists in areas where there is no facility for packaging, storing, and transportation. In such cases, fish decay is the rule, providing opportunities for financial loss to farmers. In developed regions where it is possible to manipulate the harvested fish for preservation and transport and there is adequate facility of transport vehicles as well as the presence of network of roads, less decay occurs. However, well-preserved fish and their transport within the shortest possible time will obviously reduce or check the decay of fish considerably.

24.11 Sorting and Grading of Fish During Transport

Grading, broadly means the sorting and classification of fish according to the various standards as fixed by the trade or as required by the customer. Grading assumes major importance in the marketing of fish in view of the fact that farm products are heterogeneous.

Most of the fish farmers have a tendency to stock fry and fingerlings in large numbers in their ponds. But this high rate of stocking especially in intensive farming systems may result in a number of extremely serious outbreak of diseases and cannibalistic behavior of stocks. Consequently, there is drastic mortality of cultured fish and the production potential of most of the farmed ponds where stocking densities are high enough, have come down to surprising levels. For this purpose, sorting and grading of fish must be taken into consideration. The main objective of sorting and grading of farmed fish is to segregate them according to their size and species. Sorting should be carried out at 3-month intervals. Sorting is very important in carp, trout, and salmon culture ponds for their normal growth. This is due to the fact that in trout cultivation, the older species consume the small-sized trout. Moreover, the presence of tilapia and other wild species in major carp culture ponds drastically reduce the number of natural food organisms considerably by consuming them to a greater degree.

Immediately after harvest, fish are sorted out according to species and size. If any considerable variation in size between species exists in small farms, sorting is done by hand. In case of large farms, mechanical graders or nets are used. However, it is advisable to sort out different size of fish and then processed according to demands of the market.

SORTING AND GRADING METHODS

Sorting and grading of fish are carried out according to their size because consumer preference and market demand depends very much on size. Various types of graders are available in the market and some of them are used in fish culture industry.

1. *Mechanical Grader*: The most simple form of grader used in farms is the adjustable bar grader. Different size groups of fishes can be segregated by adjusting the distance between the bars.

2. *Graders Made of Screens*: Grading of fish may also be done by using large superimposed net with variable mesh sizes. The nets are immersed in storage tanks or ponds. Fish are placed on the top-most net which has the largest mesh-size. Small-sized fish are passed through the top nets and are collected by the lower net which has the smaller mesh-size.

3. *Use of Table for sorting*: Sorting is done on a smooth sorting table. The floor of the table is covered with a aluminium sheet with elevated sides of about 15 cm in height. One side of the table is terminated in a narrow neck. The table is held in position on mobile wheels or on a support. Immediately after sorting, detritus, mucus, and mud are removed from the table with clean water.

The graded and sorted fish are then divided into species and sizes and kept in containers filled with clean water. The containers are then transferred to the growing, storage or wintering ponds.

BENEFITS OF GRADING

Some of the important benefits of grading fish are as follows:

* Grading gives incentive to farmers for marketing quality fish, since such produce may fetch better price in the market.

* Grading eliminates discarded or unwanted fish species.

* Grading simplifies marketing system by making it possible for buyers to produce fish that satisfy their requirements.

* Grading enables consumers to purchase particular consignment according to their purchasing power which varies greatly.

* Grading facilitates future trading. In the absence of grading, it would not be possible to guarantee the delivery of fish of a particular species of fish.

* Grading dispense with the need for personal inspection of fish/products by buyers. If a graded product is well-established and recognized, it can be internationally traded. Trading in such products will reduce the risk of fraudulent practices in marketing.

To sum up, sorting of fish during transport to long distances is extremely important. Fish culturists always try to purchase healthy seeds from the suppliers for high survival and yield in their ponds. Therefore, it is necessary to sort out dead or weak fish from live ones during transport. Weak fish are not likely to survive/withstand long distances. After proper sorting, only live and healthy fish should be considered for packaging and transport. At the same time utmost care should be taken to transport more or less same-sized fish in the containers with an adquate supply of clean and oxygenated water. It is also advisable to aerate the water of the container carrying sorted fish.

24.12 Storing the Fish

After sorting and grading, fish should be stored according to species for either long or short period. Storing of fish for short period is done immediately after fish are harvested. After segregation, fish stocks can be sent immediately to their destination for marketing. In other situations, fish are stored for either short or long period.

Extreme caution must be taken to ascertain that fish stocks do not face any difficulty or danger by invading parasites, enemies or infectious diseases.

PROVISIONS FOR STORING

Before storing of graded fish, the following provisions would have to be kept in mind.

1. Loss of fish weight should be least.
2. Fish should be kept in good condition.
3. The quality of fish flesh should be improved. This is particularly important in the case of bottom-dwelling fishes. Storage in good conditions may improve the quality of flesh.
4. If fish have to be kept for short duration, feeding should be stopped.
5. The water should be fresh and cold (not below 6°C). This condition helps reduce the metabolism of fish.
6. Fish should be protected from their enemies and infectious diseases.
7. For better management, the stored fish should be kept very close to the residence of the farmers.

METHODS OF STORING

Generally, small quantities of fish may be stored in floating boxes, aquaria or small holding nets. But storing of fish in large numbers require nets fixed in storage tanks. If storing is done for long time, fish must be kept in ponds.

1. *Nets*: For short time storing, fish stocks can be conveniently kept in nets which are suspended in clean and well-oxygenated water. This type of storing is very effective when ponds are drained. If several compartments are necessary, pockets are made by means of bamboo poles on which nets are placed. If clean water is available, a large number of fish can be stored.
2. *Small Ponds*: In such ponds, only delicate species may be placed. These ponds are also used to store fish stocks for a long time.
3. *Cement Tanks*: These tanks are made up of cemented brick or concrete structure and the walls are smooth in order to avoid damage to the fish. Cement tanks

should be fed with abundant supply of clean water. Fish may be kept for long or short period.

4. *Floating Boxes*: Floating boxes are made of long, thin and narrow pieces of wood or metal or perforated metal sheet on which wire nets are wrapped around the frame. Each box is provided with a lid that can be locked in order to avoid theft and the slip of the fish. In each box, only small quantity of fish is kept. These are widely used by farmers harvesting in rivers and lakes. Sometimes a large open-work pan of wooden bars or wicker or cane is used for storing and protecting fish. These pans are attached by a chain to a bamboo or wooden pole immersed into the pond water. These pans have handles which help to carry them from one place to another.

5. *Small Holding Nets*: This net is made up of metal but is always pliant and is used by anglers. They keep their catches in this net. A superior container of the same design is prepared by nylon netting wrapped on metal hoops and provided with a rubber tube distended by air. The basket floats but partially immersed in water which helps prevent fish stocks from slipping out. Fish can be kept in better condition in these small nets.

6. *Aquarium*: Glass aquaria of varying sizes may be used to store fish in better condition. This permits the buyers to prefer fish according to their need. If there is water exchange facility, there is no problem for oxygenation, otherwise aquarium water must be charged with air by using compressor machine. If the aquaria are heavily stocked with fish, frequent oxygenation is badly needed. Most aquarium fish are stocked in this way.

24.13 Conclusion

The demand of broodfish and fish seeds in fish farming industry continues to be high in view of the rapid expansion of fish culture in the tropics and sub-tropics. Since the supply of fry, fingerlings of farmed fish and inputs (fertilizers and feeds) are essential for significant production, it is inevitable that their availability must be ascertained in due time.

Lack of adequate transport facilities of live and freshly-killed fish in most regions of the world have significant impact on fish culture and marketing. Affluence of the consumers and the scientific understanding as well as technological developments in handling and packaging techniques have provided an added impetus to the prospects of commercial transportation of stocking materials and inputs for culture. It is important that fish farmers always try to transport these ingredients properly by any means so that they can better cultivate in their own/leased ponds and bring their produce into the market as quick as possible. Transport facilities are, however, very critical to farmers who live in remote/ upland areas. The rate of fish mortality during transport can be greatly minimized if different management procedures as mentioned in this chapter are adopted. The importance of transportation must receive primary attention in fish culture industry.

References

Black, E. C. and I. Barrett. 1958. Increase in levels of lactic acid in the blood of trout following handling and transportation. *Can. Fish Cult.* 20 : 12-24.

Copeland, T. H. 1947. Fish distribution units *Prog. Fish-Cult.,*9(4): 193-202.

Hult, M. 1986. Text Book of Fish Culture. Fishing News Books, Oxford.

Hutchinson, G. E. 1957. A .Treatise on Limnology. Vol. I. Wiley, New york. 1015 PP.

Lagler, K. F. 1956. Freshwater Fisheries Biology. Brown, Dubuque, Iowa, 434 pp.

Mc Farland, W. N. and K.S. Norris. 1958. The control of pH by buffers in fish transport. *Calif. Fish Game.* 44 : 291-310.

Nemoto, C. M. 1957. Methods for air transport of live fish. *Prog. Fish-Cult* 19 (4) : 147 - 157.

Norris, K. S. et al. 1960. A survey of fish transportation methods and equipments. *Calif. Fish Game.* 46 : 5-33.

Pillay, T. V. R. 1990. Aquaculture : Principles and Problems. Fishing News Books, Oxford. 575 pp.

Questions

1. What is transportation? Why transportion is so essential in fish culture industry ?

2. What are the designs of trading the stocking materials and transport systems which are essential for fish culture?

3. Discuss different methods of fish transport. Among different methods, which one is best?

4. How would you determine the packing load of fish during transport?

5. State different rules which are considered during transport of stocking materials.

6. What are the causes responsible for the mortality of stocking materials during transport? Discuss.

7. Discuss the role of drugs, antiseptics, and antibiotics in fish transport.

8. What are the principles of aeration in fish transport? Discuss.

9. How fish are handled during transport? What are the factors which are generally considered during transport? Discuss.

10. Why sorting and grading are necessary during fish transport?

11. Discuss how freshly-killed fish are transported.

12. Write notes on the following: (a) Open system of transport, (b) Closed system of transport, (c) Role of pulverized soil during transport, (d) Patro's model, (e) Mammen's model, (f) Use of antiseptics and antibiotics during transport, (g) Effects of over-stretch in transport, (h) Sorting and grading of fish.

25

Fish Marketing and Trade

A book on fish culture will remain incomplete unless a discussion on fish marketing and trade produces a clear and defined image. The fish market, with its vast size and demand base, offers great opportunities to marketers. Consumers of most of the countries reside in rural areas and substantial amounts of the national income are generated there. It is only natural that the rural market forms an important part of the total market of any particular country. After the second world war (1939-1945), the fish marketing started in full potential. The sixties and seventies witnessed its steady development. And the eighties, nineties, and the beginning of the 21 st century have been the full blooming of the fish marketing and export. This chapter has discussed the overall aspect of marketing, international marketing strategies, and some other items partaining to fish marketing as well as trade.

25.1 What is Marketing

Marketing deals with customers. Creating customer value and satisfaction are the core fact of modern marketing system and practice. Although marketing has been practised in one form or the other since the day of Adam and Eve, today most management thinkers regard marketing as the most important function in any business. Although a good number of definitions of marketing have been proposed in different marketing management books, perhaps the simplest definition is this one : Marketing is the economic process by which products and services are exchanged between the producer and the consumer and their values are determined in terms of money/dollar prices.

Sound marketing is critical to success of fish farms whether it is Iarge or small, domestic or global. Marketing of any product whether it is agricultural, aquacultural, or industrial, is well practised in many countries of the world, but in some other parts of the world where marketing had no reputations, political and social changes have created there new opportunities for marketing.

25.2 Selling and Marketing

Marketing is much wider than selling and more dynamic. Selling is, however, virtually synonymous with marketing. In reality, there is a great deal of difference between the selling and marketing. While the former involves around the needs and interests of the sellers, marketing involves around the needs and interests of the buyers.

Selling begins with the existing products of the aquafarm and views business as a task of somehow pushing the products. On the other hand, marketing starts with the customers, present and potential, and views business as a task of meeting the needs of the customers by producing and supplying products that would meet the needs of the customers. Selling seeks profits by applying open and subtle pressure on the customers to make him part with his money in exchange for the farm's products. Customers' satisfaction is secondary to the basic aim of making quick sales and quick money.

Selling is a more tactical, routine activity with a short-term perspective while marketing has strategic implications with long-range objectives such as growth, viability, sustained customers' loyalty, survival, competitive competence and so on. The difference between selling and marketing cannot be described better, however. A suspect might arise as to how selling could be entirely dispensed with under the marketing approach. selling is definitely a part of marketing what is sought to bring into prominence by playing down of selling the vanity of pushing that is congenital to the sales approach.

25.3 Functional Classification of Fish Marketing

The functional classification looks at marketing from the view point of the functions performed in the course of transporting the farm products from the place of harvesting / processing units to the customers' hands. The following activities should be recognized precisely when fish marketing is taken into consideration: culture/capture, harvesting, sorting and handling, processing, preservation, warehousing, transportation, wholesaling, retailing, delivery, selling, research, financing, and profit. All these functional activities are directly related to each other to make the marketing and trade policy more effective.

25.4 Importance and Scope of Fish Marketing

The main important role of fish marketing is the economic development of farmers' community in any developing country which is undoubtedly cardinal at present. In addition to the transport of fish, whether live, fresh or preserved, rural fish markets also help farmers to earn money. Current trends in the development process have also given an impetus for the development of rural fish marketing. These trends include (1) increase in real incomes of rural an urban people due to economic development, (2) high fish production following the adoption of modern technology, (3) adoption of integrated fish farming, and (4) increasing awareness among progressive fish farmers towards the awareness of modern life.

25.5 Fish Marketing: Developed Versus Developing Countries

DEVELOPED COUNTRIES

After the second World War, both the concepts and functions of fish marketing have given a serious attention in profit-related activities. Since fisheries and aquaculture have not become highly industrialized in almost all fishing nations, the production and new marketing techniques must have to be taken into account to produce and sale for income as more fish of commercial importance as possible. Sophisticated fish marketing systems exist in many developed countries and play an important role in meeting the demands of customers. Trout and salmon potential in the United Kingdom is, for example, considerable and the consumer trends towards fresh and healthy products provide a great opportunity to farmers. However, a strong underlying demand for fish will help sustaining farmed fish industries. The marketing systems of trout and salmon have been discussed in section 25.27.

Many developed countries have the improved methods of fish marketing where there is availability of cold storage facilities, ice plants, and good processing units. Fish markets are well equipped with proper drainage system, electricity, water supply, storage, and sanitation. Improved fish marketing systems have been developed along with the development of fisheries and aquaculture. Highly organized fish marketing systems provide remunerative prices to the primary producer though the interest of the buyers is also considered.

DEVELOPING COUNTRIES

In contrast to developed countries, many developing nations have traditional system of fish marketing. The methods and practices generally adopted in marketing system are traditional and still remain unchanged over several decades. Fish marketing in this case involves collection of fish stocks after sorting from production units and transport them to marketing places by van rickshaws (a three-wheeler carrier with inflated tyres). When the catches are small, fish are sold to the landing sites but when catches are substantial, as in the case of big farmers, mechanized transport vehicles are used to reach the marketing place. Except in some marketing systems, all the markets in many Asian countries require modernization.

In developing countries, fish marketing needs modernization. In Philippines and Singapore, fish marketing system are solely controlled by private marchents. In Sri Lanka, though regulated by private traders, government has extended some supports to co-operative enterprises. Severe exploitation of fish producers by fish marchents in Phillippines, Malayasia, and Indonesia are not uncommon. In some countries such as India, bidders exhibit some gestures with fingers for their opinion for buying or for the auction of fish, some vernacular words such as *Kaki* or *Kadi* are used.

In many developing countries, fish marketing is conducted on traditional pattern and is not efficient in practice and exploitative in feature. This pattern of fish marketing in India, however, is far from desirable and leaves much scope for improvement. This becomes more crucial in view of the adoption of new aquacultral strategy, which has resulted in increased yield of fish. Before the farmer gets disillusioned by the fact that the big traders are flourishing at their cost, something must be done to improve the marketing system.

25.6 Defects of Fish Marketing

If it is necessary to implement adequate marketing policies, we must find out the defects in the present marketing system. The most important defects of fish marketing are noted below.

SUPERABUNDANT MIDDLEMEN AND MALPRACTICES

A large army of middlemen functioning in almost all types of marketing systems become obese at the cost of the poor farmers. These middlemen act at various steps in the process of collection and distribution of fish stocks. The existence of middlemen reduces the share of the consumer's price obtained by the farmer. Nobody can refuse the activities being executed by these middlemen.

Several malpractices are also adopted by these middlemen. Some of these are as follows:-

1. Weights and sale are managed against the seller.

2. Deductions are made on religious and other purposes depending on the will.

3. The settlement of price is made under cover. Often farmers do not know at all that they are being cheated by the middlemen and consquently, unless appointed otherwise, they prefer to keep themselves away from the middlemen.

EXPENSIVE SYSTEM OF TRANSPORTATION

The means of transportation available to the farmers are not adequate. Most of the farmers have no well-equipped transport vehicles. They simply carry their produce over long or short distances by van rickshaws or bi-cycles. Since fishes are touchable to diseases, shocks and other injuries, resulting in mass mortality of fish during transport, mechanized and quick transport systems should be developed to carry fish seed, larvae, fry or fingerlings and even broodfishes as and when necessary. In many cases, transport of fish over long distances are so costly that they are beyond the reach of the small and marginal farmers.

LACK OF MARKET INTELLIGENCE

Most of the farmers do not know about the commercial system of marketing since they have no adequate knowledge on the market situation. They entirely depend on the message provided by the middlemen. This message is prejudiced in favor of the market traders and therefore, farmers restrain from understanding returns for their crops.

25.7 Co-operative Marketing

Marketing of aquaculture products on a co-operative basis is considered as one of the best methods so far the aquaculture production point of view is taken into account. Several advantages that follow can be derived from the co-operative marketing system.

1. Co-operatives protect the farmers from exploitation of traders.

2. It helps reduce the marketing margin.

3. It can generate the holding power with farmers by sanctioning cheap and easy loan to them.

4. Co-operative marketing may have a vigorous impression on market trend and it will assist to stabilize the market price.

PROGRESS OF CO-OPERATIVE MARKETING

Co-operative marketing as a separate institution came to be developed around Second Five Year plan. On the basis of the recommendation of All India Rural Credit Survey Committee, the Government of India established the National Co-operative Development and Warehousing Board (NCDWB) in 1956 under a central legislation. Since the primary responsibility for developing co-operatives rested with the sub-national governments, the board did not deal directly with co-operatives but provided assistance to sub-national governments for giving an effective support to co-operatives. However, in March 1963, the National Co-operative Development Corporation (NCDC) was established by replacing the NCDWB. This corporation has rendered a yeoman's service to co-operatives in the field of its functioning.

CO-OPERATIVE NETWORK

It is beyond doubt that co-operatives play serious role in the successful development of fish culture industries. Under the impression of policies afforded in the plan and with the support of the national and sub-national governments and the NCDC, the co-operative movement has made a spectacular progress for the last four decades of planned development. A well-weave co-operative structure has come up in the fisheries and aquaculture sectors, performing various functions such as fish culture, fishing, processing, transport and marketing of aquuaculture products, storage and management of fishing crafts and gears. A co-operative is, however an association of individuals or businesses joined together to achieve a great success. According to one estimate, there has been organized up to March 2002, 7098 fishermen co-operative societies in the country. Since all these co-operatives are functioning well, farmers are being encouraged to accrue benefits and to maximize the return for its members, and there seems to be a great scope for enrolling individual farmer to the membership of co-operative societies. This will assist to build up a strong network for marketing societies and to improve their financial resources so necessary for extension in business.

25.8 Problems and Suggestions of Co-operative Marketing

It should be pointed out that the co-operatives have manifold objectives such as (1) develop the confidence of their members, (2) provide credits to members in times and in need and other facilities, and (3) they convince them that they can not only emancipate the farmer from exploitation but also can provide remunerative prices to him. The most common problem faced by a co-operative is the lack of efficient management and the manager is the key to success.

There is need to realize the millions of farmers about the advantage of marketing their produce through co-operative societies. Most of the farmers tend to go to the trader instead of a co-operative marketing society for the sale of his produce. He is entirely dependent on the trader for several traditions so and the flexibility with which the trader is able to afford credit and other services to farmer.

25.9 Fish Marketing Strategies

As a mechanized and modernized fish farm grows and transfoms itself into a business entity, aquaculture management in general face vastly different strategic requirements at the individual farm levels. In this case, progressive fish farm managers are always concerned with the strong meeting competition in the concerned business, protecting their market shares and benefits, assuring short-term prosperity and long-term health of their business. Their job is fabricating the right competitive strategies in the market, as their fight is always in the market place. The kaleidoscope variety of strategies and strategic stances that are being expressed about in books related to marketing strategy of aqua-products adopted at the business level. Whereas the strategic demands at the farm level involve around the growth of the entire aquaculture organization with the best utilization of the aquaculture resources across the existing farms, the farm level strategies revolve around the task of smartly marketing their products and assuring profits and growth. The aquaculture organization as an entity provides the sap and succour on which the individual farms and their strategies thrive.

As a matter of fact, the discussion on business level strategies/marketing strategies is beyond the perview of this chapter since it is the subject matter of another book on the Marketing Management as a whole. We have chosen a brief structure and presented an outline of the overall strategic planning process of fish marketing and trade in this chapter so that we can provide the proper perspective for the planning task at the business level. We have seen in this chapter how the fish markting and trade organizations take certain decisions regarding the growth objectives and relative priorities of its diverse businesses. The overall plan is the strategic agenda from which each business unit takes its direction and commences its planning and strategy formulation jobs both for domestic and overseas markets.

25.10 International Fish Marketing Strategies

International fish marketing strategy involves all the activities that form part of domestic marketing. A prospective fish farming industry engaged in international marketing has to correctly identify, assess and interpret the needs of the overseas customers and carry out integrated marketing to satisfy those needs. There are several characteristics that are unique to international marketing. When the marketing strategy crosses the national boundaries, it becomes more complex. The resulting problems transcend those of marketing finance and production. A range of legal sociological, political and cultural dimensions enter the picture. While all these factors contribute to the complexity of international marketing there is one environment and cultural dynamics of global markets. The difference between domestic and international marketing is essentially environmental and cultural in character. Cultural diversity continues despite the world getting closer. Modern communication and transport systems have no doubt brought nations closer, but the cultural differences continue. The cultural approach accordingly will provide the correct insight into global markets.

MAIN TASK INVOLVED IN INTERNATIONAL FISH MARKETING

1. *Market Selection* : Proper market selection is an integral part of the strategies for international business. The opportunities afforded by the various overseas market must be carefully evaluated keeping in view of the resources and constraints of the farms. Through careful market selection, the opportunities can be fully exploited and the risk involved in international business can be minimized.

2. *Product Selection* : Product selection is also important as market selection. In fact, all aspects of the product such as product components, packaging, package size, labelling and usage instructions may need adaptation depending on the situation. They have to be matched with the need of the specific export markets. International marketing farms must pay due attention to this requirement.

3. *Entry Strategy for global Marketing* : If an enterprise wants to sell his products in a foreign country, it must determine the best mode of entry. Its choices are exporting, joint venture and direct investment. The entrepreneur exports its product from time to time or it may make an active commitment to expand exports to a particular market.

(i) *Joint Venture* : Joint ventures contribute to an important route for carrying on international market. Most of the farms have set up a number of joint ventures in several countries of the world. Setting up joint venture in foreign markets is a useful strategy gaining entry to global markets. Several projects on deep sea fishing and aquaculture farms have taken up to accelerate production and marketing in collaboration with other countries. Advanced fishing nations such as Japan, Korea USA, and Phillippines, for example, have joined hands with the Indian entrepreneurs for development of deep sea fishing in the Indian waters.The joint venture concept should also be taken in farmer roots in Inland water resource in the future; more enterprises should be picked it up as a strategy for realizing their globalization ambitions; not only has the number of joint ventures grown in both aquaculture and fishery sectors, the investment in individual ventures should also be increased substantially.

(ii) *Direct Investment*: The biggest investment in a foreign market comes through direct investment, the development of foreign-based production facilities. If an enterprise has gained experience in exporting and if the foreign market is large enough, production facilities offer many advantages. The farm may improve its image in the host nation because it creates employment opportunities. Generally, a farm develops a deeper relationship with customers, local suppliers, and distributors, allowing it to better adapt its products to the local market. And finally the farm keeps full control over investment and hence can develop marketing policy that serve its long-term international objectives.

Entrepreneurs start with indirect exporting which involves less investment because the farm does not require an overseas sales force. It also involves less risk. Domestic-related export agents and co-operative organizations bring know-how and services to the relationship.

Entrepreneurs may eventually move into direct exporting,whereby they handle their own exports and where the investment and risk are greater and so is the potential return. In general, farming industries may set up a domestic export department that carries out export activities.

4. *International Marketing Research*: It is an another important task of export marketing. It should be kept in mind that the planning of marketing should rest on the foundation of marketing research. Without reliable marketing information gathered through elaborate marketing research, major decisions in international marketing cannot be taken. It is very difficult to gather information about the various foreign markets, highly expensive but there is no other alternative for this step.

5. *International Marketing Communication*: In contrast to domestic marketing, an international marketer should venture into a foreign market only after acquiring a sound understanding through advertising, trade promotion and high level personal selling has to be adapted to suit the foreign markets.

6. *Mastering the Procedural Complexities*: The international enterprise is required to master a variety of procedural complexities covering a number of areas such as import-export licences, customs, foreign exchange, mode of payment, shipping freight procedures, quality and packaging regulations.

7. *Packaging the product*: In the field of exports, in addition to product quality, attention needs to be given to the quality of packaging as well. Improved and more attractive packaging will no doubt enhance export opportunities.

Great studies have taken place allover the world, in the area of packaging. New materials and new production technologies have revolutionized packaging practices. In the case of products for which packaging is essentially meant to give protection, effective protective packages have emerged. In cases where packaging is meant to add charm to the product, entirely new concepts and styles of packaging have emerged. It is cardinal that fish farming industries keep abreast of these developments. The requirements of packaging for the export market must be adequately appreciated by the fishing industries and exporters.

(i) *Package Materials*: Over the years, great changes have taken place in package materials. Previously, wooden boxes and baskets were the main materials used for packing. But this practice is discontinued because of the risk of pollution and subsequent putrefaction of the products. During transport, fishes are packed in containers made of bamboo slits charged with ice. However, this slowly gave place to paper and paper-board, due to the shortage in wood supplies. High quality paper coated with easy material is used for packing frozen shrimps, lobster tail, crabs, etc.

Metal containers are also used as a packaging medium. Metal containers constitute an excellent packaging medium for processed shrimp, small fish, and fillets. The acute shortage of tin in some countries makes metal packaging rather costly. In many Asian countries, jute bags lined with polythene sheets or synthetic bags are extensively used for marketing the sun-dried products.

With the growth of the petro-chemical industry, a new range of package materials has entered the export marketing scene. The new family of synthetic package materials have several advantages such as water-proof and moisture-proof property, greater resistance to sun exposure, light weight, alkali and acid proof property, transparency and attractiveness. They tend themselves to attractive branding/printing on them.

The continuous search for improved types of packing has led to a stream of innovation in this field. These innovations have embraced package materials, package design, and package convenience. Some of the innovations have been brought in with a view to reducing the costs of packaging.

(ii) *Package Design*: Similar to the size and shape of the package, design too have a communicative role. A good package design is definitely attractive to consumers. A bad package design can harm the entire product message. The product might be a quality product, but it may not get accepted in the market if the package design does not succeed in evoking a favorable response. Package design and color have to be harmonized rhythmically to make the package communication effective. Labels, pictures, and other illustrations on the package increase its communicative value.

8. *Product Quality*: Apart from the essential elements of product policy, the product management essentially involves many more issues. Product quality is one such issue and as vital as it dictates the ultimate fate of the enterprise in the market place.

It should be kept in mind that an aquaculture industry cannot attain a good and lasting reputation unless the product quality is assured. Aquaculture entrepreneurs in several advanced countries have emerged as export leaders only through their quality assurance. Rigorous control on the production process makes their products as zero defect and consquently globally acceptable.

Aqua-products must always be exported after assuring the quality of products and at the same time the quality has to be monitored throughout the life-cycle of the product *i.e* from production to the final consumption stage by the users. If there is any complaint regarding quality, the farms should have to tackle all complaints.

9. *Pricing*: Pricing is one of the most important decision areas of marketing. As a generalization, price and sales volume together decide the revenue of any business. As the sales volume in itself is dependent on price, pricing really becomes a vital decision area on account of certain other factors besides its important role bringing revenues and profits to the business.

An aquaculture farm will have a number of objectives in the area of pricing. Profit is one of the major objectives in pricing. Most farms adopt profit optimization rather than profit maximization as the pricing objective as they consider an optimum levels of profits over a long period as a more sound objective of pricing than maximum profits. Hence, optimum is an ambiguous term. Its definition will differ from farm to farm. The farm must evolve a clear idea of the optimum from its preception of aquaculture business realities.

It is important to note that no farm can remain pleased with a single objective in pricing. The various objectives sought to be conceived through pricing are noted below:

(i)   Entering new markets.

(ii) Target sales volume.

(iii) Minimum return on investment.

(iv) Providing services at prices that will stimulate economic development of farmers.

(v)   Stabilizing prices and margins in the market.

25.11 Fish Marketing Strategies for Indian Farms

In the preceding sections we have had an idea of the fundamentals of international marketing. Here, we shall discuss the issues that are specific to India and the Indian farms in their effort at global marketing, and the possible strategies that the Indian aquaculture industry can employ for the global market place. The discussion will be carried out under three sub-sections that follow.

ACHIEVING AQUACULTURAL COMPETITIVENESS

Achieving aquacultural competitiveness is an essential prerequisite for expanding exports. No country achieves a high level of export without achieving a high level of competitiveness. Aquacultural industries as production entities have to make their products internationally competitive both in quality and price if they have to realize their export ambitions. The poor competitiveness is the effect of gaps in a variety of vital areas such as technology, productivity, quality, and infrastructure. Aquacultural competitiveness involves a number of steps that follow.

1. *Conductive Policy Package*: So far the export context is considered, the government has taken several steps under the new trade policy and economic reforms such as

stable import-export policy and a conductive framework for export growth, simplifying customs procedures, allowing exporters to use a reasonable percentage of their net foreign exchange earnings for export promotion and exempting export units from any kind of domestic duties.

2. *Improving the Infrastructure*: Poor infrastructure is one of the main constraints for exports. Transport and warehousing faciliies, farm mechanization, adequacy of water and hatchery units are all inadequate and hence the exports are handicapped due to these infrastructural inadquacies. The government has to ensure that all these inadquacies will be removed in the shortest possible time.

3. *Systematic Promotion*: Systematic development and promotion of market should form an integral part of India's strategy in international marketing. Much emphasis has been given on export promotion but the result is not commensurate with the expenditure. However, the government has to sharpen its focus of the promotional effort and get better results out of the export promotion expenditure. Market promotion can generate better results if the various export promotion agencies are engaged in this task.

4. *Bridging the Technology Gap*: If the nation and the enterprises have to achieve global aquacultural competitiveness, it is cardinal for them to bridge the technology gap in the production. Aquaculture industry should produce what the overseas markets need and produce them to their technological standards. The aquaculture industries have to do by opting for the latest technology, adapting it, and by committing to higher investment for technology upgradation. In spite of changes in trade policies, a lack of latest technological dynamism in aquaculture industry has prevented a rapid increasing in export.

5. *Productivity Improvement*: Besides upgradation of technology, productivity improvement is the crucial role for achieving aquaculture competitiveness. Unless productivity is improved, the industries will find it difficult to compete in the international markets. However, upgradation of technology will contribute substantially to productivity improvement. The industries must exploit the aquatic resources of the nation to its full potential and ensure that their productivity is kept high and their products remain price competitive.

6. *Improving Product Quality, Packaging and Delivery*: Immediately after harvesting, fish farmers are always in search of packaging and delivery to market place. However, these three terms are related to each other. Weakness in anyone of these factors is sufficient to make a country's position in world markets vulnerable. Therefore, these weaknesses, if present, should be overcome.

7. *Selectivity in Products*: The type of products to be exported is a fundamental part of product-market strategy. It is obvious that the aqua-farms cannot seek to export all products that they make. Products that have a high potentiality for export have to be identified and exported.

THRUST PRODUCTS FOR EXPORT

The aquaculturists and the government have identified a number of aqua-products as high focus export items. The dictates of current competitive advantage definitely indicate that the aqua-industry engaged in export marketing would be better by concentrating on a few selected products which are assured winners. There is a lot of wisdom in adopting such an export basket. Apart from the export of modern terms, the country's export is accounted by agricultural and aquacultural products. Several aqua-items have registered a phenomenal growth in recent years. Fish and fish products cannot be expected to execute a sudden prodigy in the export scene. A lot of efforts over a relatively longer period may be required to make the product export winners for the country. The aqua farms have to move through major mechanization and efficiency improvement to make this occurrence. Industries have to ensure the maximum possible value in addition to their export of the aqua-products.

STRONG COMPETITIVE ADVANTAGE IN AQUA-PRODUCTS

Though the country has a strong competitive advantage in aqua-products, it was never exploited in full potettial. Within the aqua-product range, the country enjoys a special competitive advantage in shrimps. By concentrating on prawn and shrimps, the country can tap a good portion of the world market for aqua-foods. The Indian sub-continent abounding in potential aquatic ecosystem provides the ideal setting for shellfish cultivation.

Inspite of several strong advantages on this aspect, there are few weaknesses too. *First*, 80 per cent of the existing exports is in the form of the commodity, not in the form of branded and packaged products. *Second*, a few industries have entered the business of shrimp exports. Large-scale exploitation of the market requires large-scale farming and modern facilities for preservation and processing. For example, industries may have to shift from block-freezing of entire consignments to individual quick freezing method, for switching over to branded and packaged marketing of finfish and shellfish. Development of infra-structure relating to crafts and gears, setting up of aquaculture and hatcheries and reprocessing facilities are apt prerequisites for a large scale of entry into the world markets. This, however, requires heavy investment in technology and equipment which large establishments alone can provide.

But, as a matter of fact, this situation is now rapidly changing. With the liberalization of the economy, the Indian aqua-industry has become ripen for foreign investment and large-scale joint venture operation. Also, restriction on foreign investment has been withdrawn from the industries. The private sector is being highly encouraged to enter this field. However, the scope for entering sizeable foreign exchange from the export of high-price shrimps, tuna and other species has been a significant factor in the development of aqua-farms in recent years. Many enterprises have set up vertically integrated semi-intensive and modified semi-intensive fish farms in different maritime states with a production range of 4 - 10 tonnes/ha/crop. Most of these projects have technical collaboration or joint venture partners from overseas and have 100 per cent export-related capacity.

1. *Issues Relating Attention*: In order to put the country's competitive advantage to proper use, the establishments actively engaged in aqua-product export have to be sorted out a few important issues relating to production and marketing. *First*, since hygiene standards are poor in India, the Indian products may not have a very good image among the health conscious consumers of the developed countries. All the farms involved must devote their full attention to this issue and enhance the hygiene standards. *Second*, fixing the delivery products to overseas customers is a fundamental marketing requirement. *Third*, poor attention to quality assurance hamper the growth of aqua-products. For example, with the formation of the unified European market, all the exporting industries will have to comply with the ISO 9000 standards and the other EU directives for exporting to Europe. Therefore, Indian exporters have to do their utmost and correct the malady in quality assurance.

Another issue demanding attention relates to branding and value addition. Most farms export only the raw products and are not able to earn greater margins which may accrue to them through value addition and branding. More than 65 per cent of Indian aqua-foods exports are unbranded. At present, however, this situation has been changed. Some farms such as Alsa Marine and Harvests, for example, have gone in for export of their products with their own brands. In general, quality branded products are sold in the supermarkets of Europe and Japan.

The competitive advantage of the country in aqua-products can be better exploited if the issues in processing and marketing are sorted out. The Indian aqua-industries can then effectively counter competition from the southeast Asian countries like China, Thailand, Taiwan and Indonesia. Aqua-products as a group has the potential to become a star item in country's exports and country's share in the global aqua-product markets can be substantially enhanced.

25.12 Quality Aqua-products is Required in Fish Markets

Quality product is the most crucial decision in the fish marketing context whether that product is sold in the market of developing countries can be supplied to the developed countries. The decision depends on the situation and the type of the product. The fish processing industries must find out what kind of product is required by the developing or developed countries. In some cases, similar product may be equally acceptable in both developing and developed countries. In other cases, the aqua-product can be the same but the package may have to be specially designed for different markets. For example, good quality packages are more popular in the developed countries. Many factors significantly contribute to this reality. In the first place, limited funds and low purchasing power force the rural consumers to prefer unpacked or bad quality packs low unit price.

Branding of aqua-products also needs skilful handling and marketing in both developed and developing world. The brand name is the assured means of conveying quality to different consumers. Besides quality, the processing units exhibit sustained interest in different products and their markets. Whether the same brand name should be used in both developing and developed countries or whether different brand names must be adopted for either developed world or developing world is a matter of conscious decision by individual industries depending on the context. In some cases, the brand name that is suited to the developed countries may not be quite suitable for the developing countries. In many cases, however, the same brand is provided cost-effective.

25.13 Tapping the Fish Markets

While the fish markets definitely offer a big attraction to markers, it would be totally native to think that fish industry can easily enter the market and walk away with a sizeable share of it. A fish industry seeking a share of this market has to work for it, since the market bristles with a variety of problems. The enterprise has to grasp with these problems and find innovative solutions to them.

PROBLEMS INVOLVED IN FISH MARKETING

1. *Transport Problems*: Transportation infrastructure is no doubt poor in many rural and urban areas. Most of the market places are not connected by rails and roads at all. Many regions have only 'Kacha' roads and many regions of the interior are entirely unconnected by roads. Because of these problems, inaccessibility, delivery of fish and fish products and services continues to be difficult in aquaculture areas.

2. *Warehousing Problems*: Fish farms find it quite difficult to get adequate godowns in many regions. And there is no warehousing agency in the interiors of rual regions. In aquaculture industry, the warehouses are mostly owned by co-operatives. The cooperative societies provide the warehousing service only to their members. Consequently, most farmers has to manage with the other network to deliver their produce to the market.

25.14 Market Segmentation in Fish Market

The fish market can be segmented either geographically or demographically.

1. *Geographic Segmentation*: When the fish market is segmented geographically, different bases that follow can be used for such segmentation.

(i) *Nutrition*: Regions endowed with favorable nutrition have the more purchasing power compared with nutritionally-handicapped regions.

(ii) *Level of Productivity*: Unproductive and productive water bodies pose different marketing environments. And within productive water bodies, continuous harvesting and stocking systems indicate varying levels of marketing.

(iii) *Level of Technology*: Different parameters such as adoption level of high yielding varieties, adoption level of high-value species, use of high quality feeds, and standard aquacultural practices will have to be reckoned with.

2. *Demographic Segmentation*: Demographically, there are many possibilities of segmenting the fish marketing. Population aggregation or population dispersion can be one base. The fish marketing can be segmented on the basis of different size classes with respect to population. Surveys have shown that the population of more than 20 crore in the age group of 25-50 years dominates the purchase of fish from the market.

Regional differences can be another base for demographic segmentation of the fish market. For example, while most South Indian people are not accoustomed with fish consumption in their daily diet, that of West Bengal and North-eastern regions prefer to purchase different kinds of fish species for daily consumption.

The consumers can also be segmented into seasonal income and regular income segments. Most of the consumers are characterized by seasonality of income. This class generally consume low-values fish and fish products (especially sun-dried) since they may not have the purchasing habits of high-values fish products.

25.15 Suggestions to Improve Fish Marketing Systens

It is an accepted fact that there is no well-organized fish marketing system in many tropical countries. Such marketing system will definitely provide security to fish farmers. Immediately after landing, fish are sold to local markets or transported to metropolitan cities through rail or road. The unsold consignments are preserved with ice for one to several days. The suggested lines of development of fish marketing system are, however, mentioned below.

CONSTRUCTION OF REGULATED MARKETS

As noted earlier, a number of malpractices prevalent in most of the fish markets take away an actual share of the price by the buyers and place the farmer and the trading class at an unfavorable condition. Fish marketing should, therefore, be governed by regulated markets where transactions are governed by various rules and regulations. Markets are generally controlled either by local bodies or under state legislation. The market associations or fish sale committee are formed by fish culturists and traders and the government and non-government contractors or any voluntary organizations look after the functioning of these markets. This marketing systems are solely responsible for the enforcement of proper trading practices, introduction of free auction system of sales, enforcement of standard weights, etc. For better marketing standards, sheds, yards, and freezing rooms should be installed.

STANDARD WEIGHT AND GRADING

Grading means the sorting of fish according to their size, species, and quality according to standards fixed by the trade or as required by the consumer. Grading is important in the marketing of fish in view of the fact that the harvested fish are not homogeneous.

1. *Benefits of Grading*:

  (i) Grading of mixed fish stocks is necessary for selling better quality of fish to the market at better prices in the market.

  (ii) Grading helps consumers to purchase fish according to their purchasing capacity and choice.

  (iii) During transport, fish are graded according to their size and species to avoid the mortality of live fish stocks and to reduce fish spoilage.

ARRANGEMENT OF BETTER TRANSPORT

The cheap and easy transport systems encourage farmers to sustain their produce to the market according to their own choice. In rural areas, roads play an effective role since there are connecting links between the farm and the marketing places.

MARKETING RESEARCH AND INSPECTION

Marketing research is often neglected and at present the research activities on this line have been totally stopped in many tropical and sub-tropical countries of the world. Research in a planned organized manner not only helps collect the needed informations in identifying problems and finding solutions for them but also to develop market-related knowledge to improve market decisions. The need to evaluate marketing methods, changing demands, prices and costs is also equally important. Routine inspection of fish markets should be done by farm managers, sale and marketing personnel.

25.16 Fish Marketing and Trade

THE NATURE OF TRADE

The nature of commercial activity of fish and fishery products between nations and regions within the country has been divided into two major types such as (1) Domestic trade (or internal trade) and (2) International trade. Domestic trade means the action of conducting business that takes place within the geographical boundaries of a region or nation. It is also known as *home trade*. International trade, on the other hand, is the transactions of products among different countries or transactions across the political boundaries. International trade, thus, refers to the exchange of aqua-products and services between one country and another. It is also termed as *foreign trade*.

CHARACTERISTIC FEATURES OF FOREIGN TRADE

  1. *Heterogeneous Markets*: In the international economy, the world markets lack homogeneity on account of differences in food habit, product preferences, customs, etc. The behavior of foreign buyers in such cases would, therefore, be different.

  2. *Different Currencies*: Foreign trade involves the use of different types of currencies. Therefore, each country has its own policy with regard to exchange rates and foreign exchange.

3. *Different National Groups*: Foreign trade takes place between differently cohered groups. The socio-economic environment differs markedly among different nations.

4. *Different National Policies and Government Intervention*: Economic and political policies greatly differ from one country to another. Policies pertaining to trade, commerce, import and export also differ widely among countries though they are more or less uniform within the country. Tariff policy, import quota system, subsidies and other controls adopted by governments interfere with the course of normal trade between one country and another.

25.17 Foreign Trade: An Indicator of Economic Development

Marketing and trade of aquacultural commodities of any nation are most important indicators of its level of economic development. However, so far as India's marketing of fish and fish products are concerned, it would be clear from various marketing statistics that a number of important changes have taken place. Among these, two important changes include (1) Rapid expansion in the value of trade and (2) A change in the composition 0f trade.

RAPID EXPANSION IN THE VALUE OF TRADE

The changes that have occurred in the value of India's foreign trade can be visualized from the Table 25.1 as noted below.

Table 25.1. Value and Quantity of India's Foreign Trade of Aquaculture Products.

Year	Quantity exported (Metric Tonne)	Value (In Crore)
1950	19,657	2.45
1960	16,337	4.02
1970	37,175	35.53
1980	74,542	218.80
1990	1,33,653	818.41
1992	1,91,314	1,581.43
1994	3,01,278	3,565.51
1996	3,53,675	3,890,01
1998	3,13,503	4,700.55
2000	3,40,003	5,000.00
2005	3,89,650	5,800.40
2010	4,20,000	6.550.00

Source: Modified after Shahin Parameswaran (2001)

It is clear from the table that India's foreign trade of aqua-products has significantly increased during the last 25 years of economic planning on aquaculture sector.

CHANGES IN THE COMPOSITION OF TRADE

The commodity composition of India's trade has undergone a notable change in the last 20 years or so. Item-wise export of aqua-products, as a matter of fact, has substantially increased both in terms of quantity and value (Table 25.2).

Table 25.2 Item-wise Export of Aqua-Products

Item		1995-1996	1997-1998	1999-2000	2004-2005	2009-10
Frozen shrimp	Q	95724	100720	109868	114245	13745
	V	2356.8	3134.5	3635.0	3928.1	4445.70
	$			844.76	1964.03	
Frozen fin fish	Q	100093	188029	129679	158758	178074
	V	372.30	726.73	532.72	685.80	1045.40
	$			123.80	342.90	
Frozen squid	Q	45025	35035	34648	38270	42465
	V	319.60	270.90	293.17	311.22	670.20
	$			68.16	155.61	
Frozen cuttlefish	Q	33845	37258	32528	36745	40415
	V	260.9	323.41	284.13	301.35	570.10
	$			66.03	150.68	
Frozen lobster	Q	1587	1289	1472	1688	1985
	V	51.06	47.80	50.28	55.30	62.45
	$			13.42	27.65	
Chilled items	Q	2773	3181	3088	3475	3890
	V	26.08	44.31	44.97	48.08	54.55
	$			10.48	24.04	
Live items	Q	1700	1755	1669	1890	2345
	V	29.34	21.31	38.00	43.05	49.60
	$			8.83	21.53	
Dried items	Q	7292	5669	6136	6670	7340
	V	40.32	33.45	42.65	45.90	53.75
	$			9.91	22.95	
Others	Q	12875	8183	22377	29080	33470
	V	87.41	52.83	224.98	235.00	350.05
	$			52.29	117.50	
Total	Q	300914	381679	341475	390821	323729
	V	3498.82	4654.89	5146.01	5653.76	7301.80
	$			1197.65	4376.68	

Q = Quantity in metric tons, V = Value in Rs. crores, $ = US Dollar million.

Source : Compiled by the author from different published literatures.

25.18 Factors Affecting Trade

Several factors account for the inception of trade which dramatically affect its volume and importance to different regions or countries and restrict its free operation. Some important factors are discussed in brief.

DIVERSITY IN AQUATIC RESOURCES

The aquatic resources of the world are very unevenly distributed as a result of variations in soil, water, climatic factors, and topography. Soil largely determines the type of both flora and fauna that can grow in any given region and also decides what type of fish biomass can be grown on a wide range of soils; but generally speaking, the ideal soil for fish cultivation should be moderately clay or loany soil so that water does not readily percolate through soils. Suitable abiotic factors are also inevitable for fish production. The most obvious differences occur, however, between the major soil zones ranging from the hot and wet tropics through sub-tropical to temperate regions. Some kinds of soils are highly effective in producing finfish and shellfish. While trout and salmon are best grown in cold regions, for example, carps can only be grown under specific conditions of rainfall and temperature particularly in the tropics. In case of aquaculture products with a wide geographical spread such as tilapia, common carp, shrimp, the quality of the product, the growth and production of any particular species may be affected by soil (and also other factors like temperature and climate). Thus the production potential in the tropics is greater than those in the cooler regions since the growth of fish depends very much on the intensity of environmental temperature.

Territorial specialization promotes trade when there is a damand for the products produced in a given region. Tropical shrimps and lobsters are regarded as main export items as western consumers acquire a mouth-watering form for these products. In some regions, some products have no demand. For example, trout and salmon are only cultured and marketed only in the cold temperate regions and these items have no demand in Asian markets.

Topographical differences govern the type of fish crops to be raised. They may affect the cost of production. For example, it is very difficult and less economical to cultivate trout under tropical conditions. The cost of production in turn affects the competitiveness of any product in the regional/international market. Moreover, water areas with much high lands have very low fish production potential while those with much low land or plains have greater fish production capacity. Mountain regions have comparatively less or no marketing advantage than in plains. Lack of adequate transport networks confined trade to very limited regions.

The availability of quality water for aquaculture around the world is very limited. Huge quantities of freshwater (and also marine waters) resources contain dangerous levels of toxic substances exceeding prescribed safety standards. This has created severe problem in producing robust fish for safe consumption. Production of diseased and insufficient quantities of fish from contaminated resources have no adequate value in the export market and even in internal market.

It is important to note that the mere existence of fish stocks is not a guarantee that they will form a valuable trading asset to the nation which possesses them. Several factors such as assured quality of products, type of species raised, and the availability of transport networks will definitely affect the competitiveness of aquatic resources considerably. Abundance of robust fish is also important as a basis for farming industry and availability of valuable species may encourage commercialization and thus allow a nation to trade in aqua-products.

POPULATION

Several countries have established a tradition of producing and exporting high-value aqua-products. Thus India, China, Thailand and Singapore produce a variety of high quality shellfish products. The tradition behind such products makes them in high demand and the new processors of the same items may not be able to compete.

The population density of a country may have far-reaching effects on trade. In general, densely-populated countries have a large volume of trade but little international trade because most of their production is used for domestic consumption. Thus, although India and China are the world's greatest carp producer, it has only a very limited amount for export. Some countries have a surplus for export because they are not at all considered fish as a principal item in their daily diet.

Another way in which population size affect trade is through the availability of labor. Aqua farming and integrated fish farming rely on large supplies of labor and cannot therefore be practised in sparsely-populated areas. Hence densly-peopled countries may have trading advantages in this respect. The value of a large labor supply again depends on the standard of living in a country. Where living standards are low, wages are low and the production costs in labor-intensive occupations can be reduced. This gives a trading advantage to certain countries because products can be made cheaply. In countries where the living standard is high, a large labor force is only an advantage if it can be employed in the making of high-value products. The effect on international trade of both population size and the living standard of the population is thus closely related to the stage of economic development of any country.

EXTENT OF FOREIGN INVESTMENT

Foreign investment, particularly in under-developed countries, can affect the volume, type and pattern of world trade. These countries have very little capital available for the development of aquacultural and processing industries. The developed countries avail themselves of economic potential either by providing foreign aid or by direct investment. By investing in the tropical countries or rendering technological assistance, they secure a ready market for their products and employment for their people. At the same time, the under-developed countries undergo substantial economic changes: derelict water areas are cleared to make way for fisheries and aquaculture, fish processing industries are developed, and a number of factories have been developed through which the exports are channelled. With each phase of the development strategy, new settlements emerge, more jobs are available to the local people and the national income rises gradually. This increases demand for imported products and thus improves markets for the developed countries, while development promotes the export earnings of the developing nation.

TRANSPORT

Lack of suitable transport facilities severely restrict trade to very limited regions. In many remote places where the transport networks are almost non-existent, harvested fish could not be traded over long distances. With the development of rail and air transport, improvement in insurance facilities, and the development of improved containers such as refrigerated ships for perishable products, world trade began to expand to its present magnitude. Due to the expansion of trade, that were used in limited areas in the past, can now be available in many parts of the world. Though distance is no longer a barrier to international trade, ever in the most inhospitable geographical areas, transport still does play a crucial role in determining the comperative trading advantages of different countries. Where good transport facilities exist, products can easily be exported which will reduce costs. Products can only enter trade if they can be transported to internal/international markets at a comperative price.

GUARANTEE OF AQUA-PRODUCTS

Stringent norms must require that every aqua-product sustain a guarantee as to its content of nutrients, vitamins, minerals, etc. If any other substances are identified as being present in the product, its guaranteed content must be listed. However, if the hygienic condition of the aqua-products is not guaranteed, their export potential will obviously be jeoperdized.

Aqua-products are sometimes grouped according to their quality. This grouping is, however, cardinal when several analyzes are offered and the comperative costs become the deciding factor as to which should be purchased.

AQUA-PRODUCT INSPECTION AND CONTROL

To ensure product quality, rules regulating the sale of all aqua- products are inevitable. Such regulations protect the customer as well as the reliable entrepreneurs by preserving products of unknown value of the market. The processing industries are generally required to print the following data on the packing box.

1. Number of net kilograms of product to a package.
2. Name, brand or trademark.
3. Hygienic condition guaranteed.

4. Manufacturing and expiry dates.

5. Name and address of manufacturer.

25.19 Role of Aquacultural Commodities in India's Foreign Trade

We have noted in brief the major trends in India's foreign trade of aquaculture products during the last 50 years. An important change that has occurred in the structure of foreign trade immediately after independence was that India tried to variegate its export structure both in aqua-culture and agriculture sectors by adding new items of exports so as to earn foreign exchange to meet the increasing import needs of the economy.

EXPORTS OF AQUACULTURE COMMODITIES

India has a vast inland and marine water resources producing a variety of aquacultural items. Being a world leader in a number of aquatic food products, India has been exporting these items since then. About one-tenth of India's total exports originate from this sector. In 1970, the total aquacultural export amounted to Rs. 35.53 crores. The value of exports from this sector dramatically increased to about Rs. 5,000 crores. These are planned to touch about Rs. 9,550 crores by 2020.

It should be kept in mind that in contrast to inland waters, the share of marine aquaculture exports has been ruling rather very high. The export contribution of shrimp culture sector of India would be 50 per cent in terms of quality and 60 per cent in terms of value. This development has counteracted the general trend of capture shrimp aquaculture sector to remain static.

A major breakthrough in the Indian shrimp culture sector in the last two decades has been the successful development of the technology of culture of tiger shrimp in freshwater tanks and ponds. Apart from the upgrading the socio-economic conditions of famers' community, the annual contribution of the shrimp culture in freshwater has attained a substantial proportion both in quantity and value of the shrimp exports.

There is an increasing tendency among inland fish culturists to switch over to composite fish culture with tiger shrimp as a main component in freshwater culture. It is beyond doubt that this eco-friendly activity will eventually spread all across the country which will have far-reaching beneficial effects both socially and commercially, particularly in respect of exports.

It is important to note that marine aquaculture products are the most common items in India's export basket. Notwithstanding the various incentives and promotional devices used, mariculture products contribute to hold their own. This dependence on products derived from marine resources is not desirable for the present and future generations. It seems quite plausible that commercial exploitation of marine water resources by highly mechanized crafts and gears has led time and again to collapse of what were known as *bountiful fish stocks.* Mackerel catch, for example, in the North Sea to a peak of one million tons in 1967 but by 1970, the stock had fallen to one-tenth of its original level. Eually shattering was the collapse of the Norwegian stock of spring-spawning herring from 10 million tons in the 1950s to almost nil in 1970. Salmon fishery is also declining due to overfishing. In fact, however, we did not take utmost care in respect of freshwater fish export particularly tilapia, Indian carps and other aquatic animals of export value. The value-added products made out of these fishes may be produced in the form of fish fillets and fish loins/steaks. It can be suggested that the country would be able to earn more foreign exchange if export-related culture of freshwater finfish and shellfish is encouraged.

INDIA'S SHARE IN WORLD AQUACULTURE PRODUCTION

The total world fish production increased from 99 million tonnes in 1990 to 130 million tonnes in 2005. The share of about 80 per cent of this comes from capture fisheries sector and the remaining from aquaculture. The value of international fish trade, however, continues to increase. In 1985, the value of international fish exports was 17 US billion dollar. In 1990, it was 35.8 US billion dollar and in 2005, exports were valued at 69 US billion dollar. India's exports have been estimated to be as 2.0 US billion dollar. India's share of the world exports of aqua-products was accordingly 2.34 per cent.

COMPOSITION OF EXPORTS

More infrastructure that the value figures are the information relating to the constituents of the basket of aqua-product exports. Frozen shrimps, fish squids, cuttlefish, octopus, lobsters, crabs, crab meat, dried cuttlefish and snails have been other aquacultural exports. Frozen shrimp, fish and cephalopods are the major items in our export basket.

Besides these, other important aqua-products that are exported are shrimp powder, cuttlefish ink, live snails and crabs, fish meal, minced fish products including surimi, algae, and sea-weeds. India's export basket is diversified, however.

25.20 Major Export Baskets

It is a matter of great pleasure that the Indian aqua-products are exported to over 70 countries of the world. The main buyers for our aqua- products are Japan, member countries of European Union, USA, Southeast Asia, and the Middle East countries. However, so far the export trend is concerned, it is evident that frozen shrimps contribute substantially to about 73 per cent of the total aqua-products export by value. Frozen shrimp, fish and cephalopod contributes to about 73, 38, and 54 per cent by value, respectively, and comes mostly from Asian regions.

There has been appreciable development in export to other countries particularly in China (20 per cent), Belgium (73 per cent), Canada (18 per cent), The United Kingdom (32 per cent), and The Netherlands (10.5 per cent). However, new markets such as Argentina, West Indies, Democratic Republic of Korea, Uruguey, Syria, Hungary, Brazil, Peru, Finland, and Kenya have recently appeared for export of aqua-products.

25.21 Current Markets and Prospects for Expansion

So far the freshwater finfish varieties are considered, it is beyond doubt that tilapia, African and Asian catfish, trout, Common carp, and Indian major carps are considered as major freshwater culture fishes. In China, the production of tilapia has increased to the tune of 6,70,000 metric tonnes per year (data for the year 2010) and one-fifths of Chinese tilapia yield finds a good market in USA and Canada. The country has proclaimed to yield more than 65 million metric tonnes of tilapia fillets each year to meet the market demand of Western world. In Thailand, different species of African and Asian catfish contribute about 10 per cent of the total freshwater fish production. However, though India and some other adjacent countries have good fish production potential from freshwater resources, major emphasis is yet to be given in the processing, marketing and export functions of high-price freshwater fish. These activities would definitely sustain integrated and streamlined production. In a large perspective processing and vis-a-vis marketing functions are the two ways to secure better terms both for the producers and consumers of aquaculture products.

There is a high demand for fishery products in Western countries where food consumption is rapidly changing especially among the elder populations due to increase in awareness of health benefits among them. Fish culture in these countries has, however, certain limitations in view of the high feed and labor costs, variable weather conditions, short growing periods, and the emergence of leisure classes. For these limitations, people of western countries have been forced to depend entirely on the production and marketing systems of Asian countries where fish farming has flourished because of year-round growing season, abundance of productive freshwater resources, cheap labor, and financial assistance by the government. However, though the market prices of several freshwater fish and fish products oscillate widely, strategies are still required for further market expansion and export of different varieties of carps. This requires the availability of raw materials, sophisticated technology for processing, training of human resource, development, and identification of markets.

25.22 Global Trends in Fish Trade

The value of international fish trade continues to increase. However, the increased volume of international trade in fishery products in the mid nineties was associated with the higher trade in low-value products such as fish meal, with the result that the value of exports increased less than the volume. In recent years, the growth in value of international fish trade has slowed down.

Developed countries accounted for more than 80 per cent of total fish imports in value terms. Japan continued as the world's largest importer of fishery products. The United States, who is the world's second major exporter of fish and fishery products, is also its second biggest importer and the European Union increased its dependence on imports for its fish supply.

In many developed countries, fish trade represents an important source of foreign currency earnings. The increase in net receipts of foreign exchange is in developing countries-deducting their imports from the total value of their exports — is impressive, rising from 5.1 US billion dollar in 1985 to 32.5 US billion dollar in 2010.

International fish trade in developed and developing countries principally depends upon the production, processing, and marketing of high-value aquatic species such as shrimp, tuna, cod, squid, sardine, mackerel, salmon, and fish meal. Trade patterns of finfish and shellfish in developed countries have encouraged by sophisticated fish processing industries geared to export/import of high-value products to markets.

EXPORT POSSIBILITIES

Fish and fishery products constitute an important component of the country's exports. The country has already emerged as a major force in the international markets in aqua-pxoducts. Aquacultural exports will continue to play a pivotal role in the export trade of India. The extent to which aquacultural exports are able to fulfil this role will be influence by the level of production and domestic consumption of different items. A sustained and energetic effort in the aquacultural sector is, therefore, crucial for the realization of targets of growth of exports.

1. *International Factors*: There are good prospects of stepping up the exports of aqua-products to different countries. In this regard, India is likely to be benefitted by certain international factors such as better trade relations with different countries. The expansion of the country's trade in aquacultural commodities and processed products would also depend upon the international action in the aquacultural adjustments contemplated to help in the long term development of exports from developing countries.

High living standards in developed countries offer large possibilities for the development of exports. A greater attention should be given to increase the harvesting efficiency of squids, sepia, loligo, and octupus for which there is a long-term export potential in these countries. The consumption of these aqua-products is very low in the country but these products are very much in vogue in European Union. Therefore, keeping the preference and taste of foreign consumers in mind, we should change our priorities.

2. *Export Promotion Efforts*: Export promotion of aquaculture products requires a well-knit strategy. The components of such a policy can be as under.

(i) Restrictions on domestic consumption may sometimes have to be resorted to for the promotion of exports of fishely products. The self-controls to be adopted on the internal market would need a specific examination. A connection between production, export, and domestic consumption of such products has to be established.

(ii) Work will have to be initiated on some aspects of export marketing such as market identification, collection of information, publicity, and statistics on foreign markets.

With improvement in the present reporting system through the trade representatives of India in a number of countries, exporters might be better informed of the possibilities of actual sales.

(iii) Export promotion efforts should encompass measures for pinpointing areas of different aquacultural products in a raw, frozen or finished form, producing these items at competitive prices, and developing the infrastructure for processing aquaculture products into an exportable form.

(iv) Funds should be earmarked for investment on integrated aqua-culture projects for production, processing and exports of suitable species which are in demand in foreign countries.

(v) Efforts should be directed to promote the export of value-added products. The most important factors that are necessary here are creation of suitable institutional frameworks, quality packaging, selectivity of export markets, inspection, and stability in government policy.

(vi) Establishment of autonomous organizations is one of the cardinal steps for increasing the exports through planning. Such organizations should be given more facilities in matter of undertaking market surveys and research. Such organizations should be provided with adequate finance for playing their due role in the export effort.

(vii) Steps should be taken to promote private enterprise in the field of aquacultural export promotion. Export-related farms of viable sizes may be set up in the potential areas of the country. There is also the need for the adequate amount of aqua-cultural inputs for high production of robust and disease-free fish.

(viii) The increase in fish production has been mostly in respect of carps which have no export value and the increase in their yield has led to diminishing returns to the farmers. For this reason, adoption of advanced and refined technology to increase export-related production of catfish, Nile GIF tilapia, and freshwater prawn has changed this situation.

(ix) Farmers should be diversified into culture of these species which has a strong export market, mainly to USA and European Union in whole as well as filleted form. Obviously, there is an urgent need and responsibility on the part of the authorities concerned to formulate the policy and promote the spread of the culture of these species allover the country.

(x) Since the Nile GIF tilapia has an enormous potential as an exportable commodity, the government has given permission to some private organizations (such as Water Base Ltd., Chennai) to import Nile tilapia (*Tilapia nilotica*) from abroad, for production of mono-sex seed. Their intensive culuure will not only for meet the domestic demand but also to cater to the surging demand mainly in USA for whole dressed tilapia and also its fillets in IQF or block-frozen form.

3. *Export Promotion Bottlenecks*: It is necessary to intensify export drive removing the bottlenecks that stands in the way of export promotion. India's markets will have to be maintained and new markets are to be explored. One of the causes for the slow growth of Indian exports — and for that matter, in the exports of all developing countries — was that the GATT (See Panel 25.1) did not make any sincere effort at enabling the poor countries to expand their export trade. While high tariff walls were not being lowered by developed countries, any effort made by the developing ones to subsidize their exports to make their prices competitive were frowned upon.

Panel - 25.1
GATT (General Agreement on Tariffs and Trade)

The aim of GATT is to equalize trading opportunities by restricting future grants of trading preferences where they do not already exist and to promote trading agreements on tariff reduction. A number of developing countries have access to the European Union (EU) market according to the trade provision of the Lome convention, an agreement between the EU and 69 countries in Africa, the Caribbean, and the Pacific (The ACP countries). Under this agreement, fishery exports from ACP countries can enter the EU tariff-free. These countries depend on the EU market, to which they export 75 to 85 per cent of their total fish exports.

The GATT Uruguay Round in 1994 concluded that customs regions need to be harmonized globally. For aqua-products, the main outcome of the negotiations was an agreement to lower customs duties by an average of 26 per cent, meaning that many importing countries will lower the tariffs for a number of products. Some developed countries such as Japan and the United States have reduced their customs duties considerably, and several of the EU tariffs for aqua-products ramain unchanged. However, important tariffs for tropical shrimps will be reduced from 18 to 12 per cent for Sub-Saharan African exporters. On the other hand, canned tuna will remain subject to an important tariff of 24 per cent. However, EU concessions are subject to re-negotiation and further reductions may be made in the future.

For countries of African continent, such development will ultimately mean increased competition from other developing countries. The margin enjoyed by the ACP countries will be eroded and, even more significant, in the spirit of the GATT principle that all trading patterns should be treated equally. However, the outcome of the negotiated international trade agreements will not have any immediate effects on many ACP countries, although in the interval stage of regulation to these business systems, the fish culture activities will require to harmonize to new arrangements in order to persist competitive.

India is to make concerted efforts at export promotion by a more effective implementation of export promotion and quality control schemes. There has been no provision for expenses on quality control and inspection and trade fairs held abroad. The hinders of our export promotion should be able to provide market information to Indian exporters at nominal charges.

The quality of our export aqua-products is not always according to specifications. The aquaculture industries are not in all cases aware of buyers' taste in foreign countries. These defects are to be remedied.

WORLD MARKETS FOR SURIMI-BASED PRODUCTS

The world demand for surimi-based products has been estimated to be around 2,65,000 MT of which about 80 per cent of the total is consumed by Japan. Other countries such as Korea, European Union, Russian Federation, China, South America, Singapore, India, Indonesia, and Vietnam produce and consume moderate (4,000 MT on an average) to high quantities (72,000 MT on an average) of surimi-based products.

1. *Japanese Market*: Since surimi-based products are very essential food commodities in Japan, their production potential is highly significant. Production potential of surimi-based products, however, varies between 7,20,000 and 8,85,000 MT. Japan is one of the leading countries of the world in surimi-based products, representing 87 per cent of the total. Several of the products consumed in the USA and EC markets account for about 20 per cent of the total products processed in Japan.

2. *US Market*: Surimi-based products (especially in the form of 'Crab imitation products', which are termed as flakes,*chunks*, *cambo*, and *sticks*) have enormous potential markets in North America. These products essentially contribute to about 95 per cent of US market.

3. *EC Market*: Since the EC market has the potential to consume about 30,000 MT per year of surimi-based products, there is great opportunity for expansion in the market if quality of the product is ascertained for the products traded. The consumption volume of surimi-based products in the European Union has been estimated to be around 67,000 MT (Data for the year 2009). A substantial quantity of products is imported from some Asian countries such as Japan and India. It is expected that the consumption rate of surimi-based products will definitely be increased in many European countries in future and at the same time, the demand for the quality products will have to be met by import and/or domestic production.

4. *India*: The time lags of yesteryear between India and advanced countries in the matter of availability and adoption of surimi-based products have increased in importance. The entry of state-of-the-art technology into the country has accelerated this phenomenon. Now the country is launching such products in the Western countries such as Korea, France, USA, and in adjacent countries such as

Thailand, Malaysia and Indonesia. Quality frozen surimi, crab imitation sticks, snow crab legs, lobster tail are apt examples. Developments of this kind are compelling business industries of the country to remain doubly alert on the new/quality product innovation front. Domestic consumption of surimi-based products is very low because consumers have experienced fishy odor in the products, though efforts are being made to make the product more attractive to consumers by combining them with the Indian spices.

5. *Korean Market*: This country has the capacity to produce surimi and hence is the second largest surimi producer in the world and the domestic consumption of different surimi- based products has increased four-folds since 1995.

6. *Market in Other Countries*: The market expansion of surimi-based products in Russian Federation, South America and China is not significant as in the case of other markets stated above, although consumption of these items has been rising to a large extent. Likewise, Malaysia, Indonesia, Thailand, and Singapore have good production potential which principally depends upon Korean surimi-based industries. Other Asian countries should follow the same development strategy in the near future.

The sum and substance of all the above discussion is that as demand for surimi and surimi-based product continues to increase and the global natural fishery catch declines, by-protects from different species of high- and low-value fishes may serve as an alternative for the surimi product. Potentiality exists for the development of both minced- and surimi-based products which may, in turn, form potential new market niches that will be beneficial to the aquaculture industry around the world.

25.23 India Should be Marketed to the World Before Marketing Aqua-Products

International fish marketing is the first and foremost medium in export promotional efforts of the country. By participating in fish marketing, the country can present to the world consumers a visual display and a better representation of her products and production capabilities. International marketing of fish and fishery products is useful for sales promotion and long-term image building. Different fishery export promotion agencies have to gain complete knowledge in different techniques involved in successfully organizing international trade of fish and fish products.

If quality aqua-products of the nation has to gain ready acceptance in the world market, the image of the nation as a slipshod producer has to be wiped out from the collective memory of these countries and the country should make an enormous effort to sell 'Product nation' as such before selling the products to the world markets. In this way, several countries have become big exporters of aqua-products in the world markets. Japan, USA, Canada, China, Russian Federation, United Kingdom and other countries are apt to give examples of this phenomenon. They have promoted a good image of themselves among the countries of the world. Several of their products that were introduced in other countries have been identified as zero defect products and hence became acceptable to other countries.

India must follow this example and market as a producer of quality market. If some Indian enterprisers could generate international respect for the 'Made in India' lebel, it would foster the process of marketing Indian aqua-products as a class.

EXPORT PROMOTION ON SUSTAINED BASIS

The country should recognize the value of sustainable export promotion that would bring good results. By promotional moves or by implementing a few schemes, the country cannot hope to ask earnestly a fairly large share of world trade. Concerted and sustained promotion can help the country gain a stand over the world markets. Once the thrust sectors of markets and products for export have been selected, the country will resolute in her effort to set up the selected Products in the selected market. Exports especially to the developed countries require concentrated and sustained marketing and promotional efforts. India should set up a hard-wearing interest in its selected overseas markets. Short-term marketing and long-term promotion do not bring the best results.

25.24 Ensuring the New Aqua-Products for Better Marketing

Considering the product, policy is the main task in product management. However, a growing aquaculture business has to undertake a constant appraisal of its existing product line. No product line is free from any flaw eternally. Changes in customer tastes and preferences, extent of competition building around — all these factors exercise some pressure on the product policy of an aqua-industry. A product may be facing the threat of functional obsolescence due to superior quality products introduced by competitors, a product might have lost its market image, or a product may be low in quality. Such instances will be to the fore only if there is a continuous appraisal of product line. And all such instances call for sound decisions. These decisions may involve the stricture on quality control in respect of another; a new packing in respect of a third product, giving an independent brand name to a product or introducing a new product altogether.

PRODUCT QUALITY CONTROL IS ESSENTIAL FOR GOOD MARKETING

Though there are a number of cardinal elements of product policy, managing the product on the practical front involves more issues. Product quality is one such issue and is vital as it decides the ultimate fate of the farm in the market place.

It is an accepted fact that a farm cannot attain a good and lasting honour through 'marketing flair' alone, only through quality products, it can fetch that honour. Developed aquaculture nations have emerged as business leaders only through their quality assurance. Strict control on the production process and the post-harvest technology make their products globally acceptable. In fact, many products have been accepted as zero defect products.

A successful product strategy will definitely need reliable and well- engineered quality products. The best aqua-farms make their products so dependable that their customers are prepared even to wait for them, if necessary.

Quality assurance efforts should not end the moment that the product leave the processing plant. Product quality has to be monitored until the organization is satisfied that the right quality has been obtained. Complaints about quality, if arise at any stage in the life cycle of the product, should effectively tackle all complaints to maintain the quality of the product. In cases where quality has really.escaped from the standard, the matter must be set right and it must be ensured that such errors regarding quality do not recur.

(i) *Import of Technology Makes Easier for Aqua-Industries to Access Western Markets*: In the past, most of the aqua-industries in the tropics did not have easy access to the latest and best technologies of the world. This was largely because of preoccupation of the country concerned with protection of domestic industry. The restrictions on royalty payments on a steady basis to suppliers of technology discouraged long-term relationship between western suppliers and buyers. The restriction made the suppliers less responsible for ensuring adequate technology transfer. The result was that while developed countries of the world was leaping ahead with newer processing technology, Asian aqua-industries lagged behind in technology and consequently in products. They were forced to stay with older technology.

Now with the liberalization of technology import, there is easier access for Asian aqua-industries to modern technology. And as a corollary, several Asian countries are producing quality aqua-products to reap the benefits offered by the vast westem market.

(ii) *Competition Faced by Small/Marginal Aqua-Farm to Access Different Markets to Gain the Benefit*: In contrast to big aqua-farms, most of the farms belonging to middle and poor classes do not possess adequate processing facilities to make thier aqua-products more attractive to western and even domestic urban markets. Their products, sadly, have been rated the shoddiest and hence have no market value. Products from high-tech aqua-industries clash directly with the relevant industries having older technology or no technology at all.

Consequently, regional aqua-products have to face the severe competition from high-tech aqua-industries. High quality products have thus become more attractive to the western user. The price of their products are moderate and better in quality.

BRANDS OF INDIAN AQUA-PRODUCTS HAVE SQUEEZED BRANDS OF ADJACENT COUNTRIES

Indian brands now have to compete with the array of other adjacent countries. Indian brands enjoy several strengths such as abilities to export western world, back up of superior technology, experience gained from operations in several markets around the globe, and superior marketing/brand-building skills. Brands of adjacent countries have been subjected to a two-folds onslaught by the Indian brands. On the one hand, they have elbowed out in the market place by the development of quality/zero-defect products and powerful global brands of India. And on the other, there are any successful Indian brands, they will become targets of takeover. In fact, any brand of outside India that could be rated as reasonably good is now being poached by the Indian brands.

ECONOMIC DARWINISM BECOMES THE ORDER OF THE DAY

Darwinism which underscores *natural selection and survival of the fittest* may be applicable to the marketing scene of any country. Aqua-farms/industries which were reared in a highly scientific manner, are now open to the vicissitudes of a highly competitive markets. It is becoming obvious that only industries that have the needed farming, processing and marketing capabilities will survive. Industries have to become fit or else quit.

HIGH QUALITY OF AQUA-PRODUCTS HAS A COST, BUT IT CAUSES GOOD ECONOMIC RETURNS

The factors mentioned earlier in this chapter set the basis and fundamental steps in ensuring quality assurance. They yield the results only when the organization adhere to these basis fanatically. several progressive farms have moved up to the 'zero defect' level in their products because they are aware of the rewards coming through the quality route as lasting and substantial. At the other extreme, there are other organizations who believe that the cost of ensuring quality is too high and therefore not practicable. They allow wide relaxations on the quality front. However, quality assurance has a cost, but it is that has to be incurred; it is a cost that can be realized back in terms of good-will and long-term business activity market.

Quite a few organizations have in their experts in higher ranks who look after quality assurance. It shows the high concern of the industries for quality assurance. Product management can never be effective in organizations where quality assurance is not strong.

1. *Quality Aqua-Products are Necessary for Making More Profits*:

    Quality products are necessary from the profit angle. If aqua-products are slipped from the standard, often have their limitations in enhancing the profit level of the industry. It, therefore, becomes essential for the industries concerned to bring in quality products to replace declining and losing products. Quality products become part and parcel of the growth requirements of the industry and in many cases, the industry reaps a good deal of benefits.

2. *Quality Products are Necessary for Combating Environmental Threats*:

    On the environmental front, aqua-industries have to combat economic, political, legal, and technological threats. These threats make some of their products highly vulnerable. To reduce the vulnerability of their market as a whole, they seek out quality products. Quality products offer lucrative avenues of growth and thus secure the overall viability of the industry. The risk also gets spread over several quality products existing as well as new products, so that the industry does not face the threat of sudden extinction. An industry is vulnerable to many factors such as changes in general economic conditions, social/political pressure, and technological changes. National and international economic trends do affect markets. Social demands on environmental protection often pose new problems to several industries especially the processing industry. Technological changes make several products obsolete over a period of time. At times, problems of supply of fresh fish also render existing products/ markets risky. Due to the environmental threats, quality product development becomes a laborious job for many industries.

25.25 A Robust Fish Marketing Environment of India

The analysis in the preceding sections (sections 25.11, 25.18, 25.19, 25.23 and 25.24) would reveal that the fish marketing environment of India has been steadily acquiring greater dynamism and robustness. Similar to agriculture and animal husbandry, the breakthrough in the aquaculture sector has also been quite significant. The infrastructure required for further advancement of the aquaculture and marketing industries is being developed right now. There has been a remarkable expansion and diversification in the production and marketing of aqua-products. And most of the rural areas have emerged as a long and growing market. At the same time, the country has witnessed the evolution of a large middle class. All these developnents have made a profound impact on the size and structure of India's markets. The opening up of the economy in the nineties, as a succeeding part to the new economic policies and progressive measures has given a new momentum to the process. With the new programs, the economic and industrial environment of India is moving further towards the forces of the market-place and away from controls and regulations.

25.26 The Pattern of World Trade in Aqua-Products

As a consequence of the factors stated in previous pages, the volume of trade, the direction of trade, and the types of products traded vary greatly between the various countries and individual countries of the world. It is not impossible to classify the following principal aqua-products entering the world trade:

1. Lobster (Frozen)
2. Shrimp and prawn (Frozen)
3. Smoking, canning, freezing, curing, fish meal, etc.
4. Tuna, Salmon, and Trout
5. Salting
6. Peeled and Breaded shrimp
7. Cod, Scallop, and catfish
8. Surimi and Surimi-based products

In terms of these items, developed nations are major importers of aqua-products from Asian countries. Developed countries also invest heavily in the developing ones where vast aquatic resources may be better utilized as food for the humankind. The developing countries of the tropics export a variety of aqua-products to the western countries in exchange of other products. This general pattern does not always hold true. However, the pattern of world trade in fish and fishery products has been discussed at length.

SOUTH PACIFIC

Export of fish products from this region consist almost wholly of tuna, which is the South Pacific's most important development resource. Canneries in African Samoa, Fiji, and the Solomon Islands export tuna at slightly higher prices than their Southeast Asian counterparts, but the latter two states benefit from African, Caribbean, and Pacific fishery products from Australia and New Zealand are highly encouraged by sophisticated fish processing industries geared to export high-value products.

EAST ASIA

During the last ten years or so, imports have expanded considerably and international trade in the region focuses on Japan, the world's major importer of fish and fishery products. About one-third of the world fish trade goes to Japan (in value terms) and Japanese imports of fishery products increased several folds. The main imports were shrimp and prawn, fresh and frozen tuna, crab, and salmon (in value) and shrimp, prawn, and tuna (in quantity).

Reports for the Republic of Korea and Hong Kong exhibit per capita food fish supplies of about 50-55 kilogram per year. In the Republic of Korea, captured fish is preferred to cultured fish and fresh fish is preferred to processed products. The Republic of Korea is both a major importer and exporter of fish.

In China, the mean per capita supply was reported as 13.5 kilogram in 2000, but since then consumption has probably increased significantly. However, important variations exist within the country; in the Southern parts, the average is 45 kilogram whereas fish consumption in the isolated areas of the Northeast is negligible. Domestic demand is met by freshwater pond products, in particular various carp species; substantial quantities of Chinese fish come from culture fisheries. The increases in disposable income in China are creating market opportunities for fish imports. Alaska pollock is the main fish imported from the Russian Federation, which is processed and re-exported to the United States and Germany. Shrimp is also imported for processing and re-exported to the United States.

EUROPE

Almost all European countries have various kinds of fish processing industries such as smoking, canning, freezing, curing, and fish meal. Production of excellent quality food from fish is the fast-growing sector of European fish processing industry and retailers in the market introduce new fish products. Supermarkets are also increasing their market share of fish sales.

In certain countries, import and export patterns have changed completely. Eastern Europe and the Russian Federation are now exporting high-value species for foreign exchange earnings. At the same time, low-value products are being imported as local production decreases.

A number of fish processing industries in many countries need modernization since these industries are not able to produce substantial quantities of fish products owing to the fall in raw material supply. Moreover, during the Soviet era marketing systems of fish and fish products were controlled by large companies which failed to survive in transition. Though the old system of trade has not yet been replaced, new private companies and large self-service market chains have appeared together with a wholesale infrastructure.

Fish consumption levels differ widely among sub-regions and countries. As a whole, average per capita fish supply varies between 19 and 24.5 kilogram/year. In general, however, consumers' apprehensions of fish are absolute and the demand for fish is likely to increase in future.

SUB-SAHARAN AFRICA

Fish is a very important aqua-product in this region and fish consumption may be more important than the per capita food fish supply figures actually suggest. According to the FAO statistics, fish consumption has declined by more than 2.5 kilogran/capita/year during the 1990s — from a per capita supply of 9 kilogram in 1985 to 4.0 kilogram in 2009. This is due to rapid population growth, drastic fall in imports aggravated by the weaker purchasing power of some countries and the smaller share of domestic production retained for local markets as artisanal fishers increasingly turn to lucrative export market. Nevertheless, most fisheries production is still utilized as food for local people.

Locally produced fish and fish products are marketed fresh, smoked-and-dried or salted-and-dried. Consumers prefer fresh products and about 50 per cent of the consumption volume consists of fresh finfish. Owing to transport difficulties, fresh products are available only near production centres. Imports are mainly of frozen products. In most Western Africa, women play an important role in processing and marketing.

High transport and storage costs, poor handling practices, limited distribution networks and a lack of harmonization, and proper enforcement of fish trade controls are some of the most important constraints to intra-regional trade. Tariff barriers and other trade limitations remain among countries belonging to customs unions.

Though the regional trade balance has been positive in values since the mid 1980s, this region remains a net fish importer in volume terms. Many countries have a small but growing export trade in fresh and frozen fish and shellfish to the European Union (EU). But the overall trade balance is based on the large export volumes of only a handful of countries. Reliance on the EU market could cause difficulties in the future as trade is liberalized and African, Caribbean and Pacific countries lose their preferential status.

SOUTH AND SOUTHEAST ASIA

Fish consumption varies throughout the region, with relatively high consumption particularly in the eastern part of the region. In Thailand, average per capita food fish supply is around 25 kilogram per year, in Malaysia almost 30 kilogram, and in Philippines and Singapore around 36 kilogram. The Maldives has the highest annual fish supply in the world — it is 128 kilogram per person. In India, on the other hand, the corresponding figure is only 6 kilogram and as a corollary, the total regional average is only about 9.5 kilogram. It should be kept in mind that consumption patterns in India vary considerably.

Although 50 per cent of the harvested fish is marketed fresh, more frozen fish is marketed. Fish utilization is also characterized by great production of a wide range of value-added products or preparations for both national and international markets. Thailand produce block-frozen; peeled and breaded shrimp by the use of improved raw materials from India, New Zealand, and the United States when there is deficient in local supplies. Dried or salted and canned fish are produced in Thailand, Indonesia, and the Philippines in large quantities. Thailand uses imported raw materials and re-exported as the final product. In general, fish processing industries in this region entirely depend heavily on imported raw materials.

Fish trend in this region has expanded significantly due to economic growth and open trade policies. Some countries such as India, Malaysia, and the Philippines are currently lowering their list of charges following the consequence of the GATT Uruguay Round. At present, Thailand has ranked as the world leading exporter of fish and fishery products since 1992. Indonesia and India ranks second and third important regional exporter, respectively. All regional countries export frozen and processed fish in value term.

1. *The Compulsion of Fish-Producing Countries in Asia to go Global*: By the beginning of the new millennium, many aqua-industries in Asia are compelled to take to export at my cost. They also have started searching for more competitive advantages. In some cases, their successful entry into western markets with moderate-priced aqua-products took away the pricing comforts. However, the new marketing policy enjoyed on several Asian quality aqua-product industries to earn foreign exchange required for importing whatever technologies they are needed for staying their production rank proceding. Once the new policy and technology came into effect, major aqua-industries started striking at the export markets.

(i) *Most Aqua-industries in Asia Find Marketing in Western World a Difficult Game*: While most aqua-industries are compelled to export in a big way, they have found that this is not a facile task. Most of them lack the required global competitiveness in cost, quality, strong brand, technology, and finance.

   * *Poor Financial Plight*: Finance in aqua-industries is one of the fundamental aspects of producing and marketing quality aqua-products. Farmers of most Asian countries are not able to invest required amount of funds in their own/leased water bodies throughout the farming period. And poor financial plight has made

their own contribution to the low productivity and it is the set-back on the export front.

* *Quality Problem*: Bridging the quality gap is another major problem. In the global context, however, the quality challenge is more formidable. Aqua-industries trying to go global have offered quality that will meet the exacting demands of western buyers. Barring a few exceptions, the quality discernment of global buyers about Asian products is generally not up to the mark. For a long time, aqua-industries in Asia had been lax in respect of quality and consequently, they were not able to upgrade all of a sudden.

* *Lack of Brand Power*: There is very little of branded products in the export basket of Asia. Asian products are sold in the western markets either as a commodity or under middleman's brand name. Brand is a must if Asian aqua-industries are to become major exporters. Brand power is the best means of high value apprehension. Export items of several Asian countries did not fetch attractive prices and a major part of the value addition went to the importing middlemen abroad mainly because they lacked strong brands.

For easy brand recall, the industry should decide to opt for the name that would promote the image of different products. It should be an easy word to remember, familiar to all different customer segments in and outside the country and should convey the meaning of the product.

A picture of the product should be provided on the pack that would definitely add an effective and attractive strategies to customers. Attractive and well-designed pictures are widely used in several aqua products in the ad industries.

Quality aqua-products will generate an overnight awareness. And this awareness may eventually extend to millions of customers around the world. Several ready-to-eat items may generate tremendous customer response.

* *Low Productivity*: Low productivity of many processing units has remained unsatisfactory. In some situations, there exists desparity between developing and developed countries and as a consequence, most aqua-industries find it difficult to compete with the western markets. It is an indicator of the gap. Asian farms have to cover to become western players. Improvement of productivity following the adoption of current technology is badly needed.

NEAR EAST AND NORTH AFRICA

Fish consumption in this region varies widely among sub-regions and countries and even within countries. The average per caput supply in the region is, however, 5 kilogram per year.

Demersal fish (particularly sea-bream), cephalopods, shellfish, and pelagic fish are dried in the sun and used as animal feed and fish meal. Sardines have been processed in Morocco since the first World War and the industry contributes significantly to the economy till today. small-sized meso-pelagic fish in the Arabian sea are used to produce fish meal.

As a generalization, the region does not contribute to international fish trade. Only a few countries export high-value fish, some cephalopods, crustaceans (both fresh and frozen forms) to European markets and Japan due to application of latest technology. Some countries have an expanding intra-regional trade to Israel, Saudi Arabia, Baharin, Qatar, and the United Arab Emirates.

NORTH AMERICA

The most popular fish products are tuna, shrimp, clams, scallops, cod, pollock, salmon, catfish, flatfish, and crabs. These large varieties of aqua-products are processed and the processing industry is technically well-developed. Development in transport and freezing techniques have taken place and as a result, the shelf-life of fresh and live fish has been extended. This indicates more trade in these items beyond local markets. Consequently, most domestic landings are consumed as a fresh or frozen form. With the development of proper distribution system, fresh finfish and shellfish are available all the year round and their availability in the market is on the increase.

The region is one of the top fish-importing and exporting countries of the world and Canada is one of the leading exporters in the world. The United States is the second largest importer after Japan and the value of the imported food fish has been estimated to the tune of 12.2 US billion dollar in 2009. On the other hand, Canada exhibits an import value of about 2.0 US billion dollar in 1994 which has increased to the tune of 7.2 US billion dollar in 2009. The most important export markets for Canada are Japan, the United States, and the European Union.

LATIN AMERICA AND THE CARIBBEAN

The levels of fish consumption vary greatly within countries and among countries. The average per caput food fish supply in this region is about 9 kilogram per year (live weight equivalent) which is below the world average of about 13 kilogram. In contrast to some Caribbean islands and Chile where fish consumption has been increasing, most of the Central African countries exhibit a decreasing trend in fish consumption and consequently, the contribution of fish and fishery products is less critical in Latin America than in any other developing countries.

About two-thirds of the total catch are converted to fish meal used as feed in the poultry and aquaculture industries. Peru and Chile use substantial quantities of the raw material destined for fish meal. Though efforts have been made to find alternative uses for human consumption, no success and economic viability have been achieved up to date due to reluctance to invest in market development.

Latin American countries export fish and fishery products particularly shrimp and fish meal. While Equador and Mexico export cultured shrimp products only to United States market, Chile and Peru export fish meal mainly to the Asian market. Recently, Equador has managed to penetrate the European shrimp market. The international trade is, however, quite limited.

25.27 Marketing Research

The analysis in the preceding pages reminds us that fish marketing and trade consist of spotting the needs of customers and meeting them in the best possible manner. Marketing research helps the industry in every component of the total marketing task. It helps the entrepreneurs acquire a better understanding of the buyer, the competition, and the marketing environment. Decisions on pricing, product quality, distribution, packaging and the like need marketing research support. To carry out these researches require specialized skills and sophisticated techniques.

It should be kept in mind that market research has to be operated as an integrated activity, integrated with all the key functions of marketing. It has to be part and parcel of the problem solving process of marketing strategy. Market research cannot be effective if it is assigned to some portions of a marketing problem for exploration in a piecemeal manner. It must be supplied with full background needed to surround and attack the entire problem involved. Likewise, if market research has to be effective, it has to be manipulated as a continuous process.

Similar to agricultural and industrial research, marketing research on fish and fishery products has gained a great deal of recognition and stature in many fish producing countries of the world. In India, fish marketing trend has, for example, started in the latter half of seventies and the post liberalization period observed a new emphasis on market research. The growth of competition has turned Indian aquaculture/fishery industries and businesses towards market research. While many larger aquaculture product industries were using the market research tool, with the intensification of competition in practically every sector and every product category, more items have started looking up to market research for help.

With the economic reforms and with aquaculture industry opening up to competition, a number of entrepreneurs are now realizing the importance of market research. The market research now has been growing at more than 25 per cent in the post-liberalization years, while the world average was only 7 per cent. And the fccus is now on value-added products. However, the demand for qualitative research is gradually increasing. Almost all aqua-product exporting agencies now have a separate research cell to facilitate easy export of their products to overseas markets in excellent quality.

25.28 A Brief Case Study on Trout and Salmon Marketing in Europe

The marketing of trout and salmon particularly in the European countries, had achieved a phenomenal success in the eighties. To day, salmon and trout marketing is the most profitable venture and the sale value of farmed salmonoids in the United Kingdom has risen from less than £ 3 million (Rs. 24 crores) in 1985 to over £ 100 million (Rs. 800 crores) in 2005 (value for the year 2005).

MARKET TRENDS OF FISH

Fish market trends in this region indicate that it does not consume significant quantities of fish in comparison with some of its European partners. Neither is the value of the market comparable with many of the other food sectors.

The sale of fresh fish declined during the 1970s and the early 1980s, this was in line with the decline in availability. Frozen fish contributed to significant growth over the same period and became available in all freezer centres in a variety of forms so providing a basis for market expansion.

The introduction of attractive wet fish service centre in the multiples is increasing the availability and convenience of fish buying. Some multiples have progressed with prepared fish using modified atmosphere packing to provide a good visual presentation and a safeguard to the product's quality. Prepared fish has the advantage of giving the consumers confidence by displaying a sell-by-date and providing full cooking instructions. Such packing also provide a hygienic leak-proof unit, free of smell, the shopping bag and the domestic refrigerators.

While about 30 per cent of the total fish supply to the United Kingdom market goes for domestic consumption, the majority of fish both for food service and for retailers goes through the wholesale markets. An increasing quantity of fish is handled by processors who purchase directly from the harvesting areas and distribute to the multiples. As fish prices have increased substantially, distributors supply a range of outlets with brine fish that can carry the extra margin and the cost of direct service.

SUPPLY OF TROUT AND SALMON

Trout and salmon represented about 20 per cent of the total fish supply to the United Kingdom in 2005. However, they represents about 22 per cent of the total value and fish market has grown steadily in line with domestic production.

1. *Trout* : The world production of trout is about 2,70,000 tonnes, produced from contries as diverse as Papua New Guinea and the United States. More than 60 per cent of the production is within Europe. Some European countries such as United Kingdom and Spain exhibit a good production potential.

2. *Salmon*: Among different species of salmon, Pacific salmon is much more of a world commodity than trout. Though the annual production totals about 7,80,000 tonnes, the vast majority is frozen or canned and therefore, the product is sold throughout the year.

The United Kingdom imports about more than 25,000 tonnes of canned salmon each year from Russian Federation, Canada, and USA. Wild Chinook and Coho salmons are considered as the best quality because their high fat content and pink flesh color affords high prices and their fresh form can be substituted for other salmon. Consequently, they are being cultured in Chili, USA, New Zealand, Canada, and Japan.

Since the production of fish has grown rapidly and is set to continue at a rate in excess of about 10 per cent per year, it is likely that these fish have enormous market potential in many European countries. But the most important factor that could seriously abridged of the expected growth is the development of market in a planned manner. If the market cannot be developed at a rate that matches the growth in supply, the current profit could be drastically reduced, so restraining from further development and investment. A more consistent supply and availability of fish of different sizes will cater a basis on which the market is developed.

TROUT MARKET

Since trout has a good value against the other farmed species, the market expansion has substantially increased and consequently, profit margins also increased. The increasing availability, the good quality, the freshness and the robust image of fish, supported and promoted by several trout associations have helped invading and developing a secured market position.

The central wholesale markets dominate the distribution of trout (range vary between 40 and 70 per cent of the total market outlet). These markets are highly flexible in the quantity and quality they are able to absorb but often at a price penalty. Due to the development of market, the customer purchases more trout from supermarkets than from traditional market-supplied fish-mongers. Hotels and Restaurants receive trout directly from farmers than from the markets. Gutted trouts are packed with modified or controlled atmosphere. This permits a good presentation, leak- and smell-proof, and provides a quality safeguard. To retain the quality and the final processed and packed product, a number of trout processing operations have grown up to meet the demand of the market and to establish the link between supermarkets and farmers.

Different trout products such as chilled and frozen forms, fillets, large trout fillets, steako, and smoked products appear in different markets especially in supermarkets; of course, such product can only be developed if raw materials are stable in supply and price.

In most regions, fresh trout and home-made products are sold through farm shops. Selling the fish stocks from farm-gate represents the highest margin sales outlet for the farmers but care must be taken so that the product quality and the service is in line with accpeted sales practice.

Very small quantities of trout are exported to European markets. Most of the European countries import trout from Denmark. And the production of trout in UK will definitely continue the development of the home markets where sales can be best controlled.

THE SALMON MARKET

The salmon market is less stable than trout. This reflects the gradual increase in production across the world. With modern and efficient distribution systems, fresh salmon is available in all the world's major markets. The price is controlled by the mechanism of demand and supply. The majority of the catch is frozen and canned and very little is sold fresh and the development of the fresh market is highly limited by the very seasonal supply of salmon. Canned salmon represents a good market than fresh or frozen but in some countries such as USA and UK, it accounts for the majority of the salmon consumed.

In many European countries, fresh farmed salmon is sold through central markets. These markets serve the fish-mongers, market stalls, restaurants, and export markets. They fix the price for different weight grades of fish to match the supply with demand. Wholesale distributors are interested in volume and the ability to sell the volume is entirely based on price. Therefore, market expansion can only be achieved through price reduction.

In the United Kingdom, about 70 per cent of the total salmon is smoked and the volume generally fluctuates between 60 and 80 per cent of the total. The growth of smoked salmon market has been stimulated by the increased supply of farmed Atlantic salmon and at constant prices. However, the increasing opulence of the consumer and the adoption of packaging techniques with low weight packs of the product have extended the market demand.

According to an estimate, more than 20 per cent of the salmon sold in the UK goes to restaurants and hotels. They purchase different items of salmon from central markets. This provides the customers greater confidence in supply, quality, and price stability.

Portions of Scottish production of farmed salmon is exported mainly to France and the Netherlands. Such export markets are inevitable as the rapid increase in production and the domestic market cannot absorb the product in full potential. Export markets are dominated by Norwegian supplies who dictate the price through the central markets. Scottish products are very excellent in quality and hence are sold in supermarkets mainly in USA, France, Belgium, Luxembourg, the Netherland, and West Germany.

MARKET DEVELOPMENT

The fish market should be developed in such a way as to stimulate the demand to fulfil the supply so that adequate margins can be maintained for sustaining the industry.

It is significant to note that it is difficult to develop the stable markets unless fish supply is ascertained. Generally, the excess production is purchased by the central markets at a reduced price while attempts have been made to maintain the continuity for the proportion of the total to the more direct markets. The trout industry has developed to a stage whereby recognition of common objectives — a continuous supply of eggs and fry — has been achieved throughout the year. And the availability of fish all the year round would definitely fulfil the market demand considerably.

Most trout farmers have developed their local sales by organizing themselves. Above this local sales, fish are sold to either the central markets or an independent processor. Channeling trout through processors helps improve efficiency and the marketing focus of farmed trout and stabilizes the markets.

Quality of the product is the another important factor to be reckoned to improve the market status, not only of the finished product to be sold but also of the service offered by farmers to the consumer. Quality of the product relates back to the selection of the stock through harvesting and packing.

ROLE OF BRITISH TROUT ASSOCIATION (BTA) AND THE SCOTTISH SALMON GROWERS ASSOCIATION (SSGA) IN MARKETING STRATEGY OF TROUT AND SALMON

These two associations have produced product specifications to increase the quality standards. The quality of trout has improved by eliminating several problems relating to the flesh, poor grading as a result of pressure from quality-conscious customers. Processing industries must be reckoned with certain basic quality criteria of aqua-products that follow as a prerequisite for success in their market endeavour.

1. Stringent quality control standards should be followed in the selection and rasing of stock.

2. Fish should have a consistent flesh, color, and texture.

3. Fish should be graded to meet the customer's demand. For the production of salmon steaks at an attractive portion size, for example, 2-3 kilogram fish are rquired.

4. Fish should be killed by a method which meets the consumer's demand.

5. Good handling and temperature control should be maintained immediately after harvesting. It is essential to have a product temperature of less than 2°C to prevent them from deterioration in quality.

6. Effective grading should be carried out the meet the consumer's demands.

7. The fish should have a good visual appearance, clean and free from any damage.

8. Ideally all fish should be gutted and gilled immediately after harvesting. This will reduce the risk of bacterial contamination and autolysis considerably. Gutting permits a more effective assessment of quality by examining the body cavity.

The product quality generally involves the physical appearance and taste of the product on which the customer decides the product. When quality of the product is ascertained, markets can be penetrated and developed. Scottish and British trout are regarded as quality product and hence command a premium price.

The BTA and SSGA are funded by a levy on the numbers of smolts surviving at the end of the year of input or indirectly via the feed manufactures. These organizations have been set up not only to provide promotional support but to tackle the problems that face the industry as well. At the same time, these two associations have been able to encourage the consumption of fresh and processed trout and salmon. By identifying the consumer trends and effectively directing the promotional activity, they stimulate the increase in demand that provides the trout and salmon industry with its current stable market position.

It should be kept in mind that the stability, availability, and price have helped develop a wide market for trout to a point where more twice as much trout is sold as fresh directly to the consumer in the UK than salmon. Trout attracts more regular consumers not because it is of higher quality, but because it is more widely available and has an easily affordable price.

DEVELOPMENT OF EXPORT MARKET

The export market is cardinal to the salmon and trout industries. Since fresh Atlantic salmon is a world commodity than trout, the price being .detached the volume of Norwegian product available related to the demand. Quality and market share is fierced and there is none of the domestic market advantages. Promotional activities in France provide and excellent support for the industry's export.

The USA is a key market for European Atlantic salmon, but it depends largely on the Norwegian supply. Although there is potential for Scottish salmon, the Norwegians have set up the price and high quality standards. Japan has a huge potential market, but Atlantic salmon imports at least more than 800 tonnes per year.

To sum up, the potential for trout and salmon in the UK and world market is considerable. The main source of fish protein is hunted which will not be able to meet the projected demands. A strong demand for fish in these regions will help sustain the farmed salmon and trout industries. The UK consumer trends towards fresh, robust and high quality products provide wide opportunities for trout and salmon farmers. Development of fish markets must be based on the active involvement of the BTA and the Salmon Promotional Organization. At the same time, market development also involves closer links with the producer, the smoker, the wholesaler, the processer, and the retailers. Without a common objective and a mutual respect, the task of developing the market will be more arduous.

Prices of trout and salmon will reduce on a cyclical basis with the corresponding opportunities for product development and market penetration. More products will definitely reduce the pressure on price in central wholesale markets. The principle should be to direct the product sale closer to the consumer and reduce the number of distribution layers the product has to pass through to reach the consumer. And this, subject to quality and flow of supply, will provide more stable pricing.

25.29 A Case Study on Shellfish Marketing Strategy in India

In the shellfish industry of India, different products achieved phenomenal success in the nineties. To day, India aqua industries is a profitable business giant with interests ranging from prawn/shrimps to cuttlefish. This case study deals with the success in aqua-product commodity exports through its well known Individual Quick Frozen products.

The growth is especially striking because it has been achieved in a traditional and mature industry such as processed products, characterized by an intense and well-entrenched competition. Indian aquaculture farms made new fortunes in the old business leading to traditional leaders way behind. How did nation achieve this success? This study endeavours to answer this challenge.

INDIAN AQUA-PRODUCTS STRIKE A NEW PATH IN EXPORT

Indian aqua-products correctly sized up this scenario and formulated a unique export strategy. Several Indian aqua-farms went in for a export strategy that ammounted to a total innovation they have deviated totally from the existing pattern of export. Practically in every aspect of export, their strategy meant the polar opposite of the existing/established pattern.

MARKETING STRATEGY OF INDIA

While the strategy formed the foundation for its growth, it was the marketing strategy of the country that actually transformed the country's vision into reality; the country carried out its market targetting in the ideal manner; and it evolved an effective strategy in marketing. As a generalization, India's marketing strategy relies on (1) Effective quality management systems, (2) Excellent customer relationship, and (3) Aggression. From these points go, the country employed these strategies and expelled virtually all its neighbouring competitors from their well-entrenched positions in the market. Its strategies on product, price, and distribution equipped it for the frontal attacks on its competitors. The strategy provided the country to stay ahead of the competitors and to secure new markets.

1. *Product Strategy*: The main elements of aqua-product strategy are noted below:

(i) *Product Based on Modern Technology*: The country shaped its aqua-product strategy around the technical superior of its product. Different aqua-products are special products, supported with the best technology in the world, that is how the nation has formulated the fundamentals of its product strategy.

Today's world is a technology-driven world. And technology has emerged as a vital factor for aqua-farm industries. While for processing and culture industries operating in frontier fields such as biotechnology and aqua-product technology, the challenge on the technology front is particularly formidable even in industries that are not so high-tech, technology is emerging as an important factor since it has an impact on aqua-product profiles. In addition to the opportunity dimension, technology encompasses a threat dimension as well. Change in technology makes products suddenly antiquated.

(ii) *Attractive Products*: India offered to the overseas consumers the most attractive aqua-products. The country avoided the humdrum products and concentrated instead on lucrative products and creative food items.

(iii) *Wide Variety and Choice*: India provided the maximum possible variety in every product line. Since variety is everything, the country consciously decided to have as wide a range as feasible in product type, taste, design of package, and finish. If applied this philosophy to every value-added forms such as live, chilled, frozen, IQF, head and headless peeled, cooked, dried and freeze-dried, ready-to-sale, and ready-to-eat. And true to its tradition, the country reaped immense benefit from all these products.

(iv) *Emphasis on Quality*: Emphasis on quality is one of the major components of the product strategy of the country. The country adopted a premium quality for exports for highly discriminating and quality conscious overseas consumers. Processed and packed to the exacting HACCP and EU standards, Indian aqua-products today command a premium in the quality-conscious export markets. This is due to the fact that the country has gone in for the best technology and the most sophisticated machinery. And it is already paying the best possible attention in quality control and quality assurance.

Indian exporters ensure stringent global standards guided by international norms. That is because quality in the Indian aqua-food industry is the basic requirement. This is one of the many reasons why Indian aqua-products are preferred around the world.

There is now a new awareness about quality, and a good deal of effort is being made by well-known aqua-industries not only to attain international quality standards but to obtain international quality certification as well. The example that follows will illustrate how Indian aqua-food industries maintain their quality standards.

Indian aqua-food industries faced the quality challenge with coercion to carry into effect a total quality solution on the basis of Hazard Analysis Critical Control Point (HACCP). Many processing units have to upgrade their facilities or face the risk of being marginalized in international markets where quality was rapidly emerging as the matter of concern to the consumers.

Recently, the detection of *Salmonella* bacteria and banned-antibiotics in processed shrimps, for example, by the United States health authorities and the detection of industrial waste substances in many consignments of low-value fish following the rejection of total consignments were some of the quality problems the industries have confronted.

Developed nations have made the HACCP based quality system is obligatory for finfish and shellfish processors also. This system requires better infrastructure, facilities such as better landing centres, hygienic handling at the processing stage, well-equipped laboratory for all units, adequate supply of quality water, and continuous power supply. Training of processing staffs for implementing the HACCP system is another important area that should be kept in mind.

INVESTMENT IN RESEARCH AND DEVELOPMENT (R & D) AND INNOVATION BECOMES INESCAPABLE

Besides basic technology, R & D and innovation have also become crucial for Indian aqua-industries. In the past, Indian aqua-industries had by and large neglected R & D and innovation for a variety of reasons. The support enjoyed by them under the regulated system was one and there was no constraint to pay attention to R & D. Several aqua-industries find their profit margins too low to permit any investment in R & D. Investment in aquaculture research and development programs in many Asian countries is still very low (it is not more than 7 per cent of scale) whereas in developed countries, it has been around 28 per cent. Several fish-producing countries of the world such as Japan, Singapore, Thailand, China, USA, etc. spend a fortune on R & D and keep bringing in improved products in a stream.

THE EXPORT STRATEGY

India struck a new path in export. When it entered the overseas market, the marketing had certain established practices in the matter of export. The export of quality products was a conscious and deliberate choice of aqua-industries towards fulfilling the export objectives of the industries. There were quite a few strong motivations, strategic compulsions, and practical considerations behind the decision of the industries to rely heavily on the export of quality products.

1. *The Marketing Industries Found the Existing Assumption and Practices of the Aqua-Product Trade*: When the marketing industries entered the export market, it was found out that the marketing system in the trade followed a pattern which had its roots in certain age-old perceptions. The industries found that the aqua-product trade hardly had any strategy for trading quality aqua-products within the country. Most of the people of the country are reluctant to purchase such aqua-products for the table. The marketing authorities of the country also observed that the product becomes so gorgeous that it would not be possible for most of the people to purchase them with degree of regularity. They also found that most of the western countries prefer different processed 'made in India' aqua-products. The country found that there was a segment which needs quality products through the western countries and they could afford the product. These countries were a well-to-do segment with a taste for the better things of life. Affordability coupled with purchasing capacity of Indian products made them a distinct class. And the segment was substantial enough for the country to concentrate on.

Many fish processing industries in western countries carried the marketing strategy and created a network of what is known as *supermarkets*. Exported Indian aqua-products are sold in these supermarkets where modern purchasing facilities exist and display a variety of quality aqua-products. And most of the supermarkets have business potential almost round the year.

PROGRESSIVE MARKETING POLICY

India's marketing strategy did not end with the exporting of aqua-products to different countries. India was aware of the fact that adquate motivation of the channel holds the key to successful marketing. Selection of countries, proper marketing, and continuous development of zero-defect aqua-products were the major elements involved in India's channel management. Several trading authorities such as Nekkanti Sea Foods Ltd, The Marine Products Export Development Authority, Aqua Star, etc. provide a wide publicity to its unique distribution strategy and converted it into product promotion. The following attractive slogan always went with aqua-products as: Superior quality does not mean excellence, it is the basic requirement.

EXPORT PROMOTION STRATEGY

It would not have been feasible for the nation to achieve its success by its superior aqua-product alone. It was India's mastery of the minds of overseas customers that enabled it to rise as the market leader in aqua-products. In fact, the Indian exporters not only always follow the state-of-the-art infrastructure facilities and technologies adopted at every stage of post-harvest strategies but also ensure stringent global standards by HACCP/EU and other international norms. These serve as the basic requirements of what creative export promotion can achieve in the Indian context.

Supported by a clear policy perspective, most of the Indian exporters carried out an innovative and result-oriented export promotional strategy. The meticulously planned and carefully executed promotional strategy enabled the exporters to meet the basic export-promotional objective of establishing the country as a synonym for exporter in the field of aqua food industry.

STRATEGY OF TARGET MARKET SELECTION

India's marketing strategy to overseas countries succeeded very well as the country had carried out its target market selection in a perfect manner. The country identified the 'well-to-do urbanity' as its target market. The high quality and high priced aqua-products pointed to this group as the target. The distinctive, sophisticated and high priced characters of Indian aqua-products in itself gave the clue to any processing industry for pinpointing the target market. The strategies on quality products, distribution, pricing, and promotion had been evolved keeping in view of the specific target market. And several processing industries have been able to fine-tune to all these strategies; and the country as a whole, has changed substantially the aqua-product business in overseas markets through its market targeting,

25.30 Conclusion

In several Asian countries, export of fish and fishery products to developed countries has expanded phenomenally during the last two decades. Export-related culture projects should aim at an integrated development of fish production, post-harvest technology, storage, processing, packaging, transportation, and marketing. The project should develop a concept of establishing a link between the producer and the consumer, the processing unit having a radiating influence on production and consumption. The export-related projects have a potential for increasing not only exports but also domestic market of the selected fish species.

Adoption of modern fish marketing is necessary so as to sell huge quantities of fresh and edible fish. According to the taste and demand of the consumers, the fish markets, marketing systems, and development of export-related aqua-products have to be designed and strengthened. The fish marketing and trade are fascinating and challenging at the same time. They offer large scope on account of their sheer size and potential and it is going steadily. Even a gradual growth of fish culture pushes up the sales substantially.

Fish market, marketing system, and trade pose several problems and hurdles. The enterprises have to be encountered these problems and put in a great deal of effort to get a sizeable share of the market.

Marketing and trade of aquacultural commodities in many countries of the world are fairly diversified. There is tremendous scope for further expansion in fish marketing and export. With the help of sophisticated technologies, value addition, upgrading of facilities and infra-structure, and financial plights, the capacity of fish production for the table from both freshwater and marine water resources can be increased to a large extent.

For export of quality aqua-products, fish processing industries need discipline and for this a code of conduct is badly needed. Most of the buyers of western countries sometimes complain about the quality of aqua-products. But they should understand that Aqua-Product Exporters' Associations in several tropical countries are always heedful of their quality and fair practices. And India, in particular, is dreaming for the posterity to gain a good recognition as one of the principal countries of the world in the field of marketing and export of quality aqua-products to western countries.

References

Banerjee, S. 1994. The World Trade Organization and India. Meharia Research Centre, Kolkata.

Shahin, N. and Parameswaran, A. 2001. Marching Chimes of Indian Marine Products Exports. *Fishing Chimes*. 20 (10 & 11) 21-33.

Shaw, S.A. and Rana, J. 1987. Market outlets and marketing practices for salmon fish, 1986. Institute of Retail Studies, Market Report No.3. University of Stirling.

Rackhan, D. R. 1984. Quality-farm to supermarkets. Scottish Fish Farming Conference, February, 1984, Oban.

Robinson, M. A. 1984. Trends and Prospects in World Fisheries, FAO Fish Circular No. 772, 25 pp.

Questions

1. Outline the relationship between regional specialization of production and world trade.

2. What are the essential differences between internal trade and international trade?

3. State the factors which affect the pattern and development of world trade in aquaculture sectors.

4. Distinguish between (i) selling and marketing, (ii) marketing between developed and developing nations.

5. Examine the pattern and extent of foreign trade of fishery products in India.

6. What should be the essential components of a proper export strategies for fishery products?

7. What is marketing? State the importance of fish marketing.

8. Mention the features of fish marketing management.

9. What types of aqua-products normally enter the international trade? With reference to any three items from the tropics, name three major countries for each of the products which import them in bulk.

10. What are the defects that affect fish marketing systems?

11. Discuss how co-operative marketing systems help the farmers' community.

12. Discuss the activities which are involved in international fish marketing strategies.

13. What are the issues that are particular to Indian aquaculture industries in their trade effort at global marketing?

14. India has a strong competitive advantages in aqua-product export. Discuss the statement. What steps should be taken to tap the world market for aqua-foods?

15. What are the problems which are involved in tapping the fish market? Why quality aqua-products are necessary in fish marketing?

16. What are the steps that are required to improve fish marketing systems?

17. What are the factors that affect the trade of fish and fishery products? Discuss.

18. Discuss the importance of aquaculture commodities in India's foreign trade.

19. Trace the global trends in fish trade.

20. Efforts of quality assurance are essential for better marketing and more profits. Discuss.

21. What are the models of world trade in aqua-products?

22. The potential for trout and salmon in the UK and world market is considerable. Discuss.

23. Discuss how Indian aqua-industries have succeeded in aqua-product commodity exports.

24. The increase in aquaculture production is important for human nutrition but export of quality aqua-products means much more than the former. Discuss your opinion on this statement.

25. State the factors which affect the pattern and development of world trade in aquaculture.

26. What do you understand by (a) Fish marketing strategy, (b) Entry strategy for global marketing, (c) Pricing, (d) Product quality, (e) Packaging, (f) Productivity improvement, (g) Thrust products for export, (h) Market segmentation in fish market, (i) Foreign trade, (j) GATT, (k) Market research.

26

Practical Manual

Determination of different nutrient elements and environmental factors of soil and water which are directly related to the productivity of any fish culture ecosystem form an integral part of efficient managanent for sustained production of fish biomass. In this chapter, an attempt has been made to describe some methods by which different parameters are analyzed for adopting regular management practices. The main objective of this chapter is to provide necessary methodology to progressiye fish culturists to take up exercise on soil and water monitoring protocols. The manual that contains a number of standard methods is a very simplified guide for use by farmers and as the simplified techniques are conquered, the farmer should understand the details as given in the chapter to be aware of the possible hidden difficulties and interferences. The data obtained from soil and water testing methodologies helps farmer to know how efficiently his farm is operating and enable him to predict and prevent troubles that are developed in fish culture. The standard methods have great practical significance when fish production levels are maintained. Consequently, the water and soil tests should be made as accurately and carefully as possible for precise management of fish culture ecosystems. An arbitrary test almost is worse than no test at all.

Detailed analysis of soil and water and their interpretation are arduous and it is only feasible in well-organized and commercial fish farms by expert technicians. But it is possible for small and marginal fish farmers to obtain some valuable informations about the nature and properties of water as well as soil by carrying out simple tests that follow.

With the help of these simple tests, one can ascertain the nature of any fish culture pond. The alkalinity or acidity, nitrogen, phosphorus, calcium, pH, dissolved oxygen, dissolved organic matter, hardness, turbidity, productivity, texture and the like all have great practical value in relation to fish production. Different methods adopted for water analysis will be discussed first and then soil analysis techniques will receive attention.

26.1 Water Analysis

In fish culture, the quality of water is defined as the suitability of water for survival and growth of farmed fish which are controlled by only a few variables. Most of these variables can be measured with appreciable degree of accuracy and the results can be used for management of the pond productivity. Assessment of water quality generally involves collection, and preservation of samples and their analysis as well as interpretation of results.

COLLECTION OF SAMPLES

Studies on the entire fish culture ecosystem will be impractical on account of limitations of time. Hence, sampling becomes inevitable (See Panel 26.1).

The collection of representative water samples is a very important part of the test concerned and hence more importance should be given to different tests. For collection of water samples, however, the site of sampling, direction of the wind, hour of sampling, and availability of sunlight must be taken into consideration. Wind action, for example, not only drifts the plankton populations away but, at the same time, also fluctuates their composition from one place to another thus largely influence some of the chamical properties of water at different regions of the pond. Since photosynthetic activity is maximum in the upper-most layer of the pond water, this zone should be given due attention at the time of sampling. During day time when the intensity of sunlight is maximum, the pH and dissolved oxygen levels generally tend to increase following decrease in the amount of free carbon di-oxide. Considering these criteria, water samples should be collected during early morning hours when dissolved oxygen and free carbon di-oxide concentrations are likely to remain in critical levels. The hour of collection may, however, be kept fixed for all regular sampling schedules since disgression from routine hour collection of water does not exhibit more or less similar results.

While glass containers (about 125 ml capacity) of different sizes are used for the collection of dissolved oxygen, different water samples with certain modifications exist which are used for ascertaining the quality of water at some particular depths. The sampling of surface water, however, involves dipping the container slowly in pond water without creating any air bubble in and around the container. As soon as the container is completely filled with water, its mouth is tightly closed and the sample is then transferred at once to the laboratory.

When samples are collected from any particular depth, the sampler is immersed into the water. The sampler simply consists of (1) a container which can be closed with the aid of a messenger and (2) an outlet tube which is fitted with clamp (Figure 26.1). After collection, the sampler is pulled up and the sample is transferred to the collection bottle through the outlet pipe. The collection bottle is then removed from the sampler proper and the sample is used for analysis or it can be preserved.

PRESERVATION OF SAMPLES

Since the quality of pond water changes quite rapidly, utmost care should be taken to carry out the required tests immediately after collection. A number of changes generally occur within the sampling bottles. These changes include (1) alteration in pH and release of metabolites by micro-organisms that lead to changes in carbon di-oxide and dissolved oxygen levels, (2) variations in alkalinity, transformation of phosphorus and different forms of nitrogen (such as ammonification, nitrification, and denitrification), and many other noticeable properties of the sample.

PANEL 26.1

SAMPLING

Sampling is used to collect data when the sources of data are far too many to be handled. Data collection is the heart of aquaculture research. It is an elaborate process through which the researcher makes a planned search for production-related data and collects the entire data necessary for the assignment. A sample is, however, only a portion of the fish culture ecosystem.

The objective of sanpling is to get the maximum information about the ecosystem with the minimum effort. Often, farmers who are not familiar with the scientific basis of sampling, have an impression that data collected through sampling is less reliable than data obtained by exhaustive covering the entire ecosystem. The impression is erroneous. If adequately performed, sampling produces representative data on the entire population.

1. Advantage of Sampling :

* Sampling saves time and cost.

* Sampling enables collection of information about the ecosystem that is 'OK' for the given purpose at a lesser time and cost.

* Sampling helps ensure the required degree of precision.

2. Terminologies Used in Relation to Sampling : Certain terminologies that are generally used in relation to sampling are noted below:

(a) Population or Universe: It is an entire group of animal/plant or non-living objects such as things having some of common basic features. In sampling, however, it means the totality of all elements that are pertinent to the scope of the problem under study. All of them possess the characteristics under investigation and hence they are sources of appropriate data. If a study is undertaken to find out the manure/fertilizer applying habits of fish culturists, one feature of pond will collectively constitute the population for the study.

(b) Factors: It refers to things that contribute to result such as oxygen, pH, dissolved oxygen, alkalinity, etc. are factors that form the productivity of a pond ecosystem.

(c) Parameter: It is a value or a statement that explains a true characteristic of an entire ecosystem. It is the value that would be obtained if the total ecosystem were covered in the measurement. In the example of fish farmers, suppose the entire ecosystem is surveyed and it is observed that the quantity of fertilizers used by them is between 130 and 150 kg/ha/year, the statement that fish farmers used this quantity of fertilizer per year, is the parameter.

(d) Statistics: It is the value of a feature obtained from a sample of the pond ecosystem. It merely provides an estimate of what would be the ecosystem's parameter

with regard to the specified feature. Normally, the statistic will vary from the parameter due to errors associated with sampling. Suppose 20 per cent sample of total fish culture ecosystem in a specified area is surveyed and it is observed that the quantity of balanced fish feed used by them falls in the range of 300-400 kg, this value is a statistic.

(e) Variables: These are the measurable features which can be numerically expressed in terms of some units such as fish length in cm, mm, inches; weight in gram, kg, pound, etc.

(f) Standard Deviation: It is the square root of the arithmetic mean square of deviation from arithmetic mean.

(g) Standard Error or Sampling Error: The sampling distribution of any statistic will have its own mean, standard deviation, etc. The sample estimates of statistic will differ from population parameter. The difference or deviation between the value of statistic of a particular sample with corresponding population parameter is termed as *standard error* or *sampling error.*

(h) Confidence Limit: An estimate (the term estimate refers to the particular numerical value obtained in a computation) from a sample of observations is subject to error. As a corollary, we calculate both an estimate and its standard error. The smaller the standard error, the greater the reliability of the estimate. The extent to which thing is likely to occur and measured by ratio of favorable cases to all cases possible which is termed as *probability,* that the parameter has exactly the value since the estimate is extremely small. If the estimate is good, there is a high probability that the true value is very close to the estimate value. It would be more satisfactory if we could make a probability statement concerning a range of values one of which the parameter is likely to have. Such an interval is termed as *confidence interval* and the end points of the interval are termed as *confidence limits.* In short, confidence limit is expressed in terms of the combination of the confidence level and confidence interval.

(i) Precision or Exactness: It refers to the degree of closeness between a statistic and a parameter. It is the direct upshot of two qualities of a sample such as its representativeness and its stability. While the former denotes to what extent the sample is close to the population, the stability of the sample is related to sample size and proportion of the population embraces. These two qualities of a sample determine the precision of the sample.

Different preservatives are available in the market that helps inhibit the changes in water qualities for long periods of time which permits adequate time to complete the tests. Some guidelines for using preservatives for different tests have been presented in Table 26.1.

Table 26.1: Different Preservatives Used in Chemical Tests of Water Samples

Parameter	Preservative	Permissible time
Dissolved oxygen (Winkler's method)	Fixed with manganous sulfate and alkaline iodide at Once	6 hours
Free carbon di-oxide	Cooling at 4°C	2 hours
Acidity and alkalinity	Cooling at 4°C	24 hours
Biochemical oxygen damand	Cooling at 4°C	6 hours
Total phosphorus	Cooling at 4°C	7 days
Chlorophyll-a	Cooling at 4°C	12 hours
Chemical oxygen demand	1 ml concentrated sulfuric acid / l of sample	7 days
Hardness	1 ml concentrated nitric acid/l of sample	7 days
Ammonia and nitrate	1ml concentrated sulfuric acid/l of sample	7 days
Soluble phosphorus	Stored in iodine-treated container	7 days

This table will definitely permits to the analyst for selecting methods that are suitable for preservation of samples.

26.2 Laboratory Equipments and Reagents

A list of laboratory equipments, reagents, and standard solutions needed to carry out the tests mentioned in this chapter are noted in Appendix I. Description of different types of equipments is beyond the perview of this chapter and the readers may consult any chamistry practical book where these equipments have been described at length.

The availability of best quality reagent chemicals make it possible to prepare sufficiently useful titrant solutions by merely dissolving the accurate weight of the chemical in the adequate volume of distilled water. The standard solution of any reagent means that the concentration is known exactly. The concentration of a standard solution sometimes referred to as its normality (abbreviated as N). It indicates the concentration of strength. Thus, the normality of the standard acid used in determining the alkalinity of water is 0.02 and can be written as 0.02N, N/50, or 1/50N, all of which have same meaning.

Though standard solutions and purchased reagents are very reliable, some of them, particularly starch solution and 0.025N sodium thiosulfate are not stable and should not be kept for long periods of time. It is better to purchase a concentrated standard solution (such as 1.0 N sodium thiosulfate) and dilute it with distilled water to last for a month at a time.

26.3 Determination of pH of Water

The pH is a numerical expression of the intensity of acidity or alkalinity. The numbers 0 to 14 are used to express pH values. The pH value of 7.0, is the neutral point. A pond water having a pH of 7.0 is neither acid nor alkaline. Numbers above 7.0 denote alkalinity, with the intensity of alkalinity increasing as the numbers become higher. Numbers between 0.0 and 7.0 denote acidity, with the intensity decreasing as the numbers decrease.

The pH of water can be measured either colorimetrically or electrometrically or simply by using litmus paper. The first method requires a less expensive equipment but suffers from severe interference contributed by color, high saline content, turbidity, free chlorine, colloidal matter, and reductants and oxidants. The indicators used for the purpose are subject to deterioration. For these reasons, this method is suitable only for rough estimation. Therefore, the electrometric method is generally considered preferable.

The pH meter is kept in 'switch on' condition for some time to warm up the instrument. About 50 ml of pH 7.0 buffer solution is taken in a clean 100 ml beaker, the glass as well as calomel electrodes are immersed and the pH reading is calibrated to 7.0. The beaker is then removed, the electrodes are rinsed with distilled water and then soaked with a filter paper. The electrodes are again immersed into another buffer solution of either alkaline or acidic range, depending on the expected pH value of the sample to be measured. If there is any fluctuation in the reading of the buffer of the second solution, the pH reading with the second buffer is calibrated and the pH reading in the first buffer solution is noted by following the same method. As a generalization, the pH meter should be calibrated with both ends of pH values (3.0 and 9.5).

After calibrating, the electrodes of the instrument are immersed in 50 ml of water sample and the pH reading is noted. After reading, the electrodes are washed with distilled water and then kept in a beaker containing distilled water.

It should be added as a note that standard buffer tablets for different pH levels are available in the market. Three buffer solutions, having pH 4.0, 7.0, and 9.2 are generally used in calibrating pH meter. One tablet is dissolved in 100 ml distilled water that produces corresponding pH level of the solution.

COLORIMETRIC METHOD OF pH DETERMINATION

In this method, a pH-dependent color on the water sample by using pH indicator of specific range is developed. The color is then compared with the standard color discs or color charts.

For colorimetric determination of pH of water, Lovibond Comparator is widely used. Before using the comparator, a rough pH value of water is estimated by dipping a strip of a wide range of pH paper. The change in color of the paper will indicate the value of pH of the sample (See Volume 1, Section 13.6). This approximate pH value will indicate which type of indicator has to be selected for the purpose.

Indicator	pH range
Bromo phenol blue	3.0 - 4.6
Bromo cresol green	3.8 - 5.4
Bromo cresol purple	5.2 - 6.8
Bromo thymol blue	6.0 - 7.6
Phenol red	6.8 - 8.4
Cresol red	7.2 - 8.8
Thymol blue	8.0 - 9.6

Ten mililitre of clear water sample is taken in two glass couvettes and 0.5 ml of the selected indicator is added in one couvette This couvette is kept in the groove of the comparator. The other couvette containing only water sample is kept in another groove. The disc has different shades of colors which are comparable to colors appeared with the indicator at different pH values. The color developed in the indicator is then matched in the sample with the shades of colors of the disc. When the color of the disc and that of the water sample matches, the corresponding reading denotes the pH of water.

The colorimetric method though involves less expenditure and less accurate, it takes more time for determination of pH than that of the electrical method. In contrast to colorimetric method, the electrical one is more accurate and the test can be made within a sort period of time. Therefore, this method is widely recommended. In fields, however, portable battery-operated pH meters are widely used than electrically-operated ones.

26.4 Salinity of Water

In freshwater fish culture ecosystems, the levels of salinity are very low and does not require estimation, though most of the fresh-water fishes thrive and grow well at salinity range between 3 and 10 ppt. The problem of salinity in freshwaters is more severe in some ponds which are situated in coastal areas and should be carefully monitored.

DETERMINATION OF SALINITY FROM CHLORINITY

The estimation of chlorine in water has been described in section 26.15. If the value of chlorine is divided by 1,000, it denotes the said value in ppt and the following relationship should be used for determination of salinity of the water sample. Salinity (ppt.) = Chlorinity (ppt.) X 1.805 X 0.03

DETERMINATION OF SALINITY FROM ELECTRICAL CONDUCTIVE

The salinity of water can be estimated from the electrical conductivity values of water by the following relationship.

Electrical conductivity (μ mhos X 0.64 cm^{-1})

26.5 Electrical Conductivity (EC) of Water

The EC value denotes the total concentration of ionized constituents of the water sample. It is closely related to the amount of total dissolved solids and is used as an index of salt content of water. Electrical conductivity is measured with the aid of an EC bridge, also called *conductivity meter bridge*. This equipment has a conductivity cell and the cell constant of which is determined first by measuring the conductance of standard potassium chloride (KC1) solution, the specific conductance of which solution is known.

For determination of the EC of the water sample, the cell of the conductivity meter is immersed in the standard solution of KCI (0.02 N) and then the value of the conductivity meter is noted. The cell constant is then calculated. The cell is then immersed in the sample water and the value of the conductivity meter is again noted. The EC can be measured by the following relationship.

EC = Measured conductivity (μ mhos cm^{-1} X cell constant

> (known specific conductance of standard KCl solution/ conductance shown by the cell)

Since the EC values are related to temperature, the EC is expressed at a standard temperature of 25°C. Most of the EC meters have temperature compensation device and consequently, direct reading at 25°C can be obtained.

Freshwater fish ponds generally exhibit low EC values which may be expressed as μ mhos cm^{-1} instead of m mhos cm^{-1}. Pond waters have EC values that varies between 30 and 1500 μ mhos cm^{-1}.

26.6 Calcium and Magnesium

Calcium and magnesium are very common polyvalent cations which are very important for pond productivity as they serve as nutrient elements. At the same time, the hardness and alkalinity of water also greatly influenced by the presence of these cations.

The most reliable method for determination of calcium and magensium in water is by complexo-metric titration using ethylene diamine tetra acetic acid (EDTA). The EDTA is an efficient complex which forms chelates with a number of polyvalent cations such as Ca^{2+}, Mg^{2+}, Fe^{2+}, Mn^{2+}, Zn^{2+}, and others. However, for determination of calcium and the total amount of calcium plus magnesium, ammonium purpurate and Eriochrome black T, respectively are used as indicators. The concentration of magnesium in the sample is then determined from the difference between the value of calcium and calcium plus magnesium.

STANDARDIZATION OF EDTA SOLUTION

Five mililitre of 0.01 N calcium solution is taken in a 100 *ml* conical flask. The solution is then diluted to about 25 ml with distilled water. Five mililitre of 4N sodium hydroxide solution and about 25 mg of murexide indicator are added. After mixing, a orange-red color will appear. It is then titrated with 0.01N EDTA solution untill the color is changed to purple. The strength of EDTA is calculated by the following relationship :

$$\text{Strength} = \frac{5}{V}, \text{ where V = Volume of EDTA solution required}$$

DETERMINATION OF CALCIUM

Five mililitre of water sample is taken in a 100 *ml* conical flask and the method is followed as in the case of standardization process.

DETERMINATION OF CALCIUM + MAGNESIUM

Five mililitre of water sample is taken in a 100 *ml* conical flask and is diluted with distilled water to about 25 *ml*. To this mixture, 1 *ml* of ammonium chloride-ammonium hydroxide buffer is added   and a few drops (3 to 4) of eriochrome black-T indicator is also added to get a wine-red color. It is then titrated against 0.01N EDTA until the color changes to blue.

1. *Calculations*:

* Concentration of Ca (mg/l) = *ml* of EDTA consumed X F X 40

* Concentration of Mg (mg/l) = (*ml* of EDTA consumed for Ca + Mg — *ml* of EDTA consumed for Ca) X F X 24

26.7 Hardness of Water

The hardness refers to the presence in water of dissolved calcium or magnesium ions and sometimes iron and aluminium ions. In general, the latter two ions are seldom occur in substantial amounts in most of the pond waters and determination of Ca + Mg definitely provides a fair idea on the hardness of water and it is expressed in terms of Calcium carbonate.

There are two broad categories of pond water such as (1) hard water (hardness results from dissolved calcium sulfate/ calcium fluoride — generally termed as *permanent hardness* and cannot be removed by boiling or hardness results from dissolved calcium hydrogen-carbonate which is formed in limestone or chalk region by the action of dissolved carbon di-oxide on calcium carbonate — generally termed as *temporary hardness* can be removed by boiling) and (2) Soft water. Such grouping has no practical significance in relation to biological productivity of fish ponds. As a generalization, hardness of pond water that varies between 60 and 120 mg/l is recommended to fish farmers.

The concentration of Ca and Mg in water sample is first determined by the method described in Section 26.6 and the calculation is made in the following way:

A. Amount of Ca (mg/l) as $CaCO_3$ equivalent = Amount of mg/l of Ca in water X Equivalent weight of $CaCO_3$ / equivalent weight of Ca = Amount of Ca (mg/l) in water X 50.04 / 20.04.

B. Amount of Mg (mg/l) as $CaCO_3$ equivalent = Amount of mg/l of Mg in water X equivalent weight of $CaCO_3$/ equivalent weight of Mg = Amount of Mg (mg/l) in water X 50.04/12.16.

Therefore, the total hardness of water = A + B = mg/l as $CaCO_3$.

The toxicity of certain heavy metals is generally influenced by the presence of hardness of water and has been found to be synergistic in action. The hardness can, however, be removed from pond water by ion-exchange using zeolites or called *permutil*.

26.8 Alkalinity of Water

The amount of acid required to titrate the bases in water is a measure of alkalinity. Several bases such as ammonia, bicarbonates, carbonates, hydroxides, silicates, phosphates, etc. equally contribute to the alkalinity of water. Their presence in water exhibits considerable influence on water qualities. Therefore, for practical purposes, the presence of carbonate ($CO_3^=$) hydroxide (OH^-), and bicarbonate (HC_3^-) are generally considered for determination of alkalinity.

For determination of the alkalinity, 100 *ml* water sample is taken in a 250 *ml* conical flask and 3 drops of phenolphthalein indicator is added. If the water turns pink, the sample is then titrated with 0.02N sulfuric acid until the pink color just disappears. The volume (*ml*) of acid consumed during titration is noted. In the same sample, 3 drops of methyl orange indicator is added; and if the sample turns yellow, it is titrated with the same acid until a faint color is obtained. During the second phase of titration, the required volume (*ml*) of acid is not further noted.

If the volume of acid used for titration with phenolphthalein is P and the total volume of acid consumed during titration with both phenolphthalein and methyl orange indicators is M, then the total alkalinity of the sample in terms of mg/l of $CaCO_3$ will be M × 10.

26.9 Determination of Chloride

The concentration of chloride ion (Cl^-) in pond water vary considerably depending on the environmental conditions. In a pond ecosystem, most of the water-soluble salts exist as chloride (Cl^-) form and therefore, the amount of Cl^- in water indicates very closely to the total amount of soluble salts present in the pond. As a generalization, the range of chloride in fish culture ponds has been recommended as 1 - 150 mg/l. It should be kept in mind that in coastal areas, the presence of large amounts of chloride in pond water definitely indicates salt-water infiltration. Sulfide and excessive alkalinity are possible sources which may interfere with the accuracy of this test.

Chloride ions can be estimated by titration with silver nitrate solution in presence of chromate ions. Silver nitrate forms silver chloride by reacting with the chloride present in water. When chlorine ions in water become exhausted, silver nitrate then reacts with the $CrO_4^=$ ions to exhibit a red color of silver chromate. This indicates that the titration has been completed.

* *Standardization of Silver Nitrate Solution*: To standardize the standard silver nitrate solution, 10 ml of standard sodium chloride solution is taken in a porcelin basin and then distilled water is added to make the volume of about 100 ml. To the solution, 0.5 ml of potassium chromate indicator is added and then titrate with silver nitrate solution till a permanent red precipitation is appeared. The factor (F) of the strength of silver nitrate solution is then standardized with the help of titre value.

$$F = \frac{10}{V},$$ where V is the volume of silver nitrate required for titration

For estimation of chloride, 100 ml of water sample is taken in a porcelin basin and the sample is titrated as above. The concentration of chloride (mg/l) is expressed by the following formula:

$$Cl^- \text{ (mg/l)} = \frac{X \times F \times 1000}{V}$$

Where   X = titre value (*ml*) of silver nitrate
          F = factor of silver nitrate
          V = volume of water sample (*ml*)

26.10 Dissolved Inorganic Nitrogen

The role of inorganic nitrogen in fish production is widely known. Generally, inorganic forms of nitrogen such as ammonium (NH_4^+), nitrate (NO_3^-), and nitrite (NO_2^-) are present in pond water. While the first two forms are predominant in farmed ponds, the last one is present in highly polluted waters. Ammonia and nitrate forms are inevitable for the growth and development of fish food organisms. For this reason, the estimation of NH_4^+ and NO_3^- nitrogen in fish culture ecosystem is badly needed.

DETERMINATION OF AMMONIA- NITROGEN

For determination of ammonia-nitrogen, 250 *ml* of filtered water sample is taken in a Kjeldahl flask and it is then fitted with the distillation plant along with the condenser. Twenty mililitre of 4 per cent boric acid is added in a 250 *ml* conical flask and kept it below the condenser so that the tip of the outlet is immersed into the boric acid solution. In the Kjeldahl flask, 10 *ml* of 40 per cent NaOH solution is added and the mouth of the condenser is fitted immediately. The process of distillation is then carried out till about 40 *ml* of distillate is collected in the receiving flask containing boric acid. The collected distillate is titrated with 0.02N H_2SO_4 till a pink color appears. However, turbidity, color, and magnesium constitute the main source of interference.

DETERMINATION OF NITRATE-NITROGEN

For detennination of nitrate-nitrogen, 20 *ml* of 4 per cent boric acid is taken in a conical flask and kept beneath the condenser tip as in the case of ammonia-nitrogen estimation. About 0.2 g Davarda's alloy is added to the water sample. The mouth of the Kjeldahl flask is closely fitted with the distillation plant and distillation is continued to about dryness of the water sample. It is then titrated as mentioned earlier. The amount of nitrogen present in the sample (N in terms of mg/l) in the form of NH_4^+ or NO_3^- or $NH_4^+ + NO_3^-$ can be expressed as follows:

$$N \text{ (mg/ l)} = \frac{A}{V} \times 280$$

Where, A = *ml* of 0.02N H_2SO_4 required for titration
V = Volume (*ml*) of water sample used

The permissible limits for NH_4^+ and NO_3^- nitrogen in pond water under Indian condition vary from 0.3 to 0.8 and 0.07 to 0.8 mg/l, respectively. These values generally indicate good productivity of the pond ecosystem. Nitrite constitutes the prime source of interference.

26.11 Phosphorus

Phosphorus is most critical nutrient element in the management of pond productivity since the element greatly influences the growth and production of fish biomass. Due to its extremely low availability in aquatic ecosystems, this element has a special importance in fish culture. Under Indian condition, however, more than 2 mg/l, 0.05 - 2.0 mg/l and less than 0.5 mg/l of P_2O_5 may be considered as optimum, moderate, and poor in pond productivity, respectively.

For determination of phosphorus, a standard curve is prepared with different concentrations of phosphorus from 2 mg/l solution. These concentrations are used for developing phosphomolybdic blue color. However, the following concentrations of phosphorus solution are used for preparation of a standard curve:

2 mg/l P solution (ml)	Concentration of P (mg/l) in 25 ml volumetric flask
0.25	0.02
0.50	0.04
1.25	0.10
2.50	0.20
5.00	0.40

The above-mentioned volumes of 2 mg/l P solution is taken in five volumetric flasks and at the same time, a blank flask is also considered containing no phosphorus solution. To each flask, 5 *ml* of ascorbic-molybdate mixture is added. The volume of each flask is made up by adding distilled water up to 25 *ml*. All the flasks are allowed to stand for 30 minutes for development of blue color. The reading of optical density in a colorimeter or spectro-photometer at 660 nm wave-length is noted after setting zero optical density with the blank solution. A standard curve is then prepared by plotting concentrations of phosphorus (mg/l) in X axis against respective optical density in Y axis.

A known quantity of water sample (5-10 *ml*) is taken in a conical flask. To the flask, 5 *ml* of ascorbic-molybdate mixture is added. The volume of content in the flask is made up to 25 ml by adding distilled water. The optical density of blue color which is developed in the mixture of the flask is determined. The concentration of phosphorus in unknown water samples is thus obtained by putting the optical density values on the standard curve. The concentration of phosphorus (mg/l) is expressed by the following formula :

$$\frac{X \times 25}{V}$$

where,   X = Concentration of phosphorus in 25 ml flask
         V = Initial volume of water sample

26.12 Dissolved Oxygen

Most probably, the dissolved oxygen (DO) concentration in water is the most important chemical parameter in fish culture ecosystems. The DO not only has direct effect on respiration of aquatic organisms but controls many other physico-chemical parameters of water as well. While wastewater contains virtually no DO unless the wastewater is fresh, natural unpolluted pond water contains oxygen that is able to dissolve in appreciable amounts (usually varies from 5 to 15 mg/l). This temperature-dependent oxygen-carrying capability is inevitable for fish and fish food organisms. Various suspended matter, sulfur di-oxide, cyanide, chromium, and residual chlorine cause interference with the accuracy of this test.

Different observations have been made on critical and optimal levels of DO in pond water. The minimum concentration of DO is, however, a function of exposure time. At 0.5 mg/l of DO level is, for example, enough to survive a fish for a few hours but not for several days. Moreover, the minimum acceptable level of DO generally varies with fish species to be reared, their size, physiological condition and other factors that are related to fish growth. According to some authorities, 3 mg/l of DO or less should be considered as hazardous to fish and for better survival, the DO level should not be less than 5.0 mg/l during at least 8 hour per day. As a generalization, for normal survival, growth, and reproduction of farmed fish, the DO levels of pond water must be kept within a range that varies between 5.0 and 8.0 mg/l. The DO should be less than 5.0 mg/l in warm water fish ponds.

PRINCIPLE

Manganous ions (Mn^{2+}) form manganous hydroxide ($MnOH_2$) in alkaline condition. This $MnOH_2$ reacts with DO of water and an equivalent amount of Mn^{2+} is oxidized to higher valency states (Mn^{4+}). In presence of iodine, the oxidized Mn^{4+} is again converted to divalent Mn^{2+} and iodine is liberated in equivalent amount of DO. The presence of this iodine, estimated by titration with the standard sodium thiosulfate ($Na_2S_2O_3$), dictates the amount of DO present in water.

*** Standardization of Sodium Thiosulfate Solution:** To standardize 0.1N $Na_2S_2O_3$ solution, 25 *ml* of 0.1 N potassium dichromate ($K_2Cr_2O_7$) solution is taken in a 250 *ml* conical flask, in which 1 *ml* of alkaline iodide solution is added and it is then acidified with the solution of 2 *ml* concentrated sulfuric acid. The liberated iodine is then titrated with 0.1 N $Na_2S_2O_3$ solution using starch as indicator. Starch converts the solution blue and at once it becomes colorless when the end point of $Na_2S_2O_3$ is achieved. The strength of $Na_2S_2O_3$ is then calculated according to the following formula :

$$\text{Strength of } Na_2S_2O_3 = \frac{\text{Volume of } K_2Cr_2O_7 \text{ used}}{\text{Volume of } Na_2S_2O_3 \text{ required}}$$

The required amount of stock solution is diluted to obtain 500 ml or any other volume of 0.025 N $Na_2S_2O_3$ by using the relationship that follows.

$$V_1 S_1 = V_2 S_2$$

Where, V_1 = Volume of stock solution required to be diluted

S_1 = Strength of stock solution of $Na_2S_2O_3$

V_2 = Desired volume of solution after dilution (500 ml in this case)

S_2 = Strength of $Na_2S_2O_3$ (0.025 N in this case)

DETERMINATION OF DISSOLVED OXYGEN IN POND WATER

The DO concentration of pond water is determined by the most widely adopted method as described by L.W. Winkler in 1888 and the method is very popular among progressive fish culturists.

For determination of DO in pond water, the sample is collected in a 100 ml glass bottle. One mililitre of manganous sulfate and alkaline potassium iodide solution are added through pipettes by slowly dipping them in the bottom of the bottle. The mouth of the bottle is closed by a stopper and the bottle is shaked well by gentle inversion. The shaking is repeated after floc has settled halfway. The floc is allowed to settle again.

The stopper is then removed and 1 ml of concentrated sulfuric acid is added by holding the pipette above the surface of the liquid. The liquid is again mixed well until no floc is visible. The bottle is allowed to stand for a few minutes.

Fifty mililitre of the solution is withdrawn into a 250 ml conical flask and it is titrated with 0.025N $Na_2S_2O_3$ until the color of the solution forms fade. A few drops of starch solution are added and the titration is continued until the blue color just disappears. The thiosulfate consumed during titration is noted and any return of the blue color should be ignored.

The DO present in pond water is expressed in terms of mg/l and is equal to the volume (ml) of sodium thiosulfate solution consumed in the titration.

26.13 Free Carbon Di-oxide

Though carbon di-oxide (CO_2) has great practical significance in the photosynthesis of phytoplankton in fish ponds, it has inter-dependence with pH and carbonate-bicarbonate equilibrium. For production of robust fish, however, the pond water must contain less than 5 mg/l CO_2; and under intensive fish cultivation, the range of free CO_2 may vary between zero mg/l during afternoon and 8 mg/l at dawn without any apparent dangerous effect on fish biomass. The most toxic effects of free CO_2 on fish may result when CO_2 level increases at the time of critically low DO levels.

Pond water having more than 8.5 pH does not contain free CO_2 in adequate amounts. Consequently, phenolphthalein indicator that has end point at this pH value is used for estimation of the amount of CO_2 in the water for titration with a standard alkali.

A stock solution of N/l0 sodium hydroxide (NaOH) is prepared by dissolving 4.0 g analytical reagent grade NaOH pellets in 1 litre distilled water. This stock solution is then standardized by titrating it against N/l0 sulfuric acid or 0.1N H_2SO_4 using phenolphthalein indicator until a faint pink color appears. One hundred ml of N/l0 NaOH solution is diluted to 440ml with distilled water to have N/44 NaOH solution.

One hundred of freshly-collected water sample is taken in a white porcelin basin. A few drops of phenolphthalein indicator is added into the sample. The appearance of pink color of the water sample indicates that the pH of water is above 8.5 and free CO_2 is almost absent. If the water sample remains colorless even after addition of the indicator, the sample is then titrated with N/44 NaOH with gentle stirring till the color turns pink.

The free CO_2 present is expressed in terms of mg/l and is equal to the volume (ml) of N/44 NaOH solution consumed in the titration × 10.

26.14 Dissolved Organic Matter

The amount of oxygen consumed during oxidation of organic matter is referred to as an index of dissolved organic matter (DOM) in water. Lentic water bodies are destined to become filled with organic deposits along with time. Though organic matter provides nutrients for the growth of micro-organisms, their gradual accumulation in the bottom sediments and the water bodies become highly productive or eutrophic and the phenomenon is termed as *eutrophication*. Though a number of parameters are involved in eutrophication phenomenon; measure of oxygen, biological productivity, and organic matter can assess the degree of eutrophication. While the presence of high organic matter in some ponds/lakes are valuable because of their ability to create favorable condition for fish production/fishing, others are dangerous because of the production of highly obnoxious algal blooms which are aesthetically objectionable. A pond soil containing much organic matter is, however, said to be rich and is ideal for growing fish.

When potassium permanganate ($KMnO_4$) is used for oxidizing soluble organic matter, the permanganate (MnO_4) ion is reduced to MnO^-_2 form and the pink color becomes faint. This change in color is used for determination of the amount of oxygen required to oxidize the organic matter.

Fifty mililitre of filtered water sample is taken in a 250 ml conical flask in which 5 ml 1 : 3 H_2SO_4 is added and the flask is shaken well. Ten mililitre of $KMnO_4$ is added to this acidified water and heated in a water bath for at least 30 minutes. To the flask, 10 ml ammonium oxalate is added. When the existing pink color of the solution will disappear, $KMnO_4$ is added very slowly until the pink color just reappears.

The amount of oxygen (mg/l) required to oxidize the dissolved organic matter is equal to the volume (ml) of $KMnO_4$ required in the titration.

26.15 Free Chlorine

In pond water, free chlorine is present in very small amounts and it is extremely toxic to fish. The presence of chlorine in pond water at the rate varying between 0.01 and 0.06

mg/l kills 70-100 per cent of fish depending on the type of species to be stocked. Therefore, the presence of even detectable concentration of free chlorine is not considered as safe for the farmed fish.

Free chlorine generally reacts with N-N diethyl - P - phenylene diamine (DPD) in presence of excess iodide to generate a red color. This color disappears when the chlorine is reduced to chloride by using ferrous ammonium sulfate.

Five mililitre each of the DPD indicator and the buffer solution is well-mixed in a 250 *ml* conical flask. To this flask, 100 *ml* of the water sample and 1g of potassium iodide crystal are added. The solution is then titrated with lN ferrous ammonium sulfate until the red color just disappears.

The free chlorine present is expressed in terms of mg/l which is equal to the volume (ml) of lN ferrous ammonium sulfate × 354.5.

26.16 Temperature

Temperature is one of the most important and dominant abiotic factors affecting the fish more conspicuously, and in more different manners than any other abiotic factor. Temperature penetrates into every region of a productive ecosystem, profoundly influences fish biomass and exerts marked effects on the metabolic rates and activities of fish as well as on their survival, growth, reproduction, and development. Though most fishes cannot adapt themselves to extremes of temperature, some fishes such as common carp, tilapia and some species of catfish are adapted to survive or even thrive at the extreme of this temperature range.

The range of temperature varies considerably in different geographical locations of ponds, at different depths of altitudes, and even in the same regions at different times of the day or season. In contrast to terrestrial ecosystems, the temperature variation in the aquatic ecosystem is comparatively less. On the other hand, shallow freshwater ponds often exhibit drastic temperature fluctuations. As a generalization, at the temperature range from 22-36°C, the warmwater fish culture has been found to be congenial. Cold water fish species are adapted to a very low temperature range that varies between 10 and 20°C.

Correct temperature records are necessary for calculating dissolved oxygen saturation values, for correlation of biological productivity of fish culture ecosystems, and for many other purposes. For determination of water temperature in a fish culture pond, a mercury-filled celsius thermometer wity a range of about 0° to 100°C is recommended. For accurate determination, the scale should be sub-divided into 0.5° or 1°C for ease in reading.

For determination of surface water temperature, the thermometer is immersed in the water to a depth of the etched circle which appears around the stem below the scale level. The thermometer is kept to the proper depth for about one minute and the temperature is recorded.

For determination of sub-surface water temperature, a reversing thermometer is used. This thermometer is kept at a desired depth under the water colum and by the help of a trigger arrangement, the thermometer is reversed and the reading is fixed. By withdrawing the thermometer from the water, the temperature reading is noted.

26.17 Biochemical Oxygen Demand

A dissolved oxygen test estimates the current oxygen status of any fish culture ecosystem. As a generalization, the DO content of water can vary considerably from day to day due to some environmental factors especially wind. and sunlight. Of more significance is the rate at which oxygen is being used by micro-organisms present in the water. If pond water contains organic materials in large amounts, then the rate of decomposition of micro-organisms may be high and the water may deplete the content of its oxygen. This has important consequences for continuous decomposition of micro-organisms and the lives of fish. If pond water becomes rich in organic materials, their complete decomposition may require a long time and therefore, a 5 day incubation period has been suggested for most practical purposes. Hence the Biochemical Oxygen Demand (BOD) value is termed as BOD_5.

DETERMINATION OF BOD VALUE

1. One litre of water sample is taken in a container and soaked it thoroughly to increase its DO level.

2. Portions of the sample are placed into three glass-stoppered bottles of 250 *ml* capacity.

3. The oxygen content of one bottle is immediately measured.

4. The remaining two bottles are incubated in the dark at a standard temperature (20°C) for 5 days. Incubation in darkness at 20°C for 5 days is the standard procedure.

5. The oxygen content of the incubated bottles is then determined.

6. The mean value for the incubated samples is substracted from the original sample. This will provide the sample BOD in mg/l unless the sample was diluted before incubation. In this case, the BOD is determined by the following formula:

$$BOD = (A - B)(a + 1) \text{ mg/l}$$

where,   A = Initial DO level in mg/l

B = Mean final DO level in mg/l

a = Volume of dilution (water to 1 volume of sample)

If the water sample is diluted, the value of BOD obtained from undiluted sample is multiplied by the number of times diluted the sample.

Note:- A badly polluted pond water might require dilution in the ratio of 1 part sample: 4 parts deionized or distilled water.

Determination of BOD value is of little or no use in fish culture where serious rate of DO consumption by micro-organisms is of more interest. Therefore, a short-term BOD estimation is made for unchanged water samples. For this purpose, the sample is incubated in pond water at a desired depth for a few hours and the BOD is determined by the formula as noted above. The BOD_5 values of pond water that vary between 5 and 20 mg/l, are suitable for fish culture.

26.18 Turbidity

The penetration of light in pond water is influenced by the optical properties which, in turn, are interfered by the impurities present in it. The turbidity makes the water dirty in the sense that light transmission is inhibited. The turbidity is principally due to the presence of numerous materials in suspended conditions. These materials essentially include organic matter, inorganic substances, plankton populations, clay particles, and silt. Turbidity is very important because of aesthetic considerations and because micro-organisms can hide in (or on) the colloidal particles. At the same time, clay particles are deposited over the gill surfaces thus gill filaments are clogged with these particles and fish suffocate.

Though light penetration in water is recorded with the help of a round disc called *Secchi disc,* the standard method of measuring turbidity is with the Jackson Candle (turbidimeter). In short, it consists of a long flat-bottomed glass tube under which a candle is placed. The sample water is placed into the glass tube until the outline of the flame is no longer visible. The centimeters of water in the tube are then measured and compared to the standard turbidity unit, which is 1 mg/l silicon di-oxide equals to 1 unit of turbidity.

Secchi disc is a circular disc of metal, generally 20 cm in diameter painted alternately black and white in radial fashion (Figure 11.1 of Volume 1). It has a weight below the disc. There is a string to lower it in water. The degree of indistinction between colors in the water determines the index of turbidity of water. The disc is slowly dipped in water with the help of the string and the depth where the black and white colors just indistinguishable is noted. The disc is again lifted upward very slowly and again the depth at which the color of the disc is just reappears is noted. The value of turbidity in terms of centimeter (cm) can be measured by the following formula:

$$\frac{D_1 + D_2}{2}$$

where, D_1 = Depth of water at which the disc disappears

D_2 = Depth at which it reappears

Adverse behavioral reactions of fish have been noted when clay turbidities exceed 20,000 mg/l. Most of the fish species survive at a clay turbidity level of 1,00,000 mg/l for several weeks but appreciable mortalities have been noted at turbidities more than 1,75,000 mg/l.

In addition to the presence of clay/silt particles in high amounts, abundance of both Phyto- and zooplankton are equally responsible for the turbidity of water; of course, it is desirable to a certain limit. However, the plankton turbidity of 30-50 cm is beneficial to fish culture.

26.19 Productivity

The productivity of any fish culture ecosystem is defined as the rate of accumulation of organic substances in living components (green plants). It may be primary (the rate at wtrich radiant energy of sun is stored by chemo- and photosynthesis of green plants which are termed as *producers* in the form of organic substances), secondary (rate of energy accumulated at consumer levels), gross primary (total rate of photosynthesis including the organic matter used up in respiration at the time of measurement), net primary (rate of accumulation of organic matter in plants in excess of that utilized at the time of measurement), and net (The rate of accumulation of organic matter not used by heterotrophs). Fish production has been found to increase with the increase in primary productivity of pond water. A well-managed fish culture pond, however, exhibits gross primary productivity value that ranges from 1,500 to 5,000 mg Cm^{-3} day^{-1}.

The measurement of primary productivity is essentially inevitable in producing robust fish from those ponds which are virtually ideal for fish culture. Insufficient light and high turbidity of water are factors that may influence the accuracy of the productivity measurement. The widespread 'light and dark bottle' method of determination of primary productivity in pond water has been adopted by many fish culturists because of the recognition that it is very simple and easy to measure the primary productivity. The method of determination of the primary productivity of water has been presented in Volume 1, Section 3.13.

26.20 Estimation of Plankton Population in a Freshwater Ecosystem

As a generalization, plankton refers to animals and plants floating in the water of seas, rivers, ponds, and lakes, as distinct from animals which are attached to or crawl upon the bottom; especially minute organisms and forms, possessing weak locomotory powers. The plankton populations, however, which are divided into phyto- and zooplankton, play a very important role in the productivity of pond water. The great bulk of planktons are unicellular floating organisms, predominantly Chlorophyceae, Cyanophyceae, Diatoms, Copepods, Rotifers and Cladocerans distributed throughout the water. Although they sometimes become abundant and impart a definite color of water, their density is generally so low that there is no gross sign of occurrence. It is the enormous quantity and quality of the earth's fresh-water which they employ that makes them the most abundant of these organisms as food for fish.

Phytoplankton populations are determined both by qualitatively and quantitatively. Qualitative and quantitative estimations of plankton require several equipments such as a cone-shaped plankton collection net, plankton samplers, haemocytometer, cover glass, filter paper, and a compound microscope.

Water samples are collected at intervals and from different depths. The collection of plankton by plankton net is a very crude method, though the plankton net is widely used for the collection of zooplanktons; of course, a small fraction of phytoplankton is retained during sampling. The plankton net is a very simple conical net made of silk and strong nylon having mesh size of 50 μm. At the lower end of the net, a glass or plastic bottle is attached (Figure 26.2) for collection of plankton populations.

For effective collection of phytoplankton, Von-Dorn water sampler and shallow water sampler are also common, useful, and easily available. With these samplers, however, sample at a desired depth can be achieved. Von-Dorn sampler (Fig. 26.3) is composed of a hollow cylinder made of polyvenyl chloride (PVC) and a device to close the ends at a depth. Two cups of PVC are fitted inside the cylinder by means of a rubber tube. The sampler is slowly immersed in water at desired level in the open position. The messenger is released which helps free the cups and consequently the cylinder gets filled with water. Due to the elasticity of the rubber tube, the cups again get attached to the ends of the cylinder. The cylinder is withdrawn from the water and the sample water is then removed. The volume of cylinder may vary from 0.5 to 5.0 litre.

A laboratory-assembled shallow water sampler is widely used in sampling of waters where the depth of pond water is very low. The sampler generally consists of a glass bottle with ground stopper fitted in a brass frame and the bottle can be immersed at any depth. The stopper is pulled up with the aid of an auxillary rope to permit the entry of water in the bottle. When the bottle is filled up, the stopper is lowered to close the bottle.

Another laboratory fabricated water sampler typically consists of a bottle which is fitted in a metallic cage. The mouth of the bottle is closed with a rubber cork having two tubes, one reaching at the bottom of the bottle. The upper end of the long tube is connected with a rubber tube which is tied with a string. The pond water enters through long tube and the air within the bottle gets displaced through the smaller tube. Displacement of air prevents the mixing of air within the water sample inside the sampler.

After collection of water sample by anyone of the methods stated above, qualitative estimation can be made to find the species composition of microflora. Phytoplanktons are mostly algae. While freshwater algae belong to classes Chlorophyceae (green), Cyanophyceae (blue-green), Xanthophyceae, and Bacillariophyceae (diatoms), the zooplanktons include mainly Copepods, Rotifera, Cladocera, the Cyclopid, and the Calanoid Copepods.

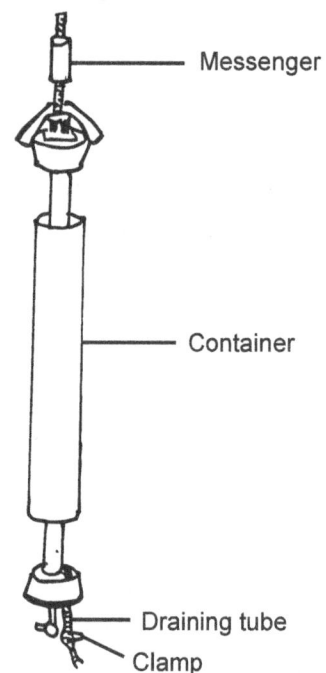

Fig. 26.1 : Structure of a water sample.

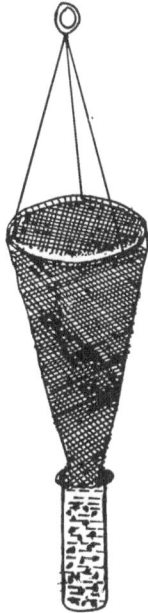

Fig. 26.2 : A simple cone-shaped net widely used for the collection of plankton from pond water.

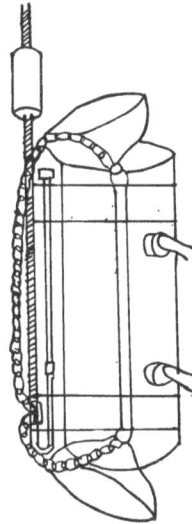

Fig. 26.3 : A Von-Dorn water sampler is also used for the collection of plankton from pond water. This sampler is very common, useful and easily available.

Quantitative estimations of these planktons are also undertaken to determine their frequency, density, and abundance. The number of organisms in unit area per volume are made. The most common method employed for counting phytoplankton is Lacky's drop method. In this method, a known volume of water (one drop) which fits below a 22 mm cover glass is placed over a glass slide. Organisms present in this drop is counted in a microfield of a compound light microscope. One microfield can be taken as One sampling unit. In this manner, the number of organisms is counted in a number of microfield. The data is recorded as follows.

Number of phytoplankton/ drop

$$= \frac{\text{Area of cover glass}}{\text{Area of microfield}}$$

= Average count per field

= 22/0.01 × 11 = 24,200

Microfield number	Number of phytoplankton	Average number of phytoplankton	Area of one microfield
1	7	55/5 = 11	0.01
2	15		
3	9		
4	8		
5	16		

If there are five drops in one mililitre, then the organisms present per mililitre

$= 24,200 \times 5 = 1,21,000$

Phytoplanktons can also be counted by the use of a haemocytometer (See Panel 26.2). It is an apparatus consisting of a special glass slide with a grid of lines imprinted on the bottom of a shallow rectangular trough so that if a coverslip is placed over the trough, the grid demarcates known volumes. Planktons collected from a well-mixed suspension are introduced into the space and the number in the grid squares is counted under the microscope.

One drop of well-agitated sample is taken on the counting chambers. A specially-designed cover slip is pressed onto these chambers. After a few minutes, counting is done in high power in central chambers (25 chambers). The procedure is repeated till at least 100 organisms of a species are counted. The data are recorded in the following manner.

$$\text{Phytoplankton}/l = \frac{\text{Count in central chamber} \times 10^{-7}}{\text{Concentration factor (C.F.)}}$$

$$\text{C.F.} = \frac{\text{Volume of water concentrated}}{\text{Volume of water made after concentration}}$$

PANEL - 26.2
HAEMOCYTOMETER

This is a counting chamber on a glass slide, in a centrally-located 'H'-shaped groove. Two prominent ridges come out of this groove, each containing five cubical chambers of the length × breadth × depth as 1 × 1 × 0.1 mm (Figure 26.4). The central chamber is divided into 25 sub-chambers, each of which is further sub-divided into 16 chambers that makes the total sub-chamber 400. Rest of the four chambers which are situated at four corners are sub-divided into 16 chambers. One of the five sub-chambers will be 1 × 1 × 0.1 mm and a value of 10^{-4} ml, while one of the five sub-chambers has 1/4 × 1/4 × 0.1 mm and a volume of 6.25×10^{-6} ml. The smallest chambers (one of the 400 sub-chambers) have a dimension of 1/20 × 1/20 × 0.1 mm and volume of 2.5×10^{-7} ml.

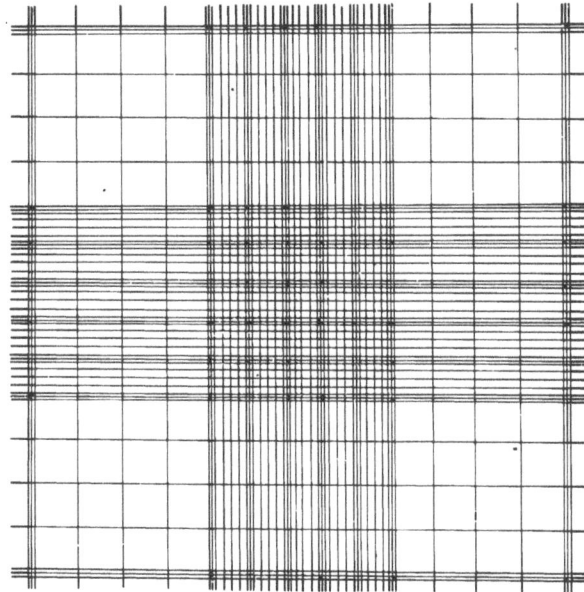

Fig. 26.4 : Counting chambers of a haemocytometer.

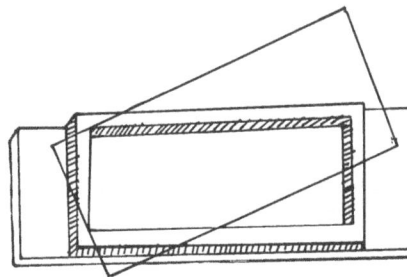

Fig. 26.5 : A Sedgewick-rafter cell is widely used for counting zooplankton populations. It is a very simple and reliable method.

The zooplankton populations are counted by a simple and reliable device termed as *Sedgewick-Rafter Cell* (Figure 26.5). It is a slide, 50 mm long, 20 mm wide and 1 mm deep. The volume of the cell is $1/cm^3$ or 1 *ml*. One mililitre of sub-sample is kept on the slide, and a special cover glass is placed over the slide. The number of zooplankton is then counted under the microscope. All organisms present in the cell (1 ml) are counted by moving the cell till the entire area is covered. Another drop may be taken and the same is repeated for several replicates.

$$\text{Number of zooplankton/ml} = \frac{\text{Number of organisms counted}}{\text{Number of replication}}$$

26.21 Toxicity Test of Chemicals on Fish (Static Bioassay Test)

Static bioassay test is performed to evaluate various chemicals as toxicant for their harmful environmental effects at tentative level. Bioassay is a test in which fish are used to detect the presence or the effects of any other physico-chemical factor or any other type of ecosystem disturbance. This test can be performed using any type of aquatic organisms.

The objective of this test is to find out the lethal concentration or effective concentration that results in mortality or other effects. They are then used for determination of safe concentration of a chemical or maxinnlm acceptable toxicant concentration. The organism is exposed to different concentrations of a toxicant for a definite period and mortality as well as behavioral changes are noted. Of the three types of toxicity tests such as static, renewal and flow-through bioassay tests, the former is very common where the organisms are exposed to the same toxicant solution for the entire experimental period.

PROCEDURE

1. *Selection and Acclimatization of Test Animals :* The fish selected for test must be healthy, disease-free and sensitive to the toxicant. Fish are collected from fresh and unpolluted waters and are acclimatized to the laboratory condition in a glass aquarium for at least 7 days in the conditions more or less similar to their natural habitat in terms of dissolved oxygen, pH, temperature, and hardness of water. Feeding is done regularly but stopped before 24 hours of the commencement of the test.

2. *Preparation of Different Concentrations of the Toxicant :* Water of the same source from where fish have been collected is used. Besides this, dilution water may also be used by using chlorine-free tap water and adjusting pH, hardness, alkalinity, dissolved oxygen, etc. The toxicant is then added to prepare various concentrations. Dissolved oxygen levels are kept at least 5 mg/l.

3. A set of glass aquaria (10 litre capacity each) of desired concentrations of the toxicant is prepared. One container is used as control (having only water but no toxicant). Replicate sets of each concentration are kept.

4. Water quality parameters such as pH, dissolved oxygen, alkalinity, and hardness in all sets are measured to ascertain the suitability of conditions.

5. Five to 10 fish of smaller size (3-6 cm long) are kept in each container.

6. The behavior of fish is noted closely for the first 4 hours. Dead fish are immediately removed from each container. A fish is said to be dead if it does not give any response after it is tapped by a glass rod.

7. Physico-chemical parameters of water are measured at 24h intervals.

8. The number of dead fish is counted in each container at every 24 hours till 96 hours. It is sometimes desirable to count dead fish at 4, 8 , 12, 16 and 24 hours

on the first day and only once a day afterwards so that a well-defined shape of the toxicity curve may be obtained. During the test period, the behavior of fish and other body changes are recorded.

CALCULATION AND INTERPRETATION OF THE RESULT

The results are recorded on a semilog graph paper with the concentration being recorded in log scale and the mortality or any other effect in arithmetic scale (Figure 26.6). A line is then drawn from 50 per cent lethality level to the graph line and is connected to the concentration. This is LC50, EC50 or 96h LC50.

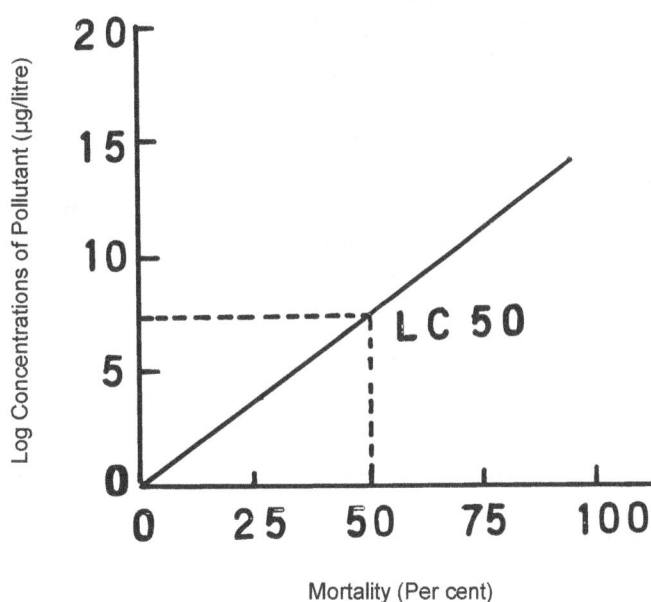

Fig. 26.6 : A toxicity curve

26.22 Soil Analysis

Pond bottom soils are very important in determining the producivity of fish culture ecosystems. Abundance of nutrient elements in a pond ecosystem largely depend on a series of biochemical and chemical reactions occurring ceaselessly in the bottom soil that help to release the content of its nutrient elements into the overlying water and their adsorption by microbial populations and soil. These phenomena greatly influence the growth and abundance of fish food organisms. Ultimately, the productivity of fish ponds significantly increased.

To determine the quantity of fertilizers to be added to soil, it is necessary to know what type of nutrient elements (or element) are deficient or excess. The total quantities of nutrients present in soils give little, if any, information on the availability of the elements to fish. Consequently, analyses suitable for measuring only the available portion of a given nutrient must be employed for aquacultural purposes. Such analyses may be termed as *partial* because only a portion of the total quantity of a soil constituent is determined. Partial analyses will now receive attention.

Similar to water analyses, the soil analyses also involve the collection and processing of soil samples and their analysis following standard methods.

PROCEDURE OF SOIL SAMPLING

A fish culture pond can be treated as a single sampling unit if the pond exhibits uniformity with respect to texture, color, depth, etc. But in cases where there are considerable variations in such properties, separate samples are collected. For each sampling unit, a number of sub-samples from different areas of the pond are collected. About 1 kilogram of sub-samples are generally collected from at least 6 inch depth of the bottom soil. This depth is the most reactive zone from the pond fertility point of view. In case of acid sulfate soils, the samples are collected from deeper horizons. After collection, all sub-samples are well mixed on a polythene sheet. The total quantity of soil sample is divided into 4 equal parts, two of which are removed and the remaining two parts are again thoroughly mixed. This amount of soil sample is considered for different tests.

PROCESSING OF SOIL SAMPLES

After collection, the soil samples are further processed. This involves air drying in shade, grinding with a wooden pestle and morter, passing throgh a strainer (2 and 80 mm mesh size), and storing in a plastic container. In different soil testing procedures, air-dried soil samples are, however, widely used.

26.23 Physical Characteristics of Soil

COLOR

The soil sample is spread uniformly over a wooden sheet. The soil particles are then matched with chips of a number of colors in the Mumseli's soil color chart. The chip with which the soil color matches, is taken out and the notation indicates on the chart is recorded which indicates the color characteristic of the soil.

TEXTURE

The soil texture is determined by the following two methods.

1. *Examination of Soil Sample under Hand Lens and Felt Between Fingers and Thumb :* The soil sample is examined under high power hand lens and small amount of it is squeezed between fingers and thumb. The texture of soil samples is observed with the help of the following chart.

2. *Using Sieves of Different Number of Meshes :* The soil sample is passed through a series of sieves with different size (diameter) of their pores (meshes) as noted in

Section 1.18 of Volume 1. Different fractions (different-sized particles) are thus separated from the sample and the relative proportion of each fraction is noted. According to this proportion, the texture of soil sample is determined.

Examination of soil and as felt between fingers and thumb	Texture
1. Individual soil grains are seen and form a cast when moist soil is squeezed.	Sand
2. Individual soil grains are seen and form a cast but fall apart when dry soil is squeezed, moist soils cast that does not break.	Sandy-loam
3. Yellow with a gritty feel; when dry soil is squeezed, it forms a cast breaking careful handling; most soils form cast that does not break.	Loam
4. Dry, appearing cloddy, with soft feeling, forming cast that does not break; on squeezing, the most soil does not form ribbons.	Silt-loam
5. Breaking into lumps, hard when dry; if moist soil is pinched, it forms thin ribbon which rapidly breaks. Moist forming cast.	Clay-loam
6. Forms very hard lumps when dry, stick when moist soil is squeezed; it forms a long and flexible ribbon.	Clay

DETERMINATION OF PERCENTAGE OF CLAY, SILT AND SAND IN A SOIL SAMPLE

1. Twenty gram of soil is taken in a 50 ml beaker and 250 ml distilled water is added to it.

2. Thirty mililitre of 6 per cent hydrogen peroxide (H_2O_2) is added to the beaker and digested on a water bath.

3. About 30 ml 2N HCl and 100 ml distilled water is added, boiled for a few minutes and the sample is allowed to stand for at least one hour with frequent stirring. Stirring makes the soil free from carbonates. The soil is filtered and washed it repeatedly with hot water till the filtrate becomes chloride free.

4. The suspension is then transferred to a 250 ml conical flask, and 5 ml of 2N NaOH is added and shaked the flask with a mechanical shaker for half an hour.

5. The content is kept in a 1,000 ml cylinder and the distilled water is added to make up the volume of the cylinder.

6. The suspension is allowed to stand for several minutes.

7. A 20 ml pipette is lowered to 10 cm depth of the suspension from which 20 ml of the content is collected. The content is then transferred to a 50 ml previously weighed beaker marked as A.

8. The volume of the 1, 000 ml cylinder is again maintained by adding distilled water. The suspension is further stirred and allowed it to stand for 6 hours.

9. Twenty mililitre of the content is again collected and transferred to a 50 ml previously weighed beaker marked as B.

10. The suspensions in beakers A and B are dried in an oven to constant weights.

11. The actual weights of dry materials in beakers are noted by subtracting the weights of the empty beakers from the values of beakers plus dried materials.

12. If the weight of dry material in beaker A is $\underline{a}$ and that in beaker B is $\underline{b}$ gram, then the percentages of clay (C), silt (Si), and sand (Sa) can be calculated as follows.

C = $\underline{b}$ x 250, Si = ($\underline{a}$ – $\underline{b}$) x 250, Sa = 100 - ($\underline{a}$ x 250).

26.24 Chemical Characteristics of Soil
DETERMINATION OF pH OF SOIL

Though the principles for estimation of soil pH are similar to that of water such as colorimetric and potentiometric methods, a soil-water suspension is used to determine the pH value of soil.

1. *Colorimetric Method:* Under this method of pH determination of soil sample, a soil-water suspension is prepared by using different soil: water ratios (especially 1 : 2). To flocculate the suspension and to make the supernatent water clear, barium sulfate is used. This supernatent is allowed to attain equilibrium with that of the soil and is determined with the help of Lovibond Comparator.

   Some amount of barium sulfate solution is taken in a 50 *ml* glass tube to form a 1 cm thick layer. Ten gram of soil sample and 20 *ml* of distilled water is added to this solution. The tube is then mixed thoroughly and is allowed to stand for about 30 minutes to settle the soil particles. About 10 *ml* of supernatent water is taken out from the mixture and the value of its pH is determined with the help of Lovibond Comparator (See Section 26.3).

2. *Potentiometric Method:* This method is more reliable and is widely used for field as well as laboratory studies. A portable battery-operated pH meter is generally used.

   About 10g of soil sample is taken in a 50 *ml* beaker. To this, 20 *ml* of distilled water is added and the soil solution is thoroughly mixed. The soil-water suspension is kept for at least 30 minutes with occasional stirring to establish an equilibrium condition and the value of its pH is determined according to the method described in Section 26.3.

DETERMINATION OF LIME REQUIREMENT IN FISH PONDS

Two methods of lime requirements (LR) are generally followed. In the first method, LR is determined from textural composition and the pH of the soil. The second method involves the use of a buffer solution.

1. *Determination of LR from the Data on Soil Texture and pH :* First, the pH and textural composition of the soil is determined following the method noted in Section 26.23. Using the following Table 26.2, the amount of lime required for ponds can be calculated.

Table 26.2: Amount of Lime (Calcium carbonate) Required For Ponds

Soil pH	Rate (Kg/hectare)		
	Clay soil	Medium soil	Sandy soil
4.0 - 5.0	3,000	2,000	1,000
5.0 - 6.0	1,800	1,200	600
6.0 - 6.5	1,500	1,000	500
6.5 - 7.5	600	400	200
7.5 - 8.0	450	250	100

Source: Modified after CIFRI (1985)

2. *Determination of LR Using Buffer Solution :* Twenty gram of soil sample is taken in a 100 *ml* beaker containing 20 *ml* distilled water and the content is stirred for 30 minutes. The pH of the suspension is measured as mentioned in Section 26.24. Twenty mililitre of buffer solution (pH 8.0) is added to this soil: water suspension and the content is allowed to stand for 30 minutes with occasional shaking. The pH meter reading is set to 8.0 with 1 : 1 mixture of distilled water and buffer solution. Under this condition, the pH reading of soil: water + buffer suspension is noted from the pH meter. The LR is calculated from the Table 26.3.

Though calcium carbonate is widely used as liming material in fish culture, other types such as calcium oxide (quick lime) and calcium hydroxide (slaked lime) are equally effective in increasing the pH value of soil. Calcium carbonate is, however, considered to be the safest and cheapest liming material.

Table 26.3: Lime Requirement in Terms of Kg/hectare of Calcium Carbonate for Fish Pond Soils

Soil pH	Soil pH: Water + Buffer								
	7.9	7.8	7.7	7.6	7.5	7.4	7.3	7.2	7.1
	Requirement of calcium carbonate								
5.7	90	180	270	360	450	540	630	725	820
5.6	125	250	380	500	630	700	880	1010	1130
5.5	200	400	600	810	1010	1210	1410	1610	1810
5.4	290	580	870	1160	1450	1740	2030	2320	2610
5.3	340	680	1020	1360	1700	2040	2380	2720	3000
5.1	440	880	1325	1760	2200	2650	3000	3530	3970
4.9	660	1310	1970	2620	3280	3930	4590	5240	59S0
4.8	670	1340	2020	2690	3360	4030	4700	5390	6050
4.7	700	1410	2120	2820	3530	4230	4940	5640	6350

DETERMINATION OF GYPSUM REQUIREMENT IN FISH PONDS

Five gram soil sample is taken in a 250 *ml* bottle and 100 *ml* of standard gypsum $(CaSO_4)$ solution is added to it. The bottle is then shaked for about 30 minutes. The filter is suspended and the Ca + Mg concentration is determined following EDTA titration method using erlochrome black T indicator as noted in Section 26.6.

1. Calculation:

(i) Meq/l of Ca + Mg in the filtrate = (*ml* of EDTA used × normality of EDTA solution × 1000)/*ml* of aliquot used.

(ii) Gypsum Requirement (Meq/100g) = Concentration of Ca in added $CaSO_4$ solution (Meq/l) – Concentration of Ca + Mg in filtrate (Meq/l) × 2.

In alkaline pond soils, sodium occurs in high levels. This results in high pH values along with several problems and needs to be recovered by using divalent cations especially calcium. Gypsum is used for replacing the exchangeable sodium and the amount of gypsum required to replace the exchangeable sodium to a safe level is termed as *Gypsum requirement* of soil.

DETERMINATION OF ORGANIC CARBOB IN POND SOIL

One gram of soil sample is taken in a 500 *ml* conical flask and a few drops of distilled water is added to it. After 10 minutes, 10 *ml* of 1(N) $K_2Cr_2O_7$ and 20 *ml* of concentrated H_2SO_4 are added. The mixture is thoroughly mixed and the flask is allowed to keep in a dark place for 30 minutes. Then 200 *ml* of distilled water and 10 *ml* of H_2SO_4 are added. About 1 *ml* of diphenyl amine is used as indicator to develop deep blue color. The standard solution of ferrous-ammonium sulfate is added drop by drop to the content of the flask until the color turns to bright green. A blank (the flask containing all the reagents but without soil sample) is carried out. The organic carbon of soil can be calculated as noted below.

$$\text{Organic Carbon (per cent)} = (B - A) \times 0.3$$

Where   B = Titration value (ml) of ferrous-ammonium sulfate in blank set.

A = Titration value (ml) of ferrous-ammonium sulfate in the sample.

Pond soil containing organic carbon in the range that varies between 1.5 and 2.5 per cent may be considered as optimum while those having less than 0.5 per cent organic carbon may be regarded as *very poor* for fish production.

DETERMINATION OF NITROGEN IN POND SOIL

In soil, nitrogen (N) occurs mostly in organic form. This form of N is gradually mineralized to inorganic ones and becomes available to producers. The mineralized forms of N (NH_4^+ and NO_3^-) are readily available to phytoplankton populations. In this sub-section, the methods of determination of easily mineralizable and mineralized forms of N and also the total N have been noted.

1. *Easily Mineralizable Nitrogen :* Fractions of organic nitrogen such as decomposable proteins, amino acids, etc. are termed as *easily mineralizable nitrogen*. Under such conditions, these fractions are transformed into available forms and contribute substantially to an index of available nitrogen status of the pond soil.

   Twenty gram of soil sample is taken in a Kjeldahal flask and 20 *ml* of distilled water is added to it. The water and soil are thoroghly mixed by swirling the flask. About 1.0 *ml* of liquid paraffin and a few glass beads are kept in the flask. This helps prevent the content from bumping. One hundred mililitre of 0.32 per cent $KMnO_4$ and 100 *ml* of 2.5 per cent NaOH solution is added to the flask. The content of the flask is distilled and about 100 *ml* of the distillate is collected in a conical flask that contains 20 *ml* of H_3BO_3 solution. The distillate is then titrated with 0.02(N) H_2SO_4 to develop a pink color.

(i) Calculation : The amount of easily mineralizable nitrogen = $\underline{a}$ × 1.4 mg/100g soil.

   Where $\underline{a}$ = The volume of 0.02 (N) H_2SO_4 required for titration.

2. *Mineralized (NH^+_4 + NO^-_3) Nitrogen :* After mineralization, organic nitrogen is transformed into ammonium and nitrate ions become available mostly in water, the ammonium ions remain adsorbed in the exchange complex of soil colloids as easily available exchangeable cations.

   One hundred gram of soil sample is taken in a 500 *ml* conical flask and 200 *ml* of NaCl is added to it. The contents are thoroughly mixed. The suspension is then filtered and the soil is washed with additional 50 *ml* of NaCl. The filtrate is taken in a Kjeldahl flask and distilled and the distillate is collected as stated above. Distillation is continued to near dryness and is tittrated the H_3BO_3 with standard 0.02 N H_2SO_4 to develop a pink color.

(i) *Calculation :* The amount of mineralized nitrogen = $\underline{a}$ × 0.28 mg/l00g soil.

   Where $\underline{a}$ = The volume of 0.02 (N) H_2SO_4 required for titration.

3. *Total Nitrogen :* Most of the soil nitrogen is present as organic form which is slowly converted into available form. Though this form of nitrogen in not directly involved in pond productivity, determination of total nitrogen is inevitable to evaluate the nitrogen supplying capacity of a soil that involves the productivity of any fish culture ecosystem.

   Five gram of soil sample is taken in a Kjeldahl flask and then 20 *ml* of distilled water is added. The flask is then allowed to stand for 30 minutes. To this sample, 35 *ml* of concentrated H_2SO_4 and 10 g of digestion mixture are added and the mixture is shaken gently to ensure effective mixing. The mixture is then digested for about 20 minutes between 360 and 410°C till the color of the mixture becomes transparent. The mixture is allowed to cool and then 250 *ml* of distilled water is added.

   The cooled digested sample is kept in the Kjeldahl flask. To this flask, a few glass beads and 1 *ml* of liquid paraffin are added. Twenty five *ml* of 4 per cent boric acid indicator is taken in a 150 *ml* conical flask which is fitted at the receiving end of the distillation apparatus. All the fittings of the apparatus is checked to avoid the possible loss of nitrogen.

About 100 *ml* of 40 per cent NaOH is slowly added to the solution and the mouth of the flask is cl osed. The flask is then heated gradually. About 150 *ml* distillate is collected from tbe receiving conical flask. Heating is stopped and the flask is removed from the apparatus.

The H_3BO_3 is then titrated with standard 0.1 (N) H_2SO_4 as noted in the case of mineralized nitrogen. If the volume of 0.1 (N) H_2SO_4 required for the titration of the sample be a, then the concentration of nitrogen in soil (per cent) can be expressed as follows:

$$\frac{a \times 1.4}{W}$$ Where W = Weight of the soil (g) taken for the purpose.

It has been suggested that the amount of the mineralizable nitrogen level below 25 mg N/100g soil and the range of 25 to 75 mg N/100g soil are considered as an index of low and moderate to high productivity respectively for most freshwater fish culture ecosystems.

DETERMINATION OF PHOSPHORUS IN POND SOIL

Phosphorus is one of the most critical and cardinal elements in maintaining pond productivity. The available form of phosphorus (P) in the water and soil phase is important for aquatic productivity because phosphate ions in pond soils form insoluble compounds with iron/aluminium and calcium under acidic and alkaline conditions respectively, rendering most of the P into unavailable forms inspite of its available in large amounts in the soil. To understand the chemistry of phosphorus transformation in fish ponds, it is therefore necessary to know the occurrence of the total amount of P and its various inorganic forms in addition to its level in available forms.

1. *Available Phosphorus :* The test for available phosphorus that follows was first developed by S.R. Olsen. This method involves the extraction of soil phosphorus by a solution of 0.5 (N) $NaHCO_3$ (sodium bicarbonate) adjusted to pH 8.5. The extraction procedure reduces the activity of calcium in soils and thus helps extract some phosphorus from calcium phosphate form. It also helps extract some phosphorus from the surface of iron and aluminium phosphate.

    Five gram soil sample is taken in a 250 *ml* bottle. To this sample, 100 *ml* of $NaHCO_3$ solution and one tea-spoonful activated charcoal are added. After shaking the suspension for 30 minutes, it is filtered through a Phosphorus-free filter paper. Ten mililitre of aliquot is taken from the filtrate in a 25 *ml* volumetric flask and then diluted with adding 15 *ml* of distilled water. A few drops (2-4) of dinitrophenol indicator (DNP indicator) is added which turns the solution yellow.

Dilute HCl (hydrochloric acid) is then added drop by drop to the solution till it turns colorless. A phospho-molybdic blue color is developed in 25 *ml* volume as noted in Section 26.11 and the optical density of the developed blue color of the solution is observed with the help of a colorimeter or spectro-photometer at 660 nm weave-length.

If the colorimetric reading for 25 *ml* volume with 10 *ml* aliquot corresponds to a mg/l of P, the concentration of available phosphorus in the soil will be a × 50 mg/l

2. **Total Phosphorus :** Though various methods are available for extraction of total phosphorus from soils, HCl extraction (Piper, 1967) and sodium carbonate fusion (Jackson, 1971) methods are most commonly employed. Of them, the former method that follows is most simple.

Various types of soil minerals are dissolved in concentrated hydrochloric acid (HCl) at the time of boiling. The concentration of phosphorus thus extracted is determined colorimetrically through development of phospho-molybdic blue color. Twenty gram of soil sample is taken out in a 500 *ml* conical flask. To this sample, 200 *ml* of concentrated HCl is added. A glass funnel is kept on the mouth of the flask and is allowed to boil for 1 hour on a hot plate. After cooling the boiled material, it is filtered through a filter paper into another conical flask. The residue is washed for several times with hot 0.5(N) HCl. The volume of the extract is made upto 500 *ml* in a volumertic flask.

Five mililitre aliquot from the extract is taken out and the phosphorus is determined following the method noted for available phosphorus estimation.

The concentration of total phosphorus (P) in soil has been calculated as follows:

$$a \times 125 \text{ mg/l}$$

Where a = Concentration of P in 25 *ml* volumetric flask in mg/l.

3. **Different Inorganic Forms :**

(i) **Saloid-Bound Phosphorus:** Two gram soil sample is taken in a 250 *ml* narrow-mouthed stopper bottle. One hundred *ml* of 1(N) ammonium chloride (NH_4Cl) is added to the bottle. The bottle is shaken well for 30 minutes and centrifuged the suspension for about 15 minutes at 2,400 rpm and the supernatent solution is decanted. If the centrifuge machine is not available, the bottle is allowed to stand for at least 6 hours which will permit the soil to precipitate at the bottom. The supernatent solution is filtered through a filter paper. The filtrate is used for estimation of saloid-bound phosphorus by the procedure noted in Section 25.11.

(ii) **Aluminium-Bound Phosphorus:** The filter paper which is used for the estimation of saloid-bound phosphorus, is placed on a funnel. The funnel is kept on the mouth of the bottle. One hundred mililitre of neutral 1(N) ammonium floride (NH_4F) is taken out by a pipette. This amount helps remove the soil particles adhering to the filter paper and pour into the bottle. This washing is replaced till the remaining NH_4F solution is exhausted. The bottle is shaken for 30 minutes. The solution is filtered which is then used for estimation of aluminium-bound P as stated earlier with the exception that 8 *ml* of 0.8 M boric acid is added to the solution in the 25 *ml* volumetric flask before development of color.

(iii) Iron-Bound Phosphorus: The filter paper sticking the soil particles is washed with about 25 *ml* of NaCl solution into the bottle containing soil. The suspension is shaken, centrifuged or placed for a few hours, decanted and discarded the supernatent solution. The soil sample is shaken with 100 *ml* 0.5 (N) H_2SO_4 and proceeded for filtration followed by washing the soil with NaCl as noted earlier. Phosphorus content in the filtered solution is determined to get the iron-bound phosphorus.

(iv) Calcium-Bound Phosphorus : One hundred mililitre of 0.1 (N) NaOH is added to the washed soil sample. The solution is then allowed to shake well for 30 minutes and the suspension is filtered. Phosphorus content in the filtrate solution is determined to get the Calcium-bound phosphorus.

The concentration of Phosphorus in each extracted solution is determined following Section 25.11. If the concentration of P in any solution be a mg/l and the aliquot taken from the solution be b *ml* to make a total volume of 25 *ml*, the amount of P in soil under each form of P will be

$$\frac{a}{b} \times 1,250 \text{ mg/l}$$

As a generalization, available phosphorus (P) below 30 mg/l, 30-60 mg/l, 60-120 mg/l, and above 120 mg/l is considered as an index of poor, average, good, and high productivity of pond water, respectively.

It should be added as a note that the value of P can be converted to phosphorus pentoxide (P_2O_5) by multiplying the data with 2.29.

To sum up, the test for a given component may either be designated in kilograms per hectare or be reported in general terms such as *medium high, high, medium* or *low.* In contrast to agricultural lands where several essential macro-and micro nutrients are necessary for crop production, tests are most commonly conducted for soils of fish culture ecosystems for those constituents such as P and Ca that are inevitable to fish growth. Although nitrogen is also an important nutrient for ecosystem productivity, the expected liberation of this element through action of microorganisms is arduous to predict.

The most significant limitation of soil test is the difficulty of proper sampling of soils. In general, only a very small sample is considered from one or several hectares of water area. The chance of error is definitely quite high. A composite sample of at least 15-20 areas of a pond from the upper 10 cm is, therefore, recommended to increase the probability that a representative portion of the soil has been obtained. The test results must be, however, correlated with fish response before dependable fertilizer/manure recommendations can be made. The recommendations are made with regard to practical knowledge of the fish crops to be reared in ponds, the characteristics of the soil under study, and other environmental conditions.

The interpretation of test data is best accomplished by experienced and skilled personnel, who clearly understand the scientific principles underlying the common field procedures. In modern laboratories, the factors to be considered in making fertilizer/manure recommendations are programmed into a computer, and the interpretation is competently printed out for use by fish culturists.

26.25 Conclusion

Though nutrient carriers play a major role in supplementing water's and soil's ability to provide nutrient for fish, deficiency or excess of nutrients definitely limit ecosystem productivity. Therefore, the adoption of suitable methods for determination of different abiotic and biotic factors which are necessary for fish yield play a major role in determining the type and amount of fertilizers to be applied. Accurate estimation of various factors of soil and water is necessary not only because of the potential for the waste of the nutrient carriers but also because of the potential to damage ecosystem quality.

In well-equipped fish farm laboratories, the estimation of soil and water parameters is made by sophisticated equipments such as pH meter, oxygen meter, salinometer, colorimeter, haemocytometer, and spectro-photometer which are economically viable for large fish farms and hence permit the best return on the money invested. A well-equipped laboratory with skilled personnel is able to handle several hundreds of samples in a month.

References

APHA. 1986. Standard Methods for the Determination of Water and Waste-water. Washington, DC.

Black, C. A. 1965. Methods of Soil Analysis. *American Soc. Agron.* Madison, Wisconsis, USA.

Boyd, C. E. 1974. Lime requirements of Alabama Fish ponds. *Anburn University Agri. Expt. Stat. Bull.* (459), pp.20.

Boyd, C. E. 1978. Water Quality in Warmwater Fish Ponds. *Agric. Expt. Stat. Auburn University.* pp. 359.

CIFRI. 1985. Carp Culture – Package of Practices for Increasing Production. Aquaculture Extension Manual (Series No.2), CIFRI, Barrackpur, India.

Dev, G. 1995. Fertilizers and their requirements in fish ponds. *In :* G.N. Chattopadhyay (Ed.). Nutrient Management in Aquaculture. Soil Testing Laboratory. Institute of Agriculture, Visva-Bharati, Santiniketan, India pp. 54-62.

Golterman, H. L. *et al.* 1978. Methods for Physical and Chemical Analysis of Freshwaters. Blackwell Scientific Publications, London, pp. 211.

Jackson, M.L. 1971. Soil Chemical Analysis. Prentice-Hall of India Pvt. Ltd., New Delhi, pp. 498.

Mondal, L.N. and G.N. Chattopadhayay. 1992. Nutrient Management in aquaculture. *In :* H.L.S. Tandon (Ed.). Non-Traditional Sectors for Fertilizer Use. FDCO. New Delhi, India, pp.1-17.

Olsen, S. R. *et al*. 1954. Estimation of available phosphorus in soils by extraction with sodium bicarbonate. *US Dept. Agric. Circ.* 939.

Parks, R. W., E. Scarsbrook and C.E. Boyd. 1975 Phytoplankton and water quality in a fertilized fish pond. *Auburn University Agric. Exp. Sta.* (Circular No. 224), pp. 16.

Santhanam, R. *et al*. 1989. A Manual for Freshwater Ecology, Daya Publishing House, Delhi, pp.134.

Subbiah, B. and G.L. Asija. 1956. A rapid procedure for estimation of available nitrogen in soils. *Curr. Sci.* 25 (8) : 259.

Theroux, F. R. *et al*. 1943. Laboratory Manual for Chemical and Bacterial Analyses of Water and Sewage. McGraw Hill Book. Co. Inc. London, PP. 274.

Vollenweider, R.A. 1969. A Manual on Methods for Measuring Primary Productivity in Aquatic Environments. IBP Hand Book No. 12. *Blackwell Sci. Publ.* Pp. 213.

Winkler, L.W. 1888. The determination of dissolved oxygen in water. *Berline Deutsch Chem. Gesellsch* 21 : 2843.

APPENDIX - I

A. Reagents and Their Preparation

1. Standard (0.02 M) Potassium Chloride (KCl) Solution: Dissolve 1.4912g of KCl in one litre of distilled water. The electrical conductivity of this solution is 2.39 m mhos Cm^{-1} at 18°C and 2.768 m mhos Cm^{-1} at 28°C.

2. 0.01 N Calcium solution: Dissolve 0.5005g dried pure calcium carbonate ($CaCO_3$) in 100 ml of 0.1 N HCl (dilute 5 ml concentrated HCl to 250 ml distilled water). Boil the solution for sometime, cool and make up the volume to one litre.

3. Ammonium Purpurate Indicator: Mix 0.2g ammonium purpurate powder with 40g potassium sulfate and store in a dark-colored bottle.

4. Eriochrome Black T indicator: Dissolve 0.5g eriochrome black T and 4.5g hydroxyl amine hydrochloride indicator in 100 ml ethyl alcohol. Add 5 ml of 2 per cent sodium cyanide solution to it.

5. 4N sodium hydroxide (NaOH): Dissolve 160g of NaOH in one litre distilled water. There is no need to standardize this solution.

6. Ammonium Chloride-Ammonium Hydroxide (NH_4Cl - NH_4 OH) buffer solution: Mix 67.5g NH_4Cl and 5.70 ml NH_4OH and make up the volume to one litre.

7. 0.01 N EDTA solution: Weigh 2.0g disodium dihydrogen ethylenediamine tetra acetic acid and dissolve it in one litre of distilled water. Standardize the solution against standard 0.01 N Ca solution.

8. Phenolpthalein indicator: Dissolve 0.5g phenolpthalein in 50 ml of 95 per cent ethyl alcohol and add it to 50 ml distilled water.

9. Methyl orange indicator: Dissolve 0.5g methyl orange in 100 ml distilled water.

10. Standard 0.02N H_2SO_4: The concentrated H_2SO_4 has a strength of 36N. To prepare a H_2SO_4 solution, this concentrated acid is used and the strength of H_2SO_4 solution is calculated by the following formula:

$$V_1S_1 = V_2S_2$$

If 35(V_2 conc. H_2SO_4 (V_1) of 36N strength (S_1) is diluted to 1,000 ml (V_2), then the strength of the diluted solution (S_2) will be 35 x 36/1,000 = 1.26N.

To get the actual strength of this diluted H_2SO_4, it should be standardized against a standard base.

Take 10 ml (V_2) of the H_2SO_4 solution (S_2) in a 250 ml conical flask. Put a few drops of methyl orange indicator, add a little distilled water and titrate with 1 N Na_2CO_3 (S_1) to a faint yellow color. If the volume of standard 1 N Na_2CO_3 solution required for titration is V_1 ml (say 11 ml), the strength of the acid may be calculated as follows:

$$S_2 = \frac{V_1 S_1}{V_2} \quad \text{or} \quad \frac{11 \times 1}{10} = 1.1$$

Therefore, the strength of the diluted H_2SO_4 is 1.1N. 1N solution of the acid can be prepared as follows:

$$V_2 = \frac{V_1 S_1}{S_2} \quad \text{or} \quad \frac{1000 \times 1}{1.1} = 909.1 \text{ ml}$$

Where, V_1 = Volume of 1 N H_2SO_4 required (say 1, 000 ml)

S_1 = Required strength of H_2SO_4 (IN in this case)

V_2 = Volume of 1.lN H_2SO_4 to be diluted (not known)

S_2 = Strength of the acid (1.1 N)

To obtain an 1 N strength, 909.1 ml of the prepared 1.1 N solution of H_2SO_4 should be diluted to 1,000 ml.

* Preparation of 0.1 N H_2SO_4 solution: This can be prepared by diluting the standard 1N solution 10 times.

* Preparation of 0.02 N H_2SO_4 solution: This can be prepared by diluting the standard 0.1 N solution 5 times.

11. Potassium chromate (K_2CrO_4) indicator: Dissolve 5g of K_2CrO_4 in about 75 ml of distilled water and add dropwise a standard solution of silver nitrate to form a red precipitate. Filter the suspension and dilute the filtrate to 100 ml.

12. Standard silver nitrate ($AgNO_3$) solution: Weigh exactly 4.791g of $AgNO_3$ and dissolve in 1 litre of distilled water. Standardize the strength of the solution by titrating against a standard sodium chloride solution.

13. Standard sodium chloride (NaCl) solution: Dissolve 1.648g dry analytical reagent grade NaCl in 1 litre of distilled water.

14. 40 per cent sodium hydroxide (NaOH) solution: Dissolve 400g NaOH in 1 litre of distilled water.

15. Davarde's alloy (Copper 50, Aluminium 45 and Zinc 5 per cent each) : It is readily available in the market.

16. Mixed indicator: Dissolve 0.5 bromo cresol green and 0.1 g methyl red in 100 ml of 95 per cent ethyl alcohol. Adjust the solution to bluish purple color, with dilute HCl or NaOH.

17. 4 per cent Boric acid (H_3BO_3) solution: Dissolve 40g boric acid in 1 litre of distilled water. Adjust the solution by adding dilute HCl or NaOH until the bluish color changes to pink.

18. Standard phosphorus solution: Dissolve 0.2195g anhydrous potassium dihydrogen orthophosphate (KH_2PO_4) in 500 ml of distilled water, add 5 ml concentrated H_2SO_4 and make up the volume to 1 litre. This forms 50 mg/l solution of phosphorus. Four mililitre of this 50 mg/l solution is diluted to 100 ml to get 2 mg/l solution of phosphorus which is used to make a standard curve for phosphorus.

19. Ammonium molybdate-Potassium antimony tartarate solution: Dissolve 1.0g anmonium molybdate [$(NH_4)_6\ MO_7\ O_{24},\ 4H_2O$] and 0.02g potassium antimony tartarate in 500 ml of distilled water. Add slowly 15 ml of conc. H_2SO_4 to this solution and make up the volume to 1 litre.

20. Ascorbic acid-molybdate tartarate solution: Weigh 88 mg ascorbic acid and add 100 ml of the molybdate tartarate solution.

21. Manganous sulfate solution: Dissolve 480g $MnSO_4.4H_2O$ or 400g $MnSO_4.2H_2O$ or 365g $MnSO_4.H_2O$ in 750 ml of distilled water. Cool, filter and make up the volume to 1 litre.

22. Alkaline iodide: Dissolve 500g sodium hydroxide and 150g potassium iodide (KI) in 350 ml of distilled water. Allow the NaOH solution to cool down and thoroughly mix the two solutions. Dissolve 10g sodium azide in 40 ml of distilled water and add to the previous solution. Make up the volume to 1 litre.

23. Standard sodium thiosulfate solution: Dissolve 24.82g of crystalline sodium thiosulfate ($Na_2S_2O_3 . 7H_2O$) in 700 ml of distilled water. Add a pellet of NaOH or 1g sodium carbonate and dilute it to 1 litre with distilled water. This stock solution (0.1 N $Na_2S_2O_3$) should be stored in a brown bottle and standardize before preparing 0.025 N $Na_2S_2O_3$ solution.

24. Concentrated sulfuric acid.

25. N/44 NaOH solution (See Section 26.13).

26. starch solution: Add 1g of starch powder to 10 of distilled water, warm to about 90°C to dissolve the starch powder in water. Allow the solution to cool and add 0.1g salicylic acid as preservative.

27. Potassium permanganate ($KMnO_4$) solution: Dissolve 0.4g $KMnO_4$ in 250 ml of distilled water.

28. Ammonium oxalate ($C_2H_8O_4N_2$) solution: Dissolve 0.888g ammonium oxalate in 1 litre of distilled water.

29. Dilute H_2SO_4 : Slowly add 100 ml of concentrated H_2SO_4 to 300 ml of distilled water.

30. DPD indicator: Dissolve 250g N-N- diethyl-P-phenylene diamine-oxalate in 150 ml of distilled water. Add 2 ml of 25 per cent H_2SO_4 (1:3; conc. H_2SO_4 : distilled water). Make the volume to 250 ml with distilled water.

31. Potassium iodide crystal: Available in the market.

32. Buffer solution: Dissolve 12g disodium hydrogen phosphate in 250 ml of distilled water and to this solution, add 10 mg mercuric chloride ($HgCl_2$). Dissolve 0.4g EDTA salt in 50 ml of distilled water in a separate beaker and add it to the former mixture. Mix the two solutions well and make up the volume to one litre with distilled water.

33. 1N ferrous ammonium sulfate [Fe $(NH_4)_2$ $(SO_4)_2$] : Dissolve 392.2g of [$Fe(NH_4)_2(SO_4)_2$. $6H_2O$J in 800 ml of distilled water, add 20 ml of conc. H_2SO_4 and dilute to one litre with distilled water.

34. Neutral barium sulfate: Dissolve 116.7g barium sulfate ($BaSO_4$) in 750 ml of distilled water. Make up the volume to 1 litre.

35. Buffer solution: Dissolve 20g p-nitrophenol, 15g boric acid, 74g potassium chloride and 10g KOH in 750 ml of distilled water. Make up the volume to one litre. This buffer solution gives a pH value of 8.0 ± 0.1.

36. Standard gypsum solution: Add 5g $CaSO_4.2H_2O$ to one litre distilled water, shake for about 30 minutes, filter and determine the calcium concentration in the filtrate by the method noted in section 26.6.

37. 1N $K_2Cr_2O_7$: Dissolve 49 g of solid potassium dichromate in distilled water and make up the volume to one litre.

38. IN $Fe(NH_4)_2$ $(SO_4)_2$: Dissolve 392.2 g of $Fe(NH_4)_2$ $(SO_4)_2.6H_2O$ in 800 ml of distilled water, add 20 cc conc. H_2SO_4 and dilute to one litre.

39. Diphenyl amine indicator: Dissolve 0.5g diphehyl amine indicator in 20cc of distilled water and add 100cc conc. H_2SO_4

40. 85 per cent Orthophosphoric acid (H_3PO_4).

41. 0.32 per cent $KMnO_4$: Dissolve 3.2g $KMnO_4$ in one litre of distilled water.

42. 2.5 per cent NaOH : Dissolve 25g NaOH in 800 ml of distilled water, cool and make up the volume to one litre.

43. Liquid paraffin (pure grade) : Available in the market.

44. 10 per cent NaCl (pH 2.5) : Dissolve l00g NaCl in 100 ml of distilled water. Add a few drops of 2-4 Dinitrophenol indicator to obtain yellow color. Add dilute HCl (HCl : Water = 1: 10) till the yellow color disappears. Add 5-6 drops of HCl and make up the volume to one litre.

45. 40 per cent NaOH : Add 40g NaOH in 100 ml of distilled water.

46. Digestion mixture: Mix 20g copper sulfate ($CuSO_4.5H_2O$), 3g mercuric oxide (HgO) and 1g selenium (Se) powder. Add this mixture to 475g sodium sulfate (Na_2SO_4) and mix thoroughly.

47. H_2SO_4 mixture: Add 3g salicylic acid to 100 ml of conc. H_2SO_4.

48. Ammonium chloride solution: Dissolve 53.5g ammonium chloride (NH_4Cl) in 750 ml of distilled water and make up the volume to one litre.

49. 0.5N neutral NH_4F : Dissolve 18.5g solid NH_4F in one litre of distilled water, adjust the pH to 7.0 with dilute ammonium hydroxide and, if necessary, dilute HCl using a pH paper.

50. Saturated sodium chloride : Put enough amount of NaCl to one litre of distilled water in a bottle so that some amount of the salt is left at the bottom.

51. 0.1 N NaOH: Dissolve 4g NaOH in one litre of distilled water

52. 0.5 N H_2SO_4 : Dissolve 15 ml conc. H_2SO_4 into 750 ml of distilled water, cool the solution and make up the volume to one litre.

53. 0.8 M boric acid: Dissolve 5g boric acid in one litre of distilled water.

54. Concentrated sulfuric acid.

55. 0.5N HCl : Dilute 50 ml concentrated HCl to one litre of distilled water.

56. 6 per cent Hydrogen peroxide solution: Hydrogen peroxide: water = 1 : 5.

57. 2N HCl : Dilute 100 ml concentrated HCl to 500 ml of distilled water.

58. 2N NaOH : Dissolve 40g NaOH pellets in 300 ml of distilled water. Cool the solution and dilute up to 500 ml with distilled water.

59. 5 per cent silver nitrate ($AgNO_3$) : Dissolve 5g $AgNO_3$ in 100 ml of distilled water.

60. 0.5 M sodium bicarbonate ($NaHCO_3$) : Dissolve 42 g $NaHCO_3$ in one litre of distilled water. Adjust the pH up to 8.5 through gradual addition of dilute NaOH using a pH paper.

61. Activated charcoal : Available in the market.

62. Different types of indicators as noted in Section 26.3

63. 1 N H_2SO_4 : Dissolve 30 ml concentrated sulfuric acid (H_2SO_4) in 1,000 ml of distilled water (stock solution).

64. 0.02 N H_2SO_4 : Dissolve 20 ml of 1 N H_2SO_4 in 1,000 ml of distilled water.

B. Apparatus and Glasswares

1. Colorimeter / spectro-Photometer.

2. Mechanical shaker.

3. Centrifuge machine.

4. Hot plate

5. Kjeldahl distillation flask with condenser

6. pH meter or Litmus paper for rough estimation of water pH

7. Conductivity meter bridge

8. Secchi disc / Jackson Turbidimeter

9. Microscope (Compound and Simple)
10. Water sampler
11. Plankton collection net
12. Lovibond Comparator
13. Salinometer
14. Pan balance and Chemical balance
15. Burette stand with clamps
16. Thermometer (0-100°C)
17. BOD Incubator
18. Haemocytometer and Sedgewick-Rafter Cell
19. Glass aquaria (10 litre capacity)
20. Strainer (2 and 80 mm mesh size) for the preparation of soil sample
21. Volumetric flask (25, 100, 250, and 1,000 ml capacity)
22. Narrow-mouthed stopper bottles (250 ml capacity)
23. Funnel (5 and 10 mm capacity)
24. Conical flask (150 and 250 ml capacity)
25. Beakers (100, 250, 500 and 1,000 ml capacity)
26. Pipette (5 and 10 ml capacity).
27. Cylinder (100, 250 and 500 ml capacity)
28. Porcelin basin
29. Filter paper (12.5 cm, Whatman No. 41)
30. Watch glasses (3 inches)
31. Brushes (medium)
32. Drop bottles (30 ml)
33. Holder, rubber, stopper
34. Evaporating dish
35. Washing bottles (500 ml)
36. Glass rod (6 mm)

Glossary

Absorption (n.) : Active movement of ions and water into the fish body as a result of metabolic processes frequently against an activity gradient.

Acidic (adj.) : Having the properties of an acid or containing acid, having a pH below 7.0. Acidification (n.); acidifying, acidified (adv.).

Acidosis (n.) : 1. An excessively acid condition of the body fluids or tissues, 2. Conditions which promote an increase in the hydrogen-ion concentration, hence a fall in the pH of pond water. Cf. *Alkalosis*.

Acid rain (n.) : Rainfall made acidic by atmospheric sulfur and nitrogen oxides from the industrial burning of fossil fuels.

Acidity (n.) : The activity of hydrogen ion in the pond water. It is measured and expressed as a pH value. Cf. @ *Alkalinity*.

Actinomycetes (n.) : A group of organisms intermediate between the bacteria and the true fungi, that usually produce a characteristic branched mycelium. Includes many, but not all, organisms belonging to the order of Actinomycetales.

Activated carbon (Also called as activated charcoal) (n.) : Charcoal that has been heated or otherwise treated to increase its adsorptive power.

Activated sludge (n.) : Aerated sewage containing aerobic micro-organisms which help to break it down.

Additive (adj.): 1.Relating to or produced by addition: 2. Relating to the reproduction of colors by the super-imposition of primary colors.

Additive effect (adj.) : Effect of genes related to the expression of a character by the super-imposition of primary characters.

Additive genetic variance (adj.) : That part of the genetic variance of a quantitative character that is transmitted and so cause resemblance between relatives.

Adhesion (n.) : 1. The action or process of adhering, 2. The striking together of particles of different substances.

Adipose fin (n.) : A small fin between the dorsal and caudal fins, most notable in tetras and other claracoids.

Adsorb (v.) : (Of a solid) hold (molecules of a gas, liquid, or solute) as a thin flim on surfaces outside or without the material. Adsorbable, Adosrptive (adj.); Adsorption (n.).

Aerate (v.) : To impregnate with air.

Aeration (n.) : 1. Technique by which air is passed through pond water for oxygenation; 2. The process by which air from the atmosphere is injected into the pond water. Poorly-aerated pond water usually contains more carbon di-oxide and correspondingly less oxygen than the atmosphere above the water.

Aestivation (n.) : Prolonged dormancy of an insect, fish, or amphibian during a hot or dry period.

Air-breathing fish (n.) : Fish which are able to breathe outside the pond/tank water for several hours with the help of specialized structures. Such as *Anabas, Clarias*, etc.

Albino (n.) : An animal having a congenital absence of pigment in the skin, scales, hair, etc.

Alevin (n.): A newly-spawned salmon or trout still carrying the yolk.

Alga (pl.: algae) (n.) : Any of a large group of simple, non-flowering plants containing chlorophyll but lacking true stems, roots, leaves, and vascular tissue, such as numerous single-celled/multi-cellular forms.

Algal bloom (n.): A rapid and often unpredictable growth of a single species of alga in an aquatic ecosystem.

Algicide (n.): Chemical substances that destroy algae.

Aliphatic (adj.): Relating to or denoting organic compounds in which carbon atoms form open chains.

Alkalinity (n.): The activity of hydroxy ion in the pond water. It is measured and expressed as a pH value. Cf. *Acidity*.

Alkalosis (n.): Conditions which promote a decrease in the H^+, hence a rise in the pH of pond water. Cf. *Acidosis*.

Allele (n.): Any one of two or more alternative forms of a gene that may occur alternatively at a given site on a chromosome. Alleles may occurs in pairs or there may be multiple alleles affecting the expression of a particular trait. If paired alleles are different, the organism is *heterozygous*; if they are same, the organism is said to be *homozygous*. A dominant allele will override the traits of a recessive allele in a *heterozygous* pairing. In some traits, alleles neither act as recessive nor dominant (called as co-dominant). An individual cannot possess more than two alleles for a given trait. All genetic traits are the result of the interactions of alleles.

Allergen (n.): A substance that cause an allergic reaction.

Allochthonous (adj.): Denoting a deposit or formation that originated at a distance from its present position. Cf. *Autochthonous*.

Allozyme (n.): Alleles for a protein encoded by a locus that can be separated electrophoretically. Because both alleles are expressed in a heterozygote, the expression is usually co-dominant.

Alluvium (n.): A deposit of clay, silt, and sand left by flowing flood water in a river valley or delta, typically producing fertile soil. Alluvial (adj.).

Alternating current (n.): An electric current that reverses its direction many times a second at regular intervals.

Altitude (n.): The height of an object or point above sea level or ground level.

Amino acid (n.): Nitrogen-containing organic acids that couple together to form proteins. Each acid molecule contains one or more amino groups (- NH_2) and at least one carboxyl group (- COOH). In addition, some amino acids contain sulfur.

Ammonia (n.) : A colorless, intensely pungent gas which dissolves in water to give a strongly alkaline solution (NH_3). A solution of ammonia used as a cleaning fluid.

Amplitude (n.) : The maximum extent or magnitude of a vibration or other oscillating phenomenon, measured from the equilibrium position or average value.

Anal fin (n.) : An unpaired fin on the underside of the body of fish, near the vent.

Analogue (adj.) : 1. Chemical substances similar in function but not necessarily alike in structural relationship, 2. Parts similar in function but not similar in genetic relationship.

Anatomy (n.) : The branch of biology concerned with bodily structure, especially as revealed by dissection. Anatomical (adj.).

Anadromous (adj.): (Of a fish) migrating up rivers from the sea to spawn. cf. *Catadromous*.

Anchor (n.) : A heavy object used to moor a ship, boat, fish cage to the bottom mud, typically having a metal shank with a pair of curved, barbed fluked.

Androgen (n.) : A male sex hormone such as testosterone.

Androgenesis (n.) : Development of an egg after entry of male germ cell without the participation of the egg nucleus. Cf. *Gynogenesis*.

Antagonism (n.) : The interaction of two chemical substances having opposite effects in a system in such a way that the action of one partially or completely inhibit the effects of the other. Antagonistic (adj.), Antagonistically (adv.). Cf. *synergism*.

Antagonistic (adj.): A substance which interferes with the physiological action of another.

Antibacterial (adj.) : Active against bacteria.

Anti-nutritional factor (n.) : Compounds which confer resistance to digestion and assimilation of food materials and this condition helps prevent nourishment of the animal concerned.

Anticeptic (adj.) : Preventing the growth of disease-causing micro-organisms.

Antioxidant (n.) : A substance that prevents or inhibits oxidation. A substance such as Vitamin C or E that removes potentially damaging oxidizing agents in a living organism.

Aromatic (adj.) : 1. Having an aroma (a pleasant and distinctive smell), 2. (Of an organic compound) containing an unsaturated ring of atoms which is stabilized by an interaction of the bonds forming the ring.

Artificial feed (n.) : Feeds that have been especially formulated to meet the nutritional needs of farmed fish.

Artificial propagation (n.) : Fish culture activity, generally involving modifications of natural spawning, incubation, or rearing environments.

Artificial selection (n.) : Fish culture activity in which breeders are chosen on the basis of heritable traits (whether purposeful or inadvertent).

Artisan (n.) : A skilled worker/farmer who makes things by hand. Artisanal (adj.).

Asphyxia (n.): A condition arising when the body is deprived of oxygen, causing unconsciousness or death. Asphyxiate (v.), Asphyxiation (n.).

Attitude (n.): A farmer's consistently favorable or unfavorable evaluations, feelings, and tendencies toward an object or idea. Attitudinal (adj.).

Attribute (v.): A quality or feature regarded as characteristic or inherent. Attributable (adj.), Attribution (n.).

Autochthonous (adj.): 1. Indigenous, 2. (Of a deposit or formation) formed in its present position. Cf. *Allochthonous*.

Autopsy (n.): Post-mortem examination of the organ and tissues of a fish's body to determine the cause of death or pathological conditions.

Auto-stocking (n.): A device or process of reproduction occurring in fish themselves with litte or no direct human control.

Back cross (v.): 1. Cross (a hybrid) with one of its parents or an organism genetically identical to one of the parents, 2. The mating of an individual to one of its parents or parental strains. In Mendelian genetics, a mating of a heterozygote to the recessive homozygote, producing a 1:1 ratio in the progeny.

Barbels (n.): Sensory growths around the mouths of various surface and bottom-dwelling fish, including catfish.

Bile (n.): A bitter greenish-brown alkaline fluid secreted by the liver and stored in the gall bladder which aids digestion.

Beel (n.) : Refers to the flood- plain wetlands which possess their temporary or permanent, association with the parent river systems.

Behavior (n.): The way in which an animal senses and reacts to its external and internal environment.

Bentonite (n.): A kind of absorbent clay formed by breakdown of volcanic ash, used especially as a filter.

Bile duct (n.): The duct which conveys bile from the liver and the gall bladder to the duodenum.

Biliary stasis (n.): A stoppage of the flow of bile within the bile duct.

Binder (n.): Substances which are used to bind the food stuffs at the time of preparation and make it insoluble in water.

Bioaccumulate (v.): (Of a substance) become concentrated inside the bodies of living things. Bioaccunulation (n.).

Bioactive (adj.): A substance having a biological effect.

Bio-availability (n.): The proportion of a drug or other substance which enters the circulation when introduced into the body and so is able to have an active effect.

Biochemical genetics (n.): The genetics of quantitative characters.

Biocontrol (n.): Any condition under which or practice whereby survival or activity of a pathogen is reduced through the agency of any other living organisms with the result that there is reduction in incidence of disease caused by the pathogen.

Biodegradable (adj.): Capable of being decomposed by bacteria or other living organisms. Biodegrade (v.); Biodegradability, Biodegradation (n.).

Biodiversity (of fish) (n.): The variety of fish in the world or in a particular habitat.

Biogenic amine (s) (n.): These are produced in fish flesh immediately after harvesting as a result of decarboxylation of amino acids due to the action of spoiled bacteria. Some common biogenic amines include tyramine from tyrosine, putrescine from arginine, cadaverine from lysine, histamine from histidine, gizzerosine from the combination of lysine adn histidine.

Biogeochemical (adj.): Relating to the cycle in which chemical elements and simple substances are transferred between living systems and the environment.

Biohazard (n.): A risk to human health or the environment arising from biological research.

Bioindicator (n.): An organism used as an indicator of the quality of an ecosystem, especially with respect to pollution.

Biolistics (n.): The firing of microscopic particles that are coated with DNA construct into living cells through a gene gun.

Biological production (n.): A process by which a number of organic substances are generated as a result of flow of energy and cycling of materials.

Biology. (n.): The study of living organisms, which includes their structure (gross and microscopical), functioning, origin and evolution, classification, inter-relationships, and distribution.

Biomagnification (n.): Also called *biological concentration*. 1. A process whereby the concentration of a pollutant, within living tissues, increases at each link in a food chain, 2. The tendency for certain chemical substances to concentrate with each trophic level. Fish preferentially accumulate certain chemicals and excrete others. When this occurs, the accumulated chemicals gradually increase as percentage of the body weight as the chemical is transferred along with the food chain.

Biomass (n.): Weight or total quantity of living organisms of one animal or plant species (species biomass) or of all the species in the community (community biomass), commonly referred to as a *unit area* or *volume* of the habitat. It is an area at a given moment is the standing crop.

Biomonitoring (n.): Direct measurement of changes in the biological component of a habitat based on evaluation of the number and/or distribution of species or organisms (before and after the change).

Bioreactor (n.): An apparatus in which a biological reaction or process is carried out.

Bioremediation (n.): The use of either naturally-occurring or deliberately introduced micro-organisms to consume and break down environmental pollutants.

Biotechnology (n.): 1. It is the application of industrial production techniques to exploit biological processes which is increasing its influence on modern life. A huge variety of products has emerged, from antibiotics to new sources of food. The six main areas in which biotechnology plays an important role are: medicine, waste management, fuel production, food production, agriculture, and aquaculture, 2. The exploration of biological processes for industrial and other purposes, especially the genetic manipulation of micro-organisms for the production of food, antibiotics, hormones, waste management, etc. Biotechnological (adj.).

Biotic potential (adj.): The inherent capability of an organism to increase in numbers under identical conditions (absence of competition, predators, and parasites).

Biotype (n.): A group of organisms having an identical genetic constitution.

Bone (n.): The main material of the endoskeleton of most vertebrates. It Consists of approximately equal volumes of collagen fibres and crystals of bone salt (containing calcium phosphate and calcium hydroxide).

Branchial organ (n.): Specialized structures which are developed in some fishes to meet the extra demand for oxygen.

Branchiostegal menbrane (n.): A membranous flap on the free edge of the gill covering, termed as *operculum*.

Brand (n.): A name, term, symbol, or design intended to identify the products of one seller or group of sellers and to differentiate them from those of competitors.

Breeder (n.): Fish that breeds.

Breeding (n.): The mating together of individuals that are related by descent. The offspring are inbred to an extent depending on the degree of relationship.

Breeding ground (n.): A place or situation that favours the development of fish.

Broadcast (v.): Scattering of fertilizers or manures or feed on the surface of the pond water.

Brood (n.): 1. A set of offspring produced at the same birth or from the same batch of eggs, 2. The offspring produced by a pair of fish: typically a group of fry that is guarded by one or both adults.

Broodfish (n.): 1. Male and female fish containing sperm and egg, respectively, which are ready for executing breeding operations. 2. Fish kept to be used for breeding.

Bubonic plague (n.): A form of plague transmitted by rat fleas and characterized by fever, delerium, and the formation of buboes.

Buccal cavity (n.): The cavity into which the mouth opens and which is, therefore, the first part of the alimentary canal. In fish, the buccal cavity is part of the respiratory as well as of the alimentary canal.

Budget (n.): 1. An estimate of income and expenditure for a set period of time, 2. The amount of money needed or available for a purpose, 3. An estimate of production and expenditure of nutrient ingredients of a fish culture ecosystem in a set period of time (nutrient budget).

Buffer (n.) : A solution containing of mixture of a weak acid and its soluble salt. It acts to resist changes in pH. Such changes can be brought about by dilution, or by addition of acid or alkali.

Bug (n.): *Entomology* : An insect of a large order having piercing and sucking mouth-parts, including aphids, leafhoppers, cicadas, and many other insects (Order : Hemiptera).

Bund (n.) : An embankment (the term mostly used in India and Pakistan).

Buoyant (n.): Able to keep afloat. Buoyancy (n.).

Burbot (n.) : An elongated bottom-dwelling fish that is the only freshwater member of the cod family (*Lota lota*).

Business market (n.): Organisation that buy products and services for use in the production of other products and for the purpose of reselling to others at a profit.

Butter fish (n.) (pl. same): Any of various fishes with oily flesh or slippery skin, especially the gunnel.

Buyer (n.): The person who makes an actual purchase.

Cage (n.): Specially-designed structure made of different materials such as bamboo, wood, nylon, etc. kept in water at a desired level for certain period of time in which fish and other aquatic animals are reared scientifically.

Cage culture (n.): Culture of aquatic animals of commercial importance in specially-designed cages made of different construction materials.

Calcium (n.): The chemical element of atomic number 20, a soft grey reactive metal of the alkaline earth metal group. (Symbol : Ca).

Cannibal (n.): An animal that eats the flesh of its own species. Cannibalism (n.), Cannibalistic (adj.).

Capital turnover (n.): Currency owned by a person, farmer, or borrowed in a particular period.

Carbon (n.): The chemical element of atomic number 6, a non-metal which has two main forms (diamond and graphite) occurs as impure form in charcoal, shoot, and coal, and is present in all organic compounds.

Carbon-di-oxide (n.): A colourless, odourless gas produced by burning carbon and organic compounds and by respiration, and absorbed by plants in photosynthesis. (Symbol: CO_2).

Carcinogenic (adj.): A substance capable of causing cancer. Carcinogen (n.).

Carnivore (n.): Any exclusively meat-eating animal. Carnivorous (adj.).

Carrying capacity (n.): (of fish) 1. Maximum number of fish that a fish culture ecosystem can support with respect to food and space, 2. The population or number of fish in a pond which can sustain during periods of lowest food availability; it may be during winter or the dry season when there may occur a drop in the water level of the pond.

Catadromous (adj.): (of a fish) migrating down rivers to the sea to spawn. Cf. *Anadromous*.

Catchment area (n.): The area from which rainfall flows into a river, lake, or reservoir.

Compatibility{n.): The degree to which fish interact favorably with one another; also, the bonding of a pair of fish prior to spawning. Compatible (adj.), Compatibly (adv.).

Capillary (n.): A vessel of the blood circulation system. Capillaries are the smallest vessels of the system with walls with one cell thick. They form extensive branching networks in all tissues of the body. Materials of all types, salts, nutrients, and water are exchanged through thin capillary wall between blood and tissues.

Cartilage (n.): A form, flexible connective tissue which is the main component of the articulating surface of joints and of structures. Cartilaginous (adj.).

Catalysis (n.):The acceleration of a chemical reaction by a catalyst.

Catalyst (n.): A substance that increases the rate of a chemical reaction without itself undergoing any permanent chemical change.

Catfish (n.): Any of about 2,500 species of scale-less, mostly freshwater fishes (Order: Siluriformes) related to carp and minnows and named for their whisker-like barbels (fleshy feelers). All species have at least one pair of barbels on the upper jaw, and some have a pair on the snout and additional pairs on the chin. Many species possess spines. They are found almost worldwide, they are generally bottom-dwelling scavengers that feed on almost any kind of plant and animal matter. Species varies from 1.5 inch to 15 ft (4 cm - 4.5 m). Many large species are used for food but small ones are popular aquarium fish.

Caudal fin (n.): The tail fin, which is often divided into lobes.

Cereal (n.): 1. A grain used for food, for example rice, 2. A grass producing much grain grown as an agricultural crop.

Chelate (n.): *Chemistry* : A compound containing an organic ligand bonded to a central metal atom at two or more points. Chelation (n.).

Chemical filtration (n.): The use of chemicals to remove dissolved wastes by adsorption.

Chemical Oxygen Demand (C.O.D.) (n.): A quantitative measure of the amount of dissolved oxygen required for the chemical oxidation of organic materials in wastewater using permanganate or inorganic dichromate salts as oxidants during a two hour test.

Chemoprophylaxis (n.) : The use of chemicals and drugs to prevent diseases.

Chemotherapy (n.) : The treatment of disease by the use of chemical substances.

Chill (n.) : 1. A feverish cold, 2. Cool in a refregerator. Chilling (adj.), Chillness (n.), Chillingly (adv.).

Chitin (n.) : A tough nitrogen-containing polysaccharide of the exo-skeleton of prawn and shrimp and other arthropods. Chitin is one of the most abundant biopolymers on earth and consists of N-acetylglucosa amine (NAG) residues in -1, 4 linkage. Chitin forms long straight chains that serve a structural role. Thus, chitin is closely related to cellulose in structure and function. Structurally, it is identical to cellulose except that the hydroxyl (–OH) group at carbon atom 2 is replaced by acetylated amino group (–NH.CO.CH$_3$).

Chlorella (n.): A common single-celled green alga, responsible for turning stagnent water an opaque green. (Genus: *Chlorella*).

Chlorine (n.): The chemical element of atomic number 17, a toxic, irritant, pale green gas of the halogen group. (Symbol: Cl).

Chlorine demand (v.): The amount of chlorine requires to reach a desired level of free chlorine residuals.

Chromatography (n.): A technique for the separation of a mixture by passing it in solution or suspension through a medium in which the components move at different rates.

Chromosome (n.): 1. In eukaryotes, the deeply-staining rod-like structures seen in the nucleus at cell division, make up of a continuous thread of DNA which with its associated proteins (namely histones) forms higher order structures termed as *nucleosomes* and has special regions — the centromere and the telomere. Normally constant in number for any species, 2. A gene-carrying structure found in the nucleus of every living cell. Genes determine the characteristics of all organisms.

Cichlid (n.): Any of more than 600 primarily freshwater fish species (Family: Cichlidae), including many popular aquarium species. They are found in the New World tropics, Africa and Madagaskar, and Southern Asia. Most species are African, appearing in great diversity in the major African lakes. They are deep-bodied with a rounded tail and grow no longer 12 inch (30 cm). They may be omnivores, carnivores, or herbivores. They are noted for their complex mating and breeding behavior. Several species are used for food and are extensively cultured in ponds in many countries of the world.

Circulatory system (n.): A system of vessels that carries nutrients, oxygen, carbon di-oxide, and waste materials around the body.

Cirrhosis (n.): A chronic liver disease marked by degeneration of cells, inflammation, and thickening of tissue.

Cladoceran (n.): A group of zooplankton used as food for rearing fry and fingerlings of different species of edible fish.

Classification (n.): A method of grouping living things to show the relationships between them.

Climate (n.): The general weather conditions prevailing in an area over a long period. Climatic (adj.).

Cloaca (n.): The common opening for the intestine and urinary tract.

Clone (n.): An organism or cell, or a group of organisms or cells, produced asexually from one ancestor to which they are genetically identical. Clonal (adj.).

Coast (n.): The part of the land adjoining or near the sea. Coastal (adj.).

Co-dominant (n.): Both alleles are expressed in a heterozygous (Aa) individual so the individual is distinguishable from both homozygous (AA and aa) types.

Coelom (n.): Also called *body cavity*. An internal space or a large fluid-filled cavity lying between the body wall and the intestinal organs; utilized as a temporary site for accumulation of excess fluids and wastes, maturation of gametes and providing space for enlargement of internal organs. Vertebrate animals have, however, a true coelom which develops as a cavity within the embryonic mesoderm which provides its lining, called *peritoneum*.

Colchicine (n.): Polypioidy is induced by treatment with aqueous solution (the concentration of this solution may vary from 0.01 to 0.50 per cent) of a drug called *colchicine*. Colchicine is an alkaloid which is extracted from seed and corn of *Colchicum autumnale*. This drug has the potency of arresting and breaking the spindle so that a cell division without cell wall formation may be severely affected leading to doubling of chromosome numbers. Since polyploids are generally more vigorous, their importance in fish stock improvement has been realized and techniques have been developed for artificial induction of polyploids.

Coldwater (n.): Coldwaters are those whose temperatures decrease within the tolerance limits. These temperatures generally range from zero to 20°C with an optimum range from 10 to 12°C. As a generalization, water bodies which are situated above an elevation of about 1,000 metres entitle for being referred to as coldwater.

Commensal (adj.): Relating to or denoting an association between two organisms in which one benefits and the other derives neither benefit nor harm. Commensalism (n.)

Community (n.) (pl. Communities): 1. A group of farmers or people living together in one place for enjoying benefits. 2. *ecology* : A group of interdependent plants or animals growing or living together.

Competition (n.): Interaction between species or organisms which share a limited environmental resource.

Composite culture (v.): Culture of different species of farmed fish in the same pond/tank.

Compost (n.): Decayed organic material used as a fertilizer for growing plants or used in fish culture ponds along with some inorganic fertilizers.

Computer (n.): An electronic device which is capable of receiving information and performing a sequence of logical operations in accordance with a predetermined but variable set of procedural instructions to produce a result in the form of information or signals.

Concentration factor (n.): The ratio of a toxicant in any aquatic organism to that of the ecosystem.

Condition factor (n.): The ratio of length (in cm) to weight (in gram) of a fish at a given condition. It is designated as *K*.

Conditional (adj.): Subject to one or more conditions being met. Condition (n.), Conditionally (adv.).

Conditioning (v.): 1. Managing of fish and their surroundings to encourage breeding, 2. Refers to the way in which water is treated it make it safe for the Pond/tank occupants.

Consumer (n.): 1. A person or thing that eats or uses something, 2. A person who buys goods and services for personal use.

Consumer market (n.): Individuals who buy products and services for personal consumption.

Consumer product (n.): Product purchased by final consumer for personal consumption.

Consumer preference (n.): The choice of fish species by consumers and accordingly, they are cultured in different systems. The people of West Bengal, for example, prefer freshwater carp and catfishes than marine water counterparts, whereas South Indian people consume live fish over carps. People of Southeast Asian countries consume milk fish, while Americans prefer catfishes than carps.

Contour (n.): An imaginary line connecting points of equal elevation on the surface of the soil. A contour terrace is laid out on a slopping soil at right angles to the direction of the slope and nearly level throughout its course.

Conservation (n.): 1. The planning and management of resources in a way so as to secure their wider use and continuous supply, maintaining their quality, value, and diversity, 2. The national/inter-national management of and care for the biosphere, in order to avoid the creation of imbalance resulting in the destruction of habitats and the extinction of species.

Conservation efficiency (n.): Fish which provide edible flesh per unit of food consumed is preferred than which provide less flesh per unit weight. Recently, fishery scientists have focussed on the protein efficiency ratio and flesh produced in protein assimilation ratio in fish for their suitability in different culture systems.

Co-operative society (n.): A fish farm or business owned and run jointly by its farmers or members with profits or benefits shared among them.

Credit (n.) 1. The ability of a customer to obtain goods or services before payment, based on the trust that payment will be made in the future, 2. An entry in an account recording a sum received. Creditable (adj.).

Crustacean (n.): A member of the large group crustacea, which comprises mainly relating to aquatic barnacles.

Cryopreservation : The action of preserving fish sperm, eggs, and embryos at low temperature (at -196°C) for storage and subsequent use.

Ctenoid (adj.): (of fish scale) having many tiny projections on the edge like the teeth of a comb, as in many bony fishes.

Cultivar (n.): A plant variety that has been produced in cultivation by selective breeding.

Cultivation (n.): A scientific farming operation used in preparing ponds for fish rearing, production, harvesting, and marketing.

Current (adj.): A body of water or air moving in a definite direction through a surrounding body of water or air in which there is less movement.

Customer satisfaction (n.): The extent to which a producer's perceived performance matches a buyer's expectations. It is the product's performance falls short of expectations, the buyer is dissatisfied. If performance matches of exceeds expectations, the buyer is satisfied.

Cyanide (n.): A salt or ester of hydrocyanic acid, some forms of which are extremely toxic.

Cycloid (n.): A curve (resembling a series of arches) traced by a point on a circle being rolled along a straight line.

Cystovarian condition (n.): Refers to condition in which oviducts of most female teleost fish are continuous with the covering of the ovaries and the ova are not shed into the body cavity. Cf. *Gymnovarium condition*.

Dace (n.): A freshwater fish related to the carp, typically living in running water. (*Leuciscus leuciscus* and other species).

Dam (n.): 1. A barrier constructed to hold back water and raise its level, forming a reservoir or preventing flooding, 2. A barrier of branches in a stream.

DDT (n.): It is the abbreviation for Dichloro-diphenyl-trichloro-ethane — a synthetic organic compound used as an insecticide but now banned in many countries.

Decanter centrifuge machine (n.): The machine chiefly consists of a rotating bowl which also consists of a conical and a cylindrical part. A conveyor is rotated inside the bowl comparately at a different motion than the bowl. This motion helps transport solid particles towards the discharge port. This process provides maximum yield of minced meat with excellent quality.

Defaulter (n.): A person or farmer who fails to repay his loan.

Degradation (n.): 1. The condition or process of degrading environment/ecosystem or being degraded, 2. The wearing down of rock or soil by disintegration.

Dehydration (n.): 1. Removal of a large amount of water from food in order to preserve it, 2. Loss of water from fish flesh. Dehydrate, Dehydrated, Dehydrating (v.).

Demersal (adj.): (of fish) living close to the sea/reservoir/river bed.

Demonstration (n.): The action of demonstrating something.

Denitrification (n.): The biochemical reduction of nitrate or nitrite to gaseous nitrogen, either as molecular nitrogen or as an oxide of nitrogen.

Denudation (n.): Combined effects of erosion and weathering in wearing away the land surface.

Depuration (n.): 1. The placing of the fish in clean water to permit cleansing from pathogens or contaminants, 2. The process of freeing from impurities. Depurative (Adj.).

Derelict Water (adj.): Any water area which is absolutely worthless for carp culture but can be utilized for culture of some air-breathing fishes such as *Anabas, Clarias*, etc. through state-of-the-art technology.

Desalinization (n.): Removal of salts from saline soil, usually by leaching.

Design (n.): 1. A plan or drawing produced to show the look and function of plan before it is built or made, 2. The purpose of planning that exists behind an action or object. (v.): Decide upon the look and functioning of (something), especially by making a detailed drawing/planning with a specific purpose in mind.

Desorb (v.): Cause the release of (an adsorbed substance) from a surface. Desorption (n.), Desorbent (adj.).

Detoxification (n.): A process by which harmful compounds such as drugs and poisonous substances are converted to less toxic compounds in the body of fish. Detoxify, Detoxifying, Detoxified (v.).

Detritus (n.): Waste or debris, in particular organic matter produced by decomposition or loose matter produced by erosion.

Diagnosis (n.): The identification of the nature of an illness or other problem by examination of the symptoms. Diagnostic (adj.)

Dialysis (n.): *Chemistry*: The separation of particles in a liquid on the basis of differences in their ability to pass through a membrane.

Diatomaceous earth (n.): A geolocial deposit of fine, gayish, siliceous material composed chiefly or wholly of the remains of diatoms. It may occur as a powder or as a porous, rigid material. Due to its high surface area, absorptive capacity and chemical stability, it has a number of uses.

Diatoms (n.): Algae having silicious cell walls that persist as a skeleton after death; any of the microscopic unicellular or colonial algae constituting the class Bacillariophyceae. They occur abundantly in fresh and salt waters.

Diffusion (n.): The movement of substances resulting from the constant random bouncing of molecules. It tends to produce a uniform distribution of each substance so that its net effect is the transport of a substance from a region of higher concentration to a region of lower one. It is a spontaneous process and does not require energy.

Dimorphism (n.): Marked differences between the males and females of a species especially differences in superficial characters such as color, size, shape, etc.

Direct current (n.): An electric current flowing in one direction only.

Direct marketing (v.): Direct communications with carefully targeted individual consumers to obtain immediate response.

Disease (n.): A disorder in animal or plant, caused by infection, diet, or by faulty functioning of a process.

Disinfectant (n.): A chemical liquid that destroys bacteria. Disinfect (v.), Disinfection (adj.).

Disintegration (n.): The breakdown of rock and mineral particles into smaller ones by physical forces such as frost action.

Distribution (n.): The way in which something is distributed. Distributional (adj.).

Ditch (n.): A narrow channel dug to hold or carry water.

Diversion (n.): An alternative route for use by fish when the usual river is temporarily or permanently closed.

Diversity (v.): 1. A measurement of the richness of fish in a given aquatic ecosystem,. 2. The number of different types of fish species in an assemblage, 3. (of a form) enlarge or vary its range of products or field of operation. Diversification (n.), Diversified (v.).

Diversity index (n.): (Of a community) the ratio between number of species and number of individuals.

Domesticate (n.): Tame (an animal) and keep it as a pet or for farm produce. Domestication (n.).

Dominant (adj.): Relating to or denoting heritable characteristics which are controlled by genes that are expressed in offspring even when inherited from only one parent. Dominate (v.). Cf. *Recessive*.

Dorsal fin (n.): The unpaired fin that runs along the centre of the back. Some fish have a divided dorsal fin.

Drainage (n.): The action or press of draining water from ponds/lakes/tanks/reservoirs.

Dropsy (n.): A condition characterized by an excess of watery fluid collecting in the cavities or tissues of the body. It is an old-fashioned or less technical term for oedema.

Drug (n.): A medicine or other substance which has a marked physiological effect when taken into the body.

Dyke (n.): An earthwork serving as a boundary or defence to prevent the pond/tank/lake from flooding.

Dry bund (n.): A shallow depression which is enclosed on three sides by an earthen embankment. Rain water and surface flow-off from the surrounding catchment area is accumulated in this depression. The latest trend is to construct masonry structures with arrangements of a sluice gate in the deepest region of the bund to facilitate complete draining. A number of brood fishes in the ratio of 2 males: 1 female are kept. They migrate to shallow areas of the bund for breeding. Adequate spawning areas may be prepared at various levels of the bund which could be suitably flooded.

Drying (adj.): A process by which fishes are preserved by evaporating the moisture from the body of fish.

Ecological energetics (n.): The energy transformations that take place within any ecosystem.

Economic development (n.): A profit-related development of any farm that can upgrade the standard of living of farmers community.

Ecosystem impact (n.): The effect of some action on the environment particularly by human beings.

Eco-toxicology (n.): The study of the harmful effects of toxic substances (toxicants) in relation to any ecosystem.

Edema (n.): A swelling of tissues due to water seeping from the blood into the interstitial fluids.

EDTA (n.): It is the abbreviation for Ethylene Diamine Tetra-Acetic Acid — the chemical formula of which is $CH_2N (CH_2COOH_2)$. A chelating agent frequently used to protect enzymes from inhibition by traces of metal ions and as an inhibitor of metal-dependent proteases because of its ability to combine with metals.

Electro-dialysis (n.): Dialysis assisted by the application of an electro-potential across the semi-permeable membrane.

Electrolytic decomposition (n.): The passage of a direct current between metallic electrodes immersed in an ionized solution. In solutions, the electric chargs are carried by electrolytic ions, each having a mass several thousand times as great as the electron. The positive ions move to the cathode and the negative ions to the anode.

Electroporation (n.): The process of introducing DNA or chromosomes into cells using a electrical pulse to create temporary pores in the cell membranes.

Electro-potential (n,) : Electric shocks that brings about alteration in outer egg membrane.

Elver (n.): A young eel.

Embankment (n.): 1. A wall or bank built to confine water and to prevent flooding by a river, 2. A bank of earth or stone built to carry a boat or railway over an area of low ground.

Embryology (n.): The branch of biology concerned with the study of embryos. Embryological (adj.).

Endangered (v.): Animal with low population numbers that are in danger of becoming extinct.

Endocrine system (n.): A network of different glands having no ducts that secretes hormone into the blood circulatory system.

Endoparasite (n.): Parasites living inside the body of the host.

Enhance (v.): Improve the quality, value, or extent of. Enhancement (n.).

Enteritis (n.): Inflammation of the intestine, especially the small intestine usually accompanied by diarrhoea.

Environmental stress (n.): An environmental factor such as temperature, water availability or acidity, which attains a level close to the tolerance limit of a species and hence cause problems for its continued survival.

Environmental sustainability (n.): The practice of adopting policies and developing strategies that both sustain the environment and profits for the farm.

Enzootic (adj.): Disease regularly affecting fish stocks in a particular district or at a particular season.

Ephemeral (adj.): A water body which flows briefly in direct response to precipitation in the immediate vicinity.

Epidemic (n.): A sudden and widespread occurrence of an infectious disease in a community at a particular time.

Epidemiology (n.): The branch of medicine concerned with the incidence and distribution of diseases and other factors relating to health. Epidemiological (adj.).

Epistasis (n.): The suppression by a gene of the effect of another gene.

Epistatic (adj.): Describes a character or a gene whose effect over-rides that of another gene with which it is not allelic; analogous to dominant applied to genes at different loci. More generally, epistasis exists when the effect of two or more non-allelic genes in combination is not the sum of their separate effects.

Erosion (n.): The process of removal of top soil by any external agent.

Etiology (n.): The cause or causes of a disease or condition. Etiologic (adj.), Etiological (adv.).

Estuary (n.)(pl. Estuaries) : The tidal mouth of a large river. Estuarial (adj.), Estuarine (adj.).

Evapotranspiration (n.): The combined loss of water from a given area, and during a specified period of time, by evaporation from the soil surface and by transpiration from plants.

Eel (n.): A snake-like fish with a slender elongated body and poorly-developed fins, proverbial for its slipperiness. *Anguilla anguilla*(Europe) and related species, order: Anguilliformes.

Exotic fish (n.): Fish originating in or characteristic of a distinct foreign country.

Erosion (n.): 1. The wearing away of the land surface by running water, wind, or other geological agents, including such processes as gravitational creep, 2. Detachment and movement of soil by water, wind, or gravety.

Esophagus (n.): That part of the digestive tract between the pharynx and the stomach.

Eurythermal (adj.): Describing an animal which is able to tolerate a wide range of temperature in its environment. *Cf. Stenothermal.*

Estrogen (n.): Female sex hormone produced by ovaries which influences secondary sexual characters, color, etc.

Euryhaline (adj.): Describing an organism which can tolerate a wide range of salinity in its environment. Cf. *Stenohaline.*

Eutrophic (adj.): (of a body of water) rich in nutrients and so supporting a dense plant population, the decomposition of which kills animal life by depriving it of oxygen. Eutrophication (n.), Eutrophicate (v.).

Evolution (n.): The process by which new species are believed to arise as a result of gradual changes that occur in populations of organisms over a long period of time.

Exploitation (n.): Make good use of resource. Exploit (v.), Exploitable (adj.).

Export (v.): Entering a foreign market by sending products and selling them through international marketing intermediaries (indirect export) or through the farm's own branch, sales representatives or agents (direct export).

Ex-situ conservation (n.): Conservation of wild animals outside their habitats by perpetuating sample populations in genetic resource centres or in the form of gene pools and gamete storage. In this type of conservation, genetic engineering and cryopreservation of gametes have been playing an important role.

Extender (v.): An inert substance added to a product to dilute or to modify its physical properties.

Extension (n.): The action or process of extending – an additional period of time given to any subject(s) to fulfil an obligation (fish culture industry, for example).

Extensive (adj.): (of agriculture/fish culture) obtaining a relatively small crop from a large area with a minimum of capital and labor. Cf. *Intensive.*

Extensive culture (adj.): Improvement of fish culture over the traditional system where selected species in pre-determined numbers are stocked. Use of fertilizers/manures and supplementary feeds is highly limited and fish depends mostly on the natural food organisms in the pond.

Extinct (adj.).: Of a species or other large group having no living members.

Extinction (n.): The irreversible condition of a species of having no living representatives in the wild which follows the death of the last surviving individual of that species. Extinction may, however, occur at a local or global level; it can result from various human activities, including the destruction of habitats or the over-exploitation of species that are harvested or hunted as a food.

Factor (n.): Any external force, condition or substance that affects the fish.

Fauna (n.): The sum total of the kinds of animals in an area at one time Cf. *Flora.*

Fecund (adj.): Producing or capable of producing an abundance of offspring or new growth; highly fertile. Fecundity (n.).

Fermentation (n.): The chemical breakdown of a substance by bacteria, yeasts, or other micro-organisms, especially that involved in the making of beers, wines, and spirits in which sugars are converted to ethyl alcohol. Fermentative (adj.), Ferment (v.), Fermentable (adj.).

Fertile (adj.): (of soil or water) producing or capable of producing abundant vegetation or crops or aquatic animals of commercial importance. Fertility (n.).

Fertilize (v.): 1. Cause (an egg, female animal or plant) to develop a new individual by introducing male reproductive material, 2. Make (soil or water) more fertile by adding fertilizer.

Fertilizer (n.): Any organic or inorganic material of natural or synthetic origin added to an ecosystem to supply certain elements essential to the growth of plankton, bottom-organisms, and fish.

(i) granulated fertilizers: Fertilizers that are present in the form of rather stable granules of uniform size, which facilitate ease of handling the materials and reduce undesirable dusts.

(ii) liquid fertilizers : Fluid fertilizers that contain essential elements in liquid forms either as soluble nutrients or as liquid suspensions or both.

(iii) mixed fertilizers: Two or more fertilizer materials mixed together. May be as dry powder, granules, pellets, bulk blend, or liquids.

(iv) fertilizer requirement: The quantity of certain nutrient elements needed for ecosystem productivity in addition to the amount supplied by the pond soil to increase fish production to a designated optimum.

Fillet (n.): A fleshy boneless piece from near the ribs/loins of fish/animals. Fillets, Fillering, Filleted (v.).

Fibrosis (n.): The thickening and scarring of connective tissue, usually as a result of injury.

Filter bed (n.): A layer of substrate through which water passes during the filtration process.

Filter medium (n.): Any material used in a filtration, system to eliminate waste, or that can be colonized by beneficial aerobic bacteria.

Filtration (n.): The removal of waste from the water in ponds.

Fin (n.) : 1. Flattened appendages which are used in swimming and are supported with bony rays. Fins control the direction of movement, 2. The projections of a fish's body that enable it to move through water. They are also often used for display and even mating purposes.

Finance (n.): The management of large amounts of money, especially by governments/banks/co-operatives. Monetary support for an enterprise. Financial (adj.).

Fish biodiversity (n.): The kaleidoscope variety of threatened and endangered fish on this planet or in an ecosystem.

Fishery (n.)(pl. Fisheries): 1. A place where fish are reared or caught in numbers, 2. The occupation or industry of catching or rearing fish.

Fish Finger (n.): A small oblong piece of flaked or minced fish coated in batter or breadcrumbs.

Fish louse (n.): A parasitic crustacean which attaches itself to the skin or gill of fish. (Many species in classes Branchiura and Copepoda.)

Fish meal (n.): Ground dried fish used as fertilizers or animal/fish feed.

Fish monger (n.): A person or shop that sells fish for food.

Fish sauce (n.): A liquid or semi-liquid substance derived from the fish consignment after passing it through rigor mortis and served with food to add flavor.

Fish silage (n.): Fish or parts of fish are compacted and stored in air-tight conditions without first being dried and used as food.

Flooding (n.): The water is released from field ditches and allowed to flood over the land.

Flood plain (n.): An area of low-lying ground adjacent to a river that is subject to flooding.

Flora (n.) : The sum total of the kinds of plants in an area at one time. Cf. *Fauna*.

Fluctuate (v.): Rise and fall irregularly in number or amount of fish in a particular pond/ tank/reservoir/lake. Fluctuation (n.).

Fluidize (v.): Cause (a finely divided solid) to acquire the characteristics of a fluid by passing against upwards through it. Fluidization (n.).

Forage fish (n.): A species of fish which is the prey of more valuable game fish.

Foul (adj.): 1. Offensive to the senses: (i) very unpleasant, 2. Polluted or contaminated. Foully (adv.)

Freezer (n.): A refrigerated cabinet or room for preserving food at very low temperatures.

Freon (n.): An organic solvent consisting of one or more of a group of chloroflurocarbons and related compounds.

Fry (n.)(pl. Fry): A young fish, especially when newly hatched.

Fungus (n.) (pl. Fungi) : Any of a large group of simple and spore-producing organisms that lack a photosynthetic pigment and feed on organic matter and include moulds, yeast, mushrooms, and toadstools. The individual cells have a nucleus surrounded by a membrane, and they may be linked together in long filaments called *hyphae*, which may grow together to form a visible body.

Gall bladder (n.) : A sec which, in vertebrates, stores the bile secreted by the liver.

Gamete (n.): A mature haploid male and female germ cells which are able to unite with another of the opposite sex in sexual reproduction to form a zygote. Gametic (adj.).

Gastritis (n.): Inflammation of the lining of the stomach.

Gear (n.): An apparatus or equipment, especially net made of nylon thread connected with floats and sinkers, by which fish are harvested

Gene (n.): The unit of hereditary information. Structural genes specify the sequence of single protein chains. Other genes contain information about when and where structural genes are to be expressed.

Gene expression (n.): The turning on or activation of a gene so that it gets transcribed and translated.

Gene flow (n.): Exchange of alleles between populations (in one or both directions).

Gene pool (n.): The stock of different genes in an inter-breeding population.

Gene regulation (n.): The various methods used by organisms to control which genes will be expressed and when.

Genetic diversity (n.): Amount of genetic variability among individuals of a single species and between species.

Genetic drift (n.): 1. The occurrence of random changes, irrespective of selection and mutation, in the genetic make-up of small isolated populations, 2. Random variation of allele frequency in a population from one generation to another that results from random errors in sampling gametes. Because such errors result from a finite sample of gametes of their parents, the genetic composition of progeny may differ from that of their parents.

Genetic engineering (n.): The deliberate modification of the characteristics of an organism by manipulating its genetic material.

Genetic integrity (n.): The degree to which the genetic composition of a population resembles its natural state, uninfluenced by anthropogenic causes.

Genetic marker (n.): An allele that may characterize a population or group of populations. The marker may be qualitative (rare or not) or quantitative (present at greater or less frequency in the marked group). Genetic markers are the basis of genetic stock identification and have been used to determine percentage.

Genetic stock indentification (n.) : Use of genetic differences occurring among populations or aggregates of populations to estimate proportionate combinations to mixtures, such as mixed stock fisheries.

Genetically modified (adj.) : (of an organism) containing genetic material that has been artificially altered so as to produce a desired character.

Genetic manipulation (n.): The use of various techniques to yield fishes containing foreign DNA as a part of their genomes. The technique involves both the isolation of DNA from another fish and its transfer to a fish. Both makes it possible to transfer genes from much less closely related species than does conventional breeding. Fishes in which foreign DNA is incorporated and expressed are called as *transgenic* or *transformed fish*. Genetic manipulation is important in research in functioning of the genome and has considerable promise in the production of new varieties.

Genetic variation (n.): The differences in genetic make-up between individuals of the same species such as brown and blue eyes.

Genetics (n.): The branch of science that deals with the way characteristics pass from one generation to the next.

Genome (n.): The complete DNA sequence of an organism.

Genus (n.): A group of closely related species. The fish 'Rohu' and 'Bata' are, for example, placed in the same genus, *Labeo* such as *Labeo rohita* and *Labeo bata*.

Geo-chemistry (n.): The study of the chemical composition of the earth and its rocks and minerals. Geo-chemical (adj.).

Genotype (n.): The genetic constitution of an individual organism. Cf. *Phenotype*.

Germplasm (n.): Either the reproductive cells of organisms as opposed to their mortal bodies, or the special region of egg cytoplasm which determines that nuclei in this region go on to make germ cells.

Gill (n.): 1. An outward extension of the body used for gas exchange. Gills are present in animals that are aquatic. 2. The main respiratory organs of a fish, located on each side of the head behind and below the eyes. Gills extract dissolved oxygen from water.

Glacier (n.): A slowly moving mass of ice formed by the accumulation and compaction of snow on mountains or near the poles.

Glaze (v.): A liquid or substance used to form smooth, thin, and shiny coating on food items for preservation of food products. Glazing (n.).

Glomerulus (n.) (Pl. Glomeruli): A cluster of nerve endings, or small blood vessels, especially of capillaries around the end of a kidney tubule. Glomerular (adj.).

Gold fish (n.): Ornamental aquarium and pond fish (*Carassius auratus*) of the carp family, but unlike most fish in this group, they lack any barbels around the mouth. They are native to East Asia but, introduced into many other areas. It is naturally greenish brown or gray, but its color varies. They may live for more than 40 years far longer than most other pond fish. Body size is highly variable, often exceeds 30 cm (12 inches). in ponds. Selective breeding has produced more than 125 breeds. They feed on plants and small animals and, in captivity, on small crustaceans and/or pelleted feed.

Gonad (n.): The special organ of multicellular animals in which the gametes are generated.

Gonado-somatic index (n.): The ratio of length (in cm) to weight (in gram) of the ovary or testes of the fish.

Gonochorism (adj.): The existence of either testes or ovaries in individual fish. Most farmed fish have this gonochoristic type of sexuality.

Gourami (n.) (pl. Gouramis) : An Asian labyrinth fish of a large group including many kinds. They are very popular in aquaria. (Osphronemidae, Belontiidae, and other families).

Grade (n.): A particular level of range, size, shape, quality, type, or value.

Grader (n.): A machine or device for sorting out fish of different sizes in a given stock.

Gradient (n.): Degree of such a slope, expressed as change of height divided by distance travelled.

Gravid (n.): Describes a female whose body is swollen with eggs or developing young.

Green house (n.) :The entrapment of heat by upper atmosphere gases such as carbon dioxide, water vapor, and methane just as glass traps heat for a green house. Increases in the quantities of these gases in the atmosphere will likely to result in global warming that may have serious consequences for animal kingdom.

Gross Domestic Product (GDP) (n.): The total value of goods produced and services provided within a country during one year.

Gross National Product (GNP) (n.): The total value of goods produced and services provided by a country during one year, equal to the gross domestic product plus the net income from foreign investments.

Ground water (n.): Sub-surface water in the zone of saturation that is free to move under the influence of gravity.

Growth factor (n.) : A subsistance such as vitamin or hormone, which is required for the stimulation of growth in living cells.

Guppy (n.): A small live-bearing brightly-coloured freshwater fish, native to tropical America and widely kept in aquaria (*Poecilia reticulata*).

Gymnovarium condition (n.):- Refers to condition in which ovaries of most female non-teleost fish open into the body cavity and the ova are conveyed through an open funnel to the oviduct. Cf. *Cystovarian condition*.

Gutted/Gutting (v.): Remove the internal organs of a fish or other animal before cooking.

Gynogenesis (n.): A phenomenon in which the irradiated spermatozoa enters into the micropyle of the fish egg and activate the development but generated without any parental genetic contribution to the individual. Hence the heplois (n) embryo that contains only maternal chromosomes is produced. Cf. *Androgenesis*.

Haemolysis (n.): The rupture or destruction of red blood cells. Haemolytic (adj.).

Haemorrhage (n.): An escape of blood from a ruptured blood vessel.

Hand- stripping (n.): Manually removal of eggs and sperm from a female and male fish, respectively.

Hardness (n.): Any multivariate cation, usually Ca^{2+} and Mg^{2+} that causes undesirable precipitation and deposition in water.

Hard water (n.): Freshwater containing a high level of dissolved calcium and magnesium mineral salts, typically in excess of 150 mg/l.

Harpoon (n.): A barbed spear-like missile attached to a long rope and thrown by hand or fired from a gun, used for catching fish, whales, and other large aquatic creatures.

Harvest (n.): The process by which agricultural crops and aquaculture species of commercial importance are collected for consumption, marketing, and export. Harvestable (adj.).

Hatchery (n.): An establishment where fish or poultry eggs are hatched.

Health (n.): The state of being free from illness or injury or disease.

Heart (n.): The muscular pump that drives the blood through the vessels of the circulatory system. In fish, however, the heart has two chambers — one auricle and one ventricle in connection with the sinus venosus and bulbous aorta.

Heavy metal (n.): A metal of relatively high density, or of high relative atomic weight.

(i) arsenic (n.): The chemical element of atomic number 33, a brittle steel-grey semi-metal with many highly poisonous compounds. (Symbol: As).

(ii) cadmium (n.); The chemical element of atomic number 48, a silvery-white metal resembling zinc. Symbol: Cd).

(iii) chromium (n.): The chemical element of atomic number 24, a hard white metal used in stainless steel and other alloys. (Symbol: Cr).

(iv) copper (n.): A red-brown metal, the chemical element of atomic number 29, which is a good conductor of electricity and heat and is alloyed to form brass and bronze.(symbol: Cu).

(v) lead (n.): A heavy bluish-grey soft ductile metal the chemical element of atomic number 82. (Symbol; Pb).

(vi) mercury (n.): The chemical element of atomic number 80, a heavy silvery-white liquid metal used in some thermometers and barometers. (Symbol: Hg).

(vii) molybdenum (n.): The chemical element of atomic number 42, a brittle silvery-grey metal. (Symbol : Mo).

(viii) nickel (n.): A silvery-white metal resembling iron, the chemical element of atomic 28. Symbol., Ni).

(ix) zinc (n.): The chemical element of atomic number 30, a silvery-white metal which is a constituent of brass and is used for galvanizing iron and steel. (Symbol : Zn).

Hepatoma (n.): A cancer of the cells of the liver.

Herbivore (n.): An animal which eats plants or parts of plants. Herbivorous (adj.).

Heredity (n.): The passing on of physical or mental characteristics genetically from one generation to another.

Heritability (adj.): A measure of the degree to which the variation of a trait is inherited. Ranges from Zero to 100 per cent. It is the proportion of additive genetic variance in the total phenotypic variance. Generally symbolized by H^2. Only when heritability is high, selective breeding is likely to alter a trait.

Hermaphrodite (n.): 1. An animal with both male and female sex organs, 2. An individual animal generating both male and female gametes.

Heterogametic (adj.): Denoting the sex which has sex chromosomes that differ in morphology. Cf. *Homogametic.*

Heteroplastic extract (n.): Extracts that are collected from different species of fish. Cf. *homoplastic extract.*

Heterosis (n.): The difference between the mean of a quantitative character in a crossbred generation and the mean of the two parental strains.

Heterozygote (n.): An individual with two different alleles at a particular locus, the individual having been formed from the union of gametes carrying different alleles. Heterozygous (adj.). Cf. *Homozygote.*

Holoblastic cleavage (n.): When the amount of yolk is less, the cleavage furrow of fertilized eggs extends through the entire egg and completely divides it.

Homogametic (adj.): Denoting the sex which has sex chromosomes that do not differ in morphology. Cf. *Heterogametic.*

Homoplastic extract (n.): Extracts that are collected from the same or closely-related species of fish. Cf. *Heteroplastic extract..*

Homozygote (n.): An individual whose two genes at a particular locus are the same allele, the individual having been formed by the union of gametes carrying the same allele. Homozygous (adj.). Cf. *Heterozygote.*

Humoral (adj.): Immunity involving the action of circulating antibodies.

Husbandry (n.): 1. The care, cultivation, and breeding of crops and animals, 2. Management and conservation of resources.

Hybridization (n.): The cross breeding of different species together.

Hybrid (n.): Offspring of a cross between two different strains, varieties, races, and species.

Hydro-edaphic (adj.): Ecology of, produced by, or influenced by the water and soil.

Hydrology (n.): The branch of science concerned with the properties and distribution of water on the earth's surface. Hydrological (adj.).

Hygiene (n.): Conditions or practices conducive to maintaining health and preventing disease, especially through cleanliness.

Hypophysation (n.): A technique by which mature male and female fish of about same weight and species are stimulated in confined waters to release sperm and eggs respectively by injecting either the extract of pituitary glands of same or related species of fish or by synthetic hormones. By this method, it is possible to fertilize eggs by sperm within very short time.

Hypothalamus (n.): Structurally small but functionally mighty area has centres for feeding, sexual maternal, and aggressive behavior.

Hypoxia (n.): 1. Deficiency in the amount of oxygen reaching the tissue, 2. Oxygen deficiency in a biotic environment.

Immobilization (n.): The conversion of an element from the inorganic to the organic form in microbial tissues or in plant tissues, thus rendering the element not readily available to other organisms or to plants.

Immune (adj.): 1. Resistant to a particular infection owing to the presence of specific antibodies or sensitized white blood cells, 2. Relating to such resistance: The immune system.

Immunization (n.): Denoting techniques making immune the body to infection. Immunize (v.).

Immuno-stimulation (n.): Stimulation of an immune response to help the survival of an organ/body after transplantation or infection with pathogens. Cf. *Immuno-suppression*.

Immuno-suppression (n.): Suppression of an immune response to help the survival of an organ/body after transplantation. Cf. *Immuno-stimulation*.

Immune system (n.): The organs and processes of the body that provide resistance to infection and toxins.

Immuno-stimulant (n.): Compounds that increase disease resistance by enhancing host defence mechanisms against pathogenic micro-organisms in the culture system.

Inbreed (v.): Breed from closely related animals, especially over many generations. Inbreeding (n.).

Inbreeding coefficient (n.): It is a measure of the degree to which an individual is inbred. Ranges from zero (when the parents are unrelated) to 100 per cent (when the parents for many generations back have been related). and is usually symboized by *F*.

Inbreeding depression (n.) : The reduction of desirable characters such as growth, survival rate, production, fertility, consequent to the homozygosis produced by inbreeding, especially those which normally outbreed.

Incubate (v.): Fish eggs in order to keep them in a hatchery tray and bring them to hatching. Incubation (n.).

Indicator species (n.): Species having a narrow tolerance range for some environmental factors. The occurrence of indicator species at a particular ecosystem or habitat indicates the corresponding habitat condition. They mainly indicate the general toxicity of the ecosystem without telling the exact chemical or physical factors responsible for this toxicity.

Indigenous (adj.): Originating or occurring naturally in a particular place; native.

Induction (n.): 1.The production of a definite condition by the action of an external factor, 2. The action or process of inducing something.

Infiltration (n.): The down-ward entry of water into the soil.

Infrastructure (n.): The basic structures such as fish farm, processing plant, cold storage, ice plant, laboratory, etc. needed for the operation of a fish farm and subsequent production, processing as well as marketing of fish products.

Innoculation (n.): The process of introducing mixed or pure culture of micro-organisms into natural or artificial culture media.

Inorganic compounds (n.): All chemical compounds in nature except compounds of carbon other than carbon mono-oxide, carbon di-oxide, and carbonates.

In-situ (Conservation of fish) (v.): Conservation of genetic resources of fish species through their maintenance within natural or even man-made ecosystems in which they occur. This type includes a system of protected areas such as fish sanctuaries. Cf. *Ex-situ.*

Insulin (n.): A polypeptide pancreatic hormone which lowers glucose levels in the blood.

Insurance (n.): 1. The action of insuring someone or something > money paid as compensation under an insurance policy, 2.A thing providing protection against a possible eventuality.

Integrated management (n.): Different functions of the aquaculture business that must be integrated with one another, keeping aquaculture/fisheries activity as a pivot.

Integrated watershed management (n.): The use of a variety of approaches to watershed management in an integrated manner. The term watershed refers to an area of land that separates waters flowing to different rivers and basins.

Intensive (adj.)* : (Of agriculture/fish culture) aiming to achieve maximum production within a limited area. Cf. *Extensive.*

Introgression (n.): The transfer of genetic information from one species to another through hybridization and repeated back-crossing. Introgression may compromise genetic integrity.

Ion-exchange (n.)** : The exchange of ions of the same charge between an insoluble solid and a solution in contact with it, used in purification and separation processes.

Ionic regulator (n.): An aquatic animal which regulates the concentrations of individual solutes in its body fluids as the environmental concentrations of these substances change.

---

* In intensive farming, ponds are heavily stocked with fish. Other management schedules involve the maintenance of water and soil qualities, frequent changing of water, water recirculation together with constant aeration and the use of formulated feeds. This system of farming is practised in raceways and in artificial tanks.

** Ion-exchange is an efficient and reliable means of ammonia removal but it is more expensive than biological filtration. Clinoptilotite – one of the zeolites used in water treatment – is an important ion-exchange material for the removal of ammonia from intensive fish culture systems. The minimum depth of an ion exchange bed is 0.60-0.75 litre/second/m^2.

Ionizing radiation (n.): Radiation consisting of particles, X-rays or gamma rays which produce ions in the medium through which it passes.

IQF (n.): It is the abbreviation for Individual Quick Frozen. A process by which individual finfish/shellfish is rapidly frozen at –40°C in a frozen chamber for long-term storage and transport.

Irradiation (n.): The process or fact of irradiating or being irradiated.

Irrigate (v.): Supply water to land or crops by means of channels. Irrigation (n.).

ISO (n.): It is the abbreviation for International Standard Organization. It is a worldwide federation of National Standard Bodies of 110 countries, has started the development of various purposes. For example, the ISO has prescribed standards under ISO 1400 and ISO 9002 for environmental management and product quality management, respectively.

Isomer (n.): Two or more different compounds which have same molecular formulas and this phenomenon is referred to as *isomerism*.

Isotope (n.): Each of two or more forms of the same element that contain equal numbers of protons but different numbers of neutrons in their nuclei.

Isozyme (n.): 1.Electro-phoretically distinct forms of an enzyme with identical activities, generally coded by different genes. 2. An enzyme which has multiple molecular forms in the same organism catalyzing the same reaction, 3. Enzymes present in more than one form distinguishable by chemical or immunological method, but have the same enzymatic activity. Isozmes are widespread in nature with over a hundred enzymes now known to occur in two or more molecular forms.

IUCN (n.): It is abbreviation for International Union for the Conservation of Nature.

Joint venture (n.): Entering foreign markets by joining with foreign establishments to produce or market of a product.

Kidney (n.): The main organ of the body of most animals including fish for the elimination of body fluid and the waste products of metabolism.

LC50 (n.): The concentration of any toxicant that cause 50 per cent mortality or which can be expected on an average.

Lab-Lab (n.): Characterized by benthic blue-green algae and diatoms, but many other forms of plants, micro-organisms and animals are associated with it and contribute to its nutritional value.

Lagoon (n.): 1. A stretch of salt water separated from the sea by a low sandbank, 2. *North America and Australia/Newzealand* : A *small* freshwater lake near a large lake or river, 3. An artificial pool for the treatment of effluent or to accomodate an overspill from surface drains.

Lake (n.): A large area of water surrounded by land.

Lake trout (n .): 1. A European brown trout of a large race, 2. A large charr living in the Grate Lakes of North America.

Larva (n.) (pl. Larvae): The post-hatching stage in an invertebrate's life cycle. Marine fish/ freshwater fish fry are often called *Larvae,* since they are poorly developed when they hatch.

Lateral line (n.): A visible line along the side of a fish, consisting of a series of sense organs which detect pressure and vibration, aiding navigation and giving early warning of impending danger.

Lateritic soil (n.): A reddish clayey material, hard when dry, forming a topsoil in some tropical regions and sometimes used for building.

Leaching (n.): Removal of materials in solution from the soil by percolating water.

Leech (n.): A parasitic or predatory annelid worm with suckers at both ends, example of which were formerly used in medicine for bloodletting [*Hirudo medicinalis*] [(medicinal leech) and other species, Class: Hirudinea].

Lethal (adj.): Sufficient to cause death. Lethally (adv.), Lethality (n.).

Lethal concentration value (LC value) (n.): The very harmful concentration of any chemical substance at which 100 per cent mortality of fish occurs. It is the concentration of a toxic compound which causes 100 per cent death or which can be expected average to cause 100 per cent death of population. It is termed as LC5, LC50, LC95, and LC100. The LC50, for example, means that cause death in 50 per cent of the fish population due to toxic action of a toxicant.

Lime requirement (n.) : The quantity of liming material needed to neutralize the acidity of the pond bottom and increase total alkalinity of pond water to above 20 mg/l.

Limnology (n.) : The study of lakes and other bodies of freshwater. Limnological (adj.)

Lipofection (n.) : The encapsulation of DNA construct in lipid vesicles and allowing them to come in contact with the plasma membrane of a target cell. This contact helps uptake the vesicles through fusion with the membrane resulting in the alteration of cellular function.

Liquid asset (n.) : Asset held in or easily converted into cash.

Live fish (n.) : Invertebrates used as fish food. They can be bred and feed to the fish alive.

Liver (n.): An organ which opens into the gut. It has several functions, including and involvement with digestion.

Loach (n.) : A small, elongated freshwater fish with several barbels near the mouth. Numerous species.

Loam (n.): The textural class name for soil having a moderate amount of sand, silt, and clay. Loam soils contain 7-22 per cent clay, 25-50 per cent silt, and 23-52 sand.

Loamy (n.): Intermediate in texture and properties between fine-textured and coarse-textural classes with the words loam or loamy as a part of the class name, such as clay-loam or loamy-sand.

Lobster (n.) : A large marine crustacean with a cylindrical body, stalked eyes, and the first of its five pairs of limbs modified as pincers. *Homarus* and other genera.

Locomotion (n.): Movement or the ability to move from one place to another.

Lymphocyte (n.) :A form of small leucocyte (white blood cell) with a single round nucleus occurring especially in the lymphatic system and having various immune functions.

Macrophage, (n.): A large phagocytic cell found in stationary form in the tissues or as a mobile white blood cell, especially at sites of infection.

Macrophyte (n.): Refers to large aquatic plants.

Mahseer (n.): A large edible carp of Northern India and the Himalayan region. (Genus: *Tor* with several species).

Manure (n.): Nutrient carrier of animal and plant origin used for fertilizing agricultural lands and fish ponds.

Market (n.): A regular gathering of people for the purchase and sale of provisions, livestock, fish and fish products and other commodities. Markets (v.), Marketed (v.), Marketability (n.).

Marketable (adj.): Able or fit to be sold or marketed.

Marketing (n.): A managerial or social process by which individuals or groups obtain what they need and want through creating and exchanging products and value with others.

Marketing environment (n.): The actions and forces outside marketing that affect marketing management's ability to develop and maintain successful transactions with its target customers.

Market development (n.): A strategy for farm growth by identifying and developing new market segments for current products.

Market potential (adj.): The upper limit of market demand.

Market price (n.): The current price of any product.

Marketing research (n.): The systematic design, collection, analysis, and reporting of data relevant to a specific marketing situation facing on organization.

Marsh (n.): Periodically wet or continually flooded area with the surface not deeply submerged. Covered dominantly with sedges, cattails, rushes or other hydrophytic plants. Subclasses include freshwater and saltwater marshes.

Maturation (n.): Final stages in the development of the germ cells; more generally, the process of becoming adult or fully developed.

Maturity (n.): 1. The time when a fish becomes sexually mature, 2. Fully-developed stage. Mature (adj.).

Mechanization (n.): Adoption of improved devices for better production.

Meroblastic (or Incomplete) Cleavage (n.): Zygotes containing large amount of yolk undergo meroblastic where only a portion of the cytoplasm is divided. The amount of cytoplasm close to the animal pole of the egg divides completely leaving most of the yolky portion of the vegetal pole remains undivided.

Mesopelagic (adj.): Fish inhabiting the intermediate depths of the sea.

Metabolism (n.): The chemical processes that occur within a living organism to maintain life. Metabolic (adj.).

Metabolite (n.): A substance formed in or necessary for metabolism.

Metamorphosis (n.): The duration change of structure and way of life as a larva takes on adult form.

Methemoglobin (n.): A stable oxidized form of haemoglobin which is unable to release oxygen to the tissue and the condition is termed as *methemoglobenemia.*

Methemoglobinemia (n.): A condition in which haemoglobin is oxidized by nitrite and forms methemoglobin. It is also called *brown blood disease.*

Micro-algae (n.) (Sin. Micro-alga) : Algae of microscopic diamensions that range from 1 to 2 microns.

Microbial degradation (n.): Breakdown of organic/inorganic compounds into toxic/non-toxic forms by micro-organisms.

Microdiet (n.): Refers to tiny nutrient particles which are surrounded by digestible materials. Three types of microdiets are generally used in fish culture/aquaculture: (i) Micro-encapsulated diet (contains tiny ingredients in solution surrounded by digestible materials), (ii) Micro-bound diet (composed of tine nutrient particles held together by a protein or carbohydrate binders), and (iii) Micro-coated diet (It comprises nutrient particles which are coated with digestible materials to form a 'skin' that is impervious to water.

Microfauna (n.): The part of the animal population which consists of individuals too small to be clearly distinguished without the use of a microscope. Includes algae, actionomycetes, bacteria, and fungi

Micronutrient (n.): A chemical element or substance required in small amounts by living organisms.

Micropipette (n.): One end of a glass micropipette is heated until the glass becomes liquified to some extent. It is rapidly stretched which forms a very fine tip at the heated end. The tip of the pipette attains to about 0.4 mm diameter that resembles an injection needle.

Micropyle (n.): A small opening in the primary egg envelope in the eggs of insects and fishes through which makes the sperm to enter into the egg at the time of fertilization.

Middleman (n.)(pl. Middlemen): 1. A person who buys goods from producers and sells them to retailers of consumers, 2. A person who arranges business deals between other people.

Migration (n.) :) 1.Long-distance movement of animals and especially the regular seasonal movements between breeding and wintering grounds or different feeding places shown by many fish, birds, and mammals, 2. Animals move from one place to another according to the seasons.

Milt (n.): The semen of a male fish.

Mince (v.): Cut-up or shred (meat) into very small pieces. Mincing (adj.), Mincer (n.), Mincingly (adv.).

Minerals (n.): Essential elements, other than carbon, hydrogen or oxygen, normally obtained as inorganic ions taken up by fish through natural food organisms.

Mineralize (v.): Convert (organic matter) wholly or partly into a mineral or inorganic material or structure. Mineralization (n.) .

Minnow (n.): A small freshwater fish of the carp family, which typically forms large shoals (*Phoxinus phoxinus).*

Mobility (n.): Spreading of toxicants from one place to another within a compartment or move to a new compartment.

Moiety (n.): Each of two parts into which a thing is or can be divided.

Molecular genetics (n.): The branch of genetics concerned with the macromolecules such as DNA and proteins.

Monitoring (v.): To measure quantitatively or qualitatively the level of a substance over a period of time. Monitorial (adj.).

Monoclonal antibody (n.): An antibody that is produced by fusing one B-cell with a myeloma cell to create a hybridoma cell. The hybridoma cell will be immortal and will produce high quantities of only one specific antibody molecule.

Morphology (n.): The branch of biology concerned with the forms and structures of living organisms. Morphological (adj.), Morphologically (adv.).

Morphometric dimension (n.): The process of measuring the external shape and dimensions of living organisms or other objects.

Mountain (n.): An elevation of the earth's surface rising abruptly and to a large height from the surrounding level.

Movement (n.): 1. Animal's activity during a particular period of time, 2. A group of animal working together to advance a shared cause.

Mouth-brooder (n.): An egg-laying fish that incubates and hatches its eggs in the mouth. Some cichlids and anabantoids are mouth-brooders.

Mucous (adj.): 1.A slimy substance secreted by the mucous membranes and glands of animals for lubrication, Protection, etc.

Mucus (n.): 2. A slimy substance which consists mainly of water with a small quantity of a protein-sugar complex which makes it very viscous.

Mucopolysaccharide (n.): A substance occurring chiefly in connective tissue and consisting of complex polysaccharides containing amino groups.

Mullet (n.): Any of various chiefly marine food fish. [Families: Millidae (red mullet) and Mugilidae (grey mullet), several species.]

Muscle (n.): A material, composed of the proteins, action and myosin, that is capable of exerting forces and of contracting. Most movements, among animals visible to the naked eye, are produced by muscles.

Mutation (n.): An unexpected change in the genetic make-up of an organism.

Mycosis (n.): A disease caused by infection with a fungus. Mycotic (adj.).

Myoglobin (n.): A red protein containing haem which carries and stores oxygen in muscle cells.

Myotome (n.): That part of the somatic mesoderm which forms striped muscles.

Necrosis (n.): The death of most or all of the cells in tissue or an organ due to disease, injury, or failure of blood supply.

Nematode (n.): A worm of the large phylum Nematoda, having a slender unsegmented, cylindrical body; such as a roundworm or threadworm.

Network (n.): Interact with others to exchange information and develop useful contacts.

Neuro-toxic (adj.): Toxic partaining to nerve.

Niche (n.): A habitat of organisms where they live and function with respect to other organisms and the environment.

Nitrification (n.): The biochemical oxidation of ammonium to nitrate, predominantly by autotrophic bacteria.

Nitrogen (n.): The chemical element of atomic number 7, a colorless, odorless unreactive gas that forms about 78 per cent of the earth's atmosphere (Symbol: N.)

Nutrition (n.): The process of feeding and subsequent digestion and assimilation of food materials.

Obstructive jaundice (n.): Jaundice caused by a mechanical impediment to the flow of bile from the liver to the duodenum. Although gallstones are the most common cause, cholangitis, cysts, and parasites in the billary ducts are responsible less frequently.

Offal (n.): 1. The entrails and internal organs of an animal used as food, 2. Decaying or waste.

Offspring (n.): The young of an animal.

Oligotrophic (adj.): 1.(of a body of water) relatively poor in plant nutrients and containing abundant oxygen in the deeper parts. 2. A lake/pond that is characterized by a deficiency in plant materials and usually by abundant dissolved oxygen in the bottom layers; its relatively small amounts of organic matter and its water is often deep.

Omnivore (n.): An animal that includes both plant and animal materials in its diet.

Oncosphere (n.): The embryonic stage of a tapeworm in which it has hooks.

Oocyte (n.): A cell in the ovary which may undergo meiotic division to form an ovum.

Operculum (n.)(pl. Opercula) : 1. A plate that closes the aperture of a gastropod mollusc's shell when the animal is retracted, 2. The gill cover of a bony fish.

Organic farming (adj.): A set of farming practice where no chemical fertilizers are used; rather, a variety of organic manures such as cattle manure, poultry litter, etc. are used for constant productivity of fish culture ecosystems.

Organic fertilizer (n.): By-product from the processing of animal or vegetable substances that contain sufficient nutrients to be of value as fertilizer.

Organic soil (n.): A soil that contains at least 20 per cent organic matter (by weight) if the clay content is low and at least 30 per cent if the clay content is as high as 60 per cent.

Organic soil material (n.): Pond bottom soils having 20 per cent or more organic carbon (by weight) if the mineral fraction is more than 60 per cent clay, more than 12 per cent organic carbon if the mineral fraction has no clay, or between 12 and 20 per cent carbon if the clay content of the mineral fraction is between zero and 60 per cent.

Organic waste skimmer (n.): A filter method in which a stream of electrically-charged air bubbles rise through a plastic tube; organic wastes stick to the bubbles and rise to the surface, where they form a thick foam. This must be collected for disposal with degree of regularity.

Organochlorine (n.): A class of organic chemicals that contain chlorine bond within the nucleus.

Organoleptic (adj.): Acting on, or involving the use of, the sense organs.

Organophosphate (n.): 1. Any organic compound whose molecules contains one or more phosphate ester groups, especially a pesticide of this kind, 2. A class of organic compounds that contain phosphorus and oxygen bonded within the broad-spectrum molecules, extremely toxic and do not persist in the environment for prolonged period.

Osmoconformer (v.): An animal which changes the osmotic concentration of its body fluids more or less in conformity with changes in the external osmotic concentration.

Osmogene (n.): A number of biodegradable substances present in many polluted waters which are odorous or volatile and incessantly lose molecules to the atmosphere. These may cause irritation in the respiratory epithelium, which finally may result in the loss of appetite, giddiness, nausia, and many other problems.

Osmoregulation (n.): The maintenance of constant osmotic pressure in the fluids of an organism by the control of water and salt concentrations.

Osmosis (n.): A process by which solvent molecules pass through a semi-permeable membrane from a less concentrated solution into a more concentrated one.

Osmotic concentration (n.): The total concentration of dissolved solutes in a solution, independent of the kinds of solutes present.

Osmotic pressure* (n.): The pressure that would have to be applied to a pure solvent to prevent it from passing onto a given solution by osmosis.

Ossify (v.): Turn into bone or bony tissue. Ossification (n.).

Overdue (adj.): Past the time when due. > not having arrieved, been born, etc. at the expected time.

Ovulate (v.): Discharge ova or ovules from the ovary. Ovulation (n.).

Oxidation (n.): Partial or complete digestion of liquid organic wastes in which the wastes are aerated by a mechanical device.

Oxidizing agent** (n.): A substance that tends to bring about oxidation by being reduced and gaining electrons.

Oxygen (n.): The chemical element of atomic number 8, a colorless, odorless reactive gas that forms about 20 per cent of the earth's atmosphere and is essential to plant and animal life. (Symbol: O).

Oxygen transfer efficiency (n.): Oxygen transfer efficiency is defined as the quantity of oxygen dissolved per energy expended (Kg O_2 KW per hour) and is determined under zero per cent saturation in freshwater at 20°C.

---

* The passage of a solvent through a semi-permeable membrane separating two solutions of different concentrations which is termed as *osmosis*. Through this membrane, the molecules of a solvent can pass but the molecules of most solutes cannot. There is a tendency for solutions separated by such a membrane to become equal in concentration. Osmosis will stop when the two solutions reach equal concentration and can also be stopped by applying a pressure to the liquid on the stronger solution on the side of the membrane. The pressure required to stop the flow is a feature of the solution and is termed as *osmotic pressure*.

** Agents such as potassium permanganate help reduce the rate of oxygen consumption by chemical and biological processes. Potassium permanganate oxidizes labile organic matter and other reduced substances and is transformed into relatively non-toxic substances which are precipitated.

Oxygenator (n.): 1. An apparatus for oxygenating pond water, 2. Aquatic plants (such as water milfoil, canadian pond weed, water crowfoot, hornwort, etc.) help to enrich the surrounding water with oxygen.

Oxygen debt (n.): A temporary oxygen shortage in the tissues arising from exercise.

Ozone (n.): An unstable, pungent, toxic form of oxygen with three atoms in its molecule, formed in electrical discharges or by ultra-violet light.

Packaging (n.): The act of designing and putting into bundles or materials or products for packing.

Pancreas (n.) (Pl. Pancreata): A large compound acinotubular gland located behind the stomach. The pancreas is both an exocrine and an endocrine organ. The exocrine glands produce digestive enzymes and empties ino the duodenum. Scattered throughout the exocrine glandular tissue are masses of cells called *islets of Langerhans*, endocrine glands that secrete hormones.

Parasite (n.): An organism which is totally dependent upon another organism (the host) for its energy, is very closely associated with its host and which often causes its reduced growth or reproduction but only rarely kills it. Parasitic (adj.), Parasitise (v.).

Parasitism (n.): The system in which one species gains a living at the expense of another, usually either within it or feeding from its surface or blood.

Parts per billion (ppb): Number of parts per one billion (10^9) parts.

Parts per million (ppm) : Number of parts per one million (10^6) parts.

Pathogen (n.): Organisms that cause diseases. Pathogenic (adj.).

Pathogenesis (n.): The manner of development of a disease.

Pathology (n.): The branch of science concerned with the causes and effects of diseases.

Pectoral fin (n.): Paired fins, one on each side of the body behind the gills.

Pedigree (n.): The record of descent of an animal showing it to be pure-bred. Pedigreed (adj.).

Pelvic fin (n.): Paired fins on the underside of the body in front of the anal fin.

Pen (n.): A small, specially-designed and restricted enclosure in which fish, sheep, pigs or other farm animals are kept for certain period of time.

Pen culture (n.): Pen culture is a method of rearing finfish and shellfish in a special type of rearing facility. The pen is a single- or double-layered enclosure made of bamboo slit, polyethylene nets (mesh size 0.5-1.0 cm), nylon ropes, poles of casuarina and palmyra and take different shapes such as circular, square, hexagonal or rectangular. The term pen, cage, and enclosures are virtually synonymous.

Perch (n.): 1. A freshwater fish with a high spiny dorsal fin, dark vertical bars on the body, and orange on the lower fins. (*Perca fluviatilis* (Europe), *P.flavescens* (Yellow Perch in

North America), and other species, 2. Used in names of similar or related species such as *Anabus testudineus* (Climbing perch in India and adjacent countries).

Perennial (adj.): Lasting for a long time, enduring or continually recurring.

Periphyton (n.): An assemblage of sessile or attached organisms grown on diverse aquatic supports.

Persistence (n.): A characteristic of a non-biodegradable material that remains essentially unchanged after being introduced into the environment. Persistent (adj.), Persistently (adv.), Persist(v.).

pH (n.): Measure of the degree of acidity or alkalinity of water in terms of the negative logarithm of hydrogen ion concentration.

Phagocyte (n.): A cell which engulfs and absorbs bacteria and other small particles. Phagocytic (adj.).

Pharyngeal teeth (n.): Projections in the throat of cyprinids and some other fish, which help to break down food.

Pharynx (n.): A part of the gut of vertebrates between the mouth and the esophagus.

Phenol (n.): A mildly acidic toxic white crystalline solid used in chemical manufacture. Phenolic (adj.).

Phenotype (n.): The observable characteristics of an organism resulting from the interaction of its genotype with the environment. Phenotypic (adj.), Phenotypically (adv.). Cf. *Genotype*.

Pheromone (n.): A chemical substance which passes through the environment from one animal to another, either as a signal which elicits a behavioral reaction in a recipient or as a messenger leading the recipient to show a physiological change. Most examples of the latter concern about the smell of the opposite sex.

Phosphate (n.): A salt or ester of phosphoric acid, typically containing the anion Po_4^{3-} or the group ---OPO $(CH)_2$.

Phosphorus (n.): The chemical element of atomic number 15, a poisonous combustible non-metal which exists as a yellow waxy solid which ignites spontaneously in air and glows in the dark (white phosphorus) and as a less reactive form used in making matches (red phosphorus) (Symbol: P).

Photochemical (adj.): Relating to or caused by the chemical action of light.

Photodegradable (adj.): Capable of being decomposed by the action of light, especially sunlight. Photodegradation (n.).

Photosynthesis (n.) : The process by which green plants use sunlight to synthesize nutrients from water and carbon di-oxide.

Physiographic (n.): A description of the physical nature of natural features; later it becomes synonymous with physical geography. Still later, the term has been restricted to the description and origin of landforms; in this sense, it is replaced by geomorphology.

Physoclistous swim bladder (n.): Refers to the absence of connection of swim bladder in some fishes with the esophagus in the adult stage. Cf. *Physostomous swim bladder.*

Physostomous swim bladder (n.): Refers to the connection of swim bladder in some fishes with the esophagus which retains throughout life. Cf. *Physoclistous swim bladder.*

Phytoremediation (n.): The action of remedying aquatic ecosystem damage with the help of macrophytes.

Pike (n.): A long-bodied predatory freshwater fish with long teeth. (*Esox lucius* and related species).

Pineal gland (n.): It is also termed as Pineal body. A pea-sized conical mass of tissue behind the third ventricle of the brain, secreting a hormone-like substance in many vertebrates.

Pituitary gland (n.)(pl. Pituitaries glands) : The major endocrine gland a pea-sized body attached to the lower surface of the hypothalamus of the brain that is important in controlling growth and the functioning of the other endocrine glands. In fish culture, however, pituitary glands are widely used in induced breeding of farmed fish.

Plankton pulse (n.): Vibration of plankton population with persistent rhythm.

Plasmid (n.): Plasmids are defined as autonomous compounds, whose genomes exist in the cell as extra-chromosomal units. They are self-replicating circular duplex DNA molecules, which are maintained in a characteristic number of copies in a bacterial cell, yeast cell or in organelles found in eukaryotic cells. These plasmids can be single copy plasmids (that are maintained as one plasmid DNA per cell) or multicopy plasmids (which are maintained as 10-20 genomes per cell). Some other plasmids are under relaxed replication control, thus permitting their accumulation in large number (up to 1,000 copies per cell). Due to their increased yield potential, they are used as cloning vectors.

Plateau (n.): An area of fairly level high ground.

Pollution (n.): Contamination of water, air, soil,etc. with harmful or poisonous substances. Pollutant (adj. & n.), Pollute (v.).

Polyacrylamide (n.): A synthetic resin made by polymerizing acrylamide, especially a water-soluble polymer used to form or stabilize gels and as a thickening or clarifying agent.

Polyandry (n.): Polygamy in which a female fish exhibits courtship behavior with more than one male. Polyandrous (adj.).

Polyculture (n.): The use of two or more species to utilize two or more trophic levels within a culture system.

Polygene (n.): Any of a group of genes influencing a quantitative trait.

Polyploid (adj.): Of or denoting cells or nuclei containing more than two homologous sets of chromosomes.

Polyploidy (n.). Artificial polyploidy, which can be induced by the use of chemicals (notably colchicine), is of economic importance in producing hybrids with desired traits.

Pond soil amendment (n.): Any material, such as lime, fertilizers, manures, gypsum, or synthetic conditioner, that is worked into the pond soil to make it more ameneble to the growth of benthic organisms, plankton, and fish.

Population genetics (n.): The branch of genetics concerned with the community of inter-breeding organisms.

Potasium (n.): The chemical element of atomic number 19, a soft silvery-white reactive metal of the alkali-metal group. It is an important plant nutrient. (Symbol : K).

Potential (adj.): 1. Having the capacity to develop into something in the future. *n.* latent qualities or abilities that may be. developed and lead to future success or usefulness,2. The possibility of something happening or of someone doing something in the future. Potentiality (n.), Potentially (adv.).

Potent (adj.): Having great power, influence, or effect. Potence (n.), Potently (adv.), Potency (n.) (pl. potencies).

Power filter (n.): An electric pump is used to drive water through the filtration media. There are two basic types such as internal power filter and external power filter. While the internal types sit inside the pond/tank, the external ones are housed outside the pond/tank and are generally used for larger ponds. An integral electric pump draws water through the filter unit, which contains one or several chambers housing filtration media inserts. After passing through the media, the clean water is returned to the main tank.

Prawn (n.): A marine/freshwater crustacean which resembles a large shrimp.

Precipitation (n.): 1. The action or process of precipitating a substance from a solution, 2. Rain, snow, sleet, or hail that falls to or condenses on the ground. Precipitate (v.), Precipitable (adj.), Precipitately (adv.).

Predator (n.): An animal which feeds upon populations of other animals (the prey). .

Preservative (n.): A substance used to preserve foodstuffs, wood, or other materials against decay. *Adj.* Acting to preserve something.

Preserve (v.): Treat (food) to prevent its decomposition. Preservable (adj.), Preservation (n.).

Prey (n.): An animal hunted and killed by another for food.

Price (n.): 1. The amount of money charged for a product or service or the sum of the values that consumers exchange for the benefits of having or using the product or service, 2. The amount of money expected, required, or given in payment for something.

Probiotic (adj.): Denoting a substance which stimulates the growth of micro-organisms, especially beneficial ones.

Process (n.): 1. A series of actions taken towards achieving a particular end, 2. Perform a series of operations to change or preserve food products.

Product (n.): Anything that can be offered to a market for attention, use, or consumption that might satisfy the need.

Productivity (n.): 1. The state or quality of producing something, 2. The effectiveness of productive effort, 3. The fertility or capacity of a given habitat or area, 4. The efficiency with which aquatic food species are produced through state-of-the-art technology. Productive (adj.), Production (n.).

Profit (n.): The excess of total revenue over total cost during a specified period of time.

Promoter (n.): A DNA region in front of the coding sequence of a gene which binds RNA polymerases and hence signals the start of a gene.

Propagation (n.): Breeding by natural processes (natural propagation) or artificial processes (artificial propagation) from the parent stock. Propagate (v.).

Prophylactic (adj.): Intended to prevent disease.

Prostaglandin (n.): Any of a group of 20-carbon compounds containing biologically active organic acids that are synthesized in the body from unsaturated fatty acids and are responsible for various biological effects.

Protandrous (adj.): 1. The maturation of male organs before the female is receptive, 2. A hermaphroditic animal having the male reproductive organs come to maturity before the female. Cf. *Protogynous.*

Protein* (n.) : Any of a class of organic compounds consisting of long chains of amino acids, having structural and functional roles in organisms and constituting an important part in the diet.

Protocol (n.): A formal record of scientific experimental observations – a procedure for carrying out a scientific experiment or a course of medical treatment.

Protogynous (adj.): 1. The maturation of female organs before the male organs liberate their contents, 2. A hermaphroditic animal having the female reproductive organs come to maturity before the male. Cf. *Protandrous.*

Pulsed direct current (n.): An electric current flowing in one direction having a single vibration.

---

* Protein molecules consist of one or a small number of polypeptide chains each of which is a linear polymer of several hundred amino acids linked through their amino and carboxylate groups by peptide bonds. The amino acid side chains have a positive or negative charge, a short aliphatic chain or an aromatic residue. Because the 20 amino acids can be arranged in nearly any sequence, the potential diversity of structure and function is enormous.

Pulverize (v.): Reduce to fine particles.

Qualitative phenotype (n.): Identification of the phenotypic characters present in an organism.Cf. *Quantitative Phenotype.*

Quantitative phenotype (n.): Measurement of the particular phenotypic characters such as length, weight, behavior, fecundity, etc. present in an organism. Cf. *Qualitative phenotype.*

Quarantine (n.): A state, period, or place of isolation for animals that have arrived from elsewhere or been exposed to contagious disease.

Raceway (n.): A Water channel, especially an artificial one for rearing fish.

Rad (n.): A unit of radiation dosage, equivalent to the absorption of 100 ergs of energy/ g of biological tissues.

Radiation (n.)*: 1. The action or process of radiating. 2. Energy emitted as electro-magnatic waves or sub-atomic particles.

Radio-isotope (n.): A radio-active isotope. An isotope is an atom with identical atomic number but different mass number.

Rare (adj.): Occurring very infrequently existing in small numbers and so of interest or value.

Recalcitrant (adj.): Compounds which are resistant to biodegradation. Recalcitrance (n.), Recalcitrantly (adv.).

Recessive (adj.): Relating to or denoting heritable characteristics controlled by genes which are expressed in offspring only when inherited from both parents. Cf. *Dominant.*

Recirculation (n.): Recurring movement of water through biofilter every 10 or 15 days in fish culture ponds to remove waste products. Recirculate (v.).

Recombinant DNA (n.): DNA that contains segments from two different species.

Recycle (v.): 1. Conversion of waste water/sewage through several treatment processes to non-toxic forms which can be reused in agriculture and aquaculture, 2. Return (material) to a previous stage in a cyclic process. Recyclability (n.), Recyclable (adj.).

Recycling (adj.): Recovery and processing of materials after they have been used, which enables them to be reused. Sewage and sludge, for example, can be treated with

---

* A process by which energy is emitted from a source through the medium. Radiation consists of a flow of atomic or sub-atomic particles or of waves. Familiar example is light (a form of electro-magnetic radiation) and can be described as waves with a range of frequencies and intensities. In fish processing industries, the electro-magnetic radiation with adequate intensities is now being widely used for long-term preservation of fish products. Ionizing radiation refers to radiation that is sufficiently energetic to strip electrons from atoms or molecules.

several processes into their constituents which form the materials for use in fish culture.

Red Data Book (n.): A compilation of data on species either threatened or endangered or rare which is maintained by the International Union for Conservation of Nature and Natural Resources (IUCN).

Refuge (n.): A place or state of safety from danger or trouble.

Remote sensing (n.): The gathering and recording of information of the earth by satellite or high-flying aircraft. These techniques involve the use of aerial photography, multispectral imagery, infra-red imagery, and radar.

Reproduction (n.): 1. The process whereby populations increase in number, 2. The natural process by which organisms produce new individuals.

Research (n.): The systematic investigation into and study of materials and sources in order to establish facts and reach new conclusions.

Research and Development (n.): (In any industry) work directed towards innovation in and improvement of products and processes.

Reservoir (n.): A large artificially constructed impoundment which is mainly used for irrigation, power generation, sport fishery, and development of fishery sector in many countries of the world.

Resource (n.): A stock or supply of aquatic food items that can be drawn in full potential when needed.

Respiration (n.): The processes in living organisms involving the production of energy, typically with the intake of oxygen and the release of carbon di-oxide, from the oxidation of organic substances.

Reverse osmosis (n.): 1. A process by which salts and other substances are removed by forcing the water through a semi-permeable membrane under a pressure that exceeds the osmotic pressure. As a result, flow occurs in reverse direction, 2. A process used by some water softening units, in which water is forced through a membrane to remove mineral salts.

Risk (n.): A situation involving exposure to danger.

Roach (n.): A common freshwater fish of the carp family.

Rotifers (n.): A Component of plankton, used as fish food for rearing fish.

Rudd (n.): A European freshwater fish of the carp family with a silvery body and red fins (*Rutilus erythropathalmus*).

Run-off (n.): The draining away of rainfall or other liquid from the surface of an area. Run-off water contains a lot of undesirable substances that cause disastrous effects on aquatic life.

Salinity (n.): A measure of the total amount of dissolved salts in water and is expressed as ppt.

Salinization (n.): The process of accumulation of salts in soil/water.

Salmon (n.): A large edible fish that matures in the sea and migrates to freshwater streams to spawn. [*Salmo salar* (Atlantic salmon) and genus *Oncorhynchus* (Pacific salmon)].

Salt fish (n.): Fish that has been preserved in common salt.

Sampling (adj.): The selection of small groups of entities to represent a large number of entities in statistics. In random sampling, each factor of an ecosystem has an equal chance of being selected as part of the sample. In stratified random sampling, the factor is divided into strata, each of which is randomly sampled and the samples from different strata are pooled. In systematic sampling, factors are chosen at fixed intervals.

Sanctuary (n.) (pl. Sanctuaries): Area constituted by competent authority in which killing, fishing, or capturing of endangered species of fish is prohibited except by or under the control of the authority in the department concerned for the management of the area.

Sanitation (n.): 1. Conditions relating to health of the culture system, 2. Conditions relating to public health.

Satellite (n.): An artificial body placed in orbit round the earth or another planet to collect information or for communication.

Saturate (v.): Cause to combine with, dissolve, or hold the greatest possible quantity of another substance.

Saturation (n.): The action of saturating or state of being saturated.

Saturation point (n.): The stage beyond which no more can be absorbed or accepted.

Scale (n.): Each of the small overlapping horny or bony plates forming the body covering of fish and reptiles. The most common types of scales include cycloid and ctenoid scales, characteristic feature of bony fish.

Scallop (n.): A bivalve having the edge of its shell in the form of a series of curves.

Seasonal (adj.): Relating to or characteristic of a particular season of the year. Seasonally (adv.).

Secondary sexual characters (n.): External features of a sexually mature fish or animal that, although no directly involved in copulation, are significant in reproductive behavior. The development of such features is controlled by sex hormones; in fishes, they are seasonal or the body color of male tilapia.

Sediment (n.): Materials carried in particles by water and deposited on the bottom of fish culture ecosystem. Sedimentary (adj.), Sedimentation (n.).

Sedimentation (n.): The settling of the solid particles through a liquid either to produce a concentrated slurry from a dilute suspension or to clarify a liquid containing solid particles.

Seedling (n.): A young plant raised from seed and not from a cutting.

Selection (n.): The process by which some individuals come to contribute more offspring than others to form the next generation in natural selection through intrinsic differences in survival and fertility, in artificial selection through the choice of parents by the breeder. Selective (adj.), Selectively (adv.), Selectiveness (n.).

Semi-intensive (adv.): Ponds are stocked with selected post-larvae reared in hatcheries, and the larvae are not dependent on the natural production of food and formulated feed as per nutritional requirement is routinely applied.

Semipermeable (adj.): Permeable only to certain substances, especially allowing the passage of a solvent but not of the solute.

Sex-determination (of fish) (n.): The method by which the distinction between males and females is established in a fish species. It is generally under the genetic control.

Sex reversal (n.): Gradual change of the sexual characters of an individual during its life time from female or vice versa.

Sexual dimorphism (n.): Marked differences between the males and females of a species, especially differences in superficial characters such as shape, size, color, etc.

Sexual selection (n.): Natural selection arising through preference by one sex for certain characteristics in individuals of the other sex.

Shellfish (n.): An aquatic shelled mollusc or crustacean, especially an edible one.

Shoal (n.): 1. A group of fish of the sane species that swims together, 2. A large number of fish swimming together. v. (of fish) form shoals.

Shock (n.): 1. An acute medical condition associated with a fall in blood pressure, caused by loss of blood, severe burn, sudden emotional stress, etc., 2. A sudden upsetting event or experience, or the resulting feeling.

Shrimp (n.) (pl. shrimps or same): A small free-swimming, typically marine edible crustacean with ten legs. .(*Pandalus*, *Crangon*, and other genera).

Sibling species (n.): The species which are morphologically similar but reproductively isolated.

Siltation (n.): The process by which fine sand, clay or other materials are carried by running water and are deposited as a sediment. Silt (n.), Silty (adj.), Silting (n.).

Silting (n.): The decomposition of water-borne sediments in stream channels, lakes, reservoirs, or on flood-plains, usually resulting from a decrease in the velocity of the water.

Sink (n.): 1. Insert beneath a surface, 2. A pool or marsh in which a river's water disappears by evaporation or percolation. technical: A body or process which absorbs or removes a particular component from a system.

Siphon (n.): A tube for removing water from a pond/tank/aquarium.

Slaughterhouse (n.): A place where animals are slaughtered for food.

Slope (n.): 1. The degree of deviation of a surface from horizontal, measured in a numerical ratio, per cent, or degrees, 2. A surface of which one end or side is at a higher level than another *v.* be inclined from a horizontal or vertical line, slant up or down. Sloppy (adj.)

Smoke (n.): A visible suspension of carbon or other particles in the air, typically one emitted from a burning substance. May be used for fish preservation. Smoking (adj. & n.), Smoked (v.).

Society (n.): 1. The aggregate of people living together in a more or less ordered community > a particular community of people, 2. (of farmer) living in a community having shared customs, laws, and farm organizations.

Socio-economic (adj.): Relating to or concerned with the interaction of social and economic factors. Socio-economically (v.).

Soft water (n.): Freshwater containing less than 100 mg/litre of dissolved calcium and magnesium mineral salts.

Soil (n.)* : The upper layer of earth in which plants grow, a black or dark brown material typically consisting of organic remains, clay, and rock particles.

Sorption (n.): Absorption and adsorption considered as a single process.

Sorting (n.): A process by which harvested fish are separated size-wise from a mixed group.

Spawning (v.): 1. The process of egg-laying and fertilization, 2.The production of eggs and sperm in those animals especially most fishes, where fertilization is external and eggs as well as sperm are just released into the water. Spawn (n.).

Spawning pit (n.): An area of substrate excavated by some species of fish, in which they spawn or guard their fry.

---

* The biologically active, porous medium that has developed in the uppermost layer of the earth's crust. Soil serves as a natural reservoir of water and nutrients, and as a participant in the cycling of carbon, phosphorus, nitrogen, and other elements through the global ecosystem. Soil has developed through the weathering of solid materials. Although the bulk of soil consists of mineral particles, fertile soils of freshwater ecosystems essentially consist of undecomposed or partially decomposed biomass as well as humus, an array of organic compounds derived from the breakdown of biomass. Such fertile soils are ideal for fish culture.

Species (n.): A group of similar living organisms whose members can interbreed to produce fertile offspring but which cannot breed with other species.

Species richness (n.): Number of species in a community of organisms. It is also termed as *ecological diversity*.

Spectrophotometry (n.): A technique for measuring the intensity of light in a part of the spectrum, especially as transmitted or emitted by particular substances. Spectro-photometrically (adv.), Spectrophotometric (adj.).

Sphincter (n.): A ring of muscle surrounding and serving to guard or close an opening.

Spoilage (of fish) (n.): The decay of fish or fish products by micro-organisms which ultimately becomes unfit for human consumption. Spoil, Spoiling (v.).

Stability (n.): The ability of an ecosystem to resist change. Stable (adj.).

Stenohaline (adj.): Describing an organism which can tolerate only a narrow range of salinity in its environment. Cf. *Euryhaline*.

Stenothermal (adj.): Describing an animal which is able to tolerate only a narrow range of temperature in its environment. Cf. *Eurythermal*.

Sterility (n.): Not able to produce offspring.

Stickleback (n.): A small freshwater or coastal fish with sharp spines along its back. (*Gasterosteus aculetus* (three- spined stick and other species.)

Stimulant (n.): A substance that acts to increase physiological or nervous activity in the body. Stimulate (v.).

Stimulus (n.) (pl. Stimuli): Refers to some aspects of the environment, either internal or external to the individual, which generates some response, although this is not always an immediate response not an easily observable one. Stimulant (n.), Stimulate (v.), Stimulation (n.), Stimulating, Stimulative (adj.).

Stock (n.): A term that varies with context and ranges in meaning from a discrete, largely reproductively isolated sub-population to an aggregation of populations managed as a unit. Alternatively, a genetic strain.

Stomach (n.): That part of the digestive tract between the esophagus and the intestine.

Strain (n.): Distinct breed, stock, or variety of an animal, plant, or other organism.

Strategic (adj.): Forming part of a long-term plan or to achieve a specific purpose.

Strategic planning (v.): The process of developing and maintaining a strategic fit between one farm's goal and capabilities and its changing marketing opportunities. It involves defining a clear farm mission, setting of supporting objectives, designing a sound business portfolio, and co-ordinating functional strategies.

Strategies (n.): A plan designed to achieve a particular long-term aim.

Strategy (n.) (pl. Strategies): A state-of-the-art technique used in fish culture or aquaculture sector designed to achieve a particular long-term aim at producing robust fish in appreciable quantities. Strategic (adj.).

Stratify (v.): Form or arrange into strata. Stratification (n.), Stratified (adj.).

Stream (n.): A small and narrow river.

Stress (n.): 1. Excess and adversive environmental factors which produce physiological responses in the individual, 2. The sum of the biological reactions to any adverse stimulus that leads to disturb an organism's physiological stability.

Subsistance (n.): Denoting or relating to production at a level sufficient only for one's own use or consumption, without any surplus for trade: *Subsistance agriculture, Subsistance aquaculture.*

Submersible (adj.): Designed to operate while submerged. Submerse (v.).

Subsistance level (n.): A standard of living (or wage) that provides only the bare necessities of life.

Sub soil (n.): The soil lying immediately under the surface soil.

Substrate (n.): The surface or material on which an organism lives, grows, or feeds.

Sump (n.): A depression in the floor of a cave in which water collects.

Super-intensive (adj.)*: (Of agriculture/Fish culture) aiming to achieve maximum production to a great or extreme degree within a limited area.

Super market (n.): Large, low-cost, low-margin, high volume and self-service store that carries a wide variety of fish products and other food products.

Surimi (n.)** : A paste made from minced fish, used to produce imitation crab-meat and lobster meat. In Japan, it is termed as *minced flesh.*

Survey (v.): 1. Examine and record of the area and features (of land/water) so as to construct a map, plan, or description, 2. Look carefully and thoroughly at.

---

* This system of farming is carried out in cement cisterns with hundred per cent water exchange. Stocking density of fish is more than that of intensive system. Continuous water exchange through biological filter system, constant aeration, stocking manipulation, and feeding of encapsulated pelleted diet are some of the important management practices generally adopted.

** A Japanese term which is termed as *wet concentrate of proteins of low-value fish muscle* (mostly of marine fish), which is mechanically deboned and water-washed fish flesh. After heating and gutting, fish meat is separated which is termed as *minced meat.* The meat is thoroughly washed with water followed by straining and dewatering and produce raw Surimi. This raw surimi is mixed with cryoprotectants and allow to freeze which is termed as *frozen surimi.* The frozen surimi is ground with salt and ingredients and after heat processing, surimi-based products are obtained. Heat processing essentially involves steaming, broiling, boiling, and deep fat frying to get the shape and develop the texture as well as appearance.

Sustain (v.): Strength or support physically or mentally. Sustainable (adj.), Sustained (v.), Sustainability (adv.).

Sustainable (adj.): Judicial management of natural resource wealth for food and trade assuring permanent and timely preparation for fish biomass.

Sustainable development (adj.): It is a notion that aims at the utilization of aquatic resources in such a manner that the utilization (but not exploitation) is well-balanced. Almost all utilization of aquatic resources for development purposes would invariably result in some negative impacts on the ecosystem.

Sustained yield (n.): A level of exploitation or agricultural/aquacultural production which is maintained by restricting the quantity harvested to avoid long-term depletion. Sustain (v.), Sustainable (adj.).

Sustenance (n.): The sustaining of someone or something in life or existence.

Swamp (n.): An area separated with water for most time in which soil surface is not deeply submerged. Swampy (adj.).

Swim bladder (n.)*: A gas-filled sac in a fish's body, used to maintain buoyancy. It is also termed as *gas bladder* or *air bladder*.

Symptom (n.): 1. A feature which indicates a condition of disease, 2. A condition of an undesirable situation.

Synchronize (v.): Cause to occur or operate at the same time or rate.

Synchronous (adj.): Existing or occurring at the same time. Synchronously (adv.).

Syndrome (n.): 1. A group of symptoms which consistently occur together, 2. A characteristic combination of behavior.

Synergism (also Synergy) (n.): The phenomenon in which the combined of two chemical substances produce a greater effect than would be expected from adding the individual effects of each substance. Synergistic (adj.). Cf. *Antagonism*.

System management (n.): Management of fish culture ecosystem relating to a set of things working together as a mechanism or inter-connecting network.

Tamarind (n.): A sticky brown acidic pulp from the pod of a tropical African tree, *Tamarindus indica*, of the pod family and used as a flavoring in Asian cookery.

Tandem (n.): A group of 2 or 3 fish species working together for several generations.

Tannin (n.) : A yellowish or brownish bitter-tasting organic substance present in some galls, barks, etc.

---

* An air-filled sac lying above the alimentary canal in most bony fish that regulates the buoyancy of fish. Air enters or leaves the bladder either via the pneumatic duct opening into the esophagus or stomach or via the capillary blood vessels, so that the specific gravity of the fish always matches the depth at which it is swimming. This makes the fish weightless, so less energy is required for locomotion.

Tapeworm (n.): A parasitic flatworm with a long ribbon-like body, the adult of which lives in the intestines (Class: Cestoda, many species).

Tariff (n.): A tax or duty to be paid on a particular class of imports or exports.

Technology (n.) (pl. Technologies) : The application of scientific knowledge for practical purposes. Technology includes the use of fish and fish products for human consumption and techniques of production to make fish culture strategy more viable and productive. Fisheries technology, however, focuses on processing and preservation of fish. Technological (adj.), Technologically (adv.).

Technology Transfer (n.): The transfer of new technology from the originator to a secondary user.

Teleost (n.): A fish of a large group (Division: Teleostei) of the sub-class Actinopterygii that comprises most bony fishes (the ray-finned fishes) apart from sturgeons, lungfishes, and some other kinds.

Telolecithal (n.): Refers to eggs where eggs contain moderate to enormous quantities of yolk, the yolk being more, is concentrated at one hemisphere termed as *vegetal pole* while the other end is called as *animal pole*.

Temperate zone (n.): Each of the two belts of latitude between the torrid zone and the northern and southern frigid zones.

Temperature (n.): The degree or intensity of heat present in a substance or object.

Tench (n.): A freshwater fish of the carp family (*Tinca tinca*), popular with anglers.

Teratogenic (adj.): Causing abnormal development of the embryo. Teratogenesis, Teratogene (n.).

Terms of trade (n.): The relationship between the export price and import price of a commodity. It may be unfavorable or favorable to a country. When the export price is greater than the import price, terms of trade are favorable and if import price is greater than the export price, terms of trade are unfavorable.

Terrace (n.): 1. A narrow level, plain bordering a lake, river, or the sea. Rivers sometimes are bordered by terraces at different levels, 2. A raised, more or less level of earth generally constructed on or nearly on a contour and designed to make the land suitable for farming and prevent accelerated erosion by diverting water from channels of concentration. Sometimes it is also called *diversion terrace*.

Testosterone (n.): A steroid hormone secreted by testes stimulating development of male secondary sexual characteristics such as body color in fishes.

Thalamus (n.): Part of fore brain acts as a central switch board for sorting out incoming sensory information and directs them appropriate areas of cerebrum.

Thermal pollution (n.): Pollution resulting from the discharge of heated waste waters at temperatures that can be harmful to aquatic life. Thermal pollution results in an increase in the temperature of the water. High temperature depletes the content of its oxygen and is very baneful to fish and other aquatic organisms.

Thermocline (n.): A temperature gradient in a lake or other body of water, separating layers at different temperatures.

Tiller (n.): An implement or machine for tilling soil; a plough or cultivator.

Titrate (v.)*: Ascertain the amount of a substance in (a solution) by measuring the volume of a standard reagent required to react with it.

Tolerance limit (n.): The extreme value of any physical or chemical substance beyond which the species cannot survive.

Topography (n.): 1.The arrangement of the natural and artificial physical features of an area, 2. Refers to the general configuration of a land surface including its relief and the position of its natural and man-made features.

Toxic (adj.): Poisonous. > Relating to or caused by poison. Toxic (n.), Toxicity,(n.).

Toxicant (n.): A toxic substance introduced into the environment, such as a pesticide.

Toxicology (n.): The branch of science concerned with the nature, effects, and detection of poisons. Toxicological (adj.), Toxicologically (adv.).

Toxicosis (n.): A condition of poison.

Toxin (n.): A poison produced by a micro-organism or other organism and acting as an antigen in the body.

Trace element (n.): Minerals, such as iron, that an organism needs in small amounts to ensure its well-being.

Trade (n.): The buying and selling of any item or products and services.

Traditional (adj.): A simplest form of fish culture with minimum or no inputs and management. Traditionally (adv.).

Trait (n.): 1. A stable and enduring attribute of an animal which varies from one individual to another, 2. Characters of an organism. .

Tranfection (n.): The process wherein a virus infects another organism and the virus genome is then expressed in the host cells. Transfect (v.).

Transcription (n.): The process by which RNA polymerase produces single-stranded RNA complementary to one strand of the DNA or rarely RNA.

---

* Titration is a method of volumetric analysis in which a volume of one reagent (the titrant) is added to a known volume of another reagent slowly from a burette until an end point is reached. The volume added before the end point reached is noted. If one of the solutions has a known concentration, that of the other can be calculated.

Transformation (n.): Introduction of a recombinant DNA of a vector into the suitable host cell for expression of foreign DNA. Transformation can occur by using $CaCl_2$ at low temperature that prevents foreign DNA from degradation by restriction enzymes present in *E. coli*.

Transgenic (adj.)* (pl. Transgenics): Containing genetic material into which DNA from a different organism has been artificially introduced. Transgene (n.).

Transposon (n.): A segment of DNA that can be translocated as a whole from a site in one genome to another site in the same genome or to a different genome.

Trauma (n.): A physical injury or wound caused by external force or violence. The majority of deaths occur in the first several hours after event.

Trematode (n.): A flatworm of the Class Trematoda, which comprises those flukes that are internal parasites.

Trench (n.): 1. A long, narrow, ditch dug by troops to provide shelter from enemy, fire, 2. A long, narrow, deep depression in the pond/ocean bed, typically running parallel to a plate boundary and marking a subduction zone.

Trickle filter (n.): The filter provides sophisticated biological and mechanical filtration. Water is drawn up from the tank/pond and is sprayed over a stack of different filter media, through which it trickles before flowing back into the pond/tank. Spraying also oxygenates the water, improving bacterial action within the filter.

Trout (n.): Any of several priced edible and game fish of the family Salmonidae, native to Northern Hemisphere but widely introduced elsewhere. Although trout species inhabit cool freshwaters, a few sea trout such as currhroat trout migrate to the sea before springs. The genus *Oncorhynchus* includes salmon and several trout species. Trout species are greatly in anatomy, color, and habitats. Most live among submerged objects or in riffles and deep ponds, eating insects, small fishes and their eggs and crustaceans.

Trophic (adj.): Relating to feeding and nutrition.

Tropic (n.): Each of two corresponding circles on the celestial sphere where the sum appears to turn after reaching its greatest declination. Tropical (adj.).

Tumour (n.): A swelling of a part of the body, generally without inflamation, caused by an abnormal growth of tissue, whether benign or malignant.

---

* Transgenic is used to describe animals such as transgenic fish which are derived from embryos into which isolated genomic DNA from another species has been introduced at an early stage of development. Such foreign genes may be incorporated into the nucleus and chromosomes by micro-injection so that the animal can express the foreign gene product. The gene for fish growth hormone, for example, can be inserted into fertilized eggs to produce fish with cells that produce fish growth hormone. Such fish offers considerable potential; for example, transgenic fish may be engineered to produce disease resistance power or extra growth hormone for improved flesh production.

Turbidity (n.): Reduction in the transparency of a pond water due to presence of silt and clay particles, organic matter, and plankton. Turbid (adj.), Turbidly (adv.), Turbidness(n.).

Turbulent (adj.): Relating to or denoting irregular and disordered flow of fluids. Turbulence (n.), Turbulently (adv.).

Undergravel filter (n.): A filter by which a colony of beneficial bacteria establishes itself on the filter bed. The bacteria break down organic waste substances produced by the fish, as water is drawn down through the filter bed.

Ureter (n.): The duct by which urine passes from the kidney to the bladder or cloaca.

Vaccine (n.): Therapeutic substance, treated to lose its virulence and containing antigens derived from one or more pathogenic organisms, which on administration to fish or other vertebrate animals, will stimulate active immunity and protect against infection with these or related organisms.

Vaccination (n.): Production of active immunity by administration of a vaccine. Vaccine is a substance injected into the body to cause it to produce antibodies and so provide immunity against a disease.

Variation (n.): A change or slight difference in condition, amount, or level.

Variety (n.) (pl. Varieties): 1. A number of things of the same general class that are distinct in character or quality, 2. A sub-species or cultivar. Varietal (adj.).

Vector (n.): 1. An organism that transmits a particular disease or parasite from an animal or plant to another, 2. A bacteiophage or plasmid which transfer genetic material into a cell.

Vent (n.): The urogenital opening which is located close to the anal fin in fish.

Vertebra (n.) (pl. Vertebrae): Each of the series of small bones forming the backbone or vertebral column.

Virgin water (n.): A water body that has not been significantly disturbed from its natural environment.

Virus (n.): A particle that is too small to be seen with a light microscope or to be trapped by filters but is capable of independent metabolism and reproduction within a living cell. A mature virus (a virion) ranges in size from 20 to 400 nm in diameter. It consists of a core of nucleic acid (DNA or RNA) surrounded by a protein coat termed as *capsid*.

Volatile (adj.): (of a substance) easily evaporated at normal temperatures. Volatilization (n.), Volatilize (v.).

Volition (n.): The faculty or power of using one's will. Volitional (adj.), Volitionally (adv.).

Volitional spawning (n.): (of fish) self-initiated or voluntary actions of fish to lay eggs and milt in the absence of applied stimuli.

Vulnerable (adj.): 1.Exposed to the risk of being attacked or harmed, either physically or emotionally, 2. Animals are under threat of actually declining in number, or animals which have been seriously depleted in the past and have yet to be recovered. Vulnerability (n.), Vulnerably (adv.).

Wage (n.): A fixed regular payment for work, typically paid on a daily or weekly basis.

Warehouse (n.): A large building where raw materials or manufacture goods may be stored. Warehousing (n.).

Waste (adj.): 1. Unusable or unwanted material. Wasteful(adj.) 2. Discarded as no longer useful to fish culture.

Waste-disposal unit (n.): An electrically operated device fitted to the waste pipe of a kitchen sink for grinding up food waste.

Waste water (n.): Discarded water as no longer required. Efforts to limit the environmental impact of waste, from the point of production through recovery processes to disposal and recycling are known as *waste water management*.

Water boatman (n.): A predatory aquatic bug that swims on its long back legs as oars.

Water hyacinth (n.): A free-floating tropical American water plant, widely introduced as an ornamental and in some warmer regions a serious weed of waterways.

Water-logged (adj.): Saturated with or full of water.

Water management (n.): The sum total of different types of operations, maintenance of water quality, farming practices, fertilizers, lime, and other treatments conducted on or applied to a water body for the production of aquatic species.

Water scorpion (n.): A predatory water bug with grasping forelegs.

Watershed (n.): 1. The total and area contributing surface or ground water to a lake, river, or drainage basin, 2. A line of division between river systems, 3. An area or ridge of land that separates waters flowing to different rivers, basins, or seas.

Weathering (n.): Wear away or change in form or appearance by long exposure of rock or other material.

Wels (n.) (pl. same): A very large freshwater catfish found from central Europe to central Asia (*Silurus glanis*).

Wet bund (n.): A perennial pond that is situated on the slope of a large catchment area of undulating terrain. Embankments are made with an inlet towards the areas or higher elevation and an outlet at the opposite end. In summer, when there is a shortage of water, the deeper area retains some amount of water containing carp breeders. In monsoon season, however, fresh rain water from the catchment areas rushes into the bund in the form of streamlets and fills it. The outlet is blocked by mud.

Wet fish (n.): Fresh fish, as opposed to fish which has been frozen, cooked, or dried.

Wetland (n.): Swampy or marshy land.

Whirling disease (n.): A disease in which fishes move rapidly round and round due to failure of nervous system.

White fish (n.): A freshwater fish of the salmon family, widely used as food. (*Coreonus* and other genera : several species).

Wholesale (n.): The selling of goods in large quantities to be retailed by others.

Wholeselling (n.): All activities involved in selling products and services to those buying for re-sale or business use.

Working capital (n.): The capital of a business which is used in its day-to-day trading operations, calculated as the current assets minus the current liabilities.

Xenobiotic (adj.): Relating to or denoting a substance that is foreign to the body or to an ecological system.

Yolk (n.): 1. The yellow internal part of a fish's or bird's egg, which is rich in protein and fat and nourishes the developing embryo, 2. The corresponding part in the ovum or larva of all egg-laying vertebrates and many invertebrates.

Yolk sac (n.): A membranous sac containing yolk attached to the embryos of reptiles and birds and the larvae of some fishes.

Zeolite (n.)* : Any of a large group of minerals consisting of hydrated alumino-silicates, used as cation exchanger and molecular sieves. Zeolitic (adj.).

---

* Zeolite is a natural or synthetic hydrated Sodium alumino-silicate $(Na_{12}[AlO_2)_{12} (SiO_3)_{12}] 27H_2O$ with an open three-dimentional crystal structure, in which water molecules are held in cavities in the lattice. The water can be driven off by beating and the zeolite can then absorbs other molecules of suitable size. Zeolites are used for separating mixtures by selective absorption and for this reason, they are often termed as *molecular sieves*.

www.ingramcontent.com/pod-product-compliance
Lightning Source LLC
Chambersburg PA
CBHW061322190326
41458CB00011B/3863